# The CHEMISTRY and FUNCTION of PROTEINS

*This volume is the second edition of CHEMISTRY AND BIOLOGY OF PROTEINS published by Academic Press Inc., 1950*

# The CHEMISTRY and FUNCTION of PROTEINS

## SECOND EDITION

## By Felix Haurowitz

Distinguished Service Professor of Chemistry
Indiana University, Bloomington, Indiana

1963

ACADEMIC PRESS • New York and London

ACADEMIC PRESS, INC.
111 Fifth Avenue, New York, New York 10003

*United Kingdom Edition published by*
ACADEMIC PRESS, INC. (LONDON) LTD.
Berkeley Square House, London W1X 6BA

LIBRARY OF CONGRESS CATALOG CARD NUMBER: 63-16962

*Third Printing, 1971*

PRINTED IN THE UNITED STATES OF AMERICA

# Foreword

The first edition of this book was published 1950. It was written as a textbook for a one-semester course on proteins for graduate students. The author's intent was to write a book which would be understood not only by graduate students of chemistry, but also by students of the biological sciences, who sometimes have a poor background in physical chemistry. The main topics discussed were (1) protein structure, (2) biological activity of proteins, and (3) biosynthesis of proteins. The book was so well accepted by the scientific world that it soon had to be reprinted. It was also translated into Russian and Japanese. These successes encouraged the author to write this second edition.

During the last decade, work on proteins has increased enormously. Chromatography, electrophoresis, and their variations have enabled biochemists to solve the amino acid sequence in a number of proteins and thus to obtain a deeper insight into structural problems. Very little was known about these methods when the first edition was written. Sanger had not yet published his classical work on the structure of insulin. The situation in protein chemistry was characterized by the following sentence (first edition, p. 116): "While the sequence of amino acids in shorter peptide chains can be determined, it would be hopeless to endeavor to fully characterize peptide chains containing 100 or more amino acids." Evidently these words were too pessimistic. Most of the space in Chapters I to VII of this book is devoted to the methods used in work on the *structure* of proteins.

A second part of the book (Chapters VIII to XV) is concerned with the properties of those proteins which have been investigated more thoroughly than others. It would have been preferable to classify these proteins systematically and to discuss them according to such a logical scheme. Since we do not yet have a satisfactory classification scheme, the system used in the first edition of this book was retained. The proteins were classified according to their occurrence in nature and their function. The latter principle is of particular value since it enables us to correlate structure with *function*. We know at present much more about the structural bases of enzyme action, of hormonal effects, and of antigen-antibody interaction than was known twelve years ago.

Our knowledge of amino acid supply and amino acid biosynthesis has

*v*

increased considerably since 1950. Work of this type is based on metabolic
experiments with isotopically labeled amino acids, their precursors and
intermediates; it is thus quite different from work on the structure of
proteins. For this reason, it was decided to limit the discussion of amino
acid metabolism to a short section in Chapter XVI.

Nevertheless, it seemed desirable to discuss in some detail the *biosynthesis* of proteins from amino acids, since the structure of proteins is
the result of the biosynthetic mechanism. In writing on biosynthesis it
was clear to the author that he was reporting on a problem which is in a
state of rapid flux. However, the correlation between structure and
biosynthesis of proteins is so important that it seemed useful to report
on our present views, realizing that they are ephemeral and that they will
change in the near future. It was the author's particular endeavor in this
part of the book to distinguish experimental facts from hypotheses which,
although frequently ingenious, still need further experimental support.

Although it was attempted to give the reader a survey on the entire
field of protein chemistry, it was obviously impossible to cite all papers
which deal with protein chemistry. It would have been easy to quote only
the most recent publications from which the interested reader can go back
to the original sources. This would have involved the omission of the
names of Kossel, Sörensen, Michaelis, and other pioneers in the development of protein chemistry. It was difficult to strike the right balance in
this respect. If good review articles on the subject under discussion were
available, the reader was frequently referred to these. Most of the papers
quoted were published during the last decade, i.e., after the publication
of the first edition of this book, and a large number of these appeared in
1960 or later.

In view of the extent of the bibliography it was decided to condense
it by using special abbreviations for those journals which were most
frequently quoted, and by listing at most two authors for each reference.
A list of the abbreviations for the journals will be found on pages ix and x.
The finding of the references cited has been facilitated by the use of bold
print for their numbers and by arranging them according to the sections
of each chapter. Further condensation of the volume of the book was
accomplished by the extensive use of small print for problems of secondary
importance. The rapid progress of research, particularly in the field
treated in the last chapter, necessitated additions and changes in the
period between the submission of the manuscript for publication and the
final completion of the page proofs. An attempt has been made to include
all of the important literature published in 1962.

Since the emphasis of this second edition of the book is on problems of
structure and function, the title was changed from "Chemistry and

Biology of Proteins" to "The Chemistry and Function of Proteins." The enormous increase in the literature on proteins made it desirable to submit most of the chapters of this book to experts in the respective fields. An appeal to several colleagues resulted in a response which far surpassed the author's expectation. The following colleagues were kind enough to read different chapters or sections and to send or give to the author their detailed written comments: Dr. R. S. Baer (Boston University), Dr. Wm. W. Bromer (Eli Lilly and Co.), Dr. H. B. Bull (University of Iowa), Dr. E. O. Davidson (Eli Lilly and Co.), Dr. J. T. Edsall (Harvard University), Dr. J. F. Foster (Purdue University), Dr. E. Gold (Indiana University), Dr. J. Gross (Massachusetts General Hospital), Dr. F. R. N. Gurd (Indiana University), Dr. R. L. Hill (Duke University), Dr. W. J. Kleinschmidt (Eli Lilly and Co.), Dr. I. M. Klotz (Northwestern University), Dr. H. R. Mahler (Indiana University), Dr. M. M. Marsh (Eli Lilly and Co.), Dr. W. F. H. M. Mommaerts (University of California, Los Angeles), Dr. E. A. Peterson (National Institutes of Health), Dr. H. V. Rickenberg (Indiana University), Dr. A. H. Sievert (Lake Bluff, Illinois), Dr. H. Wm. Sievert (Abbott Laboratories), Dr. H. A. Sober (National Institutes of Health), Dr. Ch. Tanford (Duke University), Dr. J. A. Thoma (Indiana University), Dr. G. Vidaver (Indiana University), Dr. D. Wetlaufer (University of Minnesota), and Dr. R. J. Winzler (University of Illinois). The author is also indebted to Mr. J. L. Groff for help in the preparation of the manuscript, and to Dr. L. Stewart for help in reading the proofs. The author's appeal to the readers of the first edition of the book elicited a number of valuable comments. Particular thanks are due to Dr. K. H. Gustavson (Stockholm) and Dr. R. Hotchkiss (The Rockefeller Institute) for detailed suggestions for improvements. Because of this generous help, use could be made of a series of most valuable critical remarks and suggestions, and many errors could be avoided.

In spite of the ample and generous aid of so many colleagues, it is unavoidable that a book on such a wide area of intensive research and rapid progress should contain errors. The only person responsible for them and for the mode of presentation is the author.

FELIX HAUROWITZ

*Bloomington, Indiana*
*April, 1963*

# List of Abbreviations

The following abbreviations are used for amino acyl residues in protein or peptide molecules.

Ala = alanine
Arg = arginine
Asp = aspartic acid
AspN or AsN = asparagine
Cys = cystine, cysteine
Glu = glutamic acid
GluN or GlN = glutamine
Gly = glycine
His = histidine
Hypro = hydroxyproline

Ileu = isoleucine
Met = methionine
Phe = phenylalanine
Pro = proline
Ser = serine
Thr = threonine
Try = tryptophan
Tyr = tyrosine
Val = valine

Other abbreviations frequently used in articles on protein chemistry are listed below.

$\text{Å}$ = angstrom unit = 0.1 m$\mu$ = $10^{-8}$ cm
A (in polynucleotides) = adenosine monophosphate residue
*ABB = Archives of Biochemistry and Biophysics*
ADP = adenosine diphosphate
ATP = adenosine triphosphate
*BBA = Biochemica et Biophysica Acta*
BGG = bovine $\gamma$-globulin
*BJ = Biochemical Journal (London)*
BSA = Bovine serum albumin
C = curie = $10^3$ mC(millicurie) = $10^6$ $\mu$C (microcurie)
C (in nucleic acids) = cytidine monophosphate residue
cal = calorie = $10^{-3}$ kcal
Cbo or Cbzo = carbobenzoxy residue
cpm = counts per minute
DAB = diaminobutyric acid
DFP = diisopropylfluorophosphate = diisopropyl-phosphofluoridate
DIT = diiodotyrosine
DNA = deoxyribonucleic acid
DNFB = 2,4-dinitrofluorobenzene
DNP = dinitrophenyl
DNPr = deoxyribonucleoprotein
DPN = diphosphopyridine nucleotide = NAD
DOC = deoxycholate
Dopa = dihydroxyphenylalanine
ETP = electron transport particle

ETS = electron transport system
$\Delta F$ = free energy change
FSH = follicle-stimulating hormone
$g$ = relative centrifugal force (see p. 71)
G (in nucleic acids) = guanosine monophosphate residue
GABA = $\gamma$-aminobutyric acid
$\Delta H$ = enthalpy change
Hb = hemoglobin
HDL = high density lipoproteins
ICSH = Interstitial cell-stimulating hormone
$JACS$ = $Journal\ of\ the\ American\ Chemical\ Society$
$JBC$ = $Journal\ of\ Biological\ Chemistry$
$k_1, k_2, \ldots$ = velocity constants
$K_1, K_2, \ldots$ = equilibrium constants
kcal = kilocalorie = 1000 cal
$\mu$g = microgram = $10^{-6}$ gram
MIT = monoiodotyrosine
mp = melting point
mRNA = messenger-RNA
MSH = melanophore-stimulating hormone
$N$ = normal (concentration)
NAD = nicotinamide-adenine dinucleotide = DPN
NADP = nicotinamide-adenine dinucleotide phosphate = TPN
NEM = $N$-ethylmaleimide
PCMB or $p$CMB = $p$-chloromercuribenzoate
pipsyl = $p$-iodophenylsulfonyl
pK = $-\log$ K (see p. 95)
PTC = plasma thromboplastin component
$R$ = gas constant = 1.98 cal per mole per degree = 8.314 joules per mole per degree = 0.08204 liter atmospheres per mole per degree
RNA = ribonucleic acid
RNPr = ribonucleoprotein
rpm = revolutions per minute
S = Svedberg unit (see page 73)
$\Delta S$ = entropy change
SDS = sodium dodecylsulfate
sRNA = soluble RNA
$T$ = temperature in degrees Kelvin = °C + 273
T (in nucleic acids) = thymidine monophosphate residue
$T_4$ = thyroxine (see p. 348)
TCA = trichloroacetic acid
TMV = tobacco mosaic virus
TPN = triphosphopyridine nucleotide = NADP
U (in nucleic acids) = uridine monophosphate residue
UV = ultraviolet

# Contents

*xi*

# The CHEMISTRY and FUNCTION of PROTEINS

# Chapter I

## Role of Proteins in Biology

Proteins have a particular significance in biology in that they constitute one of the indispensable components of living matter. To be sure, living organisms also contain carbohydrates and lipids, frequently in even greater abundance than proteins. Thus, green plants, which are rather poor in proteins, are rich in cellulose, a carbohydrate. There are, however, essential differences between proteins and most of the other cellular constituents. Wherever the phenomena of *growth* and *reproduction* are seen, proteins and nucleic acids are primarily involved. In animal and plant cells multiplication is initiated by the nucleus, in which proteins and nucleic acids are closely associated. In bacteria, where there is no typical nucleus, the bulk of the living substance is likewise formed by proteins and nucleoproteins. Both components are also found in viruses, the simplest of which are free of lipids and carbohydrates.

As the name of the proteins indicates (Greek *protos*, the first), they have been considered for many years the primary component of living matter. The great importance of the nucleic acids was recognized much later and has been in the foreground during the last decade. Many scientists seem to consider nucleic acids more important than proteins, and attribute to them an almost exclusive role in the genetic determination of inheritable characteristics. We must not forget, however, that growth and reproduction, like all processes which involve primarily the formation or degradation of proteins and nucleic acids, depend on the presence of certain *enzymes*, and that all these enzymes are proteins. Accordingly, nucleic acids cannot be formed in the absence of enzyme proteins. The presence of both proteins and nucleic acids is required for growth and multiplication.

The basic principle of the *chemical structure* of proteins is quite simple. They consist of long chains of amino acids linked to each other by peptide bonds. Complications arise, however, (1) from the presence of about 20 different types of amino acid residues in the peptide chains, (2) from the great length of these chains, which may consist of several hundred amino acids, and (3) from a particular *conformation* of the peptide chains, i.e., their specific folding, which results in a definite three-dimensional pattern. Even if all proteins consisted of straight peptide chains, free of folds and turns, an almost infinite number of proteins could be formed merely by changes in the order of the 20 amino acids in the long chains. Since each of these chains could assume an unlimited number of conformations, it is not surprising that each animal or plant species has its own species-specific protein. In some instances we are even able to find variations among the proteins of individuals of the same species. Only during the last decade have details of the amino acid sequence and of the conformation of some of the proteins been revealed.

It is understandable that the *conformation* of a protein is of the greatest importance for its chemical and physical properties as well as for its biological functions. The solubility of the proteins, their serological behavior, and their enzymatic and hormonal activities depend on the molecular groups present within the *surface* of the molecules, and hence on the spatial arrangement and the mode of folding of the peptide chains. It is, therefore, one of the principal endeavors of protein chemistry to elucidate the three-dimensional structure of the protein macromolecules and the distribution of functional groups within the molecules.

If the natural conformation of a protein molecule is destroyed or modified by physical or chemical operations, many of the chemical and physicochemical properties are changed, and the biological functions of the original *native protein* are altered or lost. We designate the modified product of such reactions as *denatured protein*. In using these terms we must be careful to avoid oversimplifications and fallacies. The designation *native* does not necessarily imply that the pure protein which we call native is identical with the form in which the same protein exists in the living cell. It is unavoidable that even the most careful methods of preparation destroy some of the weak bonds which link the protein molecule to neighboring molecules of other types inside the living cell. It is quite possible that some of the proteins occur in living organisms as giant "supermolecules" whose peptide chains form bridges between adjacent cells or extend over a series of cells. The pure crystalline protein preparations in a labeled bottle may not always have much in common with the protein molecules present in living tissues. It is also clear that denaturation can take place in many different ways and that we are usually dealing with a mixture of denatured products formed from one and the same native protein.

In spite of the dangers involved in the *preparation* and *purification* of proteins, protein chemists have succeeded in isolating pure, frequently crystalline, proteins which still manifest the same enzymatic or hormonal activities as the crude extracts of tissues or organs. Therefore we have every reason to assume that enzymes such as crystalline urease, isolated first by Sumner (1926), and crystalline pepsin, prepared by Northrop (1929), are essentially identical with the enzyme molecules in the living cells, and that the same is true for the pituitary hormones vasopressin and oxytocin isolated from the posterior lobe of the hypophysis by du Vigneaud (1952). Although it is quite possible that even these well-defined molecules occur in the cells more or less loosely bound to other proteins, their isolation does not seem to affect their biological activity significantly.

General problems of protein structure are discussed in the first part of this book (Chapters II–VII). A general survey of the isolation and purification of proteins (Chapter II) is followed by description of their degradation to amino acids (Chapter III) and of methods for the determination of amino acid sequences (Chapter IV). The physical chemistry of the proteins is discussed in the subsequent chapters which deal with the size and shape (Chapter V), electrochemistry and hydration (Chapter VI), and internal structure or conformation (Chapter VII), of the proteins.

In the second part of the book (Chapters VIII–XIV) the most important types of proteins are discussed. Since an enormous number of proteins is known, and since their properties vary considerably, many attempts have been made to classify proteins systematically. Most of these *classifications* are unsatisfactory and have been abandoned. One of the classifications, proposed by a committee of American and British biochemists, was based on the solubility of proteins in different solvents. Proteins soluble in 50%-saturated ammonium sulfate solution were called *albumins;* those precipitated by the same solvent were referred to as *globulins.* The latter class was subdivided into euglobulins, insoluble in salt-free water, and pseudoglobulins, soluble under the same conditions. However, the solubility of proteins in salt solutions depends not only on the concentration of the salts, but also on pH, temperature, and other factors. We do not yet know which of the chemical structures in the protein molecules are responsible for the differences in solubility.

In view of the present lack of a structural basis for the classification of proteins it seemed best to arrange the proteins in this text according to their biological distribution and function. The first group is formed by the soluble proteins of the blood plasma, milk, eggs, and seeds, and by the protamines and histones (Chapter VIII). Another group contains the structural proteins of the connective tissue, epidermis, and muscle (Chapter IX). Chapter X presents a survey on intermolecular forces acting between proteins and other molecules, and is written as an introduction to Chapter XI, which deals with the so-called conjugated proteins. These are complexes formed by the combination of proteins with nonprotein substances. Although hemoglobin and some of the other conjugated proteins occur as such *in vivo*, we still do not know whether this is also true for the nucleoproteins. They may be formed *in vitro* by combination of proteins with nucleic acids isolated from the same biological material. Proteins with enzymatic activity are discussed in Chapter XII, protein hormones in Chapter XIII, and toxins in Chapter XIV. It is unavoidable that Chapters VIII–XIV overlap to a certain extent. For example, many of the enzymes belong to the conjugated proteins. Moreover, it is impossible to draw a sharp borderline between some of the genuine proteins which contain very small amounts of carbohydrate and the typical conjugated glycoproteins.

Chapter XV is devoted to the antibody proteins. Analyses of antibodies and of their mode of formation have provided us with important knowledge on the mechanism of immunological reactions and have also given us valuable information on the specificity and the biosynthesis of animal proteins.

The two final chapters of the first edition of this book dealt with protein biosynthesis. In the first of these, the problem of the supply of indispensable *amino acids* and the mechanisms of amino acid biosynthesis were discussed. During the 12 years which separate this edition from its predecessor, however, the principal problems of amino acid supply have been solved. Most of the pathways of amino acid biosynthesis have been considerably clarified. Today the problems of protein chemistry are quite different from those of the chemistry of the amino acids. For this reason amino acid biosynthesis and amino acid supply will not be treated in this text. Readers interested in amino acid chemistry are referred to Greenstein's fundamental treatise on amino acids (1) and to an authoritative book on their metabolism (2).

The last chapter of the preceding edition of this book was devoted to a discussion of the mechanism of protein biosynthesis. Our views have undergone dramatic changes since 1950 and are at present in a state of flux. Each month brings new reports on important discoveries, particularly on the mutual interaction of proteins and nucleic acids. The problem which is currently the focus of interest is the role of the nucleic acids in protein biosynthesis. Most biochemists believe that the "four-letter code" of the nucleic acids, which contain only 4 different nucleotides, determines the specific sequence of 20 different types of amino acids present in the protein macromolecules. The nucleic acids are considered to be the material which transmits genetic characteristics from generation to generation. Whatever the mechanism of this process may be, it involves the presence of enzymes which act as catalysts in the synthesis of both proteins and nucleic acids. As far as we know, all these enzymes are proteins. The presence of these enzymes is just as indispensable for growth and replication as is the presence of nucleic acids. We are still far from an understanding of these intimate relations between nucleic acids and proteins. Therefore, our interpretations are only tentative and their validity is ephemeral.

In Chapter XVI an attempt has been made to draw a picture of the rapidly changing views on the mechanism of protein biosynthesis and on its dependence on the genetic role of the nucleic acids. This picture is based on a limited number of experimental facts and on their interpretation by means of ingenious theories and speculations. Since the latter undergo rapid changes, it is unavoidable that parts of this picture will be obsolete when this book is published.

The development of protein research in the near future can be predicted with more confidence. In the next few years we will learn the complete amino acid sequence of most of the well-known "pure proteins." We will find many more examples of genetically determined deviations in the amino acid sequence. We will learn whether the primary amino acid sequence alone determines the conformation of the peptide chains or whether proteins of identical primary structure can occur in two or more different conformations, and can have different types of folding of their peptide chains. Much more insight into the three-dimensional structure of the peptide chains will be gained. This, in turn, will give us more information on the "active sites" or "active patches" on the surface of globular enzymes, hormones, antigens, and antibodies, and also more information about the basis of their biological activities. There is hardly any doubt that these predictable discoveries will open new vistas and will thus lead to advances into unknown areas of molecular biology where the boundaries between biochemistry, biophysics, and the morphological sciences disappear.

### REFERENCES

**1.** J. P. Greenstein and M. Winnitz, "Chemistry of Amino Acids," Wiley, New York, 1961. **2.** A. Meister, "Biochemistry of the Amino Acids," Academic Press, New York, 1957.

# Chapter II

## Purification, Isolation, and
## Determination of Proteins[*]

### A. Methods of Isolation and Separation from Nonprotein Material

Since most proteins are extremely sensitive to heat, acids, bases, organic solvents, and, in some instances, even to distilled water, the methods generally employed for the isolation of other types of organic compounds can hardly be applied in protein chemistry.

The *insoluble cellular proteins*, frequently designated as *scleroproteins*, can easily be prepared by extracting the cell with water and with organic solvents to remove fats, carbohydrates, and the soluble proteins. Since some of these soluble proteins are dissolved only in the presence of neutral salts, an extraction by dilute solutions of NaCl or sodium bicarbonate is frequently necessary. If the insoluble protein is resistant to proteolytic enzymes a further purification is achieved by treatment with pepsin at pH 1–2 or with trypsin at pH 8–9. In this way keratin, the insoluble protein or cornified tissues such as horn and hair, can be prepared. Difficulties may arise when the insoluble proteins are accompanied by insoluble carbohydrates such as chitin or cellulose.

Keratin may be solubilized by reduction of its dithio bonds with sulfides or sulfhydryl compounds, collagen by treatment with collagenase or by boiling with water, fibroin by concentrated solutions of lithium bromide (A-1) or by dichloroacetic acid (A-2). The solubilized products are, however, not identical with the original material; they are partially degraded. All of the insoluble proteins can also be solubilized by heating with dilute solutions of sodium hydroxide whereas most of the carbohydrates are resistant to this treatment. Obviously, the proteins are drastically degraded by NaOH and are converted into a mixture of low molecular weight products. Owing to their insolubility the scleroproteins cannot be purified by crystallization. It is impossible to ascertain, therefore, whether the preparations obtained by the usual methods are of a uniform nature or are mixtures of several different proteins.

[*] See reference (A-19).

The isolation and purification of *soluble proteins* involves their extraction from the cells by suitable solvents and their precipitation by altering the concentration of salts and/or hydrogen ions, or by adding organic solvents. In many instances it has been possible to obtain *crystalline* precipitates by these procedures. In order *to avoid denaturation* of the protein to be isolated one must work at low temperatures, since solutions of many proteins are subject to denaturation even at room temperature. This is especially true for solutions of proteins in salt-free water. The rate of denaturation is reduced by the addition of neutral salts. For example, crystalline trypsin can be stored in a saturated solution of magnesium sulfate (A-3). The rate of denaturation is also reduced by storing the protein solution at *low temperatures*. Refrigerators, refrigerated centrifuges, and cold rooms, therefore, form an important part of the equipment required for the preparation of proteins.

At low temperatures a second important danger is also reduced, that of *bacterial decomposition*. Protein solutions form an excellent nutritional medium for bacteria and invariably become infected and are destroyed if kept at room temperature. Owing to the thermolability of protein solutions, they may not be sterilized by heat. Bacterial infection and growth can be inhibited by the addition of disinfectants, but these disinfectants may form compounds with the proteins and alter their physicochemical properties, or may denature them. Whenever possible, bacteria should be removed from protein solutions by centrifugation at high speed or by filtration through Seitz or Berkefeld filters or through filter pads. The disadvantage of the filtration is that small portions of the protein are adsorbed to the porous filter mass and are lost by denaturation.

The best method for storing protein solutions over long periods of time is to freeze them and to keep them at approximately $-10°C$ to $-20°C$ or at lower temperatures. At these temperatures bacteria cannot multiply, making it unnecessary to sterilize the solutions. Immune sera and enzyme solutions have been stored in the author's laboratory in the frozen state for many months, some of them for many years, without an appreciable loss of their biological activity. Inactivation or denaturation may occur, however, when some of the more labile proteins (lipoproteins, purified antibodies, pure ovalbumin) are repeatedly frozen and thawed.

Since the cellular membranes are impermeable to the massive protein molecules, the destruction of these membranes is a prerequisite for the *extraction of soluble proteins*. The destruction of cells can be accomplished mechanically by grinding them with sand or with kieselguhr. In these methods a part of the protein may be adsorbed to the silicate particles, and some denaturation may occur. It is preferable, therefore, to destroy the cells by suitable mills (Latapie mill, Potter-Elvehjem homogenizer).

The cellular structure is also destroyed by the action of organic solvents such as alcohol, acetone, or glycerol. If the concentration of *glycerol* does not exceed 85% a large portion of the soluble protein passes into the glycerol extract. In this way hydrolytic enzymes can be extracted from the pancreas and other organs. Glycerol extracts are rather stable at room temperature. It would appear that the rate of denaturation is reduced by the loose association of glycerol molecules with the protein, due to the polar hydroxyl groups of glycerol. If *acetone* is used for the destruction of the cells, it is necessary to mince the organ to be extracted or to grind it in a meat chopper and then to place the pulp under vigorous stirring into 5–10 volumes of acetone. In this manner a medium containing 80–90% of acetone is obtained in which proteins are not so easily denatured as in lower concentrations of acetone. The advantage of this procedure is not only that the cellular membranes are destroyed, but also that most of the lipids are extracted by the acetone. They can be removed quantitatively by a subsequent treatment with peroxide-free ether. The extracted residue is dried by spreading it over filter paper and is then extracted with water or with dilute solutions of salts or buffers. While many proteins and important enzymes withstand the action of acetone, other less stable proteins are denatured by the action of this solvent.

The simplest and best method for the *destruction of the cellular membranes* is disintegration by repeated *freezing and thawing.* Since pure ice particles are formed, the concentration of the salts in the cellular liquid increases during freezing, and the cellular membranes burst owing to the heightened osmotic pressure and to mechanical lesions of the cellular membranes by the ice particles. Destruction of the cells is also achieved by grinding cells or tissues with dry neutral salts.

The extracts prepared by any of the methods described above are centrifuged to remove insoluble particles. Salts, glycerol, and other substances of low molecular weight can be removed by *dialysis* against distilled water. The euglobulins and many of the plant proteins are insoluble in salt-free water and are obtained as precipitates on prolonged dialysis. They can be redissolved in isotonic salt solutions. Crystalline oxyhemoglobin is obtained when oxygen gas is passed through a solution of the dialyzed, salt-free, reduced horse hemoglobin (A-4,5). This phenomenon demonstrates impressively that the solubility of proteins depends to a great extent on very small changes in their molecules in the example given, on the replacement of a water molecule by an oxygen molecule as one of the ligands of the iron atom.

In the past dialysis was performed in *membranes* of animal origin (gut), in parchment paper, or in collodion membranes. The disadvantage of animal membranes is their inhomogeneity, while that of parchment paper and of collodion is their high content of sulfuric and nitric acid ester groups. Many proteins are denatured by adsorption to these groups. The best material at present for dialyzers is cellulose, which is obtainable commercially as *Cellophane*. The small size of its pores renders the rate of diffusion very low, increasing the time required for dialysis and hence the danger of bacterial infection. It is advisable, therefore, to perform the dialysis in the refrigerator. The pore diameter of Cellophane can be increased by treating the membrane with aqueous solutions of zinc chloride (A-6). It would considerably facilitate laboratory work involving dialysis if Cellophane membranes of higher permeability were commercially obtainable. The most convenient material for small-scale laboratory experiments is Cellophane tubing; dialyzers of different sizes can easily be prepared by tieing off the tubing at one or at both ends to form sacs of the desired size. Dialysis time can be reduced considerably by stirring the outer liquid or by rocking devices.

If the purpose of dialysis is merely to *remove salts*, the following method is used in the author's laboratory. The dialysis tubing which contains the protein solution is suspended horizontally a few millimeters under the surface of a large volume of distilled water (Fig. II-1) and is kept in a refrigerator. Owing to their

  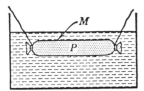

FIG. II-1. Three types of dialysis membranes. *G*: glass tube; *M*: membrane; *R*: rubber ribbon; *P*: protein solution. (From F. Haurowitz, "Biochemistry, An Introductory Textbook," p. 184. Wiley, New York, 1955.)

higher density, the salts diffuse rapidly downward to the bottom of the jar and the dialysis bag remains surrounded by salt-free distilled water. The high concentration gradient causes rapid dialysis and a concentrated solution of the dialyzable material accumulates at the bottom of the jar. After 10–15 hours the jar is emptied, rinsed, and filled again with fresh distilled water and this process is repeated until the salts are removed from the protein solution. Obviously this method cannot be used for dialysis against salt or buffer solutions. An increase in the diffusion rate of electrolytes can be achieved by *electrodialysis;* however, one has to avoid acid and alkaline reactions near the surface of the membranes which separate the protein solution from the anodic and the cathodic liquids respectively. A considerable portion of the protein is denatured when the reaction becomes strongly acid or alkaline.

Dialysis of proteins can frequently be replaced by filtration through columns of cross-linked dextran gels which are commercially obtainable under the trade name Sephadex (Pharmacia, Uppsala, Sweden; Box 1010, Rochester, Minnesota). These gels form a sponge-like structure which traps in its holes and cavities the small dialyzable ions and molecules and thereby retards their elution, whereas the large protein molecules pass into the first portions of the eluate (A-18). The dextran gel is stable between pH 2 and 10, and is therefore a very valuable tool for the purification of proteins.

### Isolation of Proteins from Their Solutions

While some proteins precipitate from their solutions upon dialysis, others are precipitated by the addition of neutral salts to their solutions ("*salting-out*" *method*). The usual procedure consists first of adjusting the pH of the solution to the isoelectric reaction of the particular protein where the solubility of proteins has its minimum value, then of adding ammonium sulfate (A-7) or sodium sulfate (A-8). Since different proteins are salted out at different salt concentrations, fractionation is possible by the stepwise increase in salt content (A-20). If the concentration of salts is very high, bacterial multiplication is sharply reduced, so that one can frequently work at room temperature. Thus, ovalbumin or serum albumin are obtained in crystals by the slow addition of small amounts of these salts to their isoelectric solution until the solutions become opalescent. The turbidity is cleared up by a small volume of water and the solution is kept at room temperature. Owing to the slow evaporation of water, crystals of the albumin are slowly formed and settle at the bottom of the beaker containing the solution. If an excess of the salt is added, amorphous precipitates of the proteins are produced. Crystals also can be obtained by dialysis of the protein solution against a saturated solution of ammonium sulfate.

### Removal of Water

While solutions of more stable substances are concentrated by evaporation on the steam bath or under reduced pressure, none of these methods can be applied in dealing with proteins. Heating on the steam bath causes denaturation of most proteins, although some of the smaller protein molecules, e.g., trypsin or ribonuclease, are resistant to short heating at definite pH and ionic strength (A-3,9). Distillation at reduced pressure cannot be used for concentrating protein solutions, since foaming occurs under these conditions. Small volumes of protein solutions can easily be *concentrated in desiccators* under reduced pressure over large amounts of water-binding substances. In order to reduce the volume of larger amounts of protein solution, the solution can be placed in a Cellophane bag which is suspended in air in front of an electric fan (pervaporation). As the water evaporates from the surface of the bag, the protein solution becomes more concentrated (A-10). However, some of the protein may undergo surface denaturation. Better results are obtained by ultrafiltration, i.e., by applying positive or negative pressure during the dialysis, or by dialysis against dry polyvinylpyrrolidone, Carbowax (polyethylene glycol) (A-11), or Sephadex.

Another method which can be applied to large volumes of protein solutions consists of freezing the solution solid and then thawing it very slowly, without stirring or shaking. Crystals of ice nearly devoid of

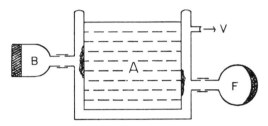

Fig. II-2. $A$: mixture of dry ice and acetone at $-70°C$; $V$: outlet to high vacuum pump; $F$: flask, $B$: bottle. The protein solutions are frozen in $F$ and $B$; their water evaporates in the high vacuum and is condensed and frozen inside the lyophilization apparatus by the low temperature in $A$. The protein is obtained as a dry powder in $F$ and $B$.

protein rise to the surface of the solution, while the bulk of the protein is concentrated in the solution in the lower part of the vessel. The concentrated protein solution is siphoned off and can be submitted repeatedly to the same procedure until high concentrations of protein are obtained (A-12a,12b). Human hemoglobin was obtained in this way in crystalline form by the author (A-12a).

The best method for concentrating large volumes of protein solutions is *freeze-drying* (lyophilization) (A-13). The protein solution is frozen in a bottle or a round-bottom flask and exposed to high vacuum in the presence of a cold area kept at about $-70°C$ by a mixture of solid $CO_2$ (dry ice) and acetone (Fig. II-2). Ice sublimes from the frozen protein solution and condenses at the cold surface. The protein is obtained as an air-dry powder and is usually in the native state. Dilute protein solutions can be concentrated very efficiently by adsorption to ion exchange columns and subsequent elution with small volumes of salt solutions (Section B, this chapter).

One of the oldest methods for the preparation of crystalline proteins is the cautious addition of *ethanol* or *acetone* to the cold protein solution. Crystalline oxyhemoglobin was obtained from the blood of many animals in this way by Hoppe-Seyler, and also by Hüfner, toward the end of the last century. It is essential to keep the temperature close to 0°C during this procedure since the organic solvents cause denaturation at room temperature. It is surprising, therefore, that serum albumin of man and various mammals remains in the native state when the precipitate produced by trichloroacetic acid is dissolved in methanol, ethanol, or acetone (A-14). Whereas the other serum proteins, after precipitation by trichloroacetic acid, are insoluble and denatured, native serum albumin can be extracted by ethanol or acetone and obtained after dialysis in the native state, free of the other plasma proteins (A-15,16,17).

## B. Fractionation of Protein Mixtures

The extracts of cells or of organs consist of a mixture of various proteins. The same is true for the blood serum and other body fluids. The classical method for the preparation of albumins and globulins from the blood serum is based on the fractional addition of ammonium sulfate. In many papers the concentration of the ammonium sulfate used is recorded as the percentage of a saturated solution. Thus a solution containing equal parts of water and of a saturated solution of ammonium sulfate is described as 50%-saturated. This mode of designation is rather inaccurate because the saturation depends to a certain degree on the temperature of the solution. It is preferable, therefore, to record the concentration of the ammonium sulfate in the more usual way, i.e., by indication of its molarity or weight per unit volume (Chapter VIII, B).

The systematic purification of proteins by means of organic solvents was introduced and successfully elaborated by Cohn and his co-workers (B-1). The methodical application of ethanol and other organic solvents is based on the fact that these solvents reduce the dielectric constant of the aqueous protein solutions. The opposite effect, an increase in the dielectric constant, is achieved by the addition of glycine. The *alcohol method* is superior to other methods when proteins are to be prepared on a large scale, since most of the alcohol is easily removed by filtration or evaporation, whereas the removal of salts by dialysis is time-consuming and laborious. The essential feature of the new method is, however, the systematic variation and close control of the ionic strength, the hydrogen ion concentration, the temperature, and the dielectric constant of the solution. By varying these factors systematically, a large number of new proteins have been isolated from the blood serum (Chapter VIII, B).

The preparation of a pure protein from its mixtures has in some instances been facilitated by the selective denaturation of the other proteins present. Thus, crude trypsin was purified by heating its solution in 0.5 $N$ hydrochloric acid to 90°C (B-2). Inert proteins were removed in this way.

During the last few years several extremely powerful methods for the fractionation of protein mixtures have been developed. The most widely used of these is the fractionation of proteins by *chromatography on ion exchange columns* (B-16). Another important method is *partition chromatography*, which had been used earlier for the separation of amino acids and peptides. Partition chromatography is based on the distribution of the solute between two only partially miscible liquid phases, such as water and butanol or water and phenol. Since many of the proteins are denatured by the organic solvents, these methods can be applied only to some of the smaller, more stable, protein molecules. The danger of denaturation has been lowered by using glycol ethers as the organic phase. Thus Porter (B-3) succeeded in fractionating γ-globulins by partition chromatography when a concentrated solution of potassium phthalate at pH 9.0 and −3.1°C was used as the aqueous phase and methylcellosolve (glycolmonomethyl ether, $CH_3OCH_2 \cdot CH_2OH$) as the non-aqueous phase. This technique is difficult and its applicability is limited.

Similar distribution of proteins between two liquid phases is accomplished by omitting the solid phase and using instead *countercurrent distribution* (CCD). As mentioned before, this method is particularly suitable for the smaller proteins which are not so easily denatured by organic solvents. It has been developed chiefly by Craig and his co-workers (B-10), who use a machine containing several hundred distribution tubes.

In this manner ribonuclease or lysozyme were fractionated in a system consisting of 2 volumes of ethanol and 3 volumes of ammonium sulfate solution (40 g ammonium sulfate + 100 ml water) (B-10). Serum albumin was fractionated by distribution in a mixture of water-butanol-ethanol (20:20:1) in the presence of small amounts of trichloroacetic acid (B-10). Serum albumin was also separated from globulins by distribution between an aqueous solution of magnesium sulfate and diethyleneglycol diethylether; the globulins had a higher affinity for the organic phase than the albumins (B-11,19).

One of the best methods for the separation of protein mixtures is their adsorption to synthetic ion exchange resins. These are usually sulfonated or carboxylated cation exchangers (B-4), or anion exchangers which contain the basic quaternary ammonium groups (B-5). Thus hemoglobin can be separated into two fractions on the weak cation exchange resins IRC-50 or XE-64 (B-4). Some of the proteins, however, undergo denaturation when adsorbed to and eluted from the synthetic ion exchange resins.

Further improvement in protein fractionation has been accomplished by the use of weakly acidic or basic *cellulose derivatives as ion exchangers* (B-6,17). One of the latter is DEAE (diethylaminoethyl) cellulose in which the OH groups of cellulose are linked by ether bonds to the OH group of diethylaminoethanol, $(C_2H_5)_2NCH_2CH_2OH$. Other basic derivatives of cellulose are triethylaminoethyl cellulose and ECTEOLA-cellulose the latter prepared by the treatment of cellulose with EpiChlorohydrin and TriEthanOLAmine). The most widely used acidic cellulose derivative is CM-cellulose (carboxymethyl cellulose) in which the OH groups of cellulose are converted into $OCH_2COOH$ groups. The proteins are bound to the acidic or basic groups of the cellulose derivatives and can be eluted by salt solutions or by changing pH by means of buffer solutions. Weakly acidic buffers can be replaced by $CO_2$ which has the advantage of being volatile (B-7). Other volatile buffers are ammonium acetate or formate. Fractionation of proteins has also been accomplished by columns of calcium phosphate (B-8). If very small amounts of protein are examined, the columns of ion exchange cellulose can be replaced by ion exchange paper, which is now commercially obtainable.

All fractionation methods on filter paper, cellulose powder, or ion exchange resins are based essentially on the same principle, the distribution of each of the proteins between two phases. If water and an organic solvent are used, the ratio of these two liquids will be different in the solid-liquid interface, and the mobile liquid phase. The affinity of the protein to the solid material depends not only on the mutual attraction of charged groups of opposite sign, but also on hydrogen bonds and interaction by van der Waals forces (Chapter VII, E). Proteins with the lowest affinity to the solid phase pass the column first. The adsorption of the proteins to the column material depends on the pH of the solution, which determines the extent of ionization of both the ion exchanger and the protein. The electrostatic attraction between ionized groups of the exchanger and charged groups of the opposite sign in the protein is lowered by salts. Hence, separation of several proteins can be accomplished by raising the salt concentration or by changing pH. If these changes are produced gradually, we speak of elution by a salt or by pH gradient, respectively (B-9, 18).

Separation of the components of a protein mixture can also be accomplished by means of *zone electrophoresis* (B-12) (see also Chapter VI, C), i.e., electrophoresis in a solid medium which can consist of starch gel, of filter paper, or of synthetic polymers such as polyvinyl chloride. The resolution of protein mixtures is particularly good when the electrophoresis is carried out in vertical columns (B-12). More powerful fractionation has been accomplished by exposure of the protein mixture to zone electrophoresis in two directions at two different pH values (B-13). This method revealed the presence of at least 21 different proteins in the normal blood serum and led to the discovery of genetically determined variations in haptoglobin (B-14) and other proteins (Chapter VIII, B). The presence of a large number of proteins in the blood serum and in other fluids of the human or animal body has also been proved by immuno-electrophoresis (B-15) in which the solution is first exposed to electrophoresis in a gel; an antiserum to the examined protein mixture is then allowed to diffuse into the gel at a right angle to the direction of electrophoresis. Each of the protein components of the mixture reacts with the homologous antibody forming an insoluble precipitate (see Fig. VI-8). Although this method is a very powerful tool for the discovery of numerous proteins in their mixture, it is less suitable for the isolation of pure proteins.

## C. *The Problem of Protein Heterogeneity*

In organic chemistry the formation of homogeneous, *uniform crystals* is generally considered conclusive evidence of the uniformity of a compound. If the crystals have a reproducible melting point which remains unchanged after recrystallization, their homogeneity is considered definitely established. This criterion cannot be applied to protein crystals, because proteins decompose when they are heated. Originally the ability of some proteins to crystallize was considered an indication of their homogeneity. This assumption is, however, invalidated by the results of the fractionation procedures described in the preceding paragraphs. Most of the crystalline proteins are mixtures of two or more compounds which can be separated from each other by chromatography, countercurrent distribution, or electrophoresis. There is no doubt, therefore, that crystallizability alone cannot be taken as a criterion for the purity of a protein. Nevertheless, crystallization is one of the best procedures for the removal of other proteins. Even if the crystals obtained are heterogeneous, they will be a good material for further purification. Alternately, one may "rest and be thankful" (C-1) for having succeeded in producing crystals.

The shape of the *solubility curve* has been used as another criterion for homogeneity. The solubility of a homogeneous substance is constant and independent of the excess of solid phase. Thus, at a given temperature a liter of water will always dissolve the same maximum amount of sodium chloride irrespective of any excess of solid NaCl. If the concentration of the solution is plotted against the added solid phase (NaCl), it increases linearly until saturation is achieved, and then remains unchanged, causing a sharp break in the curve, which from this point on appears as a horizontal straight line (Fig. II-3). It is hardly possible to apply this method to solutions of proteins in pure water because the solubility is highly dependent on traces of electrolytes and of hydrogen ions. The influence of these factors can be eliminated by using concentrated salt solutions as solvents for proteins. When crystalline ovalbumin or carboxyhemoglobin were examined in this way by Sørensen (C-2), he found that their solubility in ammonium sulfate solutions increased with the amount of the insoluble phase. Evidently, the protein molecules were not homogeneous. Sørensen attributed this behavior to the reversible dissociation of large

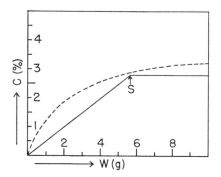

FIG. II-3. Relation between $W$, the amount of added solid substance to a constant volume of solvent, and $C$, the concentration of the dissolved substance. The solid line shows the behavior of a pure substance. Its concentration increases linearly until saturation is attained at $S$. The broken line is typical of a protein.

protein complexes into smaller units. Indeed, it has been found that insulin, serum albumin, hemocyanin, and other proteins occur in their solutions as mixtures of monomers with dimers or higher polymers (see the chapters dealing with these proteins). Even though chymotrypsinogen (C-3), ribonuclease (C-4), and a few other crystalline proteins behave in the solubility test as homogeneous substances would behave, they are heterogeneous as shown by other methods of investigation.

Of particular interest are those cases in which heterogeneity can be attributed to genetic determinants. Thus, we can differentiate two forms of $\beta$-lactoglobulin, called $\beta_1$ and $\beta_2$, or A and B, respectively. In the milk of some cows either one or the other, or a mixture of both, of these two forms occurs. The occurrence of each of these forms is determined by the Mendelian laws of genetics and depends accordingly on the occurrence of the corresponding genotypes in the parents of the animals (C-5, 6). Similarly, human blood sera contain different types of haptoglobin (C-7; see also Chapter VIII, B). The differences in chemical structure which are responsible for the different behavior of the two components of $\beta$-lactoglobulin in chromatography or electrophoresis are very small. The dependence of their occurrence on genetic determinants is, however, of great importance, since it excludes the possibility that the observed small differences are caused by manipulation of these proteins during their isolation. It also demonstrates quite clearly that the biogenesis of these proteins depends on a mechanism which is controlled by the genes of the respective individuals and thus also depends on the genes of their parents and earlier ancestors.

The steady improvement in our methods of fractionation and also in our techniques of analysis has increased enormously the number of proteins which can be separated and differentiated from each other. It will be shown later in this text (Chapter XI, G) that the differences between two forms of a protein in many instances are due to the replacement of one of a hundred or more amino acids by another single amino acid. In such cases the two proteins differ from each other in their *primary structure*, i.e., in the composition of their covalently linked chains of amino acid residues. We still do not know whether the *conformation* of the protein molecules, i.e., the three-dimensional arrangement of their peptide chains in space, is strictly determined by the primary structure. Such correlation between primary structure and conformation may exist in the smaller protein molecules; it is less probable in molecules whose peptide chains consist of several hundred amino acids.

It is well known that the dithio groups, —S—S—, of the cystine residues in proteins react *in vitro* slowly with the SH groups of sulfhydryl compounds according to the exchange reaction:

$$R^IS^IS^{II}R^{II} + R^{III}S^{III}H \rightleftharpoons R^IS^IS^{III}R^{III} + R^{II}S^{II}H$$

If this reaction were to take place *in vivo* between SS and SH groups *within* a single protein molecule, different types of this protein would be formed, differing from each other by their *secondary structure*, the position of the dithio cross-links between the peptide chains. It is not yet clear whether such changes take place during the biological lifetime of a protein molecule. In other words, we do not know whether proteins "age" in this or some other manner, or whether "old" protein molecules differ in their primary or secondary structure from "young" molecules of the same type. Problems of this type can be investigated by injecting an organism with isotopically labeled amino acids. Obviously, only those protein molecules which are formed *after* injection of the labeled amino acid will contain the labeled amino acids. Consequently, if the animal is killed shortly after injection, the "young" protein molecules contain more of the labeled amino acids than the "old" protein molecules. Investigations of this type, carried out with $S^{35}$-labeled rat serum albumin, demonstrate that "young" and "old" rat serum albumin molecules are metabolized in the rat at the same rate. Evidently the rat organism cannot differentiate the old from the young molecules and degrades them randomly, at identical rates (C-8). In similar experiments with rabbits it was found that fetal and adult serum albumin are metabolized in adult rabbits at the same rates (C-9).

Fifteen or twenty years ago, hemoglobin, $\beta$-lactoglobulin, serum albumin, ovalbumin, ribonuclease, edestin, and other proteins were considered well-defined molecules which, in their solutions and under certain conditions, might associate to form larger complexes but were otherwise homogeneous. At present we know about 20 different human hemoglobins (see Chapter XI, G), at least 4 fractions of human serum albumin (C-10), about 15–20 different human serum globulin fractions (B-15, C-11), and 10 or more various cathepsins, all of them occurring in the spleen (C-12). Heterogeneities have also been discovered in the $\beta$-lipoproteins of human blood serum (C-13), in the protamines clupein and salmine of fish sperm (C-14), in Bence Jones protein of the human urine (C-15), in edestin (C-16), and in numerous enzymes (see Chapter XII, C).

Will further refinements of our techniques reveal an increasing number of heterogeneities and variations? Is there a limit of variations or should we abandon the notion of completely uniform protein molecules and assume that the large protein molecules consist of a population of closely related individual molecules of the same primary structure but of different conformations? Are we, in this sense, dealing with a continuous "microheterogeneity" (C-17)? We do not yet know the final answer to these intriguing questions. However, even if protein populations were never quite homogeneous, it would not lessen the great importance of the new methods of fractionation and isolation described in the preceding paragraphs. It would merely mean that proteins, as a result of their large size and the numerous cross-links between their peptide chains (see Chapter VII, E), can exist in numerous possible conformations, and that they persist in some of these for periods of many hours, days, or months. Analogous conformational changes in small tautomeric molecules may take place continually and so rapidly that we cannot isolate any one of the numerous short-lived forms, but always end up with an equilibrium mixture which has constant properties and thus behaves as a "homogeneous uniform substance." According to this view, the difference between the large protein molecules and the small molecules of other substances would be merely a difference in the lifetime of certain conformations of their molecules but not a fundamentally new type of difference.

## D. *Qualitative Tests for Proteins*

The textbooks of biochemistry enumerate various *color reactions* of proteins.

The phenolic side chain of tyrosine is responsible for the yellow color produced by nitric acid in the xanthoproteic reaction, for the red color given by Millon's reagent, for the diazo test, and for the blue color given in an alkaline solution of phosphomolybdates. The color reactions brought about by dimethylamino-benzaldehyde (Ehrlich) or glyoxylic acid (Hopkins-Cole) are due to the presence of tryptophan; the red color reaction with hypochlorite and $\alpha$-naphthol is a reaction of arginine (Sakaguchi), while the dark coloration formed when alkaline solutions of lead acetate are heated with proteins is due to the presence of cystine or cysteine. These and other color reactions are discussed in greater detail in Chapter III, D, and Chapter VII, F.

The *biuret reaction* performed by adding a dilute solution of copper sulfate to a strongly alkaline solution of the protein produces a purplish-violet color. Since the same reaction is given by all polypeptides, it is due to the presence of peptide bonds, —CONH—, in the protein molecule. It is assumed that a complex is formed wherein the copper ion is coordinated with the peptide bond to form a copper-containing ring of the probable structure (I) or another similar coordination compound. At low

(I)

pH values the cupric ions combine also with carboxyl groups, at higher pH values with amino groups of the proteins (D-1). The biuret reaction is frequently used for the quantitative determination of proteins in the blood serum. However, it is not very sensitive, and a negative reaction does not prove the absence of proteins. Owing to their content of numerous amine groups the proteins give also the ninhydrin reaction, i.e., a deep purplish color with ninhydrin (see Chapter III).

*Proteins are precipitated* by the salts of heavy metals (Cu, Pb, Hg, $UO_2$, Fe), by a series of organic acids including trichloroacetic acid, salicylsulfonic acid, and picric acid, and by colloidal acids (e.g., tungstic acid, tannic acid) and bases (ferric hydroxide). These precipitants, in general, cannot be used for the preparation of pure proteins because many proteins are denatured by their action. However, native proteins have also been prepared from their compounds with zinc, provided that the zinc salt is added to the protein solutions cautiously and an excess is avoided (D-2). If an excess of the precipitating reagents is used, most proteins are quantitatively precipitated. This method finds extensive use in clinical laboratories in the deproteinization of biological fluids, in the quantitative determination of proteins in the blood serum and similar fluids, and as a qualitative test for the presence of proteins in urine and other biological solutions.

While it is very difficult to separate the precipitated denatured protein from the precipitating agents mentioned above, the protein can be obtained easily in the denatured state by heat coagulation or by Sevag's method (D-3), in which the protein is denatured and rendered insoluble by shaking with chloroform and amyl alcohol or octanol. In tissue sections proteins can be localized by impregnation with phosphotungstic acid and detection of the latter by X-rays which are absorbed by the heavy tungsten atoms (D-4).

## E. Quantitative Determination of Proteins (E-1)

The precipitate produced by the reagents referred to in the preceding section can be used for the quantitative determination of proteins by Kjeldahl analysis. In this procedure it is assumed that the precipitate contains only protein nitrogen. This assumption is not quite justified because some high-molecular polysaccharidic acids and certain lipids are precipitated along with the proteins. The Kjeldahl determination consists of converting all the nitrogen into ammonium sulfate by boiling the protein with concentrated sulfuric acid and potassium sulfate in the presence of a catalyst such as cupric sulfate, selenium, or mercury salts. The average nitrogen content of proteins is 16% and the conversion factor from nitrogen to protein equal to 6.25 (100/16). Actually this value varies considerably. Thus ovalbumin contains 15.75% nitrogen and edestin 18.7% (E-2). It is necessary, therefore, to accurately determine the nitrogen content of the protein to be analyzed. The conversion factors of the principal fractions of human blood serum are: 6.63, 6.78, and 6.52 for the $\alpha$-, $\beta$-, and $\gamma$-globulins, respectively, and 6.53 for the albumins (E-3).

Although the Kjeldahl method is one of the standard methods employed in biochemical laboratories, the nitrogen values obtained in the customary procedure are frequently too low. Reliable values have only been obtained with the use of mercury as a catalyst (E-4). The heating of the protein with sulfuric acid, potassium sulfate, and the catalyst should be continued for 8 hours (E-2). When these precautions are observed, precise results are obtained. Attempts have been made to simplify the method by avoiding the distillation of ammonia. The ammonium sulfate formed can be oxidized directly by NaOBr and the excess of the hypobromite titrated iodometrically (E-5), or ammonia can be determined colorimetrically by means of Nessler's reagent.

A simplifying modification was introduced by Conway, who devised the distillation of ammonia at room temperature in a closed "unit" (Fig. II-4) consisting of two concentric compartments (E-6).

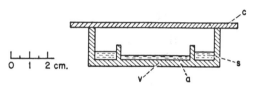

Fig. II-4. Cross section through a Conway unit (E-6); *v*: reaction vessel; *c*: glass cover; *a*: inner compartment containing acid; *s*: solution of ammonium salts made alkaline by potassium carbonate. Diameter of unit, 70 mm.

If sufficient amounts of protein are available, good results are obtained by the colorimetric evaluation of the biuret reaction (E-4,7). The sensitivity of the reaction can be increased by using a measured amount of cupric ions, isolating its excess which in the alkaline solution precipitates as cupric hydroxide, and determining it colorimetrically with diethyldithiocarbamate (E-8). Most, but not all, proteins can be precipitated quantitatively by boiling or by adjusting pH to the isoelectric pH value; the precipitate so formed can be dried and weighed. The main difficulty here consists in removing salts and other soluble compounds, since protein precipitates frequently give rise to colloidal solutions when they are washed with salt-free water. This difficulty can often be avoided by washing with dilute acetic acid or with methanol-water (1:1), followed by acetone and ether.

Since the specific gravity of the blood serum is largely determined by its protein content, the latter has been evaluated by measuring the specific gravity of the serum in a gradient tube containing a mixture of bromobenzene and kerosene (E-9). Standardization of the gradient tube is accomplished by using potassium sulfate solutions of differing densities. Copper sulfate solutions have also been used for gradient tubes. Neither of these methods is free from error (E-10), due to the partial dependency of the specific gravity of the sera on their content of substances other than proteins.

If it is desired to determine very small amounts of protein use can be made of the intensive ultraviolet absorption of tyrosine and tryptophan at 280–290 mμ (E-11) or of the blue color caused by the reducing action of proteins on Folin's phenol reagent (i.e., phosphomolybdic acid) (E-12). Both methods are very simple and rapid. The intensity of the ultraviolet absorption and of the blue color reaction depend on the content of tyrosine and the other aromatic amino acids; their percentage varies from protein to protein. It is necessary, therefore, to use a sample of the protein to be determined for standardization.

Owing to the frequent use of paper electrophoresis for the separation of proteins, attempts have been made to determine the protein bands on the paper strips by staining with bromphenol blue, amido black, or other dyes and to evaluate the color densitometrically or by colorimetry after extraction of the dye from the separated bands. The values obtained in this manner are not very accurate because of dependence of the results on temperature and other conditions (E-13) which may lead to trailing of some of the proteins and overlapping of their bands (E-14).

Particular problems arise in the determination of the *radioactivity of isotopically labeled proteins*. The labels most frequently used are $C^{14}$, $S^{35}$, and $H^3$. The β-radiation emitted by radioactive carbon and sulfur is very soft and is easily absorbed by organic substances. Consequently, a considerable self-absorption of proteins is observed and it is necessary to construct a self-absorption curve by measuring the activity (counts per minute) of increasing amounts of the protein. If sufficient amounts are available, one can measure always at infinite thickness where the specific activity of the sample is strictly inversely proportional to its weight. The energy of the β-emission of tritium ($H^3$) is an order of magnitude lower than that of $C^{14}$ or $S^{35}$. Accordingly the β-radiation emitted by tritium is easily absorbed even by very thin layers of organic material; not more than 10% of the radiation can penetrate a layer of 1 μ thickness (E-15). This precludes in most cases counting of dry powders of $H^3$-protein. Very small amounts of tritiated protein samples can, however, be analyzed by scintillation counting provided the sample can be dissolved in toluene, which is used as a solvent for the scintillating materials. Dissolution of the protein samples in toluene can be accomplished by means of some cationic detergents such as the quaternary ammonium base Hyamine, a substituted cresoxy alkoxy alkyl benzyl ammonium hydroxide (E-16). Although the β-radiation emitted by tritium cannot penetrate organic material, the light emitted by the scintillating "phosphors" penetrates easily. The instrument, by means of a photomultiplier, counts light flashes instead of the β-particles which cause these flashes.

## REFERENCES

### Section A

**A-1.** P. P. von Weimarn, *Ind. Eng. Chem.* 19: 109(1927).   **A-2.** E. J. Ambrose, *et al.*, *Nature* 167: 264(1951).   **A-3.** J. H. Northrop and M. Kunitz, *J. Gen. Physiol.* 16: 267(1932).   **A-4.** F. Haurowitz, *Z. physiol. Chem.* 136: 147(1924).   **A-5.** M. Heidelberger, *JBC* 53: 34(1922).   **A-6.** W. B. Seymour, *JBC* 134: 701(1940).   **A-7.** S. Sørensen and M. Høyrup, *Z. physiol. Chem.* 103: 16(1918).   **A-8.** G. S. Adair and M. Robinson, *BJ* 24: 993(1930).   **A-9.** R. J. Dubos and R. H. S. Thompson, *JBC* 124: 501(1938).   **A-10.** L. Farber, *Science* 82: 158(1935).   **A-11.** J. Kohn, *Nature* 183: 1055(1959).

**A-12a.** F. Haurowitz, *Z. physiol. Chem.* 186: 141(1930). **A-12b.** T. Shapiro, *Science* 133: 2063(1961). **A-13.** E. W. Flosdorf, *J. Chem. Educ.* 22: 470(1945). **A-14.** S. Levine, *ABB* 50: 515(1954). **A-15.** G. W. Schwert, *JACS* 79: 139(1957). **A-16.** E. Kallee *et al.*, *Z. Naturforsch.* 12b: 777(1957). **A-17.** S. Fleischer and F. Haurowitz, *Arzneimittel-Forsch.* 10: 362(1960). **A-18.** J. Porath, *BBA* 39: 193(1960). **A-19.** P. Alexander and R. J. Block, "Analytical Methods of Protein Chemistry." Pergamon Press, London, 1961. **A-20.** M. Dixon and E. C. Webb, *Adv. in Protein Chem.* 16: 197(1961).

### Section B

**B-1.** E. J. Cohn *et al.*, *JACS* 68: 459(1946); 71:541(1949). **B-2.** J. H. Northrop, "Crystalline Enzymes," p. 268. Columbia Univ. Press, New York, 1948. **B-3.** R. R. Porter, *BJ* 59: 405(1955). **B-4.** N. K. Boardman and S. M. Partridge, *Nature* 171: 208(1953); N. K. Boardman, *BJ* 59: 543(1955). **B-5.** H. G. Bowman and L. E. Westlund, *ABB* 64: 217(1956). **B-6.** H. A. Sober and E. A. Peterson, *in* "Ion Exchangers in Organic and Biochemistry" (C. Calmon and T. R. E. Kressman, eds.) p. 318. Interscience, New York, 1957; H. A. Sober *et al.*, *JACS* 78: 756(1956). **B-7.** M. A. Mitz and S. S. Yanari, *JACS* 78: 2649(1956). **B-8.** A. Tiselius *et al.*, *ABB* 65: 132(1956). **B-9.** S. Ellis and M. E. Simpson, *JBC* 220: 939(1956). **B-10.** L. C. Craig *et al.*, *JACS* 80: 2703; 3366(1958); P. V. Tavel and R. Signer, *Adv. in Protein Chem.* 11: 238(1956). **B-11.** A. Tiselius and P. Flodin, *Adv. in Protein Chem.* 8: 461(1953). **B-12.** J. Porath, *BBA* 22: 151(1956). **B-13.** O. Smithies *et al.*, *BJ* 71: 585(1959); 72: 121(1959); *Adv. in Protein Chem.* 14: 65(1959); M. D. Poulik and O. Smithies, *BJ* 68: 636(1958). **B-14.** G. E. Connel and O. Smithies, *BJ* 72: 115(1959). **B-15.** P. Grabar, *Methods of Biochem. Anal.* 7: 1(1959); P. Grabar and C. Williams, *BBA* 10: 193(1953). **B-16.** S. Keller and R. J. Block, "Analytical Methods of Protein Chemistry," p. 65. Pergamon Press, London, 1961. **B-17.** E. A. Peterson and H. A. Sober, *in* "Methods in Enzymology" (S. P. Colowick and N. O. Kaplan, eds.), Vol. 5, p. 1. Academic Press, New York, 1962. **B-18.** E. A. Peterson and E. A. Chiazze, *ABB* 99: 136(1962). **B-19.** P. v. Tavel, *Helv. Chim. Acta* 45: 1576(1962).

### Section C

**C-1.** A. G. Ogston, *Ann. Rev. Biochem.* 24: 183(1955). **C-2.** S. Sørensen, *Compt. rend. trav. lab. Carlsberg, Sér. Chim.* 18: No. 5; 19; No. 11(1933). **C-3.** M. Kunitz and J. H. Northrop, *Cold Spring Harbor Symposia Quant. Biol.* 6: 325(1938). **C-4.** M. Kunitz, *J. Gen. Physiol.* 24: 15(1940). **C-5.** A. G. Ogston and M. P. Tombs, *BJ* 66: 399(1957). **C-6.** R. Aschaffenburg and J. Drewry, *BJ* 65: 273(1957). **C-7.** O. Smithies and N. F. Walker, *Nature* 178: 694(1957). **C-8.** H. Walter and F. Haurowitz, *Science* 128: 140(1958). **C-9.** S. Fleischer and F. Haurowitz, *Arzneimittel-Forsch.* 10: 362(1960). **C-10.** A. Saifer and H. Craig, *JBC* 217: 23(1955). **C-11.** J. W. Williams *et al.*, *JACS* 74: 1542(1952); F. W. Putnam, "Plasma Proteins." Academic Press, New York, 1960. **C-12.** E. M. Press *et al.*, *BJ* 74: 501(1960). **C-13.** D. Gitlin, *Science* 117: 591(1953). **C-14.** H. M. Rauen *et al.*, *Z. physiol. Chem.* 292: 101(1953). **C-15.** F. W. Putnam, *JBC* 203: 347(1953). **C-16.** D. A. I. Goring and P. Johnson *ABB* 56: 448(1955). **C-17.** J. R. Colvin *et al.*, *Chem. Revs.* 54: 687(1954).

### Section D

**D-1.** I. M. Klotz *et al.*, *J. Phys. & Colloid Chem.* 54: 18(1950); 55: 101(1951). **D-2.** E. J. Cohn *et al.*, *JACS* 72: 465(1950). **D-3.** M. Sevag *et al.*, *JBC* 124: 425(1938). **D-4.** A. Engström and M. A. Jakus, *Nature* 161: 168(1948).

**Section E**

**E-1.** P. L. Kirk, *Adv. in Protein Chem.* 3: 139(1947). **E-2.** A. C. Chibnall *et al., BJ* 37: 354(1943). **E-3.** F. W. Sunderman, Jr., *et al., Am. J. Clin. Pathol.* 30: 112(1958); *FP* 17: 319(1958). **E-4.** A. Hiller *et al., JBC* 176: 1401(1948). **E-5.** F. Rappaport and G. Geiger, *Mikrochemie* 18: 43(1935). **E-6.** E. J. Conway *et al., BJ* 27: 419(1933); 36: 655(1942). **E-7.** J. Fine, *BJ* 29: 799(1935); H. A. Stiff, *JBC* 177: 179(1949). **E-8.** H. Nielsen, *Acta Chim. Scand.* 12: 38(1958). **E-9.** O. H. Lowry and T. Hunter, *JBC* 159: 564(1945). **E-10.** J. Harkness and R. Wittington, *Anal. Chim. Acta* 1: 249(1947). **E-11.** H. Eisen, *J. Immunol.* 60: 77(1948). **E-12.** O. H. Lowry *et al., JBC* 193: 265(1951). **E-13.** W. P. Jencks *et al., BJ* 60: 205(1955). **E-14.** G. T. Franglen and N. H. Martin, *BJ* 57: 626(1954). **E-15.** J. S. Robertson *et al., J. Appl. Radiation and Isotopes* 7: 33(1959/1960). **E-16.** M. Vaughan *et al., Science* 126: 446(1957).

# Chapter III

# The Amino Acid Composition
# of Proteins

## A. Cleavage by Hydrolysis and Other Methods

Acids, bases, and enzymes are the agents used for the *hydrolytic cleavage* of proteins. In most cases, boiling *hydrochloric acid* has been utilized to effect hydrolysis. Since the constant boiling acid contains only 20.5% HCl, the commercial concentrated acid, which contains 35% HCl, is diluted before use. The dry protein is mixed with approximately 10 volumes of the 20.5% acid and kept in a boiling water bath for 30–60 minutes. By this procedure the formation of froth during hydrolysis is avoided. Total hydrolysis is then achieved by refluxing the protein for 12–70 hours with the boiling hydrochloric acid.

The hydrochloric acid used for hydrolysis should be free of traces of iron since ferric chloride catalyzes the oxidative destruction of some of the amino acids (see below). This can be prevented to some extent by hydrolysis with a mixture of concentrated HCl and formic acid (A-1). The best results are obtained when tubes containing the protein and glass-distilled HCl are evacuated, sealed, and then heated until hydrolysis is complete. In all these procedures the excess HCl is removed in vacuum over NaOH. If formic acid was used it is removed at the same time. The residue contains the amino acids as the hydrochlorides. They can be converted into the free amino acids by being passed through a column of an anion exchange resin. When it is intended to prepare free amino acids, it may be advantageous to hydrolyze with 8 *N* sulfuric acid which can be removed from the hydrolyzate quantitatively by the addition of the equivalent amount of barium hydroxide. The heavy precipitate of barium sulfate must subsequently be extracted several times with boiling water in order to remove the considerable amounts of amino acids which have been adsorbed. The filtrates and washings are pooled and evaporated, leaving a mixture of the amino acids.

The advantage of acid hydrolysis is that racemization is avoided, the amino acids being obtained as unchanged L-amino acids. Most of the amino acids are resistant to boiling mineral acids. *Tryptophan*, which is completely destroyed by the action of boiling acid, can be isolated from enzymatic hydrolyzates. The decomposition products of tryptophan are converted into a dark brown substance called *humin* which is probably formed by the condensation of the indole nucleus of tryptophan with small amounts of aldehydes produced during the hydrolysis (A-3). During acid hydrolysis considerable amounts of the sulfur amino acids are oxidized to various products, and a small portion of the hydroxyamino acids, serine and threonine, undergoes oxidation to the corresponding α-keto acids (A-2). Asparagine and glutamine are hydrolyzed by boiling HCl to aspartic and glutamic acid. The liberated ammonia is found in the hydrolyzate as ammonium chloride.

The best method for the quantitative determination of the sulfur amino acids *cystine* and *cysteine* is oxidation of the protein by treatment with cold performic acid (prepared from $H_2O_2$ and formic acid) and subsequent hydrolysis with HCl. Cystine and cysteine are converted into cysteic acid, $HO_3S \cdot CH_2 \cdot CHNH_2 \cdot COOH$, which is resistant to the action of boiling HCl and can be determined in the hydrolyzate (A-4). The oxidation of *hydroxyamino acids* during HCl hydrolysis takes place very slowly. The true content of these acid-sensitive amino acids in the protein can be determined by hydrolyzing different protein samples for various periods of time, measuring in each of the hydrolyzates the amount of serine and threonine, and calculating the initial amount of these amino acids by extrapolation to zero time.

Complete hydrolytic cleavage can also be accomplished by boiling the proteins with 4 *N barium hydroxide* (A-5) or with alkali hydroxides. Preference is given to barium hydroxide since the excess can be removed by an equivalent amount of sulfuric acid. The alkaline hydrolyzates are colorless and free of humin. The disadvantages of the alkaline hydrolysis are that the amino acids undergo racemization, that some of them are deaminated, that arginine is converted into ornithine and urea, and that cystine and cysteine are destroyed.

Complete hydrolysis can also be achieved by the action of *proteolytic enzymes* under very mild conditions (A-13). In this manner quantitative conversion of several proteins into amino acids has been accomplished. The enzymes used were papain, leucine aminopeptidase, and carboxypeptidase (A-13). The enzymatic hydrolyzate contains not only tryptophan but also glutamine and asparagine. Hence both amides can be isolated from the hydrolyzate (A-6). Enzymatic hydrolysis is especially useful when it is intended to obtain peptide intermediates by partial hydrolysis. In view of the great importance of this method it will be discussed separately in Chapter IV.

*Partial hydrolysis* can be accomplished also by concentrated HCl at 37°C (A-7). The presence of specific peptides in such hydrolyzates indicates that the susceptibility of various peptide bonds to HCl is different; for example, peptide bonds formed by aspartyl residues (A-12) or by serine and threonine are very easily hydrolyzed. The —CO·NH— bond formed by the amino groups of the hydroxy acids is easily converted into an ester linkage —CO·O— in which the hydroxyl group of serine or threonine combines with the carboxyl group of an adjacent amino acid. The ester bonds formed by this "acyl shift" are more susceptible to hydrolysis than the peptide bonds. In contrast to these easily hydrolyzable peptide bonds, those formed by valine are particularly stable (A-7,8). Partial hydrolysis of proteins has also been achieved by treatment with 25 $N$ sulfuric acid at 40°C (A-9) or by boiling the protein with Dowex 50, a sulfonated ion exchange resin (A-10).

When proteins are heated with anhydrous hydrazine, $N_2H_4$, the peptide bonds are cleaved by *hydrazinolysis;*

$$R·CO·NH·R' + N_2H_4 \rightarrow R·CO·NH·NH_2 + H_2N·R'$$

Akabori (A-11), who described this reaction, utilized it for the determination of the terminal amino acid which has a free COOH group. In contrast to all other amino acids in the peptide chain, which are converted to their hydrazides, the terminal amino acid is found in the hydrolyzate as free amino acid.

## B. Determination of Rate of Hydrolysis

The hydrolysis of proteins consists of the cleavage of peptide and amide bonds according to the following reaction:

$$R·CO·NH·R' \xrightarrow{+H_2O} R·COOH + R'·NH_2$$

The extent of hydrolysis can be ascertained at any point by measuring the increase in carboxyl, in amino groups, or in both.

The *increase in amino groups* can be measured by titration with alkali in the presence of formaldehyde or ethanol. The *formol titration* was first applied by Sørensen (B-1) and was originally explained by the formation of $N$-methylene compounds of the structure $R·N{=}CH_2$. Most probably mono- or dimethylol compounds, $R·NH·CH_2OH$ or $R·N(CH_2OH)_2$, respectively, are formed (B-2). The apparent basicity of the amino groups is thereby greatly reduced (see Chapter VI). The $N$-hydroxymethylamino acids can be titrated with alkali and phenolphthalein in the same manner as fatty acids. Accordingly, each equivalent of alkali consumed corresponds to one equivalent of peptide bond cleaved. The *ethanol titration* is rendered possible by a shift of the pK value of phenolphthalein used as an indicator; while the color change in aqueous solution occurs over a wide pH range near pH 9 where many of the amino groups are still protonated, in alcoholic solution it takes place near pH 12 within a narrow pH range.

At this high pH value the reaction: $R \cdot N^+H_3 + OH^- \rightarrow R \cdot NH_2 + H_2O$ goes to completion (see Chapter VI) (B-3).

Because amino acids are weaker acids than are peptides (see Chapter VI, Table 2) the enzymatic hydrolysis of peptides is accompanied by a slight decrease in acidity. If carbon dioxide and bicarbonate are present, absorption of carbon dioxide results and the decrease of the carbon dioxide pressure is a measure of the extent of hydrolysis. In this manner the rate of enzymatic hydrolysis can be measured manometrically in the Warburg apparatus (B-4).

The *increase in amino groups* during hydrolysis can also be determined by the method of Van Slyke (B-5), which is based upon the reaction of nitrous acid with primary amino groups to form nitrogen gas:

$$R \cdot NH_2 + HONO \rightarrow R \cdot OH + N_2 + H_2O$$

While the $\alpha$-amino groups of amino acids and the terminal $\alpha$-amino groups of peptides react very rapidly with nitrous acid, the $\epsilon$-amino groups of lysine react very slowly. A slow formation of nitrogen gas is also caused by ammonia, which is produced by hydrochloric acid from the amide groups of asparagine and glutamine (B-5). The volume of the nitrogen formed may be measured at atmospheric pressure or manometrically under reduced pressure. The Van Slyke method provides excellent results with most of the amino acids and peptides; however, consideration must be taken of the fact that nitrous acid itself is converted into nitrogen by strongly reducing substances such as cysteine. Amounts of nitrogen in excess of the theoretical are also produced by glycine and its peptides (B-6).

While the methods outlined above enable us to determine the rate of the liberation of amino and carboxyl groups during the hydrolysis, they do not provide a means of distinguishing between *free amino acids* and peptides. Such a differentiation is accomplished by using reagents which react with the carboxyl and the $\alpha$-amino group of the same amino acid molecule at the same time. The most important of these reagents is triketohydrindene hydrate, the trade name of which is *ninhydrin*. If ninhydrin is heated with an aqueous solution of an amino acid, the amino acid is oxidized to $CO_2$, ammonia, and an aldehyde which contains one carbon atom less than the amino acid. At the same time the hydrated keto

group, $\diagdown{}C(OH)_2\diagup{}$, of ninhydrin is reduced to an alcohol group, $\diagdown{}CHOH\diagup{}$.

Ninhydrin

This reduced form of ninhydrin combines with the excess of the hydrated keto form and ammonia to give a purplish or blue dye whose concentration can be measured colorimetrically (B-7). If the heating with ninhydrin is performed at pH 1–5, the amount of free amino acid can be determined by measurement of the $CO_2$ gas evolved (B-8). Although proteins and peptides, like ammonia and amines, give a blue color with ninhydrin, they do not form $CO_2$ when heated with ninhydrin. Hence, free amino acids can be determined by this method in the presence of proteins or peptides (B-8,-9).

## C. Fractionation and Isolation of Amino Acids (C-1)

Methods for the separation and isolation of all amino acids have become available only recently. Previously it was necessary to be content with the separation of certain classes of amino acids. Thus the *basic amino acids* arginine, lysine, and histidine were precipitated by phosphotungstic acid and were separated from each other by precipitation with silver sulfate at different pH values (C-2). The principal procedure used for the fractionation of the *monoamino acid* mixtures was the fractional *distillation of their esters* as described in the classical papers of Fischer (C-3). The distillation involved great losses, so that the results were far from quantitative. But it was by this method that the peptide-like structure of the proteins and their content of amino acids were conclusively established.

In the last few years, new methods for fractionating protein hydrolyzates have been developed. Synge (C-4) has made use of the variations of the *partition* coefficient of amino acids between water and butanol, phenol, or collidine, to achieve their resolution. Even if the solubility of several amino acids in water or in the organic solvent is almost the same, their distribution between water and the organic solvent will be different. If filter paper, starch, or another hydrophilic solid phase is used, adsorption phenomena will take place at the solid-liquid interface. Moreover the ratio of organic solvent to water at the interface may be different from the same ratio in the bulk of the mobile liquid phase. Therefore different amino acids move with different velocities over the surface of the paper or through chromatographic columns.

In paper chromatography (C-5), a small amount of the protein hydrolyzate is placed onto a line about 6 cm from the end of the paper strip. This end is then dipped into the solvent mixture and the entire system placed in a closed glass jar or another closed chamber. It is important that the atmosphere in this closed system be saturated with the vapors of water and the organic solvent. Excellent results with complete separation of all amino acids have been obtained when chromatography was carried out under rigorously controlled conditions (C-24). In ascending paper chromatography (C-6) the paper is suspended from the top and its lower end dipped into a dish with the solvent mixture. Large paper sheets can be folded to form a cylinder which can be dipped into the solvent mixture without the necessity for suspension. In descending paper chromatography the upper end of the paper strip or sheet is dipped into a trough

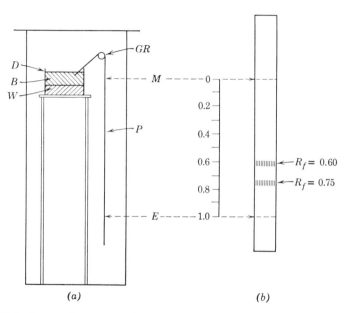

*(a)*          *(b)*

Fig. III-1. Descending paper chromatography $D$: dish; $W$: water; $B$: butanol; $M$: mixture of amino acids at beginning of migration; $P$: paper; $E$: front of solvent; $GR$: glass rod. The symbol $R_f$ is explained in the text. (From F. Haurowitz, "Biochemistry, an Introductory Textbook," p. 70. Wiley, New York, 1955.)

with the solvent mixture which is allowed to flow downward (Fig. III-1). Excellent and rapid resolution is also accomplished by thin-layer chromatography of the amino acids on various gels (C-20). In this method the mixture of amino acids is placed on glass plates which are coated by a homogeneous layer of silica gel or aluminum hydroxide (C-23). Resolution in ascending chromatography is accomplished within 30–100 minutes.

To fractionate larger amounts of amino acids, columns of starch (B-7, C-7), silica gel (C-8), or ion exchange resin (C-9) are used. Since distribution of the amino acids between mobile and stationary phase is different, they are eluted at a different rate. In this way they are separated from each other. If filter paper or thin-layer plates are used as support, further fractionation can be achieved by turning the support around by an angle of 90° and chromatographing with a second solvent system. In this manner the mixture of amino acids is resolved into "spots" which can be made visible by the ninhydrin test as shown in Fig. III-2. Each of these spots consists in general of a single amino acid so that by this method separation of most of the amino acids is possible.

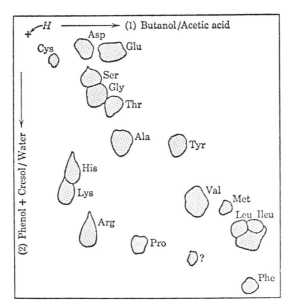

FIG. III-2. Paper chromatogram of a protein hydrolyzate. The hydrolyzate was placed on the point marked *H* and first run with a mixture of 10 % acetic acid and butanol, then turned by an angle of 90° and run with a mixture of phenol plus cresol and water. The amino acids, made visible by spraying with ninhydrin, are denoted by the first three letters of their names, except for isoleucine (ileu). (From F. Haurowitz, "Biochemistry, an Introductory Textbook," p. 135. Wiley, New York, 1955.)

The rate of migration, $v_a$, of an amino acid in a given solvent system depends on $v_s$, the flow rate of the solvent mixture. The ratio $v_a/v_s$ is a constant typical for each of the amino acids (C-1,4). This ratio has been designated by the symbol $R_f$ (ratio of flow rates). To determine the $R_f$ value of an amino acid we measure $d_a$, the distance of migration of the respective amino acid, and divide it by $d_s$, the distance between the origin and the front of the solvent mixture (Fig. III-1). The ratio $d_a/d_s$ is equal to $R_f$. Obviously, $R_f$ is always smaller than 1.0 since the amino acids cannot migrate beyond the liquid front. Care must be taken that the measurement is made before the liquid front reaches the end of the paper since this would render impossible the determination of the distance from the origin to the front of the solvent.

After the discovery of the high resolving power of paper chromatography, a large number of solvent systems containing two or more components has been used for the improved resolution of amino acid mixtures (C-10,18). Further resolution is frequently accomplished when one-dimensional paper chromatography is combined with high-voltage electrophoresis (C-11,12) in a direction perpendicular to that of the chromatography. In another modification, the cellulose fibers which form the stationary phase in paper chromatography are converted into DEAE-cellulose, or another ion exchange paper similar to the ion exchange resins mentioned earlier (Chapter II, B).

The invisibility of the amino acids on the untreated paper is a disadvantage common to all these methods. If we make them visible by ninhydrin, they undergo decarboxylation. Several attempts have been made to use colored derivatives of the amino acids whose migration can be followed visually during chromatography. Good results have been obtained with the intensely yellow dinitrophenylamino acids, prepared by treatment of the amino acid mixture with 2,4-dinitrofluorobenzene; the DNP-amino acids are easily separated by two-dimensional paper chromatography (C-13). Since the DNP-amino acids are light-sensitive, the chromatography must be carried out in a dark room or under an opaque cover. Similar results have been obtained with the colorless phenylthiohydantoin derivatives of the amino acids (see Chapter IV, B) in the presence of dimethylformamide (C-19). The phenylthiohydantoins, because of their ability to absorb ultraviolet light, are visible as dark spots on the paper if the latter is made fluorescent by exposure to ultraviolet light in the absence of visible light (C-14). Attempts have also been made to convert the amino acids into the analogous amino alcohols by treatment with lithium aluminum hydride, $LiAlH_4$, or with $LiBH_4$ (C-15), and to separate these or their DNP-derivatives by paper chromatography (C-16). If radioactive amino acids containing $C^{14}$, $S^{35}$, or $H^3$ are fractionated, their location on the paper can be determined by scanning the paper with a counting device or by autoradiography. The $R_f$ values of the "heavy" $C^{14}$-amino acids are somewhat lower than those of the $C^{12}$-amino acids (C-17).

The amount of amino acids which can be fractionated by paper chromatography is usually much less than 1 mg, depending on the thickness of the paper. For larger amounts of amino acids, column chromatography is used. This extremely powerful method makes it possible to determine amino acids quantitatively as will be shown below (Section E). Since the L-amino acids identified by column chromatography account for more than 99% of the total nitrogen of the examined proteins, no appreciable amounts of other substances can be present in the molecules of these proteins. That one must beware of generalizations, however, is illustrated by the finding of ethanolamine, diaminobutyric acid, and the "unnatural" D-amino acids in certain bacterial polypeptides (see Chapter XIV).

Since amino acids are not volatile, their mixtures cannot be analyzed as such by the very powerful method of gas chromatography. Gas chromatography has been used, however, after conversion of the amino acids into their volatile N-trifluoroacetyl derivatives or their methyl esters (C-21,22).

## D. *Amino Acids: General Properties and Specific Reactions (D-1, 4 to 7)*

The amino acids are zwitterions, easily soluble in water and insoluble in organic solvents such as ethanol. Only the imino acids proline and hydroxyproline are soluble in ethanol. Because of their similar physicochemical behavior, the amino acids cannot be separated by fractional addition of alcohol or of neutral salts; however, the solubility of cystine in neutral solutions and that of tyrosine in slightly acid solutions is so low that the bulk of these two amino acids can be precipitated by adjusting the pH to the proper value. Other amino acids can be precipitated by the addition of suitable precipitating reagents. None of these methods is fully satisfactory, because all the precipitates are soluble to a certain extent.

The amino acids obtained by total hydrolysis of the proteins are *optically active* if racemization is avoided during the hydrolysis. Although some of the amino acids are dextrorotatory (*d*-amino acids) and other amino acids levorotatory (*l*-amino acids), all the amino acids isolated from the well-known proteins have the same configuration, i.e., the same spatial arrangement of the four substituents of the α-carbon atom. All these natural amino acids are designated L-amino acids. Their mutual relations have been proved by the conversion into identical derivatives (D-1). The original designation of the natural amino acids as L-amino acids was quite arbitrary. Much later it was found that their configuration happens to be the same as that of L-glyceraldehyde, which is the reference substance for the L-sugars (D-2). Among the twenty amino acids which occur in proteins only glycine is devoid of an asymmetric carbon atom and therefore is optically inactive.

Although the separation of D-amino acids from their L-isomers can be accomplished by fractional crystallization of their salts with asymmetric bases such as brucine, quinine, or fenchylamine (D-3), better yields are obtained by enzymatic methods in which enzymes are used which attack only one of the two isomers. Thus a mixture of D- and L-amino acids can be separated by acetylation and subsequent exposure of the mixture of D- and L-acetylamino acids to the enzyme acylase from hog kidney. Only the L-acetylamino acids are hydrolyzed by the enzyme (D-4). The unhydrolyzed D-acetylamino acids can be extracted from the aqueous solution by organic solvents whereas the free L-amino acids remain in the aqueous layer. The D-amino acids are easily obtained by hydrolysis of their acetyl derivatives with hydrochloric acid.

D- and L-amino acids can also be differentiated by means of *microbiological methods. Leuconostoc mesenteroides,* mutant strains of *Neurospora,* or different strains of lactobacilli are used in these tests. The microorganisms are cultivated on a medium containing all amino acids except the one to be detected in an unknown sample. If a strain is used for which this amino acid is indispensable, the cells will not multiply. The growth of the culture is measured by the turbidity produced by the growing microorganisms, by determination of the lactic acid formed, or by weighing the mycelium of the mold culture (D-5,6; Section E, below).

Some of the amino acids give typical *color reactions.* Many of these reactions have been used for their identification and colorimetric determination. They are at present rarely used for this purpose since chromatography with ninhydrin renders possible the identification and determination of all amino acids, not only those which give specific color reactions. The colorimetric methods can be used, however, to obtain some insight into the conformation of the peptide chain of a protein, i.e., into the mode of folding. Obviously only those amino acid residues which are in or close to the surface of a globular protein macromolecule are accessible to color reagents and will give positive color reactions even if they are protein-bound. Amino acid residues which are buried inside the globular protein molecule are inaccessible to the reagents and therefore will not react in the native protein. They react, however, when the protein molecule is unfolded by denaturing reagents as described later in Chapter VII (Sections F and H). In the same chapter it will be shown (Section B) that the polar side chains of the amino acids are of great importance for the conformation (i.e., folding and tertiary structure) of the native protein molecule. For all these reasons it may be useful to describe in the following paragraphs those properties of the amino acids which are of importance in their role as building stones of the proteins. More details on the chemistry, the physical chemistry, and the metabolism of amino acids will be found in special texts on these topics (D-4 to 7).

*Glycine* (German: *Glykokoll*), $NH_2 \cdot CH_2 \cdot COOH$, can be determined by a color reaction with *o*-phthalaldehyde (D-8) or by its oxidation to formaldehyde and subsequent colorimetric determination of the aldehyde with chromotropic acid (1,8-dihydroxynaphthalene-3,6-disulfonic acid) (D-9). *Leuconostoc mesenteroides* has been used for its microbiological determination (D-10). With nitrous acid, glycine evolves more than one equivalent of gas since some $CO_2$ and $N_2O$ are formed in addition to $N_2$ (D-11).

*Alanine* is converted by nitrous acid to lactic acid, $CH_3 \cdot CHOH \cdot COOH$, which is determined in the usual way by oxidation to acetaldehyde (D-12).

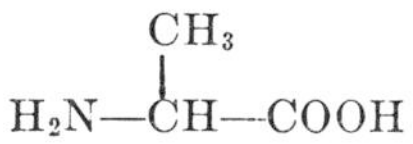

$$CH_3$$
$$|$$
$$H_2N-CH-COOH$$
Alanine

Alanine can also be determined by means of *Streptococcus faecalis* (D-10). On heating with ninhydrin (see Section B, above) alanine is converted into acetaldehyde which is distilled off and determined by titration (D-13).

*Valine, leucine,* and *isoleucine* also furnish volatile aldehydes under these conditions and can be determined similarly (D-13). Each of these three

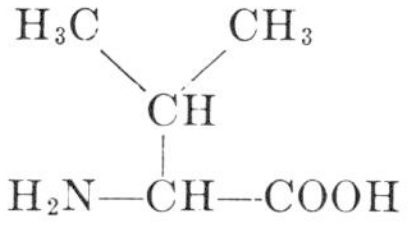

$$H_3C \quad CH_3$$
$$\backslash \quad /$$
$$CH$$
$$|$$
$$H_2N-CH-COOH$$
Valine

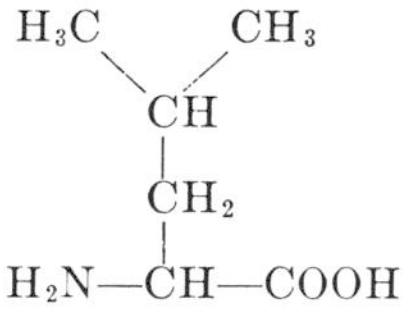

$$H_3C \quad CH_3$$
$$\backslash \quad /$$
$$CH$$
$$|$$
$$CH_2$$
$$|$$
$$H_2N-CH-COOH$$
Leucine

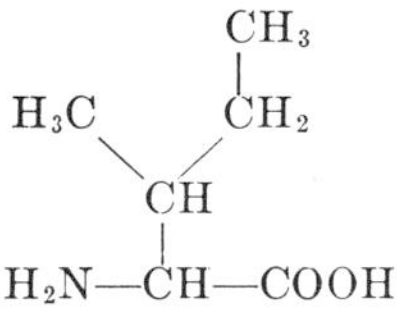

$$CH_3$$
$$|$$
$$H_3C \quad CH_2$$
$$\backslash \quad /$$
$$CH$$
$$|$$
$$H_2N-CH-COOH$$
Isoleucine

amino acids can be determined microbiologically (D-5,14). Leucine, owing to its low solubility, can be determined easily by isotopic dilution (D-15) (see Section E). The nonpolar side chains of leucine, isoleucine, and valine give rise to the formation of hydrophobic bonds in protein molecules (see Chapter VII, E).

*Aspartic acid* and *glutamic acid* can be precipitated as calcium or barium salts from alcoholic solution (D-16). Both amino acids have also been

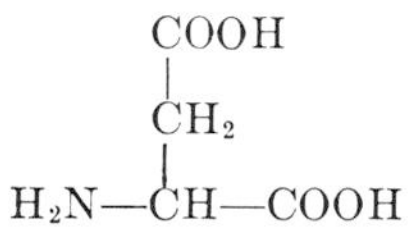

$$COOH$$
$$|$$
$$CH_2$$
$$|$$
$$H_2N-CH-COOH$$
Aspartic acid

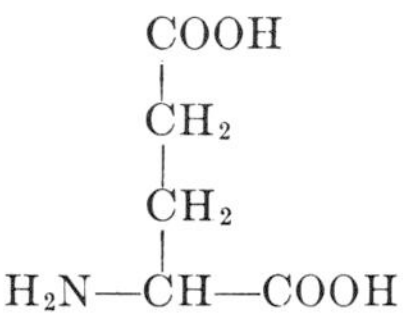

$$COOH$$
$$|$$
$$CH_2$$
$$|$$
$$CH_2$$
$$|$$
$$H_2N-CH-COOH$$
Glutamic acid

determined microbiologically (D-5,10). Glutamic acid, upon oxidation with Chloramine T ($CH_3 \cdot C_6H_4SO_2 \cdot NHCl$), is converted into succinic acid, $COOH \cdot CH_2 \cdot CH_2 \cdot COOH$, which is determined manometrically by means of succinoxidase (D-17). The amount of glutamic acid present in a protein hydrolyzate can also be estimated from the decrease in amino nitrogen on heating of the hydrolyzate at pH 3.3 under pressure. Under these conditions asparagine and aspartic acid are stable whereas glutamine and glutamic acid are converted into pyrrolidonecarboxylic acid (D-18,20) which has no free amino group. While some of the aspartic and glutamic acid residues are present in protein molecules as the dicarboxylic acids with free carboxyl groups, a portion of these groups is substituted by ammonia and converted into an acid amide group, —$CONH_2$. In enzymatic hydrolyzates asparagine as well as glutamine can be determined by chromatography (D-19).

$$H_2C \overline{\quad\quad} CH_2$$
$$H_2NOC \quad\diagdown \quad CH-COOH$$
$$\diagup$$
$$H_2N$$
Glutamine

$$H_2C \overline{\quad\quad} CH_2$$
$$OC \diagdown_{\underset{H}{N}} \diagup CH-COOH \; + \; NH_3$$
Pyrrolidonecarboxylic acid

If proteins are heated with hydrazine (*hydrazinolysis*), the aspartyl and glutamyl residues are converted into free α-monohydrazides of aspartic and glutamic acid. Under these conditions, the residues of the amides asparagine and glutamine undergo hydrazinolysis not only of their α-peptide bonds, but also of their β- or γ-amide bonds and are thus converted into their dihydrazides (D-21).

$$CO \cdot NH_2$$
$$|$$
$$CH_2 \quad \xrightarrow{\;N_2H_4\;}$$
$$|$$
$$-NH \cdot CH \cdot CO-$$
Asparagine residue

$$CO \cdot NH \cdot NH_2$$
$$|$$
$$CH_2$$
$$|$$
$$H_2N \cdot CH \cdot CO \cdot NH \cdot NH_2$$
α,β-Dihydrazide of aspartic acid

When the carboxyl groups of aspartyl and glutamyl residues are esterified and subsequently reduced to primary alcohol groups by treatment with $LiBH_4$, the amide groups of asparagine and glutamine residues in the protein chains remain unchanged. If the reduced protein is hydrolyzed by hydrochloric acid, the residues of the amides asparagine and glutamine are converted into aspartic and glutamic acid, whereas aspartyl and glutamyl residues of the protein are found in the hydrolyzate as α-amino-γ-hydroxybutyric acid and α-amino-δ-hydroxyvaleric acid, respectively (D-22).

*Hydroxyamino acids* can be determined by means of oxidation with periodic acid $HIO_4$ (D-23). By this procedure *serine* is converted to

$$CH_2OH$$
$$|$$
$$H_2N-CH-COOH$$
Serine

$$CHOH-CH_3$$
$$|$$
$$H_2N-CH-COOH$$
Threonine

formaldehyde, and threonine to acetaldehyde, one equivalent of ammonia being liberated. Ammonia is also formed from *hydroxylysine*, an amino acid found in noticeable amounts only in collagen (D-24). For the determination of hydroxylysine, the protein hydrolyzate is precipitated with phosphotungstic acid and the precipitate treated with periodate (D-24). Some of the protein-bound hydroxyamino acids react with alkyl fluorophosphates and are involved in the catalytic activity of esterases (see Chapter VII, F; Chapter XII, B).

$$CH_2—CH_2—CHOH·CH_2NH_2$$
$$H_2N—CH—COOH$$

Hydroxylysine (D-24,-25)

*Cystine* is easily reduced by thioglycol, thioglycolic acid (D-26), or other reducing agents, to yield *cysteine*. In the absence of reducing substances, cystine is stable whereas cysteine solutions, at alkaline pH values and in the presence of oxygen, undergo oxidation to cystine. In acid solutions, cysteine is stable. It can be oxidized to cystine by iodine (D-27), Folin's reagent (phosphomolybdic acid, D-28), or porphyrindin (D-29). Each of these methods may be used for the quantitative determination of cysteine. The sulfhydryl groups of cysteine can also be titrated amperometrically with silver or mercury salts (D-30) or determined colorimetrically by means of the intensely colored 4-(*p*-dimethylaminobenzeneazo)-phenyl-mercury acetate (D-31). Many of these reactions, particularly that with PCMB (*p*-chloromercuribenzoate), have been applied to detect protein-bound cysteine residues. In some of the proteins the sulfhydryl groups of cysteine react with added iodine to yield sulfenyl iodide groups: $RSH + I_2 \rightarrow RSI + HI$ (D-32).

$$CH_2SH$$
$$H_2N—CH—COOH$$

Cysteine

$$CH_2—S—S—CH_2$$
$$H_2N—CH \qquad CH—COOH$$
$$COOH \qquad NH_2$$

Cystine

One of the most sensitive reactions for the detection of traces of cystine or cysteine on paper chromatograms is Feigl's iodine-azide test, in which the dark blue iodine-starch complex is reduced by alkali azide to colorless iodide: $I_2 + 2N_3^- \rightarrow 2I^- + 3N_2$ (D-33). The reaction is catalyzed by compounds containing the groups —S—, —SS— or —SH, and allows us to detect $10^{-3}$ µg of cystine or cysteine.

By means of isotopically labeled cystine and cysteine it has been shown that in their aqueous solutions the following *exchange reactions* take place (D-44b):

$$RSSR + R'S^*H \rightleftharpoons RSS^*R' + RSH \quad \text{and} \quad RSSR + R'S^*S^*R' \rightleftharpoons 2RSS^*R'$$

Evidently, the dithio bond, —SS—, is easily and reversibly cleaved. The intermediates are most probably not the radicals RS• but the sulfide ion RS⁻ and the sulfenium ion RS⁺ or its hydrated form (D-34). Similar exchange reactions take place in proteins either inter- or intramolecularly (see Chapter VII, E and H). Protein-bound cystine and cysteine can be determined by such an exchange reaction with the intensely yellow bis(dinitrophenyl)cystine (D-44c).

The dithio bonds of cystine can undergo cleavage by oxidation, reduction, sulfitolysis, or by the action of cyanides. The first of these reactions is usually brought about by performic acid (A-4) or peracetic acid at 0°C; it yields the stable compound cysteic acid. The oxidation of cystine or cysteine to cysteic acid has also been carried out on paper during paper chromatography (D-35). Reductive cleavage of the dithio bond is accomplished by addition of an excess of mercaptoethanol ($CH_2OH \cdot CH_2SH$), thioglycolate, or other sulfhydryl compounds, or by treatment with sodium borohydride, $NaBH_4$.

$$CH_2{-}SO_3H \qquad CH_2{-}S{-}SO_3H \qquad CH_2{-}S{-}CH_2{-}CH_2{-}NH_2$$
$$H_2N{-}CH{-}COOH \qquad H_2N{-}CH{-}COOH \qquad H_2N{-}CH{-}COOH$$

| Cysteic acid | Cysteine-S-sulfonic acid | S-Aminoethylcysteine |
|---|---|---|

Treatment of cystine with sulfite results in the formation of one molecule of cysteine and one of cysteine-S-sulfonate (D-36):

$$RSSR + H_2SO_3 \rightarrow RSSO_3H + RSH$$

In the presence of oxidizing agents the RSH formed is reoxidized to cystine and thus all of the cystine is finally converted into cysteine-S-sulfonic acid (D-37). Cyanides convert cystine into one molecule of a thiocyanate and one of cysteine: $RSSR + HCN \rightarrow RSCN + RSH$.

The conversion of cystine into cysteine or cysteic acid is accompanied by an enormous change in the optical rotation since cystine has a specific rotation which, depending on pH, varies from approximately $-200°$ to $-300°C$, whereas the specific rotations of cysteine and cysteic acid under the same conditions vary from $-5°$ to $+8°C$. Homocystine, in which another $CH_2$ group is inserted between the SS group and each of the two asymmetric carbon atoms, does not have the strong levorotation shown by cystine (D-38). Evidently, the high $[\alpha]_D$ value of cystine is caused by the proximity of the two asymmetric carbon atoms to the dithio group (D-39).

Cysteine and its S-sulfonate are labile compounds, easily oxidized to cystine or cysteic acid. Stable derivatives of cysteine are obtained by alkylation of the sulfhydryl group, for instance by iodoacetic acid which yields carboxymethylcysteine: $RSH + CH_2I \cdot COOH \rightarrow RS \cdot CH_2 \cdot COOH + HI$. In analogous reactions the sulfhydryl groups react with iodoacetamide, $ICH_2 \cdot CONH_2$ (D-40), or bromoethylamine. Reaction of cysteine with the last-named compound converts the molecule into S-aminoethylcysteine, which is an analog of lysine. It differs from lysine merely by the replacement of a $CH_2$ residue by a sulfur atom. Peptides of S-aminoethylcysteine, like those of lysine, are hydrolyzed by the enzyme trypsin (D-41). The SH groups of cysteine also react specifically with $N$-ethylmaleimide as shown below (D-42):

$$RSH + \begin{array}{c} CH{-}CO \\ \| \qquad \qquad NC_2H_5 \rightarrow \\ CH{-}CO \end{array} \qquad \begin{array}{c} RS{-}CH{-}CO \\ | \qquad \qquad NC_2H_5 \\ CH_2{-}CO \end{array}$$

If proteins are hydrolyzed with alkali, their cystine residues lose one of the sulfur atoms and are converted into lanthionine: $S(CH_2\cdot CHNH_2\cdot COOH)_2$ (D-43). Cystine and cysteine are destroyed by hydrazinolysis, which reduces their sulfur to hydrogen sulfide (D-44a).

For the quantitative determination of *methionine*, advantage is taken

$$CH_2—S—CH_3$$
$$|$$
$$CH_2$$
$$|$$
$$H_2N—CH—COOH$$
Methionine

of its property of reacting to form methyl iodide when boiled with hydriodic acid. The methyl iodide evolved is determined in the usual way, titrimetrically (D-45). Methionine gives a red color with nitroprussides in acid solution (D-46). During HCl hydrolysis, some of the sulfur atoms of the thioether group undergo oxidation to SO or $SO_2$; consequently, the hydrolyzate contains small amounts of methionine sulfoxide and methionine sulfone (D-47).

*Arginine* is one of the basic amino acids (hexone bases) which are precipitated from the protein hydrolyzate with phosphotungstic acid. All these amino acids (arginine, histidine, lysine, and hydroxylysine) can be isolated from the precipitate as hydrochlorides by acidification with HCl and extraction of the phosphotungstic acid by ether and amyl alcohol. In dealing with small amounts of these bases it is simpler to remove them from the hydrolyzate by a short column of a cation exchange resin which at neutral pH values retains only these basic acids (see below, Section E). In the proteins the basic side chains of the hexone bases neutralize the COOH groups of aspartic and glutamic acid (see Chapter VII, E).

*Arginine* can be separated from the other basic amino acids by precipitation with flavianic acid (1-naphthol-2,4-dinitro-7-sulfonic acid) (D-48). Boiling with alkali destroys arginine, two molecules of ammonia being formed for each molecule of arginine. The enzyme arginase hydrolyses arginine, cleaving it to form urea and ornithine. A very sensitive color reaction given by arginine with sodium hypochlorite and $\alpha$-naphtol or 8-hydroxyquinoline (Sakaguchi reaction) can be used for the quantitative determination of arginine (D-49).

$$CH_2—CH_2—NH—C(=NH)—NH_2$$
$$|$$
$$CH_2$$
$$|$$
$$H_2N—CH—COOH$$
Arginine

$$CH_2—CH_2—CH_2—CH_2NH_2$$
$$|$$
$$H_2N—CH—COOH$$
Lysine

*Lysine*, another basic amino acid, has been separated from the other bases by precipitation with picric acid. It can be determined microbiologically with *Leuconostoc mesenteroides* or by bacterial decarboxylation (D-50). Some of the ε-amino groups of protein-bound lysine residues are free and can be substituted by alkylation with methyl iodide, DNFB (2,4-dinitro-fluorobenzene), or other alkyl halides. Similarly, the ε-amino groups can be substituted by acylation (e.g., with acetic anhydride or benzoyl chloride). They also react with phenylisocyanate or phenylisothiocyanate, to give phenylhydantoins or phenylthiohydantoins respectively (see below, Section E). The ε-amino groups form condensation products with aldehydes. They also react with aromatic diazo compounds to give colored azo derivatives (D-51). Amidination of the protein-bound lysine residues results in the formation of homoarginine residues.

$$NH_2 \cdot C(:NH) \cdot NH \cdot (CH_2)_4 \cdot CHNH_2 \cdot COOH$$
Homoarginine

*Hydroxylysine*, the 5-hydroxy derivative of lysine, has been discussed with the other hydroxyamino acids. *Ornithine*, ($\alpha,\delta$-diaminovaleric acid), the lower homolog of lysine, has been found in some of the toxic bacterial peptides. Other unusual amino acids found in these peptides are $\alpha$- and $\gamma$-aminobutyric acid, $\alpha,\gamma$-diaminobutyric acid and $\beta$-alanine (D-52) (see Chapter XIV). Many of the bacteria and molds contain $\alpha,\epsilon$-diaminopimelic acid, $HOOC \cdot CHNH_2 \cdot (CH_2)_3$—$CHNH_2 \cdot COOH$, which by decarboxylation of one of its carboxyl groups yields lysine (D-52).

CH—NH
‖        ＼
         CH
C — N ⁄
|
CH₂
|
H₂N—CH—COOH

Histidine

*Histidine*, the third basic amino acid, gives a red color with diazotized sulfanilic acid, and thus can be determined colorimetrically (D-53). If proteins are exposed to intense light in the presence of methylene blue, their histidyl residues are destroyed by photooxidation (D-54). The reaction is highly specific for histidine, and only small amounts of methionine undergo similar destruction (D-55). All other amino acid residues are resistant to photooxidation under these conditions.

The imino acids *proline* and *hydroxyproline* differ from the amino acids by their solubility in ethanol. Since the imino acids do not lose their NH

H₂C——CH₂              HOHC——CH₂
 |        |                  |        |
H₂C＼   ⁄CH—COOH       H₂C＼   ⁄CH—COOH
    N                        N
    |                        |
    H                        H

Proline                  Hydroxyproline

groups after exposure to nitrous acid, they can be determined by their color reactions with ninhydrin after deamination of the other amino acids of the hydrolyzate (D-56). In contrast to the amino acids which give a blue or purplish color reaction with ninhydrin, the two imino acids give yellow or reddish colors on the paper chromatogram. Hydroxyproline, after oxidation with ninhydrin, gives an intense color reaction with isatin (D-57). It can also be determined colorimetrically by oxidation with $H_2O_2$ or Chloramine T and subsequent condensation with dimethylamino-benzaldehyde (D-58).

*Tryptophan* is precipitated from its solution in sulfuric acid by the addition of mercuric sulfate. It can be determined spectrophotometrically by measuring the pronounced absorption in the ultraviolet region (D-10). Tryptophan gives intense color reactions with aldehydes such as formaldehyde, dimethylaminobenzaldehyde, or glyoxylic acid, which have been used for the colorimetric determination of tryptophan (D-59). As men-

$$\text{Tryptophan structure: indole ring—} CH_2\text{—CH(NH}_2)\text{—COOH}$$

Tryptophan

tioned earlier, tryptophan is destroyed during the acid hydrolysis of protein, causing the formation of dark brown humin substances. Some of the tryptophan residues are also destroyed by exposure of the proteins to performic or peracetic acid; in this reaction colorless oxidation products are formed. The peptide bonds formed by the amino groups of tryptophan residues are cleaved selectively by $N$-bromosuccinimide (D-60).

*Phenylalanine*, like other aromatic compounds, has an absorption band in the ultraviolet spectral region. However, its absorbancy is weaker than that of tyrosine or tryptophan. For colorimetric determination, phenylalanine is converted to 3,4-dinitrobenzoic acid (D-61) or to dinitrophenylalanine, which is reduced by zinc powder to diaminophenylalanine and coupled with naphthoquinonesulfonic acid to give a dye suitable for colorimetry (D-62).

Phenylalanine                    Tyrosine

The water solubility of *tyrosine* is so low that it precipitates upon neutralization of the protein hydrolyzate. Tyrosine has an intense ultraviolet absorption and gives a series of color reactions: (1) the diazo test, a red color given after coupling with diazo compounds (D-63), (see Chapter VII, F); (2) the Millon test, a red color produced by a solution of mercury in nitric acid; and (3) a blue color resulting from the reducing action of tyrosine on phosphotungstic or phosphomolybdic acids (D-28). The intensity of these reactions is higher in denatured than in native proteins (see Chapter VII, H). Tyrosine, when treated with iodine in slightly alkaline solution, is iodinated and converted to diiodotyrosine and thyroxine (D-64). A similar reaction occurs when proteins are treated with iodine.

## E. Quantitative Determination of Amino Acids in Proteins

As mentioned in the preceding section of this chapter, some of the amino acids can be determined in the proteins without the necessity of hydrolyzing the protein. However, most of the amino acids fail to give a characteristic color reaction and can be determined only after total hydrolysis of the protein. The quantitative determination of all amino acids in the protein hydrolyzate was an unsolved problem until about 1945. At that time, Brand and his co-workers (E-1) succeeded in determining more than 90% of the amino acids in human serum albumin and in ovalbumin. In some instances they had to resort to microbiological methods, in others to precipitation methods which required relatively large amounts of material. In 1949 Stein and Moore (E-2) designed a chromatographic separation of the amino acids which allowed the quantitation of all of the amino acids in yields frequently exceeding 99% of the total amino acid content. In the last few years this method has been further improved and refined in such a manner that the amino acid analysis of the protein hydrolyzate can be performed automatically. Before discussing this development in detail, it may be useful to review briefly the older quantitative methods since these are valuable for the determination of one or a small number of amino acids.

One of these methods is the *isotope dilution method* (D-15). To determine a certain amino acid in a hydrolyzate, a small amount of the same amino acid containing an isotopic element is added to the hydrolyzate. Precipitation is then achieved by a suitable treatment and the amount of the isotope in the precipitate and in the supernatant is determined and expressed in counts per minute (cpm). Since:

$$\frac{\text{isotope in supernatant}}{\text{isotope in precipitate}} = \frac{\text{amino acid in supernatant}}{\text{amino acid in precipitate}}$$

the amount of the respective amino acid in the original hydrolyzate can be calculated. If $C_o$ is the specific activity (cpm/mg) of the amino acid added, and $C_s$, the specific activity of the specimen of amino acid isolated, then $B$, its amount in the original hydrolyzate, is:

$$B = A(C_o/C_s - 1)$$

where $A$ is the amount of amino acid added. The method can be applied even when the precipitation is incomplete since the value of $B$ in the equation given above does not depend on the amount of precipitate obtained.

If a good precipitating agent is not available, the amino acid can be converted into some derivative and precipitated in the presence of the same radioactive derivative. Thus glycine has been determined in a mixture of all other amino acids by treating the mixture with $p$-iodophenyl sulfonyl chloride (pipsyl chloride) containing radioactive iodine, then adding a very large excess of nonradioactive pipsyl glycine and determining the specific activity of the isolated pipsyl glycine precipitate (E-3). The ratio of the specific activities of $I^{131}$ in the pipsyl chloride used and in the pipsyl glycine isolated is a direct measure of the dilution of the glycine present in the hydrolyzate by the added nonradioactive pipsyl glycine. Further refinement is possible by using $S^{35}$-containing pipsyl glycine instead of the nonradioactive materials.

Another valuable quantitative method is based on the *microbiological evaluation* of the amino acid required for the growth of certain bacteria or molds (see preceding Section D). For the quantitative determination of an amino acid, it is necessary to compare the growth response or the metabolic activity of the microorganisms with the response or activity in media containing varying amounts of the examined amino acid for which the assay is made. If the extent of growth or the amount of lactic acid, $CO_2$, or any other typical metabolic product is plotted against the amount of amino acid added, a standard curve is obtained which makes it possible to determine the amino acid content of the unknown sample by graphic interpolation (D-5). The principal advantage of the microbiological methods is their ability to differentiate L- from D-amino acids. However, the results must be interpreted with caution since it is difficult to exclude stimulation or inhibition of the response by contaminating substances present in the unknown material.

The most powerful methods of quantitative amino acid analysis in protein hydrolyzates are based on *column chromatography*. In the first experiments of this type a starch column was used (E-2). The adsorbed amino acids placed at the top of the column were eluted with a mixture of $n$-butanol, $n$-propanol, and 0.1 $N$ hydrochloric acid (1:2:1). Although the total yield of recovered amino acids was very high, this method did not separate leucine from isoleucine, and gave only poor resolution of methionine from tyrosine and of alanine from glutamic acid. An important im-

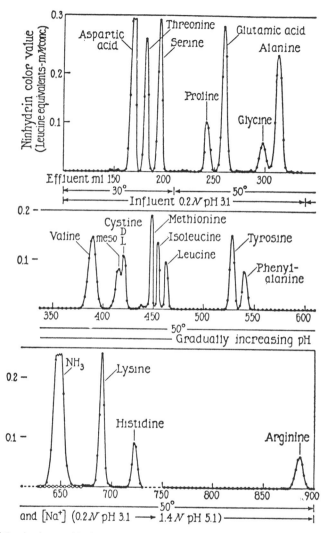

Fig. III-3. Amino acid determination of the hydrolyzate of ribonuclease. The abscissa shows the total volume of effluent in ml; the ordinate of each point indicates the absorbancy in the ninhydrin test. From Hirs *et al.* (E-4).

provement was accomplished by using the cation exchange resin Dowex 50 instead of starch (E-4). The acidic sulfonate groups of this resin remain negatively charged even at pH 3–4 where the amino groups of the amino acids are converted into positively charged ammonium groups. When the hydrolyzate of about 2–3 mg of protein is dissolved in 0.2 *N* phosphate-citrate buffer mixture and slowly passed through the column, the positively charged ammonium groups of the amino acids are bound by the negatively charged groups of the resin and are retained in the column. At pH 3.1 and 30°C, aspartic acid is eluted first, and is followed by threonine and serine as shown in Fig. III-3. If the temperature is raised to 50°C, while gradually increasing pH to 5.1 and the concentration of the buffer mixture to 1.4 $N$, the amino acids are eluted in the sequence proline, glutamic acid, glycine, alanine, valine cystine, methionine, isoleucine, leucine, tyrosine, and phenylalanine. The last fractions of the eluate contain ammonia, lysine, histidine, and arginine, since these basic substances are strongly bound by the acidic resin (see Fig. III-3).

The effluent is collected in small portions in test tubes which are moved automatically so that each of them contains an equal volume, e.g., 2.0 ml. About 400–800 test tubes are required for a complete analysis of the hydrolyzate. Each of the eluates is then mixed with the ninhydrin reagent and heated in a water bath under standard conditions, i.e., at a definite pH value and temperature for the same length of time. The intensity of the color is then measured in a photometer. Since the color value is different for each of the amino acids, it is necessary to determine the absorbancy for each of them along with the factor for the conversion of absorbancy into concentration. The measured absorbancies are then plotted against the numbers of the test tubes, i.e., against the total volume of the eluate. Each point on the curve shown in Fig. III-3 indicates the result of the photometric analysis in the content of one test tube. The values which form a single peak are added to each other to give the total amount of the respective amino acids in the hydrolyzate.

Since elution must be slow in order to avoid turbulence and changes in the sequence of amino acids eluted, it requires considerable time, usually two days, and requires the colorimetric analysis of several hundred eluates. For this reason an automatic amino acid analyzer has been developed in which the pH gradient and the temperature gradient are electrically regulated; the eluates are automatically mixed with ninhydrin, heated, and the absorbancy automatically recorded. Although this equipment is rather expensive, it saves man power and renders possible routine analyses of protein hydrolyzates (E-5,10). If only very small amounts of a protein are available, a micromodification of the chromatographic method can be used (E-6), or one can resort to two-dimensional paper chromatography. The intensity of the ninhydrin spots can be evaluated by densimetry or by elution of the amino acids and photometric determination of their concentration. *Radioactive amino acids* containing $C^{14}$ or $H^3$ can be detected on the paper chromatogram by a suitable counting device (E-7). The entire paper is scanned and the result recorded graphically; curves similar to those shown in Fig. III-3 are obtained. If the specific activity (cpm per milligram) of each of the amino acids is known, their concentration can be calculated. Similar results are obtained with radioactive derivatives of nonradioactive amino acids, e.g., with $N$-acylamino acids produced by acylation with $C^{14}$-acetic anhydride (E-8) or $I^{131}$-pipsyl chloride (E-3).

The results of the amino acid analysis can be expressed in different ways (E-9), as shown in Table III-1. The first column of this table lists the gram of amino acid per 100 g of protein. The sum of these values always exceeds 100 g since the protein contains the amino acyl residues —NH·CHR·CO— and not the entire amino acid molecules $H_2N·CHR·$-COOH whose molecular weight is higher by 18, the molecular weight of water. In a good analysis the sum of all amino acyl residues should obviously be 100 g. Lower values indicate analytical losses. Evidently, the yield expressed in amino acyl residues (column 2 in Table III-1) is more useful than that expressed in amino acids (column 1). The most

TABLE III-1
AMINO ACID COMPOSITION OF RIBONUCLEASE[a]

| Amino acid | Amino acid (g/100 g protein) | Amino acid residues (g/100 g protein) | Nitrogen (% of total N) | Number of residues per protein molecule[b] | | |
|---|---|---|---|---|---|---|
| | | | | Calculated | Nearest integer | Found in peptides |
| Aspartic acid | 15.0 | 13.0 | 8.86 | 15.8 | 16 | 15 |
| Glutamic acid | 12.4 | 10.9 | 6.64 | 11.8 | 12 | 12 |
| Glycine | 1.64 | 1.25 | 1.72 | 3.05 | 3 | 3 |
| Alanine | 7.67 | 6.12 | 6.77 | 12.0 | 12 | 12 |
| Valine | 7.49 | 6.34 | 5.03 | 8.95 | 9 | 9 |
| Leucine | 2.02 | 1.74 | 1.21 | 2.15 | 2 | 2 |
| Isoleucine | 2.67 | 2.30 | 1.60 | 2.85 | 3 | 3 |
| Serine | 11.4 | 9.44 | 8.54 | 15.0 | 15 | 15 |
| Threonine | 8.90 | 7.56 | 5.88 | 10.45 | 10 | 10 |
| Cystine ($\frac{1}{2}$) | 7.00 | 5.95 | 4.58 | 8.15 | 8 | 8 |
| Methionine | 4.00 | 3.52 | 2.10 | 3.75 | 4 | 4 |
| Proline | 3.94 | 3.32 | 2.69 | 4.79 | 5 | 4 |
| Phenylalanine | 3.51 | 3.13 | 1.67 | 2.97 | 3 | 3 |
| Tyrosine | 7.60 | 6.84 | 3.30 | 5.87 | 6 | 6 |
| Histidine | 4.22 | 3.73 | 6.41 | 3.80 | 4 | 4 |
| Lysine | 10.5 | 9.22 | 11.3 | 10.0 | 10 | 10 |
| Arginine | 4.94 | 4.43 | 8.92 | 3.97 | 4 | 4 |
| Amide $NH_3$ | 2.07 | — | 9.58 | 17.0 | (17)[c] | (17)[c] |
| | 116.97 | 98.79 | 98.79 | — | 126 | 124 |

[a] From Hirs *et al.* (E-11).

[b] The first calculation of the peptide residues was based on a molecular weight of 14,000. In the peptide mixture prepared by partial hydrolysis of ribonuclease only 15 instead of 16 molecules of aspartic acid and only 4 instead of 5 proline molecules were found. The molecular weight of ribonuclease, based on the amino acid analysis, is only 13,683.

[c] The number of amide groups is not included in the total number of amino acid residues since it merely indicates conversion of a residue of aspartic or glutamic acid into one of asparagine or glutamine, respectively.

meaningful method of recording is shown in column 3, which lists the nitrogen of each of the amino acids in per cent of the total nitrogen. One of the advantages of this method of presentation is its independence of contamination of the protein by small amounts of carbohydrates. Many of the proteins contain a small per cent of sugars and also amino sugars. These carbohydrates cannot be removed by purification methods since they are bound to the protein by covalent bonds. They are either free of nitrogen or so poor in nitrogen that their nitrogen content can be neglected. For this reason they do not significantly affect the nitrogen values shown in column 3 of Table III-1, but they contribute distinctly to the total weight of the protein and lower the values shown in columns 1 and 2.

If the molecular weight of the protein is not known, it is impossible to calculate the number of amino acyl residues per protein molecule. The same situation arises if the molecular weight of the protein is considerably higher than 100,000, because the number of some of the amino acids per protein molecule is then higher than 30. Since the analytical uncertainty in the photometric amino acid determination is $\pm 3\%$, we are not able to differentiate 30 from 29 or 31 amino acids. The difficulties increase with higher numbers of amino acids per molecule. It is customary in such instances to indicate merely the number of amino acyl residues per $10^5$ g of protein.

If the molecular weight of the protein is known, the number of molecules of each of the amino acids per protein molecule is easily calculated from the values given in columns 1 or 2 of Table III-1. If the calculated numbers shown in column 4 are close to an integer, this integer (column 5) indicates the number of the respective type of amino acid molecules in one molecule of the protein. If the calculated value is very different from an integer, it means that either the analysis is wrong or that the molecular weight is not correct. The latter is frequently the case since the analytical error in molecular weight determinations is rarely less than about $\pm 3\%$. This circumstance complicates calculations of the number of amino acids per molecule. Another difficulty arises from the fact that cysteine, during the analytical procedure, is converted to cystine. Consequently, these two amino acids cannot be differentiated by chromatography. Since oxidation of a cysteine molecule gives one-half of a cystine molecule, it is customary to represent the results by indicating the number of cystine half-molecules (Cys, $\frac{1}{2}$). The number of preformed cystine residues can be ascertained by the methods described in the preceding Section D of this chapter.

The values shown in Table III-1 can be easily converted into each other. Thus the value 15.0 g of aspartic acid (column 1) multiplied by the factor 115/133 gives 13.0 (column 2); here 115 is the molecular weight of the aspartyl residue —NH·CH(CH$_2$·COOH)·CO— and 133 that of aspartic acid H$_2$N·CH(CH$_2$· COOH)·COOH. Since aspartic acid contains only one nitrogen atom, 15.0 g of aspartic acid contain $15.0 \times 14/133 = 1.58$ g nitrogen. The nitrogen content of 100 g of ribonuclease is 17.8 g; hence, 1.58 g of aspartyl-$N$ are

$$100 \times 1.58/17.8 = 8.86\%$$

of the total nitrogen (column 3). Assuming for ribonuclease a molecular weight of 14,000, we find that 13.0% of aspartyl residues correspond to

$$13 \times 14,000/100 = 1820 \text{ g}$$

of aspartyl residues or $1820/115 = 15.8$ residues per molecule of ribonuclease (column 4). When analogous calculations are made for the other amino acids, it must be remembered that lysine contains 2, histidine 3, and arginine 4 nitrogen atoms per molecule. Therefore, their contribution to the values of column 3 is higher than to those of columns 1 or 2 of the table.

In most proteins, the average value for the equivalent weight of the amino acid residues is approximately 115. Hence the number of amino acid residues in a protein of the molecular weight $M$ is approximately $M/115$, or close to 870 amino acid equivalents per 100,000 g of protein.

In the wake of the development of the new methods of quantitative amino acid determination, more than 100 papers on complete amino acid analyses have been published. Some of these analyses will be presented in the chapters in which the respective proteins and their properties are discussed. Many more analyses of this type will certainly be published in the next few years.

### *REFERENCES*

**Section A**

**A-1.** G. L. Miller and V. du Vigneaud, *JBC* 118: 101(1937).   **A-2.** B. Franck and J. Knoke, *Ber. deut. chem. Ges.* 90: 2450(1957).   **A-3.** R. A. Gortner and M. J. Blish, *JACS* 37: 1630(1915).   **A-4.** G. Toennies and R. P. Homiller, *JACS* 64: 3054(1942); F. Sanger, *Nature*, 160: 295(1946); *BJ* 39: 507(1946).   **A-5.** A. R. C. Warner, *JBC* 142: 741(1942).   **A-6.** M. Damodaran *et al.*, *BJ* 26: 1704(1932).   **A-7.** R. Acher *et al.*, *BBA* 5: 493(1950).   **A-8.** P. Desnuelle and G. Bonjour, *BBA* 7: 451(1951).   **A-9.** F. Haurowitz, *Z. Physiol. Chem.* 162: 41(1926).   **A-10.** J. C. Paulson and F. E. Deatherage, *JACS* 76: 6198(1954).   **A-11.** S. Akabori *et al.*, *Bull. Chem. Soc. Japan* 25: 214(1952).   **A-12.** S. Blackburn and G. R. Lee, *BJ* 58: 227(1954).   **A-13.** R. L. Hill and W. R. Schmidt, *JBC* 237: 389(1962).

**Section B**

**B-1.** S. Sørensen, *Biochem. Z.* 7: 45(1907).   **B-2.** M. Levy and D. E. Silverman, *JBC* 118: 723(1937); D. French and J. T. Edsall, *Adv. in Protein Chem.* 2: 278(1945). **B-3.** R. Willstätter and E. Waldschmidt-Leitz, *Ber. deut. chem. Ges.* 54: 2988(1921); L. Michaelis and M. Mizutane, *Biochem. Z.* 147: 7(1924).   **B-4.** H. A. Krebs, *Biochem. Z.* 220: 283(1930).   **B-5.** D. D. Van Slyke, *JBC* 9: 185(1911); 23: 407(1915); 83: 425(1929).   **B-6.** T. L. McMeekin and R. Warner, *Ann. Rev. Biochem.* 15: 126(1946); P. Cristol *et al.*, *Bull. soc. chim. biol.* 38: 639(1956).   **B-7.** S. Moore and W. H. Stein, *JBC* 178: 53, 79(1949); 211: 907(1954).   **B-8.** D. D. Van Slyke *et al.*, *JBC* 141: 627(1941).   **B-9.** A. Markovitz and D. Steinberg, *JBC* 228: 285(1957).

## Section C

**C-1.** A. J. P. Martin and R. L. M. Synge, *Adv. in Protein Chem.* 2: 1(1945). **C-2.** A. Kossel *et al.*, *Z. physiol. Chem.* 31: 165(1900); 110: 241(1920). **C-3.** E. Fischer, *Z. physiol. Chem.* 33: 151(1901). **C-4.** R. L. M. Synge *BJ* 33: 1913, 1918(1938); *Symposia Biochem. Soc.* 3: 90(1949). **C-5.** C. E. Dent, *BJ* 41: 240(1946); R. Consden *et al.*, *BJ* 37: 79(1943). **C-6.** R. S. Williams and H. Kirby, *Science* 107: 481(1948). **C-7.** W. H. Stein and S. Moore, *JBC* 176: 337(1948). **C-8.** A. H. Gordon *et al.*, *BJ* 37: 79(1943). **C-9.** S. Moore and W. H. S. Stein, *JBC* 211: 893, 907(1954). **C-10.** L. F. Wiggins and J. H. Williams, *Nature* 169: 279(1951); C. H. Edwards *et al.*, *J. Chromatog.* 2: 188(1959). **C-11.** W. Grassmann and K. Hannig, *Z. physiol. Chem.* 292: 32(1953). **C-12.** N. Gross, *Nature* 176: 72(1955); F. Turba *et al.*, *Z. physiol. Chem.* 296: 97(1954). **C-13.** A. L. Levy, *Nature*, 174: 126(1954). **C-14.** J. Sjöquist, *BBA* 41: 20(1960). **C-15.** C. Fromageot *et al.*, *BBA* 5: 283(1950); A. C. Chibnall and M. W. Rees, *BJ* 68: 105(1958). **C-16.** W. Grassmann and A. Riedel, *Z. physiol. Chem.* 312: 206(1958). **C-17.** K. A. Piez and H. Eagle, *Science* 122: 968(1955). **C-18.** I. M. Hais and K. Macek, "Papirova Chromatografie" (in Czech), Czechoslovak Academy of Sciences, Prague, 1954. **C-19.** A. N. Cherbuliez *et al.*, *Helv. Chim. Acta* 44: 319(1961). **C-20.** M. Brenner *et al.*, *Experientia* 17: 145(1961); *Helv. Chim. Acta*, 44: 2022(1961). **C-21.** F. Weygand *et al.*, *Z. physiol. Chem.* 322: 38(1960). **C-22.** H. A. Saroff and A. Karmen, *Anal. Biochem.* 1: 344(1960). **C-23.** E. Stahl, *Chem. Zeitung* 82: 323(1958). **C-24.** C. S. Hanes *et al.*, *Can. J. Biochem. and Physiol.* 39: 119(1961).

## Section D

**D-1.** A. Neuberger, *Adv. in Protein Chem.* 4: 298(1948). **D-2.** P. Brewster *et al.*, *Nature* 166: 178(1950). **D-3.** L. R. Overby and A. W. Ingersoll, *JACS* 73: 3363(1951). **D-4.** J. P. Greenstein, *Adv. in Protein Chem.* 9: 122(1954). **D-5.** M. Dunn, *Physiol. Revs.* 29: 219(1949); E. E. Snell, *Adv. in Protein Chem.* 2: 85(1945). **D-6.** J. P. Greenstein and M. Winnitz, "Chemistry of Amino Acids" (3 vols.). Wiley, New York, 1961. **D-7.** A. Meister, "Biochemistry of the Amino Acids." Academic Press, New York, 1957. **D-8.** A. R. Patton and E. M. Foreman, *Science* 109: 339(1949). **D-9.** B. Alexander *et al.*, *JBC* 160: 51(1945). **D-10.** E. Brand *et al.*, *JACS* 67: 1524(1945). **D-11.** A. T. Austin, *JCS* 1950: 149. **D-12.** A. I. Kendall and T. E. Friedemann, *J. Infectious Diseases* 47: 171(1930). **D-13.** A. Virtanen and N. Rautanen, *BJ* 41: 101(1947). **D-14.** F. Ryan and E. Brand, *JBC* 154: 161(1944). **D-15.** D. Rittenberg and G. L. Foster, *JBC* 133: 737(1940); 159: 431(1945. **D-16.** F. W. Foreman, *BJ* 8: 463(1914). **D-17.** P. P. Cohen, *BJ* 33: 551(1939). **D-18.** H. S. Olcott, *JBC* 153: 71(1945). **D-19.** R. L. Hill and W. R. Schmidt, *JBC* 237: 389(1962). **D-20.** J. Melville, *BJ* 29: 179(1935). **D-21.** K. Ohno, *J. Biochem (Japan)*, 41: 345(1954). **D-22.** P. Jolles and C. Fromageot, *BBA* 9: 287(1952). **D-23.** L. A. Shinn and B. H. Nicolet, *JBC* 138: 91(1941). **D-24.** D. D. Van Slyke *et al.*, *JBC* 141: 681(1941); S. Bergström and S. Lindstedt, *ABB* 26: 323(1950). **D-25.** J. C. Sheehan and W. A. Bolhofer, *JACS* 72: 2469(1950). **D-26.** A. E. Mirsky and M. L. Anson, *J. Gen. Physiol.* 18: 307(1935). **D-27.** Y. Okuda, *J. Biochem. (Japan.)* 5: 217(1925). **D-28.** O. Folin and J. M. Looney, *JBC* 51: 421(1922); 69: 519(1926); O. H. Lowry *et al.*, *JBC* 193: 265(1951). **D-29.** R. Kuhn and P. Desnuelle, *Z. physiol. Chem.* 251: 14(1938). **D-30.** R. E. Benesch *et al.*, *JBC* 216: 663(1955); I. M. Kolthoff *et al.*, *Anal. Chem.* 26: 366(1954); P. D. Boyer, *JACS* 76: 4331(1954); R. Brdička, *Mikrochemie* 15: 167(1934). **D-31.** M. G. Horowitz and I. M. Klotz, *ABB* 63: 77(1956). **D-32.**

L. W. Cunningham and B. J. Nuenke, *JBC* 234: 1447(1959). **D-33.** H. Holter and S. Løvtrup, *Compt. rend. trav. lab. Carlsberg* 27: 72(1949). **D-34.** R. E. Benesch and R. Benesch, *JACS* 80: 1666(1958). **D-35.** N. R. Ling, *Nature* 178: 1054(1956). **D-36.** R. Cecil and U. E. Loening, *BJ* 66: 18P(1957). **D-37.** J. M. Swan, *Nature* 180: 643(1957); J. F. Pechère *et al., JBC* 233: 1364(1958); H. Würz and F. Haurowitz, *JACS* 83: 280(1961). **D-38.** L. F. Fieser, *Rec. trav. chim.* 69: 410(1950). **D-39.** J. E. Turner *et al., in* "Sulfur in Proteins" (R. Benesch, ed.), p. 10. Academic Press, New York, 1959. **D-40.** R. E. Benesch and R. Benesch, *BBA* 23: 643(1957). **D-41.** H. Lindley, *Nature* 178: 647(1956). **D-42.** J. E. Moore and W. H. Ward, *JACS* 78: 2414(1956); R. Benesch, *Science* 123: 987(1956). **D-43.** W. R. Cuthbertson and H. Phillips, *BJ* 39: 7(1945). **D-44a.** S. Akabori and T. Fujiwara, *Bull. soc. chim. biol.* 40: 1983(1959). **D-44b.** E. V. Jensen, *Science* 130: 1519(1959). **D-44c.** A. N. Glazer and E. L. Smith, *JBC* 236: 416(1961). **D-45.** H. D. Baernstein, *JBC* 115: 25, 33 (1936). **D-46.** T. E. McCarthy and M. X. Sullivan, *JBC* 141: 871(1941). **D-47.** J. Pikkarainen and E. Kulonen, *Ann. Med. Exptl. et Biol. Fenniae* (*Helsinki*) 37: 382(1959) (*Chem. Abstr.* 54: 22746). **D-48.** A. Kossel and E. Gross, *Z. physiol. Chem.* 135: 167(1924). **D-49.** S. Sakaguchi, *Japan. Med. J.* 1: 278(1948) (*Chem. Abstr.* 44: 1158). **D-50.** E. F. Gale, *BJ* 38: 232(1944). **D-51.** A. N. Howard and F. Wild, *BJ* 65: 651(1957). **D-52.** E. Work and D. L. Dewey, *J. Gen. Microbiol.* 9: 394(1953). **D-53.** M. Hanke, *JBC* 66: 475(1926). **D-54.** L. Weil and T. S. Seibler, *ABB* 54: 368(1955). **D-55.** W. J. Ray *et al., JACS* 82: 4743(1960). **D-56.** P. B. Hamilton and P. J. Ortiz, *JBC* 187: 733(1950). **D-57.** M. G. Kolor and H. R. Roberts, *ABB* 70: 620(1957). **D-58.** R. E. Neumann and M. A. Logan, *JBC* 184: 299(1950); H. Stegemann, *Z. physiol. Chem.* 311: 41(1948). **D-59.** J. R. Spies, *Anal. Chem.* 22: 1447(1950). **D-60.** A. Patschornik *et al., JACS* 80: 4747(1958); L. K. Ramachandran and B. Witkop, *JACS* 81: 4028(1959). **D-61.** R. Kapeller-Adler, *Biochem. Z.* 252: 185(1932). **D-62.** W. C. Hess and M. X. Sullivan, *ABB* 5: 165 (1944). **D-63.** H. Pauly, *Z. physiol. Chem.* 94: 284, 426(1915); M. Hanke, *JBC* 79: 587(1928). **D-64.** P. von Mutzenbecher, *Z. physiol. Chem.* 261: 253(1939); E. P. Reineke and C. W. Turner, *JBC* 149: 555(1943).

*Section E*

**E-1.** E. Brand *et al., JACS* 67: 1524(1945); E. Brand, *Ann. New York Acad. Sci.* 47: 216(1947). **E-2.** W. H. Stein and S. Moore, *JBC* 178: 79(1949). **E-3.** A. S. Keston *et al., JACS* 71: 249(1949); 72: 748(1950). **E-4.** C. H. W. Hirs, W. H. Stein, and S. Moore, *JBC* 211: 941(1954). **E-5.** D. H. Spackman, W. H. Stein, and S. Moore, *Anal. Chem.* 30: 1185(1958). **E-6.** J. E. Eastoe, *BJ* 79: 652(1961). **E-7.** F. Turba, *in* "Symposium on Protein Structure" (A. Neuberger, ed.), p. 116 Wiley, New York, 1958. **E-8.** J. K. W. Whitehead, *BJ* 68: 653, 662(1958). **E-9.** G. T. Tristram, *in* "The Proteins" (H. Neurath and K. Bailey, eds.), Vol. I, p. 181. Academic Press, New York, 1953. **E-10.** K. A. Piez and L. Morris, *Anal. Biochem.* 1: 187 (1960). **E-11.** C. H. Hirs *et al., JBC* 211: 941(1954); 219: 623(1956).

# Chapter IV

## Amino Acid Sequence and
## Primary Structure of the Proteins

### A. General Technique

The term *primary structure* is frequently used to designate the "chemical" formula of the proteins which shows in which sequence the amino acids are linked together by peptide bonds. It does not take into consideration electrostatic interaction between positively and negatively charged groups of the proteins nor interaction by van der Waals forces. The dithio bonds of cystine which can form cross links between different parts of a peptide chain or between different peptide chains are not as stable as carbon-carbon bonds or even as stable as peptide bonds. They undergo disulfide exchange with each other or with sulfhydryl groups (see Chapter III, D); hence their role in structure is intermediate between that of the stronger covalent bonds and the aforementioned weaker bonds. Since the dithio bonds cause difficulties in the sequential analysis of proteins, they are frequently cleaved by oxidation with performic acid (see Chapter III, D) before the order of the amino acids in the peptide chain is investigated.

The first step in the investigation of the primary structure of proteins and peptides is the determination of the *N-terminal amino acid*, i.e., the amino acid which carries the free α-amino group. If a method is available to remove only the *N*-terminal amino acid, it can be isolated and identified. Repetition of this procedure sometimes permits stepwise degradation of the peptide chain from its *N*-terminus and elucidation of its sequence. One limitation of the method arises from the inability of certain amino acids to react with the reagents used, and also from the unavoidable losses in each of the steps. These limitations prevent repetition more than 5–10 times. Essentially, the same procedure can be applied to the *C-terminal amino acid*, which carries the free α-COOH group, provided that methods for the specific degradation of this amino acid are available. Even when this is accomplished, not more than a few amino acids at the *N*-terminus and a few others at the *C*-terminus of the peptide chain can be determined by these methods.

Further information on the amino acid sequence is obtained after the *partial hydrolysis* of the protein, fractionation of the peptide mixture, isolation of the different peptides, and analysis of each by the just-described stepwise degradation from the $N$- or $C$-terminus. Even if the precise amino acid composition of each of the peptides has been elucidated, it is still necessary to know the order in which these peptides occur in the large protein molecule. Using different methods of partial hydrolysis, for instance hydrolysis by various enzymes, we can split the original peptide chain at different points and can then reconstruct the original sequence.

Let the structure of a polypeptide be: $A \to B \to C \to D \to A \to E \to F \to C \to A \to G \to H \to F \to G \to B \to C \to A$, where each of the letters stands for a definite amino acid, and where each arrow indicates a peptide bond of the type $CO \cdot NH$ (and not $NH \cdot CO$). Hence the $N$-terminus is at the left, the $C$-terminus at the right side. If an enzyme is used which cleaves peptide bonds formed by the carboxyl groups of A, the following fragments are obtained: A, $B \to C \to D \to A$, $E \to F \to C \to A$, and $G \to H \to F \to G \to B \to C \to A$. The order of the split products in the original polypeptide is still unknown. If now another sample of the same protein is hydrolyzed by an enzyme which splits the peptide bonds formed by the carboxyl groups of D and F, the following peptides will be obtained: $A \to B \to C \to D$, $A \to E \to F$, $C \to A \to G \to H \to F$, and $G \to B \to C \to A$. Comparing the two results, we find that they can be superimposed only in the following way:

$$A \quad B \to C \to D \to A \quad E \to F \to C \to A \quad G \to H \to F \to G \to B \to C \to A$$
$$A \to B \to C \to D \quad A \to E \to F \quad C \to A \to G \to H \to F \quad G \to B \to C \to A$$

In this manner the structure of the entire polypeptide can be clarified (A-1,2). In the following sections of this chapter we will discuss each of the four essential steps of this method, namely: (1) determination of $N$-terminal amino acids, (2) determination of $C$-terminal amino acids, (3) hydrolytic cleavage of the protein by two or more different enzymes, (4) fractionation of the peptides obtained and clarification of their position in the original peptide chain of the protein.

## B. Determination of the N-Terminal Amino Acids

The terminal $\alpha$-amino group of proteins or peptides is easily acylated or alkylated. *Acylation* can be accomplished by treatment of the proteins in slightly alkaline solution, at pH 8–9, with benzoyl chloride or arylsulfonyl chlorides such as toluene or naphthalenesulfonyl chloride. Acetylproteins are produced by heating the dry protein with acetic anhydride or by acetylation in aqueous solution with ketene (B-1) or acetic anhydride and sodium acetate (B-2). In all these procedures the acylating agent also reacts with the hydroxyl groups of the hydroxyamino acids. However, the ester bonds formed by this reaction are easily cleaved upon acid hydrolysis whereas the acid amide bonds are more resistant to hydrolysis. The $N$-terminal amino acid can then be separated from the free amino acids by chromatography or by extraction with an organic solvent. Acylation has not been used extensively for the determination of the $N$-terminus because better results are obtained by the $N$-alkylation technique (see below). Acylation is mentioned here because of the use of pipsyl chloride (p-iodophenylsulfonyl chloride) as a versatile acylating reagent which can be labeled by radioactive $I^{131}$ or $S^{35}$. If the radioactive pipsyl peptides or proteins are hydrolyzed, the $N$-terminal amino acid is obtained as a radioactive pipsyl derivative and is easily identified (B-3).

Acylating and alkylating reagents react not only with the $\alpha$-amino group of the terminal amino acid but also with the $\epsilon$-amino groups of lysine. For this reason, the hydrolyzate of the acylated or alkylated protein contains also $\epsilon$-$N$-acyl or alkyl derivatives of lysine. These can be separated from the $\alpha$-substituted $N$- terminal amino acid because of the basic properties of the free $\alpha$-amino group of the $\epsilon$-substituted lysine derivative. Whereas the $\alpha$-substituted $N$-acyl and $N$-alkyl derivatives of the monoamino acids are extracted by organic solvents from their aqueous acid solution, the mono-substituted $\epsilon$-derivatives of lysine remain in the aqueous layer.

The great importance of the $N$-alkylation for the determination of the $N$-terminal amino acid was recognized by Sanger who alkylated proteins in aqueous solution under mild conditions with 2,4-dinitrofluorobenzene (A-1). The $N$-alkyl residues are highly resistant to acid hydrolysis and are obtained in very good yield after complete hydrolysis of the peptide bonds. Other investigators have methylated proteins by means of dimethyl sulfate or diazomethane. Better results are obtained when the milder reductive methylation with formaldehyde and palladium is applied (B-4). However, *dinitrophenylation* is simpler than these methods and yields intensely yellow DNP (dinitrophenyl) amino acids which are easily detected on the paper chromatograms. For dinitrophenylation the proteins or peptides are dissolved in a dilute aqueous solution of sodium bicarbonate; a small excess of dinitrofluorobenzene, dissolved in methanol or ethanol, is added, and the mixture is stirred or shaken. The yellow DNP-protein is then hydrolyzed with HCl and the hydrolyzate, after removal of the excess of HCl, is extracted with ether or other organic solvents. The free amino acids and $\epsilon$-DNP-lysine remain in the aqueous layer whereas the DNP-derivative of the $N$-terminal amino acid passes into the organic solvent. It can be identified by its $R_f$ value in paper chromatography.

The dinitrophenylation shares with other alkylation methods the disadvantage that isolation of the $N$-terminal residue requires total hydrolysis of the protein. Hence, the method cannot be used more than once, nor can it give any information on the sequence of the amino acids at the $N$-terminus. To overcome this difficulty the protein-bound DNP group can be reduced by ammonium sulfide to a 2-amino-4-nitro-phenyl group whose amino group forms a lactam with the carboxyl group of the $N$-terminal amino acid and can be isolated as a 3-substituted 7-nitro-3,4-dihydro-2-oxyquinoxaline (B-5).

The $N$-terminal amino acid sequence can be elucidated by substitution of the $N$-terminal $\alpha$-amino group with phenylisocyanate (B-6) or preferably with phenylisothiocyanate (B-7). The first product formed in this reaction is a phenylthioureide which through an unstable 2-anilino-5-thiazolinone undergoes rearrangement to a stable 3-phenyl-2-thiohydantoin.

The peptide bond between the substituted $N$-terminal amino acid and the adjacent amino acid is cleaved by mild hydrolysis, for instance by barium hydroxide or by HCl in nitromethane (B-7). The cleavage can be carried out on the paper chromatogram of a peptide (B-8). After removal of the liberated phenylthiohydantoin, the residual peptide can be coupled again with phenylisothiocyanate and the procedure can be repeated several times. It is limited by unavoidable losses on repetition of the process, and also by the inability of certain amino acids to yield the typical phenylthiohydantoins. The phenylthiohydantoins of the amino acids are colorless substances but can be detected on paper chromatograms by their ability to absorb ultraviolet light (B-9), by the iodine-azide reaction (see Chapter III, D-33), or by the blue color given with Grote's reagent (sodium aquoferricyanide) (B-10).

The terminal $\alpha$-amino groups of proteins and peptides also react with *carbon disulfide*, $CS_2$, to give the sulfur analogues of carbamino derivatives (B-11). The $N$-terminal amino acid can be split off as a 2-thio-thiazolide-

5-one which on hydrolysis with HCl gives the free amino acid:

$$X \cdot NH \cdot CO \cdot CHR \cdot NH_2 \xrightarrow{+CS_2} X \cdot NH \cdot CO \cdot CHR \cdot NH \cdot CS \cdot SH \rightarrow$$

$$XNH_2 + \overset{\boxed{\phantom{x}S\phantom{x}}}{CO \cdot CHR \cdot NH \cdot CS} \rightarrow HOOC \cdot CHR \cdot NH_2$$

The $N$-terminal amino acid of proteins is also released by the action of the enzyme *leucine-aminopeptidase* (B-12). The action of the enzyme however, does not stop at the first peptide bond but proceeds further through the peptide chain. Some information on the amino acid sequence can be gained by interrupting the action of the enzyme at different times and determining the quantity of each of the liberated amino acids. Kinetic analysis can then reveal the sequence in which the amino acids were released.

The protein of tobacco mosaic virus (TMV) and some of the other proteins react neither with dinitrofluorobenzene nor with phenylisothiocyanate at their amino end. For a time it was assumed that these proteins form loops in which the amino end of the peptide chain is bound to one of the carboxyl groups of aspartic or glutamic acid. In the meantime it has been found that the $N$-terminal serine residue of the TMV protein occurs in the native protein as $N$-acetylserine (B-13). *Acetylated or otherwise substituted N-terminal amino acids* also occur in other proteins. In some of the proteins the $N$-terminal amino acid is formed by cystine which is a dia-mino-dicarboxylic acid. In these proteins one of the two amino groups is free, the other linked to a carboxyl group of an adjacent amino acid residue (B-14). This results in a loop which is closed by a dithio bond.

$$HO(CO \cdot CHR \cdot NH)_x \cdot CO \cdot \overset{\displaystyle NH(CO \cdot CHR \cdot NH)_y \cdot CO}{\underset{}{CH \cdot CH_2 \cdot S \cdot S \cdot CH_2 \cdot CH \cdot NH_2}}$$

## C. Determination of the C-Terminal Amino Acids

In Chapter III, Section A, it has already been mentioned that, according to Akabori, the $C$-terminal amino acid is obtained as free amino acid after *hydrazinolysis* whereas all other amino acids are converted into their hydrazides. Dinitrophenylation of the hydrazinolyzate converts the amino acids hydrazides into DNP-amino acid hydrazides which can be extracted from the weakly alkaline solution with ethyl acetate. The alkaline aqueous layer, after repeated extraction with ethyl acetate, contains those molecules which have still a free carboxyl group, viz., the DNP-derivative of the $C$-terminal amino acid and the DNP-derivatives of the $\alpha$-hydrazides of aspartic and glutamic acid (C-1). Separation of the $C$-terminal amino acid from other substances is then accomplished by chromatography of the yellow DNP-derivatives.

Attempts have been made to *reduce the C-terminal carboxyl group* by means of lithium aluminum hydride, LiAlH$_4$, to a primary alcohol group —COOH → —CH$_2$OH, and to isolate the $C$-terminal residue in the form of its ether-soluble amino alcohol (C-2) or DNP-amino alcohol (C-3). However, some of the peptide bonds are also reduced by LiAlH$_4$ and converted into secondary amines: —CO·NH— → —CH$_2$·NH— (C-4). Better results are obtained when the carboxyl groups of the protein are first esterified by methanol or ethanol and then reduced by means of sodium or lithium borohydride (C-5). The reduction to alcohols is not quantitative when carried out at 0°C; at higher temperatures some of the peptide bonds are reduced (C-6).

If peptides are heated in acetic anhydride with ammonium thiocyanate, their $C$-terminal amino acid undergoes *condensation* with the *thiocyanate* residue and a thiohydantoin ring is formed (C-7).

$$X·NH·CHR·COOH + NH_4CNS \rightarrow X\text{—}N\text{———}CHR$$

The $C$-terminal thiohydantoin can be split off by the action of alkali (C-7) or hydrochloric acid (C-8) and can be separated from the amino acids and peptides by extraction with nitromethane or by chromatography. In evaluating the results of this method it is necessary to keep in mind that many of the proteins bind far more than 1 mole of thiocyanate. This has been demonstrated by using isotopically labeled thiocyanate containing $C^{14}$ or $S^{35}$ (C-9). Evidently, thiocyanate reacts with more than the $C$-terminal amino acids; it is well known that the thiocyanate reacts also with the carboxyl groups of $\gamma$-bound glutamyl or $\beta$-bound aspartyl residues (C-9). However, the occurrence of such residues in proteins has not yet been proved convincingly.

Other chemical methods for the determination of the $C$-terminal amino acid are its condensation with carbodiimide and conversion by alkali into an amino acyl urea (C-10), or anodic oxidation in methanol to a methyl ether:

$$—CO·NH·CHR·COOH \xrightarrow{-CO_2} —CO·NH·CHR·OCH_3$$

which on treatment with HCl gives —COOH + NH$_3$ + R·CHO + CH$_3$OH (C-11).

Since none of the chemical methods for the determination of the *C*-terminal amino acid are very satisfactory, resort is frequently had to the enzymatic action of *carboxypeptidase*. For obvious reasons, the action of the enzyme proceeds beyond the terminal amino acid (C-12). Kinetic analysis of the rate of appearance of the amino acids may reveal the identity of the terminal amino acid. The method is particularly valuable for the elucidation of the structure of peptides or DNP-peptides isolated from partial hydrolyzates (C-13). If the hydrolysis of the protein or peptide by carboxypeptidase is carried out in water containing $H_2O^{18}$, the isotopic $O^{18}$ is incorporated into the carboxyl groups of all amino acids except the *C*-terminal amino acid (C-14). This permits the identification of the *C*-terminal amino acid unambiguously. The method is based on the fact that the carboxyl groups in aqueous solution do not exchange their oxygen atoms with oxygen atoms of the solvent, but that the latter is incorporated on cleavage of the peptide bond:

$$X \cdot CO \cdot NH \cdot Y + H_2O^{18} \rightarrow X \cdot CO \cdot O^{18}H + H_2N \cdot Y$$

## D. Partial Degradation of Proteins and Separation of Peptides

*Partial hydrolysis* of proteins is usually accomplished by incubation of the protein with one of the proteolytic enzymes. Contamination by the applied enzyme can be avoided by using insoluble enzyme-carrier preparations (see Chapter XII, F). Although partial hydrolysis can also be brought about by the action of concentrated HCl at 37°C or by boiling with dilute hydrochloric acid, enzymatic hydrolysis is much more specific since each of the enzymes can split only certain types of peptide bonds.

Since many of the native proteins are not attacked in their native state by proteolytic enzymes, it is necessary to denature them by heating their aqueous solutions. In most instances it is also advantageous to cleave first the dithio bonds by oxidation with performic acid (D-1) or by reduction with sulfhydryl compounds (see Chapter III, D) or sodium borohydride (D-18). The excess of $H_2O_2$ after treatment with performic acid can be destroyed by catalase (D-2). Reoxidation after treatment with reducing agents is prevented by alkylation of the SH groups with iodoacetic acid and conversion into S-carboxymethyl groups (D-18). Denaturation as well as cleavage of the SS bonds cause unfolding of the peptide chain and make the susceptible peptide bonds more accessible to the enzyme molecules. The peptides produced by the action of subtilisin on heat-denatured hemoglobin are, however, different from those produced by the same enzyme from the native protein (D-19). Evidently, different bonds are hydrolyzed in the native and the denatured hemoglobin.

The most specific enzyme action is that of *trypsin*, which attacks only peptide bonds involving the $C{=}O$ groups formed by the basic amino acids lysine and arginine. Consequently, hydrolysis by trypsin results in the formation of peptides which have a $C$-terminal arginine or lysine residue, the only exception being the $C$-terminal peptide which may have any one of the amino acids in $C$-terminal position. The maximum number of peptides formed by tryptic hydrolysis is equal to $n + 1$ where $n$ is the total number of lysyl and arginyl residues in the protein molecule. A smaller number of peptides may be due to abnormal, resistant bonds or to the formation of free lysine or arginine. Thus, ribonuclease, which has 10 lysyl and 4 arginyl residues, yields only 13 peptides after tryptic hydrolysis (D-3). If the lysyl residues are converted into homoarginyl residues by amidination or if they are deprived of their basicity by carbobenzoxylation or trifluoroacetylation only the peptide bonds formed by the arginyl residues are hydrolyzed by trypsin (D-4). The opposite effect, increase of the number of bonds hydrolyzed, can be brought about by treatment of the protein with 2-bromo-ethylamine which converts the residues of cysteine (I) into those of S-aminoethylcysteine (II) whose structure is analogous to that of lysine (III) (D-5). Peptide bonds formed by (II) are hydrolyzed by trypsin.

$$
\begin{array}{ccc}
\overset{\displaystyle |}{N}H & \overset{\displaystyle |}{N}H & \overset{\displaystyle |}{N}H \\[2pt]
\overset{\displaystyle \cdot}{C}H{\cdot}CH_2{\cdot}SH & \xrightarrow{\;BrCH_2CH_2NH_2\;} & \overset{\displaystyle \cdot}{C}H{\cdot}CH_2{\cdot}S{\cdot}CH_2{\cdot}CH_2{\cdot}NH_2 & \overset{\displaystyle \cdot}{C}H{\cdot}CH_2{\cdot}CH_2{\cdot}CH_2{\cdot}CH_2{\cdot}NH_2 \\[2pt]
\overset{\displaystyle \cdot}{C}O & \overset{\displaystyle \cdot}{C}O & \overset{\displaystyle \cdot}{C}O \\[2pt]
| & | & | \\
(I) & (II) & (III)
\end{array}
$$

The action of other enzymes is less specific than that of trypsin. Chymotrypsin hydrolyzes preferentially peptide bonds formed by the carboxyl groups of tyrosine, phenylalanine, methionine, and tryptophan, but also some of those formed by other amino acids (D-6). Pepsin splits predominantly peptide bonds formed by the amino group of tyrosine, but also some of the bonds in which the amino group is contributed by glutamic acid, glycine, or alanine (D-6). Peptide bonds which are not hydrolyzed by any of the animal or plant enzymes are frequently hydrolyzed by bacterial enzymes, e.g., by subtilisin, an enzyme from *Bacillus subtilis* (D-7).

If two or more different proteolytic enzymes have been used, the hydrolyzate contains a mixture of peptides and free amino acids. The latter can be destroyed by heating with ninhydrin while peptides are not attacked by this reagent (D-8). Specific cleavage of the peptide chains can also be caused by nonenzymatic reagents (D-21). For instance, peptide bonds formed by the carboxyl groups of tryptophan, tyrosine, or histidine residues are specifically cleaved by aqueous solutions of *N*-bromosuccinimide or bromine (D-9,21). Methionyl peptides are split by treatment with cyanogen bromide (BrCN) or iodoacetamide ($ICH_2CONH_2$) and subsequent heating in aqueous medium (D-21,22).

The methods used for the fractionation of the peptide mixtures of the partial hydrolyzate are essentially the same as those used for the fractionation of proteins (see Chapter II). Since the molecules of the peptides are smaller than protein molecules, they cannot be salted out as insoluble precipitates, nor can they be separated from the salts of the buffer solutions by dialysis. For this reason it is advantageous to use volatile buffers which can be removed by evaporation at low temperatures. The acetates and formates of ammonia (D-10) or of triethylamine (D-11), or mixtures of acetic acid with pyridine or collidine (2,4,6-trimethyl-pyridine) (D-12) have been used for this purpose. These salts are volatilized during the process of freeze-drying. The mixture of peptides is then fractionated by chromatography on columns of cation exchange resins, e.g. Dowex 50 × 2 (D-3), or on anion exhange resins such as Dowex 1 × 2 (D-12). Acidic peptides are bound to the basic resins with higher affinity than basic peptides. Consequently, the latter are eluted first. Basic resins have also been used to separate the peptides from free amino acids (D-13). Since the peptides quite generally are more acidic than the amino acids, only the amino acids are eluted from the basic anion exchange resins by ammonium acetate buffer at pH 8.65, whereas removal of the peptides requires elution with acetic acid (D-13). Fractionation of the peptides can also be accomplished by high voltage electrophoresis (D-14) or by passage through Sephadex (D-20).

The $R_f$ value of a peptide depends on the $R_f$ value of the amino acids which it contains. The correlation between these two values is shown by the equation:

$$RT \ln \left( \frac{1}{R_f} - 1 \right)_p = (n - 1)A + B + \sum RT \ln \left( \frac{1}{R_f} - 1 \right)_{aa}$$

where $R$ = gas constant, $T$ = absolute temperature, $A$ and $B$ are constants, and the subscripts $p$ and $aa$ refer to peptides and amino acids, respectively (D-15). A list of 2,000 peptides prepared during the years 1950–1956 has been published and can be used as reference material (D-16).

The methods described in the preceding paragraphs have led to the complete elucidation of the structure of insulin by Sanger (D-1). Since then the structure of other large peptides or proteins has been clarified by means of the same techniques. The number of proteins and peptides whose amino acid sequence is known has grown from year to year and comprises at present oxytocin and vasopressin, the melanophore-stimulating and the adrenocorticotropic hormone of the pituitary gland, glucagon from the pancreas, ribonuclease, hemoglobin, cytochrome c, and the protein of tobacco mosaic virus.

The structure of these proteins will be discussed in connection with their properties in other chapters of this book. It should, however, be emphasized here that the importance of these results goes far beyond the information which they give us on the structure of proteins as a class of particularly large organic molecules. They are the first approach to the fundamental problem of the assembly of amino acids in a definite sequence. As more sequences of amino acids in proteins are elucidated, it will be possible to compare sequences of analogous proteins in different organisms and also to compare different proteins manufactured in one cell, one tissue, or one organ. Attempts in this direction have been made with the limited material available at present (D-17). They have revealed certain regularities in the sequence of amino acids, similarities in sequences occurring in the chains of trypsin and chymotrypsin (see Chapter XII), and also striking similarities between the amino acid sequence of various hemoglobins (Chapter XI, G). Further analyses of this type will give us more insight into the mechanism of protein biosynthesis and will clarify the influence of ontogenetic and phylogenetic evolution on the amino acid sequence in the peptide chains of the proteins.

## *REFERENCES*

### Section A

**A-1.** F. Sanger, *BJ* 39: 507(1946); *Adv. in Protein Chem.* 7: 1(1952).    **A-2.** G. Braunitzer *et al.*, *Nature* 190: 480(1961).

### Section B

**B-1.** R. M. Herriott, *J. Gen. Physiol.* 19: 283(1936).    **B-2.** H. S. Olcott and H. Fraenkel-Conrat, *Chem. Revs.* 41: 151(1947).    **B-3.** S. Udenfriend and S. F. Velick, *JBC* 190: 733(1951); A. S. Keston, S. Udenfriend, and M. Levy, *JACS* 72: 748(1950). **B-4.** R. E. Bowman, *J. Chem. Soc.* 1950: 1349; V. M. Ingram, *JBC* 202: 201(1953). **B-5.** E. Scoffone *et al.*, *Gazz. chim. ital.* 87: 354, 1348(1957).    **B-6.** S. J. Hopkins and A. Wormall, *BJ* 28: 2125(1934).    **B-7.** P. Edman, *Acta Chem. Scand.* 4: 277(1950); *Nature* 177: 667(1956).    **B-8.** H. Fraenkel-Conrat, *JACS* 76: 3606; 6058(1954).    **B-9.** J. Sjöquist, *BBA* 16: 283(1955).    **B-10.** W. Landmann *et al.*, *JACS* 75: 3638(1953). **B-11.** J. Leonis and A. Levy, *Bull. soc. chim. biol.* 33: 779(1951).    **B-12.** R. L. Hill and E. L. Smith, *BBA* 31: 257(1959).    **B-13.** K. Narita, *BBA* 28: 184(1958).    **B-14.** K. Titani *et al.*, *J. Biochem. (Tokyo)* 43: 737(1956).

### Section C

**C-1.** K. Ohno., *J. Biochem. (Tokyo)* 40: 621(1953); C. Niu and H. Fraenkel-Conrat, *JACS* 77: 5882(1955).    **C-2.** C. Fromageot *et al.*, *BBA* 6: 283(1951).    **C-3.** W. Grassmann *et al. Z. physiol. Chem.* 296: 208(1954).    **C-4.** P. Karrer and B. J. R. Nicolaus, *Helv. Chim. Acta* 35: 1581(1952).    **C-5.** A. C. Chibnall and M. W. Rees, *BJ* 48: XLVII(1951).    **C-6.** J. C. Crawhall and D. F. Elliott, *Nature* 175: 299(1955); D. M.

Meyer and M. Jutisz, *Bull. chim. France* 1957: 1211. **C-7.** T. B. Johnson and B. H. Nicolet, *JACS* 33: 1963(1911); P. Schlack and W. Kumpf, *Z. physiol. Chem.* 154: 125(1926). **C-8.** J. Tibbs, *Nature* 168: 910(1951); J. T. Edwards and S. Nielson, *Chem. & Ind.* 1953: 197; V. H. Baptist and H. B. Bull, *JACS* 75: 1727(1953); S. W. Fox *et al.*, *JACS* 77: 3119(1955). **C-9.** F. Haurowitz *et al.*, *JBC* 224: 827(1957); F. Haurowitz *et al.*, *FP* 18: 244(1959). **C-10.** H. G. Khorana, *J. Chem. Soc.* 1952: 2081. **C-11.** R. A. Boissonas, *Nature* 171: 304(1953). **C-12.** P. Desnuelle, *Ann. Rev. Biochem.* 23: 55(1954). **C-13.** E. Waldschmidt-Leitz and K. Gauss, *Ber. deut. chem. Ges.* 85: 352(1952). **C-14.** A. Kowalsky and P. D. Boyer, *JBC* 235: 604(1960).

### Section D

**D-1.** F. Sanger, *BJ* 44: 126(1949). **D-2.** R. L. Hill *et al.*, *Ann. Rev. Biochem.* 28: 97(1959). **D-3.** C. H. W. Hirs *et al.*, *JBC* 219: 623(1956); S. Moore and W. H. Stein, *Adv. in. Protein Chem.* 11: 191(1956). **D-4.** C. B. Anfinsen *et al.*, *ABB* 65: 156(1956); *Biochemistry* 1: 401(1962). **D-5.** H. Lindley, *Nature* 178: 647(1956). **D-6.** R. Acher *et al.*, *BBA* 19: 97(1956). **D-7.** K. Linderström-Lang, *Compt. rend. trav. lab. Carlsberg, Sér. chim.* 26: 403(1949). **D-8.** A. Markovits and D. Steinberg, *JBC* 228: 285(1957). **D-9.** A. Patchornik *et al.*, *JACS* 82: 5923(1960); E. J. Corey and L. F. Haefele, *JACS* 81: 2225(1959). **D-10.** C. H. W. Hirs *et al.*, *JBC* 195: 669(1951). **D-11.** J. Porath, *Nature* 175: 478(1955). **D-12.** V. Rudloff and G. Braunitzer, *Z. physiol. Chem.* 323: 129(1951). **D-13.** P. R. Carnegie, *BJ* 78: 687(1961). **D-14.** C. B. Anfinsen *et al.*, *JBC* 221: 385; 405(1956). **D-15.** A. B. Pardee, *JBC* 190: 757(1951); T. B. Moore and C. J. Baker, *J. Chromatog.* 1: 513(1958). **D-16.** M. Goodman and G. W. Kenner, *Adv. in Protein Chem.* 12: 465(1957). **D-17.** F. Šorm *et al.*, *Collection Czechoslov. Chem. Communs.* 26: 531; 1174; 1180(1961); H. Gibian, *Z. Naturforsch.* 16b: 18(1961); F. Lanni, *Proc. Natl. Acad. Sci. U.S.* 47: 261(1961). **D-18.** H. Fasold *et al.*, *Biochem. Z.* 334: 255(1961). **D-19.** M. Ottesen and W. A. Schroeder, *Acta Chem. Scand.* 15: 926(1961). **D-20.** V. Stepanov *et al.*, *Z. Naturforsch.* 16b: 626(1961). **D-21.** B. Witkop, *Adv. in. Protein Chem.* 16: 221(1961). **D-22.** I. Bernier and P. Jolles, *Compt. rend.* 253: 745(1961).

# Chapter V

## Size and Shape of Protein Molecules*

### A. Introduction. The Molecular Weight of Proteins

The molecular weight of small molecules may be determined by dissolving them in suitable solvents and measuring the lowering of the freezing point, the elevation of the boiling point, or the decrease in the vapor pressure of the solvent. Satisfactory values are obtained by these methods if the concentration of the solute is at least 0.01–0.1 $M$. It would be physically impossible, however, to prepare a 0.01 $M$ solution of a protein of molecular weight 100,000, since such a solution would have to contain 1000 g protein per liter.

Some proteins, such as certain albumins, are highly soluble; with these it is possible to prepare solutions containing 300 g or more of the protein per liter of water. It is hardly feasible, however, to determine molecular weights using such solutions, because traces of contaminating low molecular substances would have an enormous influence on the result. Thus, a millimole of carbon dioxide and a millimole of serum albumin will each depress the freezing point of water to the same extent; but whereas a millimole of carbon dioxide is only 0.044 g, a millimole of serum albumin is 68 g.

The first estimations of the molecular weight of proteins were based on the *chemical determination of elements or of amino acids* occurring in the protein in only *small* amounts. The classical example of this procedure is the estimation of the molecular weight of hemoglobin from its iron content. The hemoglobin of mammals contains 0.34 g iron in 100 g hemoglobin. Since the atomic weight of iron is 56, 0.34 g is equal to $0.34/56 = \frac{1}{165}$ of an equivalent, and the amount of hemoglobin bound to 1 gram equivalent (56 g) of iron is $165 \times 100 = 16{,}500$. This value is the minimum molecular weight of hemoglobin and was long considered to be the actual molecular weight. Physicochemical methods give a "molecular weight" of 68,000 for hemoglobin in solution, corresponding to a particle formed by four of the subunits determined by the iron analysis.

*See references (A-3,8).

Similarly, minimum molecular weights of proteins can be calculated from their content of those amino acids which occur in the protein molecules in very small amounts only, or from quantitative determinations of the terminal $\alpha$-amino groups by means of dinitrophenylation (A-1). For instance, if 1.0 g of a protein contains $0.025 \times 10^{-3}$ moles of terminal $\alpha$-amino groups, its minimum molecular weight is $1.0/(0.025 \times 10^{-3}) = 40,000$.

The *physicochemical methods* used for determining the molecular weights of proteins depend not only on the size and the mass of the protein molecule, but also on its electrical charge and its shape. The latter factor is particularly important when measuring the rate of molecular movements such as the diffusion rate or the rate of sedimentation in a gravitational field. While spherelike molecules behave normally in such experiments, the diffusion rate of the threadlike, elongated molecules of fibrous proteins depends on their shape. Deviations from the spherical shape give rise to an increase in the frictional coefficient and tend consequently to reduce the diffusion rate. In concentrated solutions of threadlike molecules further complications arise by the mutual collisions and by temporary adhesion of the molecules. The results of these *dynamic* methods are also affected by the hydration of particles, since the motion of a protein through its solvent will be retarded if its cross-sectional area is increased by hydration. Interference by these factors need not be feared, however, when we examine protein solutions by means of *static* methods in a state of equilibrium, e.g., by measuring their osmotic pressure or by measuring their concentration gradient in the centrifugal field of an ultracentrifuge.

Difficulties in the interpretation of the results arise from the phenomenon of reversible association of protein units. If $n$ soluble protein units of the equivalent weight $M$ combine with each other to form larger particles, the equilibrium $nM \rightleftarrows (M)_n$ depends on pH, ionic strength, and also on the concentration of the protein. In general the particle weight increases when the protein concentration is increased. The equivalent weight of the smallest protein unit in the aqueous solution can be estimated by measuring the "molecular weight" at different concentrations and extrapolating to the concentration $c = 0$. We may then designate the smallest unit as molecule and the larger units as dimers, trimers, tetramers, or polymers. This procedure fails, however, when a large molecule dissociates into subunits of different types. Thus, hemoglobin of the molecular weight 68,000 dissociates on alkalinization into two subunits formed by $\alpha$-chains and two other units formed by $\beta$-chains (see Chapter XI). This dissociation does not involve any cleavage of peptide or other covalent bonds. Upon neutralization the original hemoglobin molecule, which has the composition $\alpha_2\beta_2$, is regenerated.

It is apparent from this example that the meaning of the term "molecular weight" is not always the same in protein chemistry. The molecular weight determined by the physical chemical methods is merely the average weight of the kinetically active protein particles in the solution. These particles may be the true molecules or aggregates of two or more molecules (A-2,3,4). As we will see later (Chapter VII, D), the association of protein units is frequently a highly specific process in which only units of the same type are linked to each other. It evidently depends on a specific shape and composition of the surface of the globular monomers.

If the analyzed protein solution contains aggregates of different sizes, the molecular weight found by physicochemical methods is an average of the different particle sizes. Two different averages are used, the number-average and the weight-average molecular weights. The *number-average molecular weight* is the arithmetic mean of the particle weights of all protein particles present in the solution. If the solution contains: $n_i$ particles of the molecular weight $M_i$; $n_j$ particles of the molecular weight $M_j$; and . . . $n_z$ particles of the molecular weight $M_z$, the number-average molecular weight is

$$M_n = \frac{(n_i M_i) + (n_j M_j) + \cdots (n_z M_z)}{n_i + n_j \cdots + n_z}$$

or quite generally,

$$M_n = \frac{\Sigma(nM)}{\Sigma n}$$

Since the product $nM$ is equal to the concentration, $C$, of the protein, we can also write $M_n = \Sigma C/\Sigma n$. In many of the physicochemical methods the results depend much more on the number of the large protein particles than on that of their subunits. Thus, the intensity of light scattering increases considerably when $n$ subunits of the size $M$ form particles of the formula $(M)_n$ although the total number of scattering particles decreases by a factor of $n$. It is customary to use under these conditions the *weight-average molecular weight*, $M_w$, which is defined by the equation:

$$M_w = \frac{(n_i M_i^2) + (n_j M_j^2) + \cdots (n_z M_z^2)}{n_i M_i + n_j M_j + \cdots n_z M_z} = \frac{\Sigma(nM^2)}{\Sigma(nM)}$$

Since $nM = C$, as shown above, $M_w = \Sigma(CM)/\Sigma C$.

Example: A solution contains an equal number, $n$, of serum albumin monomers (molecular weight = 68,000) and dimers (molecular weight 136,000). The number-average molecular weight, $M_n$, is $\dfrac{68,000n + 136,000n}{2n} = 102,000$. The weight-average molecular weight, $M_w$, is $\dfrac{(68,000)^2 n + (136,000)^2 n}{68,000n + 136,000n} = 113,000$. If the solution contains molecules of only one size, $M_n$ and $M_w$ are identical.

As mentioned earlier, the dynamic methods for the determination of molecular weights depend not only on the size of the protein particles but also on their shape. At present all calculations are based on the assumption that the diffusing, sedimenting, or otherwise moving particles are spheres, ellipsoids of rotation, rods, or coils; other more complicated shapes are not tractable (A-5). The true shape of the protein molecules does not correspond to any of the above-named simple geometrical bodies, and therefore all calculations are only approximations. The true shape of protein molecules can be investigated by means of X-ray crystallographic analysis. Important progress in this area has been accomplished (A-6) as will be shown in Chapter VII, B, and one can reasonably assume that the true shape of a number of protein molecules will be known within a few years.

## B. Osmotic Pressure of Proteins (B-1)

The osmotic pressure is measured with an osmometer, which consists essentially of a semipermeable membrane containing the protein solution

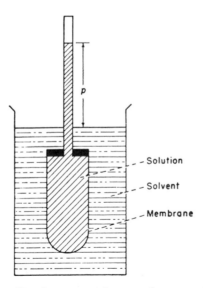

FIG. V-1. Osmometer (p: osmotic pressure).

and a capillary serving as a manometer (Fig. V-1). If the semipermeable membrane containing the protein solution is placed in water, water will flow into the protein solution and the level in the capillary will rise until the hydrostatic pressure, $p$, is equal to the osmotic pressure of the protein solution ; $p$ is usually expressed in atmospheres .

Equilibrium is attained rapidly when the volume of the osmometer is small. However, more precise values are obtained with larger volumes which require longer periods of equilibration. Since this involves the danger of bacterial growth in the protein solution, attempts have been made to shorten the time required for equilibration by *dynamic osmometry* in which $v$, the rate of flow of water into or out of the osmometer, is measured at different external pressures, $p_e$, applied to the capillary. If $v$ is plotted against $p_e$, interpolation for $v = 0$ gives the osmotic pressure, $p$, of the protein solution.

The osmotic pressure is approximately proportional to the molar concentration, $C$, of the protein and to the absolute temperature, $T$, or, mathematically expressed, $p = KCT$, where $K$ is a constant. Van't Hoff found that $K$ is identical with the gas constant, $R$, the value of which is 0.08207 liter-atmospheres per degree; the molar concentration, $C$, is equal to $c/M$, where $c$ is the concentration in grams per liter and $M$ the molecular weight of the protein. By substituting these values in the above equation we obtain $p = RTc/M$ and $M = RTc/p$, which is van't Hoff's equation for the osmotic pressure. It is identical in form with the Gay-Lussac law for gases; the osmotic pressure is evidently equal to the pressure which the same substance would exert if it were present in the form of a gas of the same molecular concentration.

Accurate measurements of the osmotic pressure of proteins showed, however, that proportionality between pressure and concentration is found only in very dilute solutions over a limited range of concentration. The deviation from ideal behavior is accounted for by Eq. V-1,

$$M = \frac{RTc}{p - Kc^2} \tag{V-1}$$

where $K$ is a constant of the system examined. The reciprocal Eq. V-2 can be multiplied by $RT$ and transformed as shown below.

$$\frac{1}{M} = \frac{p - Kc^2}{RTc} \qquad \text{and} \qquad \frac{p}{c} = \frac{RT}{M} + Kc \tag{V-2}$$

If $p/c$ is plotted against $c$, the slope gives the factor $K$. At very low concentrations of the solute, $c^2$ becomes so small that the term $Kc^2$ can be neglected. The molecular weight can therefore be determined by plotting the osmotic pressure, $p$, against the concentration, $c$, and extrapolating the curve to $c = 0$. The intercept on the $p$-axis is considered the osmotic pressure of an ideal solution and is used to compute $M$ according to van't Hoff's law. The constant $K$ corrects for the volume occupied by the protein molecules themselves. Deviations from ideal behavior are particularly large in solutions containing threadlike molecules which immobilize molecules of water.

The osmotic pressure has frequently been attributed to the impact of the protein molecules on the membrane. A more adequate concept (B-1) describes the osmotic pressure as that which arises from the difference in the activities of the water molecules on both sides of the membrane. While the water in which the osmometer is immersed has the full vapor pressure and activity of free water, the vapor pressure and activity of the water inside the membrane is reduced by hydration of the proteins and also by the immobilizing action of threadlike protein molecules. From the thermodynamic standpoint, the flow of water into the protein solution can also be viewed as a tendency toward an increase in entropy.

The osmotic pressure of protein solutions is affected by the pH of the solution. In acid solutions the proteins are present as cations, in alkaline solutions as anions, so that dialyzable anions and cations, respectively, are required to balance the charge of the proteins, and an unequal distribution of these dialyzable ions inside and outside the membrane results (Donnan effect). Due to this phenomenon the osmotic pressure of acid or alkaline protein solutions is higher than that of isoelectric solutions. Thus the osmotic pressure of a 1.2% solution of hemoglobin at pH 5.4, 6.5, 7.2, and 10.2 was found to be 13.4, 3.2, 5.0, and 21.4 mm Hg, respectively (B-2); the osmotic pressure had its minimum value near the isoelectric point of hemoglobin at pH 6.9.

Although the equipment required for osmometry is simple (Fig. V-1), accurate determinations are difficult. In order to avoid bacterial decomposition of the protein, the determinations must be made at low temperatures and over short periods of time. In order to reduce Donnan effects and to eliminate errors arising from traces of salts present in the protein solution, a concentrated salt solution is employed as solvent and the osmometer membrane is immersed in a salt solution of equal concentration. The measurement of membrane potentials permits the calculation of that part of the pressure, $p$, which is due to differences in the ion activities.

Some of the molecular weights found by osmometry are: ovalbumin, 45,000 (B-3); serum albumin, 72,000 (B-4); hemoglobin, 67,000 (B-2); $\beta$-lactoglobulin, 37,000 (B-5). Lower values of $p$ are obtained when protein mixtures are measured, indicating that protein complexes are formed in such mixtures (B-6). Formation of such complexes has been observed not only between acidic and basic proteins but also between serum albumin and the $\gamma$-globulins of the blood serum (B-7). The plant proteins amandin and excelsin, which have molecular weights of 206,000 and 214,000 respectively in aqueous solutions, give values of 30,300 and 35,700 in concentrated solutions of urea (B-8). Evidently, they dissociate into smaller subunits in the presence of urea. Ovalbumin, serum albumin, and gliadin have the same molecular weights in water and in urea solution.

Osmometric determinations of molecular weights between 20,000 and about 150,000 are more accurate than other methods of molecular weight determination. The osmotic pressure does not depend on the shape and hydration of the protein molecules. However, the method does not provide us with any basis for deciding whether the protein in the solution examined is homogeneous or whether it is a heterogeneous mixture of proteins of different sizes. If the solution contains more than one kind of particle, the value computed from the osmotic pressure is the number-average molecular weight, $M_n$.

## C. Diffusion Rate of Proteins (C-1)

When a protein solution is placed in contact with the protein-free solvent, diffusion of protein molecules into the solvent will occur. The quantity, $dS$, of protein diffusing through an area $Q$ in time $dt$ is, according to Fick's law:

$$dS = -DQ(dc/dx)dt \qquad \text{(V-3a)}$$

where $dc$ is the difference in protein concentration over the distance $dx$, and $dc/dx$ is the concentration gradient. $D$ is called the diffusion coefficient. Its order of magnitude is $10^{-5}$ cm$^2$ sec$^{-1}$ for amino acids, and $10^{-6}$ to $10^{-7}$ cm$^2$ sec$^{-1}$ for proteins. The diffusion rate is proportional to the absolute temperature and inversely proportional to a quantity called the frictional constant, $f$. Stated more concisely:

$$D = RT/Nf \qquad \text{(V-3b)}$$

where $R$ and $T$ are the gas constant and the absolute temperature, respectively, and $N$ the Avogadro number ($6.02 \times 10^{23}$). For spherical molecules of radius $r$ the frictional constant is proportional to the radius and to $\eta$, the viscosity of the solvent:

$$f = 6\pi\eta r$$
$$D = RT/Nf = RT/6\pi\eta rN \qquad \text{(V-3c)}$$

Since the volume of a sphere of radius $r$ is $4r^3\pi/3$, the molecular weight of a spherelike protein molecule can be calculated by means of the equation $M = 4r^3\pi N\rho/3$, where $\rho$ is the density of the dissolved protein particle. Substituting $r$ from Eq. V-3c we obtain:

$$M = (4\pi N\rho/3)(RT/6D\pi\eta N)^3$$

Since most of the protein molecules are nonspherical, their frictional constant, $f$, is larger than $f_0$, the frictional constant of a spherical molecule of an equal molecular weight. The frictional ratio, $f/f_0$, is therefore greater than one. The magnitude of the frictional ratio can be calculated for regularly shaped particles such as cylinders, rods, and ellipsoids. If, however, the shape of the molecule is not known, one is unable to calculate the molecular weight from the diffusion rate.

In order to provide a relatively simple model, it is often assumed that the protein molecule is an ellipsoid of revolution with either one long axis, $a$, and two short axes, $b$ (prolate ellipsoid), or with $a$ shorter than the two $b$ axes (oblate ellipsoid). If $M$ and $f/f_0$ are known, the ratio $a/b$ can be calculated. The frictional constant of a protein depends, however, not only on its molecular weight and its shape but also on its hydration, because swelling of the dissolved particle through binding water of hydration clearly results in an increase of the friction. If the protein has a spherelike shape, the frictional ratio depends solely on the degree of hydration. This can be calculated from the relationship:

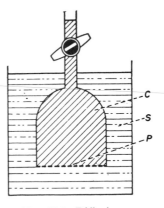

FIG. V-2. Diffusion apparatus. $C$: colloid solution; $S$: solvent; $P$: porous disk.

$$w = (f/f_0 - 1)\rho/\sigma$$

where $w$ is the weight of water bound by 1 g of protein and $\rho$ and $\sigma$ are the densities of the solution and of the dissolved protein, respectively. In computing the molecular weight of nonspherical molecules a value of 0.3–0.5 is usually assumed for $w$.

The diffusion constant is determined experimentally by forming a sharp horizontal boundary between protein solution and solvent in a vertical tube and measuring the rate at which the boundary spreads by means of one of the optical methods used in sedimentation and electrophoresis (see Chapter VI, C). If the protein investigated is homogeneous, the concentration gradient $dc/dx$ will decrease according to a Gaussian distribution curve; in protein mixtures the curve of the gradients will not fit the Gaussian curve (C-2).

Since the formation of a sharp boundary between protein solution and solvent is difficult to accomplish experimentally, the two solutions have often been studied when separated by a porous disk of sintered glass (C-3) or by a dialysis membrane (C-4). The rate of diffusion is measured by determining the amount of protein diffusing into the protein-free solution through the disk or the membrane. A simple form of the porous disk method is shown in Fig. V-2. By placing the heavier protein solutions above the solvent, $S$, the diffusion is accelerated by gravity. The diffusion tube is calibrated with solutions of known diffusion coefficient. If the substance used for standardization is a protein of known molecular weight $M_1$, with diffusion coefficient $D_1$, the molecular weight, $M_2$, of the unknown protein is:

$$M_2 = M_1 \left(\frac{\sigma_2}{\sigma_1}\right) \left(\frac{D_1 f_1}{D_2 f_2}\right)^3$$

If the frictional ratios and the densities of both proteins are the same the molecular weight is: $M_2 = M_1(D_1/D_2)^3$. The results obtained by this method are somewhat less reliable than those obtained by means of the free-boundary method. The precision of this method has been increased considerably by the use of interferometric techniques (C-5) and the availability of membranes with well-defined pore sizes (C-6).

Molecular weights calculated from diffusion rates are of the same order of magnitude as values obtained by other methods. Owing to the variable factors introduced by hydration and by deviations of the molecules from the spherical shape, determinations of the diffusion rate are rarely used for the direct calculation of molecular weights. Frequently they are combined with the sedimentation velocity method.

## D. Sedimentation Equilibrium (D-1)

If a protein solution is centrifuged at high speed, the protein molecules, because of their high specific gravity, will tend to sediment, with the result that their concentration increases from the center of the centrifuge to the periphery. If $\sigma$ is the density of the protein and $\rho$ the density of the solvent, the centrifugal force is $\varphi = m\omega^2 x(1 - \rho/\sigma)$, where $m$ is the mass of one protein molecule, $\omega$ the angular velocity, and $x$ the distance of the protein molecules from the center of the rotor. The amount of protein flowing through a cross section $Q$ of the cell is equal to $cQ(dx/dt)$ where $c$ is the concentration of the protein, $Q$ the area of the cross section of the cell, and $dx$ the distance of protein migration during the time $dt$. This outward sedimentation is counteracted by backward diffusion due to the higher concentration of the protein in the periphery, which causes the protein molecules to move from regions of high concentration toward the more dilute central part of the system. The extent of backward flow is given by $-DQ(dc/dx)$ where the diffusion coefficient (see Eq. V-3b) is $D = RT/Nf$. Since the velocity of sedimentation is $dx/dt = \varphi/f$, the flow of protein through the area $Q$ is

$$dm/dt = Q(cdx/dt - Ddc/dx) = Q(\varphi c/f - Ddc/dx)$$

Replacing $f$ by $RT/ND$ we obtain

$$dm/dt = Q(\varphi cDN/RT - Ddc/dx) = QD(\varphi cN/RT - dc/dx)$$

At equilibrium the flow by sedimentation is equal to the opposite flow by diffusion. Hence $\varphi cN/RT = dc/dx$. As shown above, the centrifugal force is $\varphi = m\omega^2 x(1 - \rho/\sigma)$. Replacing $\varphi$ by this term and multiplying by $N$ in order to convert $m$ to $M$, we finally obtain:

$$dc/dx = \varphi cN/RT = \frac{\omega^2 x M(1 - \rho/\sigma)c}{RT}$$

If $dc$ is the difference between the concentrations $c_1$ and $c_2$ at the distances $x_1$ and $x_2$ from the center of rotation, integration of the above equation yields:

$$M = \frac{2RT \ln (c_1/c_2)}{\omega^2(1 - \rho/\sigma)(x_1{}^2 - x_2{}^2)} \tag{V-4a}$$

The molecular weight can thus be calculated by measuring the protein concentrations at $x_1$ and $x_2$.

The term $\rho/\sigma$ can be replaced by $\rho V$, where $V$ is the partial specific volume of the protein. For most proteins $\sigma$ is approximately 1.33 and $V$ close to 0.75 (see Chapter VI, F). The angular velocity, $\omega$, is equal to $v/x$ where $v$ is the velocity of the centrifuged solution and $x$ the distance from the center of the rotor. Obviously $v$ is different for different layers of the centrifuged solution, depending on $x$, whereas the angular velocity $\omega$ is the same for all layers. If the number of revolutions per second is $z$, the velocity of the centrifuged solution is $2r\pi z$, and $\omega$ is equal to $2\pi z$. It is customary to indicate the velocity in revolutions per minute ($rpm$). Hence, $z = rpm/60$. Since $2\pi = 6.28$, $\omega = 6.28 rpm/60 = 0.105\ rpm$. The centrifugal force, $G'$, is equal to $mv^2/x$ where $m$ is the mass of the centrifuged material. Instead of calculating this value, we can use the *relative centrifugal force*, $g$, which is equal to $G'/G$, where $G$ is the gravitational force of 980 $m$. We write, therefore, $g = G'/G = mv^2/980mx = v^2/980x$. Remembering that $v = x\omega = 6.28xz$, we substitute this expression for $v$ and find that

$$g = \frac{(6.28xz)^2}{980x} = 4.024\ 10^{-2}xz^2$$

Replacing $z$ by $rpm/60$ we obtain $g = 1.12 \times 10^{-5}x(rpm)^2$. Evidently, the centrifugal force is proportional to the radius, but increases with the square of the angular velocity.

The attainment of the high speeds necessary for experiments of this kind was accomplished by Svedberg, who constructed an oil turbine "ultracentrifuge." Since the sedimentation equilibrium would be disturbed by stopping the centrifuge, the observation has to be made during the centrifugation. For this purpose the protein solutions are placed in transparent chambers and the gradient of the protein concentration is measured by a refractive index method, similar to that used in electrophoresis (Chapter VI, C), or by other optical devices (D-2). The heavy oil-driven centrifuge has been replaced by lighter air-driven or electrically-driven ultracentrifuges. Although the speed attainable in the ultracentrifuge is very high, lower velocities are used for determinations of the sedimentation equilibrium than for determinations of the sedimentation velocity (see Section E). If very high velocities were used in equilibration, all protein would be packed in a narrow region at the bottom of the cell and the precise determination of the values of $c_1$, $c_2$, $x_1$, and $x_2$ would become difficult.

The principal disadvantage of the sedimentation equilibrium method is that a very long time, as much as several days, may be necessary for equilibrium to be attained. This difficulty has been overcome to a certain extent by a technique proposed by Archibald (D-4). It is based on the fact that the concentration of the protein solution during centrifugation decreases at the meniscus of the solution and increases at the bottom of the centrifuge tube, but that no net flow of protein molecules occurs through these two cross sections. The concentrations at the meniscus and at the bottom of the tube can be calculated from the measured refractive index gradients. If such measurements are made at different times of centrifugation, the equilibrium concentrations of protein $c_m$ and $c_b$ in the meniscus and at the bottom of the tube, respectively, can be calculated by extrapolation without waiting for equilibrium. If $x_m$ and $x_b$ are the distances of the meniscus and bottom from the center of the rotor, the molecular weight can be computed from an equation similar to Eq. (V-4a). The Archibald technique has become widely used and may displace the older methods (D-1). It is particularly valuable for the investigation of small protein molecules which sediment very slowly. If small cells are used, ultracentrifugation may not take more than 45–70 minutes (D-5).

## E. Sedimentation Velocity of Proteins (D-1)

If protein solutions are spun in an ultracentrifugal field at very high speed, the protein particles sediment at a high rate. If the molecular weight is high, the rate of diffusion can be neglected since the rate of sedimentation increases and the diffusion rate considerably decreases with increasing molecular weight of the solute studied in such experiments. In the preceding section it has been shown that the centrifugal force for a protein molecule is $\varphi = m\omega^2 x(1 - \rho/\sigma)$ and that $dx/dt = \varphi/f$. Since $f = RT/DN$ (Eq. V-3b), we can write

$$(dx/dt)(RT/DN) = m\omega^2 x(1 - \rho/\sigma)$$

Multiplying by $N$ and thereby converting $m$ into the molecular weight $M$ we obtain

$$M = \frac{RT}{D(1 - \rho/\sigma)} \left( \frac{dx/dt}{\omega^2 x} \right)$$

Replacing the second term by $s$ we find $M = RTs/[D(1 - \rho/\sigma)]$ where $s$ is called the sedimentation coefficient. The *sedimentation coefficient* is defined as the rate of sedimentation in a centrifugal field of unity. It is equal to

$$s = (dx/dt)/\omega^2 x$$

where $x$ is the distance of the moving boundary from the center of the rotor at the time $t$, and $\omega$ the angular velocity. The dimension of the sedimentation coefficient is seconds, and the unit of $10^{-13}$ sec has been designated as one svedberg $(1S)$. The sedimentation coefficients of proteins are of the order $10^{-13}$ to $10^{-12}$ sec. If the sedimentation constant is corrected to a standard state (water, 20°C), it is denoted by the symbol $s_{20}$ or $s_{20,w}$. If $s$ was determined at the temperature $t$, $s_{20}$ is defined by the equation:

$$s_{20} = s \frac{\eta(1 - \rho_{20}/\sigma_{20})}{\eta_{20}(1 - \rho/\sigma)}$$

where $\rho_{20}$ and $\sigma_{20}$ are the densities of solvent and solution and $\eta_{20}$ the viscosity at 20°C. The diffusion coefficient $D$, which is required for the calculation of the molecular weight from $s$, is determined by the diffusion methods described in Section C. The velocities needed for sedimentation measurements are much higher than those needed for the sedimentation equilibrium.

The sedimentation velocity, like diffusion, depends on the shape and the hydration of the protein molecules whereas the sedimentation equilibrium is independent of these factors and depends only on the molecular weight. Nevertheless, most of the molecular weight determinations during the last few years have been made by means of the sedimentation velocity method, combined with diffusion measurement. The principal reason for this is the shorter time of centrifugation, which eliminates the danger of bacterial growth and decomposition of the proteins. The sedimentation velocity method does permit one to decide whether or not the protein under investigation is homogeneous since protein mixtures form more than one boundary in the centrifuged solution. Difficulties in the interpretation of results may arise from dissociation and aggregation phenomena and from the formation of protein-protein complexes (D-3; E-6). The sedimentation of the protein and the formation of one or more boundaries in the protein solution can be measured by means of photometry, refractometry, or interferometry. Although photometry is in princi-

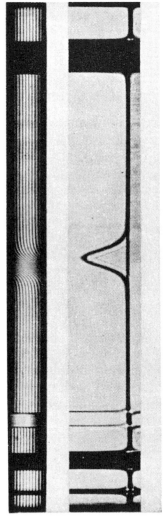

FIG. V-3. Optical pattern in the determination of sedimentation velocity. The upper pattern shows Rayleigh interference fringes, the lower pattern, the schlieren phenomenon. Both were obtained during the ultracentrifugation of bushy stunt virus. From Schachman (E-1).

ple the simplest method, it has rarely been used since it necessitates the use of quartz windows for the measurement of the ultraviolet absorption of the colorless proteins. Only a few colored proteins (hemoglobin, hemocyanin, cytochrome, or the flavoproteins) absorb visible light. Recently, devices for the measurement of ultraviolet absorption during ultracentrifugation have been developed. It seems that this photometric method will soon replace refractometry (E-7).

At the present time, most sedimentation analyses are made by refractometry. This method is based on the deflection of a beam of light in a boundary which separates two media of different refractivity. The refractive index of a protein solution increases linearly with an increase in the protein concentration. If during the ultracentrifugation the protein sediments, a zone of high protein concentration (and high refractive index) will be formed in the bottom portion of the protein solution, and a zone of low protein concentration (and low refractive index) in the top portion of the compartment. Since the deflection of the light beam depends only on differences in refractivity and not on the magnitude of the refractive index, no deflection occurs initially in the homogeneous protein solution. As soon as the sedimentation begins, slight deflection over the wide area of changing concentration is observed. After prolonged centrifugation, when a distinct boundary has been formed, the light beam is strongly deflected in the narrow zone of this boundary (see Fig. V-3) (E-1) where the gradient of refractivity has its maximum. For the same reason, interference fringes undergo a strong deformation in the region of a high gradient of refractive indices (E-5).

Although the sedimentation rate depends also on the electrical charge of the protein molecules, this factor can be largely eliminated by the addition of salts (E-2,3). Difficulties of interpretation arise in protein mixtures due to the mutual interaction of the proteins and the formation of complexes (D-3; E-3,6). Svedberg had originally assumed that all molecular weights of proteins are multiples of 17,000. Later, when the precision of the technique increased, deviations from this rule were found in numerous instances. In spite of the improvement of methods, however, the limit of error is still 1–3% of the molecular weight.

A particular situation arises when the density of the protein particles, $\sigma$, approaches the density of water. This occurs when lipoproteins are investigated. Some of these contain 50% or more of lipids whose density is lower than the density of serum or of physiological salt solution. If the density of the solvent is increased by the addition of more salts, increasing amounts of the lipoproteins, on ultracentrifugation, rise to the surface of the solution. Instead of sedimentation we observe flotation. The sedimentation coefficient, $s$, has then a negative sign. The result of such measurements is recorded in $S_f$ values, where $1S_f = -1S$. Several classes of lipo-proteins have been differentiated by their $S_f$ values, which range from about $10S_f$ to more than $100S_f$ (E-4). (See Chapter VIII, B, and Chapter XI, B.)

## F. Viscosity of Protein Solutions (F-1,13)

It is a familiar observation that solutions of some proteins such as gelatin, are extremely viscous, while solutions of others such as ovalbumin or the serum proteins flow much more easily, even if their concentration is much higher than that of gelatin. In pure solvents, high viscosity is caused by strong intermolecular attraction which lowers the free motility of the molecules. For this reason the viscosity of water, whose molecules are bound to each other by hydrogen bonds, is higher than that of ether or ethanol, although the molecular weight of water is lower. For the same reason glycol or glycerol have much higher viscosities than ethanol or propanol. In protein solutions the increase in viscosity is caused by *intermolecular* electrostatic interaction between protein molecules (F-2) and also by strong deviation of certain protein molecules from the spherelike shape and *intramolecular* interaction between their peptide chains. The high viscosity of gelatin is due to the threadlike shape of its molecules which, owing to Brownian motion, occupy a larger volume of the solvent than spherical molecules of the same molecular weight would occupy.

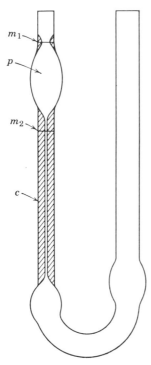

Fig. V-4. Ostwald viscometer.

Measurements of viscosity do not involve any experimental difficulties. This may be one of the reasons for their frequent application. The type of viscometer most commonly used is the Ostwald viscometer (Fig. V-4). The aqueous solution of the protein is allowed to flow from the pipet ($p$) through the capillary ($c$), and the time ($t$) required for the level to drop from $m_1$ to $m_2$ is noted. If the viscosity of the protein solution does not differ too greatly from that of water, and if the dimensions of the capillary are appropriate, the viscosity, $\eta$, is directly proportional to $t$, the time of outflow, and to $d$, the density of the solution:

$$\eta = Cdt$$

If $\eta_0$ is the viscosity of water and $\eta$ that of the protein solution, the relative viscosity: $\eta_r = \eta/\eta_0 = Cd_p t_p/Ct_w = d_p t_p/t_w$ where $d_p$ is the density of the protein solution, and $t_p$ and $t_w$ are the time of flow of the protein solution and of water, respectively. The relative viscosity, $\eta_r$, is evidently independent of the constant $C$, which is a function of the dimensions of the capillary. Table V-1 shows the relative viscosities of solutions of ovalbumin at 25.2°C (F-3) and of gelatin at 37°C (F-4). At lower temperature

TABLE V-1

RELATIVE VISCOSITIES OF AQUEOUS SOLUTIONS OF OVALBUMIN AND GELATIN[a]

| Ovalbumin | | | | | | | |
|---|---|---|---|---|---|---|---|
| Concentration, %......... 3.02 | 8.88 | 14.53 | 20.12 | 28.15 | | | |
| Relative viscosity......... 1.22 | 1.57 | 2.21 | 3.60 | 9.99 | | | |
| Gelatin | | | | | | | |
| Concentration, %...   1 | 2 | 3 | 4 | 5 | 6 | 8 | 10 |
| Relative viscosity... 2.39 | 3.44 | 4.54 | 5.78 | 7.12 | 9.06 | 14.2 | 22.0 |

[a] From Chick and Lubrzynska (F-3); Kunitz (F-4).

the viscosity of gelatin cannot be measured because its solutions form a gel. If the shape of the protein molecules is very elongated, their solutions do not follow Newton's law of flow when investigated in capillaries (F-5). This has been designated as non-Newtonian flow.

For very dilute suspensions of spherelike particles Einstein formulated the equation:

$$\eta = \eta_0(1 + 2.5Nv/V) \tag{V-4b}$$

where $N$ is the number of suspended particles, $v$ the volume of each particle, and $V$ the total volume of the solution. If we know $N$, the number of protein molecules dissolved, we can calculate the volume occupied by each particle and can thus determine the extent to which it is hydrated. On the other hand, if the hydration is known, we can calculate from Eq. (V-4b) the number of molecules, $N$, and the particle weight, which is equal to the total weight of the protein divided by $N$. Most of the proteins have a nonspherical shape, so that this method is of very limited applicability.

Since the viscosity, $\eta$, of a protein solution is always higher than $\eta_0$, the viscosity of water, the relative viscosity $\eta_r = \eta/\eta_0$ is always greater than one. The increase in relative viscosity, which is equal to $\eta_r - 1$, has been called the specific viscosity, $\eta_{sp}$. According to Staudinger (F-6), the specific viscosity of solutions of threadlike macromolecules of a homologous series is directly proportional to the concentration $c$ and to $M^\alpha$ where $M$ is the molecular weight of the dissolved macromolecule; $\alpha$ assumes values between 0.5 and 1.5 (F-8). In the equation: $\eta_{sp} = KcM^\alpha$, $K$ is a constant for a given series of homologous chain molecules of the same structural pattern but of different lengths.

If chain molecules are dissolved, their molecules will only rarely be present in the fully extended shape. They will fold and unfold continually, owing to free rotation around the valence bonds, and will assume different

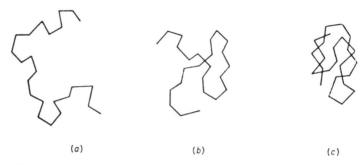

(a)                    (b)                    (c)

Fig. V-5. Different conformations of a threadlike macromolecule.

constellations at different times (Fig. V-5). According to statistical considerations (F-7) the average chain length of the molecule, i.e., the distance between the two ends of the chain, is proportional to $\sqrt{L}$, where $L$ is the chain length of the fully extended molecule. Since $M$ is proportional to $L$, the average chain length is also proportional to $\sqrt{M}$.

The viscosity of the solution depends not only on this average chain length, but also on the shape resistance of the molecule, i.e., on the resistance of the chain molecule to a change in conformation.

If viscosity measurements are performed at different concentrations of the macromolecular solution, different values of $\eta_r$ or of $\eta_{sp}$ are obtained. By plotting the specific viscosity against the concentration of the solution and by extrapolation to the concentration $c = 0$, the *intrinsic viscosity*, $[\eta]$, of a molecule can be determined. It is (F-1):

$$[\eta] = (\eta_{sp}/c)_{c\to 0} = (\ln \eta_r/c)_{c\to 0}$$

The intrinsic viscosity of gelatin has been found to be 0.6. The analogous values for other proteins are: ovalbumin, 0.04; gliadin, 0.11 (F-9a); tobacco mosaic virus, 0.6 (F-9b).

While viscosity measurements allow the determination of the molecular weight of chain molecules such as those of rubber and of cellulose esters, the situation in solutions of proteins is complicated by the electrostatically interacting anionic and cationic protein side chains and by their action on water molecules. The viscosity of a protein solution will depend, therefore, on the pH of the solution. Increasing acidity causes expansion of the molecules of bovine serum albumin and of human $\gamma$-globulin whereas insulin, $\beta$-lactoglobulin, trypsin, chymotrysin, ribonuclease, ovalbumin, and other proteins do not show this phenomenon (F-10). Electrostatic interaction between ionic groups can be eliminated to a certain extent by the addition of neutral salts.

Although it is not possible to calculate molecular weights from viscometric data alone, viscometry in combination with other methods gives valuable information on the size and shape of protein molecules. Thus it has been found that the product $D^3M$ $[\eta]$ is constant and is approximately $8 \times 10^{-16}$ (F-11). This would allow us to calculate the molecular weight, $M$, from the diffusion coefficient, $D$, and the intrinsic viscosity.

If the rodlike particles of a solution are very long and if they possess sufficient rigidity, the phenomenon of *thixotropy* is observed (F-12). This consists of the formation of a gel during prolonged standing of the solution, and of the liquefaction of the gel upon agitation. Thixotropy has been observed in gels of myosin, the contractile muscular protein.

## G. Flow Birefringence (G-1)

If a solution of a macromolecular substance such as a protein is placed between two crossed Nicol prisms, the field will remain dark. If, however, the solution is made to flow, a light field is observed in solutions of rodlike or threadlike proteins, while solutions of spherelike proteins remain dark. The appearance of light in the dark field of the crossed Nicol prisms is due to the same phenomenon as the birefringence of crystals and other anisotropic structures. It is believed that rodlike protein molecules undergo a uniform orientation in the streaming solution (Fig. V-6). Quantitative measurements of flow birefringence are carried out by means of a device in which the protein solution is placed between two concentric cylinders. While

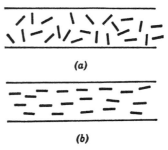

FIG. V-6. Flow birefringence. (*a*) Particles at rest; (*b*) particles oriented owing to flow.

one of the cylinders is fixed, the other one rotates, so that the rod-like protein molecules tend to assume a position tangential to the direction of flow. The angle between the long axis of the rods and the direction of flow decreases as the speed of the moving cylinder increases and as the axial ratio of the rods increases; this ratio can, then, be calculated from the angle of orientation (G-2,3). In the rotating system a dark cross, the cross of isocline, is visible; it is similar to the cross observed in starch grains under a polarizing microscope. The average angle of orientation of the rods is equal to 90° minus the angle of isocline. By means of this device it has been found that the molecules of myosin from muscle (G-3), fibrinogen from blood plasma (G-4), and tobacco mosaic virus (G-5) are elongated, threadlike molecules with high axial ratios, while those of $\gamma$-globulin are less asymmetric (G-4); $\beta_1$-globulin was found to be highly symmetric. Attempts have been made to get more information about the shape of proteins in their solutions by evaluating the sedimentation rate, the intrinsic viscosity, and the flow birefringence in the same solution (G-6).

## H. Polarization of Fluorescence (H-1,2)

It has been known for a long time that fluorescent molecules, after absorption of light, remain for a certain time in the excited state before they emit fluorescence. This period between absorption and emission is of the order of $10^{-8}$ sec. If the molecules are irradiated with polarized light, the emitted light should also be polarized. However, molecules of low molecular weight undergo very rapid Brownian motion during the time interval of $10^{-8}$ sec, and rotate so rapidly that randomization takes place and the emitted fluorescent light is depolarized. Large molecules, however, move much more slowly than the small ones, and also rotate at a lower rate. Consequently, the absorption of polarized light causes large molecules to emit polarized fluorescent light. This phenomenon has been utilized for the estimation of molecular weights (H-2,3). Since most proteins are colorless and do not emit fluorescent light on irradiation, it is necessary to render them fluorescent by coupling with a fluorescent dye. Weber (H-2) used for this purpose aromatic sulfonyl chlorides such as 1-dimethylamino-naphthalene-5-sulfonyl chloride. Another useful dye is fluorescein isocyanate. The mean relaxation time of proteins, i.e., the time during which the molecules remain oriented in a definite direction, is considerably longer than the lifetime of the excited state. Consequently, the protein-dye complex will emit *polarized* light after absorption of polarized light. The smaller the molecule is, the more randomization takes place. No measurable polarization of the emitted fluorescent light is observed when the molecular weight is less than approximately 1000. The relaxation time of the molecules depends not only on their size but also on their shape. It increases with an increase in the axial ratio and with deviations from the spherical shape. The mean relaxation time of lysozyme is $2.8 \times 10^{-8}$ sec; it is $9.0 \times 10^{-8}$ sec for ovalbumin, and $\sim 13$ sec for bovine serum albumin (H-2). The molecular weights calculated from these data agree well with those determined by other methods. Good results have been obtained with a simplified modification of the apparatus used for the determination of the polarization of the emitted light (H-4).

## I. Light Scattering by Proteins (I-1)

The fact that colloidal solutions such as those of proteins are turbid, and that they scatter part of the incident light, an effect called *Tyndall phenomenon*, has long been known. Much less scattering is observed in solutions of small molecules or ions. However, even liquids and gases scatter a small part of the incident light. The intensity of scattering can be determined either by measuring the decrease in the intensity of transmitted light which has passed through the investigated solution, or by measuring the intensity of light scattered at a right angle or any other suitable angle. Since the intensity of the scattered light decreases with the square of $r$, the distance from the scattering medium to the observer, the reduced intensity at $90°$ is $R_{90} = I_s r^2/I_0$ where $I_s$ and $I_0$ are the intensities of the scattered and the incident light respectively. The scattered light is more intense in a protein solution than in the pure solvent. The difference in the intensity of light scattering is denoted as the turbidity of the solution and is designated by the symbol $\tau$. The intensity of the scattered light increases with the number and size of the protein molecules. This is evident from the equation:

$$1/M = Hc/\tau \tag{V-5}$$

where $M$ is the molecular weight, $c$ the concentration in grams per ml, $H$ a proportionality constant, and $\tau$ the excess turbidity. If $\mu$ is the refractive index of the solution, $\mu_0$ that of the solvent, and $\lambda$ the wave length of the light used, $H$ can be determined according to the following equation:

$$H = \frac{32\pi^3}{3} \frac{\mu_0^2}{N\lambda^4} \left(\frac{\mu - \mu_0}{c}\right)^2$$

where $N$ is the Avogadro number. The equation shows that the intensity of the turbidity, $\tau$, depends on $\mu_0$ and on the difference between the two refractive indices $\mu$ and $\mu_0$, and that it is inversely proportional to the fourth power of the wave length, $\lambda$. If solvent and solution have the same refractive index, the last term becomes zero and no scattering will occur.

Equation (V-5) is valid only for ideal systems in which there is no interaction between the molecules. Experimental results are closer to Eq. V-6, which contains a term similar to the one used in the calculations of molecular weights from the osmotic pressure. $B$ is an empirical constant. If $Hc/\tau$ is plotted against $c$, a straight line is obtained, the intercept of which with the ordinate is $1/M$. In this way the molecular weight, $M$, can be determined.

$$\frac{1}{M} = \frac{Hc}{\tau} - 2Bc \qquad \text{(V-6)}$$

The light scattering method has the great advantage of rapidity. Its principal difficulty is the great influence of traces of dust and other

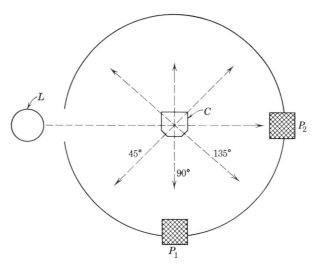

FIG. V-7. Light-scattering apparatus. $L$: light source; $C$: cuvette containing protein solution; $P_1$, $P_2$: photoelectric cells in position to measure scattering.

particles which, because of their large size, cause very intense light scattering. Some of the molecular weights determined by light scattering are: ovalbumin, 45,000–50,000, the highest values being found at the isoelectric point (I-2); serum albumin, 74,000; hemocyanin, 6.34 millions (I-3); and tobacco mosaic virus, 40 millions (I-4).

The intensity of light scattering is usually measured by a photocell (Fig. V-7) mounted on an arm which can be rotated around the center of the system into which the cuvette with the protein solution in placed (I-5). It is customary to measure the light which is scattered at an angle of 90° to the direction of the incident light. If the size of the scattering molecules is smaller than the wave length of the incident light, the intensity of the scattered light is symmetrically distributed in the forward and backward direction. Hence, identical intensities will be found at 45°

and at 135 (90 + 45)°. If however, the size of the protein molecule is so great as to be comparable with the wave length of the incident light, interference occurs and the intensity of the scattered light is no longer symmetrically distributed about the right-angle position. The intensities ($\dot{I}_1$ and $\dot{I}_2$) of light scattered at 45° and 135°, i.e., forward and backward from 90° direction, are found to be different. The value $q = (\dot{I}_1/\dot{I}_2) - 1$ has been called the dissymmetry coefficient. With increasing dilution of the solution, $q$ approaches a limiting value called the "intrinsic dissymmetry coefficient"; this value is a measure of the spatial extension of the colloidal molecule and can be used to determine directly the size of the molecule in solution (I-7). In this way the length of the tobacco mosaic virus was found to be 2750 Å, in agreement with electron microscope measurement which furnished the value 2800 Å (I-6).

## J. Other Methods Used for the Determination of Molecular Weights and the Shape of Proteins in Their Solutions

Attempts have been made to obtain information on the approximate size of protein molecules by X-ray analysis of their solutions. Since the protein molecules in their aqueous solutions are not regularly arranged, as are those of solid protein crystals, but are randomly oriented in space, the method cannot reveal more than the radii of gyration, i.e., the radii of the spheres which result from the rotation of the molecules around their centers of gravity. Investigations of this type give for human serum albumin a radius of gyration of 31.0 Å, for its dimer (mercaptalbumin Hg-complex) 37.2 Å, and for bovine serum albumin 29.8 Å (J-1). A similar investigation (J-2) gave for $\beta$-lactoglobulin the radius of gyration 24.6 Å, for lysozyme 16.0 Å, and for ovalbumin 24.0 Å. The respective axial ratios were 3.6, 2.7, and 2.9 (J-2). Different values were obtained from measurements of the electro-optic effect, i.e., the birefringence produced in solution of bovine serum albumin by an electrical field. From the measured relaxation times it was concluded that the dimensions of the molecule are 235 and 33 Å and the axial ratio close to 7 (J-3).

Molecular weights and axial ratios of proteins have also been calculated from measurements of the dielectric constants of their solutions at different wave lengths. Since the results of this method can be interpreted in various ways, they will be discussed in connection with the dipole moments and the electrochemistry of proteins (Chapter VI, H). Finally, molecular weights of dissolved proteins have been determined from the pressure of protein films in the interphase water/air or water/nonaqueous solvent. The method will be treated in connection with other properties of protein films (Chapter VII, J).

## K. Electron Micrography of Dry Proteins (K-1)

Electron micrography is based on the diffraction of electrons by the molecules. Since the electrons of an electron beam are diffracted not only by proteins but also by water and other molecules present, it is necessary to investigate the dry dehydrated protein. Dehydration may change the shape of the protein molecules. This can be prevented to a certain extent by careful fixation with formaldehyde, ethanol, or other reagents. The method allows us to magnify the diameter of the protein molecules approximately $10^5$ times. We can see in electron micrographs not only the size but also the shape of the largest protein molecules. The method thus

Fig. V-8. Electron micrograph showing the surface of edestin crystals. 135,000 ×. From Hall (K-6).

gives the most direct information on these properties. Particularly beautiful pictures are obtained by first shadowing the particles, i.e., by exposing them to the oblique evaporation of chromium or of gold. The metal is deposited on one side of the particle, giving the suggestion of a shadow and imparting depth to the picture obtained (K-2). The first electron micrographs of proteins were produced by von Ardenne (K-3), who found a molecular diameter for hemocyanin of 220 Å. Each of the hemocyanin particles is apparently a bundle of four subunits (K-4). Similarly, four subunits can be seen in ferritin molecules (K-5). The protein molecules, whose molecular weights are less than 100,000, are too small to reveal details of their structure. Electron micrographs of microscopic crystals of these proteins show, however, quite distinctly (Fig. V-8) how the protein molecules are arranged in the crystal lattice (K-6). If the total amount of protein present in the electron micrograph is known, the molecular weight of very large particles can be determined directly by counting the number of visible molecules in the electron micrograph; this method gave for tobacco mosaic virus a "molecular weight" of $49 \times 10^6$ (K-7). For the smaller protein molecules this method is not suitable. The approximate molecular weight of a smaller protein molecule can, however, be determined from the two dimensions in the electron micrograph and the length of the "shadow," from which the height in the third dimension can be calculated (K-8).

If instead of being shadowed the proteins are directly treated with tungstate, they appear on electron microscopy light on a dark background. When investigated by this method, $\gamma$-globulins appear as small spheres whereas their physicochemical properties in solutions indicate an axial ratio of $5:1$ (K-9). Evidently, there is little relation between the shapes observed in electron micrography of the tungstate-treated proteins and those deduced from the hydrodynamic behavior of the same proteins.

## L. X-Ray Analysis of Protein Crystals (L-1)

Although X-ray analysis has been applied to proteins for more than twenty years, the great power of this method has been recognized only recently. It has been shown, particularly by Perutz (L-2), Kendrew (L-3), and their co-workers, that X-ray analysis under favorable conditions is able to reveal all details of the conformation of the peptide chain(s) inside a globular protein molecule. This application of X-ray analysis and its results will be treated in Chapter VII, B. Here we will discuss X-ray analysis merely as a method for the determinations of the molecular weight and shape of proteins.

When crystalline proteins were examined by X-rays, it was at first found that the interference pictures obtained were less typical than those obtained with protein fibers. The reason for this disappointing result was limitation of the older methods to wide-angle scattering by means of which only those periodicities which correspond to short intervals between structural elements can be detected. Periodicities of a higher order can be determined only by small-angle scattering. Another difficulty which had to be overcome was the change which wet protein crystals underwent when they were dried. Even under the mildest conditions of drying, the structure collapses, frequently irreversibly, and with disorientation. Much better results are obtained when the wet crystals are investigated in their mother liquor (L-4).

The regular periodicity of the X-ray diffraction maxima indicates that the crystals are indeed formed by protein molecules which are identical or almost identical in size and shape. Analysis of a crystal in two or three different orientations makes it possible to determine the size of the unit cell which is usually represented by the length of its three axes $a$, $b$, and $c$, and the interaxial angles $\alpha$, $\beta$, and $\gamma$. The equivalent weight of the unit cell, whose size is determined by these dimensions and angles, can be calculated provided the density and composition of the protein crystals is known. The equivalent weight of the unit cell is not necessarily identical

TABLE V-2
X-RAY MOLECULAR WEIGHTS

| Protein | Dimensions of unit cell (Å) | | | | Mol. weight | References |
|---|---|---|---|---|---|---|
| | a | b | c | $\beta°$ | | |
| Ferritin (wet) | 131.5 | 131.5 | 186 | | 545,000 | (L-7) |
| (dry) | 109 | 109 | 154 | | | |
| Horse CO-hemoglobin (wet) | 109 | 63.2 | 54.4 | 111° | 66,700 ± 5% | (L-6) |
| (dry) | 102 | 56 | 49 | 134 | | |
| Human serum albumin (wet) | 178 | 54 | 166 | 95 | 65,000 ± 2% | (L-6) |
| (dry) | 168 | 38 | 134 | | | |
| Horse metmyoglobin (wet) | 57.3 | 30.8 | 57.0 | 112 | 17,000 | (L-8) |
| (dry) | 51.5 | 28.0 | 37.0 | 100 | | |
| Ribonuclease (wet; 12.5% $H_2O$) | 30.90 | 38.80 | 54.06 | 106 | 13,400 | (L-5) |
| (dry) | 29.10 | 30.08 | 51.03 | 114 | | |
| Zinc insulin (wet; 5.4% $H_2O$) | 83 | | 34 | | 37,600 | (L-5) |
| (dry) | 74.8 | | 30.9 | | (hexamer) | |

with the molecular weight of the protein, but can be a multiple of it. Thus the unit cell of rhombohedral zinc insulin, which corresponds to a molecular weight of 37,600 (L-5) consists of 6 molecules of insulin. Some results of the X-ray analysis of protein crystals are shown in Table V-2.

## M. Comparison of Molecular Weights and Shapes Found by Different Methods (M-1)

In the introduction to this chapter the difficulties in defining a protein molecule were emphasized. Most of the physicochemical methods for the determination of the molecular weight of proteins measure the size of the particle which moves in diffusion, sedimentation, or rotation as a kinetic unit. This unit may consist of a single protein molecule or of an aggregate of molecules. This is exemplified by the divergence between the *molecule* and the *kinetic units* of insulin. The molecular weight of insulin, determined by complete elucidation of its amino acid sequence and structure, is approximately 6000 (M-2), yet the size of its diffusing, sedimenting, or osmotically active particles in solution is 48,000 in concentrated solutions; on dilution it decreases to approximately 12,000 (see Table V-3). Evidently, insulin is present in dilute aqueous solutions as a dimer, in concentrated solutions as a hexamer or octamer.

Difficulties are also encountered by the fact that many of the phenomena discussed in the preceding sections of this chapter depend not only on the molecular weight, but also on the shape of the protein molecules and on the extent of their hydration. In a number of the equations quoted above there appears the term $\sigma$, the density of the protein particles in solution. This value cannot be determined experimentally. It is sometimes replaced by $1/\bar{V}$, where $\bar{V}$, the partial specific volume, is the increase in volume of a large volume of solvent brought about by the dissolution of 1 g of the protein. The inadequacy of this procedure will be discussed in Chapter VI, D. Finally, it must be kept in mind that protein molecules combine easily with almost any anions or cations, with numerous uncharged substances, and also with other protein molecules (E-6). The molecular weights determined experimentally are affected by these phenomena. In view of all these uncertainties one would expect wide divergencies between molecular weights determined by means of different methods. Table V-3 shows, however, a surprisingly fair agreement between most of the experimental values. This result may be attributed in part to the fact that all proteins seem to have a similar amount of water of hydration, and that the error in replacing $\sigma$ by $1/\bar{V}$ is the same in most of the techniques used.

Although the molecular weights, as shown by Table V-3, do not deviate too much from each other, great differences exist between the axial ratios calculated from viscosity or diffusion on the one hand, and from X-ray analysis on the other (M-3). Thus axial ratios, $a/b$, of 3–4 have been calculated for serum albumin, hemoglobin, and for lactoglobulin in solution, while X-ray analysis shows that the axial ratio in the wet crystals is less than 3 for lactoglobulin and less than 2 for hemoglobin. Similar divergencies were found when axial ratios calculated from dielectric increments (see Chapter VI, H) were compared with those obtained by other methods. The poor agreement between axial ratios obtained by different methods is not surprising since calculations of the axial ratios are based on the simplifying assumption that the protein molecules are ellipsoids of revolution. This assumption is contradicted by the results of X-ray diffraction (Chapter VII, B) and electron micrography (this Chapter, Section K). It has become clear during the last few years that too much confidence has been placed on simple shapes and that we are rather ignorant concerning the true shapes of proteins in solution (D-5). The most reliable

TABLE V-3

MOLECULAR WEIGHTS OF PROTEINS DETERMINED BY DIFFERENT METHODS

| Protein | Molecular weight ($\times 10^{-3}$) determined by: | | | | |
|---|---|---|---|---|---|
| | Osmometry | Sedimentation equilibrium | Sedimentation velocity | X-ray analysis | Light scattering |
| Insulin | 12[a] | 35[d] | 6.3–47.8[d] | 39.5[j] | 12[c] |
| Ribonuclease | — | 13[d] | 12.7[f] | 14.5[l] | — |
| Cytochrome c | — | — | 15.6[d] | — | — |
| Myoglobin | — | 17.5[d] | 16.9[g] | — | — |
| Lysozyme | 17.5[a] | 14–17[d] | — | — | 14.8[c] |
| Chymotrypsinogen | 36[a] | — | 38[d] | — | — |
| Pepsin | 36[a] | 39.0[d] | 35.5[d] | 39.3[m] | — |
| β-Lactoglobulin | 36[a] | 38.0[d] | 41.5[d] | 40[k] | 36[c] |
| Ovalbumin | 44[b] | 40.5[d] | 44.0[d] | — | 46[c] |
| Bovine serum albumin | 69[a] | 68[e] | 68[g] | 82.8[m] | 77[c] |
| Hemoglobin | 67[a] | 68[m] | 68[m] | 66.7[m] | — |
| γ-Globulin (man) | | | 176[d] | | |
| Excelsin | 214[a] | — | 295[m] | 306[m] | 280[c] |
| Catalase | — | — | 250[d] | — | — |
| Phycoerythrin | — | 290[d] | 290[d] | — | — |
| Fibrinogen | — | — | 330[d] | — | 540[h] |
| Thyroglobulin (pig) | — | 650[d] | 630[g] | — | — |
| Hemocyanin (*Helix pomatia*) | 1800[a] | 6700[d] | 8900[d] | — | 6340[c] |
| Tobacco mosaic virus | — | — | 40,000[g] | — | 40,000[i] |

[a] J. T. Edsall, *in* "Proteins" (H. Neurath and K. Bailey, eds.), Vol. 1B, p. 594. Academic Press, New York, 1953.

[b] H. Gutfreund, *Nature* 153: 406(1944).

[c] J. T. Edsall, *in* "Proteins" (H. Neurath and K. Bailey, eds.), Vol. 1A, p. 608. Academic Press, New York, 1953.

[d] J. T. Edsall, *in* "Proteins" (H. Neurath and K. Bailey, eds.), Vol. 1A, pp. 634–645. Academic Press, New York, 1953.

[e] Horse serum albumin.

[f] A. Rothen, *J. Gen. Physiol.* 24: 203(1940).

[g] T. Svedberg and K. O. Pedersen, "The Ultracentrifuge." Oxford Univ. Press, London, 1940.

[h] P. Putzeys and T. Brosteaux, *Trans. Faraday Soc.* 31: 1314(1934).

[i] G. Oster *et al.*, *JACS* 69: 1193(1946).

[j] D. Crowfoot and D. Riley, *Nature* 144: 1011(1939).

[k] F. Senti and R. Warner, *JACS* 70: 3318(1948).

[l] I. Fankuchen, quoted in "Proteins, Amino Acids and Peptides" (E. J. Cohn and J. T. Edsall, eds.), p. 382. Reinhold, New York, 1943.

[m] P. Johnson, *Ann. Repts. on Progr. Chem.* 43: 53(1948).

axial ratios are those of the very large protein molecules which can be determined directly by electron micrography.

## REFERENCES

### Section A

**A-1.** A. R. Battersby and L. C. Craig, *JACS* 73: 1887(1951). **A-2.** K. O. Pedersen, *Cold Spring Harbor Symposia Quant. Biol.* 14: 140(1950). **A-3.** J. T. Edsall, *in* "Proteins" (H. Neurath and K. Bailey, eds.), Vol. 1, p. 554. Academic Press, New York, 1953. **A-4.** R. F. Steiner, *ABB* 49: 71(1954). **A-5.** A. G. Ogston, *Ann. Rev. Biochem.* 24: 181(1955). **A-6.** M. F. Perutz *et al.*, *Nature* 185: 416(1960); J. C. Kendrew *et al.*, *Nature* 185: 422. **A-7.** G. M. Edelman, *JACS* 81: 3155(1959). **A-8.** C. Tanford, "Physical Chemistry of Macromolecules," Wiley, New York, 1961.

### Section B

**B-1.** D. W. Kupke, *Adv. in Protein Chem.* 15: 57(1960); W. J. Badgley and H. Mark, *Frontiers in Chem.* 6: 75(1949); E. J. Cohn and J. T. Edsall, eds., "Proteins, Amino Acids and Peptides," p. 382. Reinhold, New York, 1943. **B-2.** G. S. Adair, *Proc. Roy. Soc.* A108: 627(1925); A109: 292(1925). **B-3.** J. R. Marrack and L. F. Hewitt, *BJ* 23: 1079(1929); H. B. Bull, *JBC* 137: 143(1941); H. Gutfreund, *Nature* 153: 406(1944). **B-4.** G. S. Adair and M. Robinson, *BJ* 24: 1864(1930). **B-5.** O. Smithies, *BJ* 56: 57(1953). **B-6.** A. G. Ogston, *BJ* 31: 1952(1937). **B-7.** G. Scatchard *et al.*, *J. Phys. Chem.* 58: 783(1954); H. Gutfreund, *Trans. Faraday Soc.* 50:628(1954). **B-8.** N. F. Burk, *JBC* 120: 63(1937).

### Section C

**C-1.** H. Neurath, *Chem. Revs.* 30: 357(1942); L. J. Gosting, *Adv. in Protein Chem.* 11: 430(1956); E. O. Kraemer, *Frontiers in Chem.* 1: 74(1943). **C-2.** J. W. Williams *et al.*, *JACS*, 74: 1542(1952); R. L. Baldwin, *J. Phys. Chem.* 58: 1081(1954). **C-3.** J. H. Northrop and M. L. Anson, *J. Gen. Physiol.* 12: 543(1929). **C-4.** L. C. Craig *et al.*, *JACS* 79: 3729(1957). **C-5.** H. K. Schachman, *Brookhaven Symposia in Biol.* 13: 49(1960). **C-6.** E. Fuchs and G. Gorin, *Biochem. Biophys. Research Communs.* 5: 196(1961).

### Section D

**D-1.** H. K. Schachman, "Ultracentrifugation in Biochemistry." Academic Press, New York, 1959; J. W. Williams *et al.*, *Chem. Revs.* 58: 715(1958). **D-2.** T. Svedberg and K. O. Pedersen, "The Ultracentrifuge." Oxford Univ. Press, London, 1940; L. J. Gosting, *Adv. in Protein Chem.* 11: 430(1956). **D-3.** J. P. Johnston and A. G. Ogston, *Trans. Faraday Soc.* 42: 789(1946). **D-4.** W. J. Archibald, *J. Phys. Chem.* 51: 1204 (1947); G. Kegeles *et al.*, *ABB* 63: 247(1956); *J. Phys. Chem.* 61: 1286(1957); *JACS* 80: 5724(1958); R. Trautman and C. F. Crampton, *JACS* 81: 4036(1959); H. G. Elias, *Angew. Chem.* 73: 209(1961). **D-5.** H. K. Schachman, *Brookhaven Symposia in Biol.* 13: 49(1960).

### Section E

**E-1.** H. K. Schachman, "Ultracentrifugation in Biochemistry." Academic Press, New York, 1959. **E-2.** A. Tiselius, *Kolloid-Z.* 59: 306(1938). **E-3.** K. O. Pedersen, *Compt. rend. trav. lab. Carlsberg, Sér. chim.* 22: 427(1938); *J. Phys. Chem.* 62: 1282 (1958). **E-4.** O. F. de Lalla and J. W. Gofmann, *Methods of Biochem. Anal.* 1: 459

(1954).  **E-5.** G. Kegeles *et al.*, cited in **D-4.**  **E-6.** J. W. Williams, *Prog. in Chem. Org. Nat. Prods.* 18: 434(1960).  **E-7.** H. K. Schachman *et al.*, *ABB* 99: 175(1962).

### Section F

**F-1.** J. T. Edsall, *Adv. in Colloid Sci.* 1: 269(1942); E. O. Kraemer, *Frontiers in Chem.* 1: 74(1943).  **F-2.** C. W. N. Cumper and A. E. Alexander, *Australian J. Sci. Research* A5: 146(1952).  **F-3.** H. Chick and E. Lubrzynska, *BJ* 8: 59(1914).  **F-4.** M. Kunitz, *J. Gen. Physiol.* 10: 811(1926/7).  **F-5.** J. T. Edsall, *in* "Proteins" (H. Neurath and K. Bailey, eds.), Vol. 1, p. 693. Academic Press, New York, 1953.  **F-6.** H. Staudinger, *Ber. deut. chem. Ges.* 65: 267(1932).  **F-7.** W. Kuhn, *J. Colloid Sci.* 3: 11(1948).  **F-8.** T. Alfrey *et al.*, *JACS* 61: 2319(1939); P. Debye and A. Bueche, *J. Chem. Phys.* 16: 573(1948).  **F-9a.** A. Polson, *Kolloid-Z.* 88: 51(1939).  **F-9b.** M. Lauffer, *Science* 87: 469(1938).  **F-10.** J. T. Yang and J. F. Foster, *JACS* 77: 2374 (1955).  **F-11.** A. Polson, *BBA* 21: 185(1956).  **F-12.** A. S. C. Lawrence, *Ann. Repts. on Progr. Chem.* 37: 101(1941).  **F-13.** J. T. Yang, *Adv. in Protein Chem.* 16: 323 (1961).

### Section G

**G-1.** J. T. Edsall, *see* **A-3,** p. 677(1953).  **G-2.** A. v. Muralt and J. T. Edsall, *JBC* 89: 315, 351(1930).  **G-3.** G. Boehm and R. Signer, *Helv. Chim. Acta* 14: 1370(1931).  **G-4.** J. F. Foster and J. T. Edsall, *JACS* 70: 1860(1948).  **G-5.** M. A. Lauffer, *JBC* 126: 443(1938).  **G-6.** H. A. Scheraga and L. Mandelkern, *JACS* 75: 179(1953).

### Section H

**H-1.** J. T. Edsall, *in* "Proteins" (H. Neurath and K. Bailey, eds.), Vol. 1B, p. 712. Academic Press, New York, 1953.  **H-2.** G. Weber, *BJ* 51: 145, 155(1952); *Adv. in Protein Chem.* 8: 439(1953).  **H-3.** R. F. Steiner and A. J. McAlister, *J. Polymer Sci.* 24: 105(1957).  **H-4.** P. Johnson and G. Richards, *ABB* 97: 250, 260(1962).

### Section I

**I-1.** J. T. Edsall, *in* "Proteins" (H. Neurath and K. Bailey, eds.), Vol. 1B, p. 602. 1953; P. Doty and J. T. Edsall, *Adv. in Protein Chem.* 6: 37(1952); J. T. Edsall *et al.*, *JACS* 72: 4641(1950); C. E. H. Bawn, *Ann. Repts. on Progr. Chem.* 47: 88(1951).  **I-2.** J. F. Foster and R. C. Rhees, *ABB* 40: 437(1952).  **I-3.** P. Putzeys and T. Brosteaux, *Trans. Faraday Soc.* 31: 1314(1934).  **I-4.** G. Oster *et al.*, *JACS* 69: 1193(1947).  **I-5.** B. A. Brice *et al.*, *J. Opt. Soc. Am.* 40: 768(1950); M. Halwer *et al.*, *JACS* 73: 2786(1951).  **I-6.** P. M. Doty *et al.*, *Trans. Faraday Soc.* 42(B): 66(1946).  **I-7.** B. Zimm, *J. Chem. Phys.* 16: 1093(1948).

### Section J

**J-1.** J. W. Anderegg *et al.*, *JACS* 77: 2927(1955).  **J-2.** H. N. Ritland, *J. Chem. Phys.* 18: 1237(1950).  **J-3.** S. Krause and C. T. O'Konski, *JACS* 81: 5082, 5507(1959).

### Section K

**K-1.** R. W. G. Wyckoff, *Adv. in Protein Chem.* 6: 1(1951).  **K-2.** R. C. Williams and R. G. W. Wyckoff, *Proc. Soc. Exptl. Biol. Med.* 58: 265(1945).  **K-3.** M. v. Ardenne, *Z. physik. Chem. (Leipzig)* 187: 1(1940).  **K-4.** A. Polson and T. G. W. Wyckoff, *Nature* 160: 153(1947).  **K-5.** L. W. Labaw and R. G. W. Wyckoff, *BBA* 25: 263 (1957).  **K-6.** C. E. Hall, *JBC* 185: 45(1950).  **K-7.** R. C. Williams *et al.*, *JACS* 73: 2062(1951).  **K-8.** C. E. Hall, *J. Biophys. Biochem. Cytol.* 7: 613(1960).  **K-9.** R. C. Valentine, *Nature* 184: 1838(1959).

### Section L

**L-1.** B. W. Low, *in* "Proteins" (H. Neurath and K. Bailey, eds.), Vol. 1A, p. 235. Academic Press, New York, 1953; I. Fankuchen, *Adv. in Protein Chem.* 2: 387(1945); F. H. C. Crick and J. C. Kendrew, *Adv. in Protein Chem.* 12: 134(1957); R. B. Corey, *Adv. in Protein Chem.* 4: 385(1948). **L-2.** M. F. Perutz *et al., Nature* 185: 416(1960). **L-3.** J. C. Kendrew *et al., Nature* 185: 422(1960). **L-4.** J. D. Bernal and D. Crowfoot, *Nature* 133: 794(1934). **L-5.** D. Crowfoot, *Proc. Roy. Soc.* A164: 580(1938). **L-6.** B. W. Low, *in* "Proteins" (H. Neurath and K. Bailey, eds.), Vol. 1A, p. 294–297. Academic Press, New York, 1953. **L-7.** D. C. Hodgkin, *Cold Spring Harbor Symposia Quant. Biol.* 14: 65(1949). **L-8.** J. C. Kendrew, *Acta Cryst.* 1: 336(1948).

### Section M

**M-1.** J. T. Edsall, *in* "Proteins" (H. Neurath and K. Bailey, eds.), Vol. 1B, p. 550. Academic Press, New York, 1953. **M-2.** E. Fredericq and H. Neurath, *JACS* 72: 2684(1950); E. J. Harfenist and L. C. Craig, *JACS* 74: 3087(1952). **M-3.** D. Crowfoot, *Chem. Revs.* 28: 215(1941).

# Chapter VI

## Electrochemistry and
## Hydration of Proteins

### A. *Amino Acids as Dipolar Ions*

It was long believed that amino acids enter water solution in the form of neutral molecules having the general formula $H_2N \cdot R \cdot COOH$. That amino acids in acidic or alkaline solution migrate to the cathode or to the anode respectively, was ascribed to the following set of reactions:

$$H_2N \cdot R \cdot COOH + H^+ \rightleftharpoons H_3N^+ \cdot R \cdot COOH \qquad \text{(VI-1)}$$
$$H_2N \cdot R \cdot COOH + HO^- \rightleftharpoons H_2N \cdot R \cdot COO^- + H_2O \qquad \text{(VI-2)}$$

It was first proposed by Adams (A-1) and by Bjerrum (A-2) that the neutral formula of amino acids, $H_2N \cdot R \cdot COOH$, has to be replaced by the dipolar formula, $H_3N^+ \cdot R \cdot COO^-$, and that only negligibly small amounts of the amino acids can be present as uncharged, neutral molecules. If we accept this view, Eqs. (VI-1) and (VI-2) have to be replaced by Eqs. (VI-3) and (VI-4):

$$H_3N^+ \cdot R \cdot COO^- + H^+ \rightleftharpoons H_3N^+ \cdot R \cdot COOH \qquad \text{(VI-3)}$$
$$H_3N^+ \cdot R \cdot COO^- + HO^- \rightleftharpoons H_2N \cdot R \cdot COO^- + H_2O \qquad \text{(VI-4)}$$

The positive charge of the ion $H_3N^+ \cdot R \cdot COOH$ and the negative charge of the ion $H_2N \cdot R \cdot COO^-$ are proved by their migration in the electrical field. For a long period of time, however, it was not clear whether the intermediate form was the neutral molecule $H_2N \cdot R \cdot COOH$ or the dipolar ion $H_3N^+ \cdot R \cdot COO^-$, since dipolar ions, owing to their positive and negative charges, are attracted with the same force by both the anode and the cathode, and therefore do not migrate as do true anions or cations. As a result, dipolar ions (zwitterions, ampholytes, or hybrid ions) fail to contribute to the conductance of the solution; they behave, in other words, as though they were neutral, uncharged molecules.

The dipolar formula of the amino acids was fully confirmed by the calorimetric determination of the *heat of ionization* (A-3). It is well known that the heat of the reaction: $R \cdot COOH \rightleftharpoons R \cdot COO^- + H^+$ in aliphatic carboxylic acids, approximately $+1000$ cal per mole, is quite different from the heat of ionization of aliphatic amines: $R \cdot NH_3^+ \rightleftharpoons R \cdot NH_2 + H^+$ which is close to $+12,000$ cal per mole. The heat of ionization of amino acids in acid solution varies from $-1300$ to $+2100$ cal and the heat of ionization in alkaline solution from $+10,000$ to $+13,300$ cal per mole. Evidently hydrochloric acid reacts with the carboxyl groups of the amino acids, sodium hydroxide with their ammonium groups.

The dipolar formula of amino acids is furthermore supported by the fact that amino acids raise the *dielectric constant* of water in which they are dissolved (see Section H below). The existence of $COO^-$ groups rather than $COOH$ groups in neutral solutions of amino acids is also indicated by their Raman spectra (A-4), which depend on the vibrations and hence on the structure of the molecular groups involved.

Finally, the dipolar structure of amino acids in their aqueous solutions is confirmed by the phenomenon of *electrostriction*, i.e., by an excessive contraction when the solid amino acid is dissolved in water. Thus a large volume of water increases by 56.2 ml when one mole (75 g) of glycolamide, $CH_2OH \cdot CONH_2$, is dissolved, but by only 43.5 ml when one mole of the isomeric glycine, $H_3N^+ \cdot CH_2 \cdot COO^-$, is dissolved. The pronounced electrostriction is due to the strong attraction and resultant compression of water molecules by the ionized groups of the amino acid (A-5).

That the *solid amino acids* also exist as dipolar ions rather than as neutral molecules is shown by their high densities and their high melting points. Both properties indicate strong electrostatic attraction between the oppositely charged ionic groups of adjacent molecules, making it more difficult for them to separate than would be the case with adjacent, neutral molecules. While the density and the melting point of glycolamide are 1.390 and 117°C, the isomeric glycine has a density of 1.607 and a melting point of 232°C (A-5).

The concept of a dipolar structure of amino acids leads to an agreement of the dissociation constants of their acid and basic groups with the dissociation constants of aliphatic acids and aliphatic amines. If the older Eqs. (VI-1) and (VI-2) are used as a basis for the calculation of dissociation constants, values are obtained which differ widely from the typical dissociation constants of the aliphatic acids and amines; thus, constants of the order of $10^{-9}$ were found for the acidic groups of amino acids, and $10^{-11}$ for their basic groups, while the dissociation constant of acetic acid is $1.8 \times 10^{-5}$, and that of ethylamine $1.5 \times 10^{-11}$. The dissociation of the carboxyl groups of amino acids, according to this old concept, would be lower than that of carbonic acid ($K_a = 4.5 \times 10^{-7}$). These contradictory results gave rise to the proposition of a dipolar formula for amino acids. The undissociated form of the amino acids, $NH_2 \cdot CHR \cdot COOH$, occurs in their aqueous solutions only in very small amounts. The ratio "dipolar form:uncharged form" is approximately $2 \times 10^5$.

Before discussing the dissociation of the acid and the basic groups of amino acids, however, we must modify Eq. (VI-4) in order to make it consistent with Brönsted's theory of acids and bases. Brönsted has defined acids as substances which are capable of giving up protons, and bases as substances which are capable of taking up protons. If we accept this definition, the groups $COO^-$ and $NH_2$ are to be regarded as basic groups because they combine with protons, $H^+$; likewise the groups $COOH$ and $N^+H_3$ are to be regarded as acid groups because they are proton donors. Accordingly, Eq. (VI-4) has to be replaced by Eq. (VI-5) or (VI-6):

$$H_3N^+ \cdot R \cdot COO^- - H^+ \rightleftharpoons H_2N \cdot R \cdot COO^- \qquad \text{(VI-5)}$$
$$H_3N^+ \cdot R \cdot COO^- \rightleftharpoons H_2N \cdot R \cdot COO^- + H^+ \qquad \text{(VI-6)}$$

While the designation of $R \cdot COO^-$ as basic and of $R \cdot NH_3^+$ as acid is free from objection, the older designation of the carboxyl group as acid and of the amino group as basic is frequently used. Care must be taken to avoid confusion in terminology.

Brönsted's theory on the nature of acids and bases was advanced in order to explain the reaction between acids and bases in nonaqueous solvents, where hydroxyl ions, $HO^-$, do not occur. In aqueous solutions water can act both as an acid, i.e., as proton donor ($H_2O \rightarrow H^+ + HO^-$), and as a base, i.e., as proton acceptor ($H_2O + H^+ \rightarrow H_3O^+$); in the latter case hydronium ions, $H_3O^+$, are formed. It is possible, therefore, to formulate the reaction of amino acids with bases in aqueous solution according to Eq. (VI-4). In the absence of water, only Eq. (VI-6) is adequate.

According to Brönsted's theory, and since acids and bases are defined respectively as proton donors and proton acceptors we can represent the dissociation of acid and of basic groups by the general formula:

$$A \rightleftharpoons B + H^+ \qquad \text{(VI-7)}$$

where A is an acid (proton donor) and B a base (proton acceptor). The dissociation constant of the acid, A, is, from Eq. (VI-7):

$$K_A = \frac{a_B a_{H^+}}{a_A} \qquad \text{(VI-8)}$$

where $a_A$, $a_B$, and $a_{H^+}$ are, respectively, the activities of the acid, the base, and the hydrogen ions. According to Eq. (VI-3), the dissociation constants of the carboxyl groups of amino acids are represented by:

$$K_a = \frac{a_{H_3N^+ \cdot R \cdot COO^-} \times a_{H^+}}{a_{H_3N^+ \cdot R \cdot COOH}}$$

and the dissociation constants of the amino groups, according to (VI-6) by:

$$K_b = \frac{a_{H_2N \cdot R \cdot COO^-} \times a_{H^+}}{a_{H_3N^+ \cdot R \cdot COO^-}}$$

In dilute solutions, where they are not very different from unity the activity coefficients can be neglected and the activity terms of the above equations replaced by the molar concentrations. The last equation can then be written as:

$$K_b = \frac{[H_2N{\cdot}R{\cdot}COO^-][H^+]}{[H_3N^+{\cdot}R{\cdot}COO^-]}$$

When the dissociation constants of carboxyl groups, $K_a$, or of ammonium groups, $K_b$, are determined employing these equations, values between $10^{-2}$ and $10^{-3}$ are obtained for $K_a$, while the $K_b$ values found for the amino groups vary from $10^{-9}$ to $10^{-10}$. The meaning of these dissociation constants becomes clearer when we consider those cases in which the concentration of the proton donor, A, is equal to that of the proton acceptor, B. When [A] = [B], the dissociation constant, K, is equal to the hydrogen ion concentration:

$$K = \frac{[B][H^+]}{[A]} = [H^+]$$

The hydrogen ion concentration is, under these conditions, a measure of the dissociation of the acid or base. It indicates the region where the ratio $[H_3N^+{\cdot}R{\cdot}COO^-]/[H_3N^+{\cdot}R{\cdot}COOH]$ or $[H_2N{\cdot}R{\cdot}COO^-]/[H_3N^+{\cdot}R{\cdot}COO^-]$ is unity. Since it is customary to indicate the hydrogen ion concentration by the negative logarithm, $pH = -\log[H^+]$, we can apply the same method to the dissociation constants and write $pK_a = -\log K_a$, and $pK_b = -\log K_b$. Evidently $pK_a$ and $pK_b$ are equal to those pH values at which 50% of the amino acid is present as dipolar ion and 50% as cation or anion, respectively. They are determined by electrometric titration of the amino acids with hydrochloric acid or with sodium hydroxide (A-6). If pH is plotted against the amount of acid or base added to the amino acid, the $pK_a$ and $pK_b$ values are equal to those pH values of the titration curve which correspond to the addition of 0.5 equivalent of acid, or of base, per mole amino acid.

The pK value does not always indicate the nature of the reacting group. Thus, for some time it was not clear whether the pK value 9.1 found in the electrometric titration of tyrosine corresponded to the dissociation of its phenolic hydroxyl group into a proton and an anion, or to the dissociation of the ammonium group $NH_3^+$ into an amino group, $NH_2$, and a proton. Since it is not possible in such cases to use the symbols $K_a$ and $K_b$, it is preferable to number the dissociation constants in sequence of increasing pK values, as $pK_1$, $pK_2$, $pK_3$, etc. This mode of indication of dissociation constants has been used in Table VI-1, which shows that the acid and the basic ionization constants of the *aliphatic monoamino acids* vary only slightly. The aliphatic side chain of the amino acids recorded in the table apparently has no great influence on the dissociation of the amino and the carboxyl groups. The table shows, moreover, that the $pK_1$ values of peptides are higher by 0.8 pH unit, the

$pK_2$ values lower by 1.4–1.7 pH units, than those of the corresponding amino acids. This means that the isoelectric point of the peptides is at a lower pH value and that the peptides are more acidic than the analogous amino acids. Therefore, hydrolysis of peptides is accompanied by a decrease in the acidity of the solution which is sufficiently marked to be manometrically measurable in an atmosphere of carbon dioxide (see Chapter III, Section B).

TABLE VI-1

IONIZATION CONSTANTS OF MONOAMINO ACIDS AND PEPTIDES

| Substance | $pK_1$ | $pK_2$ | Isoelectric point |
|---|---|---|---|
| Glycine | 2.35 | 9.78 | 6.1 |
| Glycylglycine | 3.12 | 8.07 | 5.6 |
| Alanine | 2.34 | 9.87 | 6.1 |
| Alanylalanine | 3.17 | 8.42 | 5.8 |
| Valine | 2.32 | 9.62 | 6.0 |
| Leucine | 2.36 | 9.60 | 6.0 |
| Hexaglycine | 3.05 | 7.60 | 5.32 |
| Serine | 2.21 | 9.15 | 5.68 |
| Proline | 1.99 | 10.60 | 6.30 |
| Tryptophan | 2.38 | 9.39 | 5.89 |

Since amino acids migrate in alkaline solutions to the anode and in acid solutions to the cathode, there is a pH value at which no migration occurs. This pH value has been called the *isoelectric point*. It can be calculated from the ionization constants according to the equation:

$$pH_I = \frac{pK_1 + pK_2}{2}$$

and is recorded in the last column of Table VI-1.

If the acid and the basic groups of an amino acid were ionized to the same extent, its salt-free water solution would have the same pH value as pure water. Since the ionization of the carboxyl groups is higher than that of the amino groups, monoamino acids are slightly acid substances whose isoelectric points are in the neighborhood of pH 6. An aqueous solution of a monoamino acid therefore, contains small amounts of hydrogen ions and of the anions $H_2N \cdot R \cdot COO^-$ in addition to large amounts of the dipolar ion $H_3N^+ \cdot R \cdot COO^-$. Since the amino acids are weak acids and weak bases at the same time, their mixtures with strong acids and bases are used as buffer solutions. Figure VI-1 shows pH values of mixtures of glycine with hydrochloric acid or sodium hydroxide.

If the amino acids contain ionized groups other than the $\alpha$-amino and the $\alpha$-carboxyl groups, the titration curve (Fig. VI-1) will be complicated by further inflection points. The ionization constants of some amino acids containing functional groups in their side chains are recorded in Table VI-2. This table shows that the aminodicarboxylic acids possess a stronger

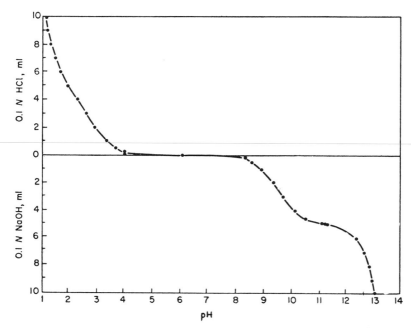

Fig. VI-1. Buffering action of glycine (A-7). If the volume of hydrochloric acid or of sodium hydroxide indicated in the diagram is $v$ ml, the volume of 0.1 $N$ glycine solution used is $10 - v$ ml.

acid group than do the monoamino acids. This is in agreement with the well-known strength of the organic dicarboxylic acids. The table shows, moreover, that the pK value for the phenolic hydroxyl group of tyrosine is approximately 10.1; it is evidently a very weak acid, being practically uncharged in neutral solutions and ionized only in alkaline solutions. The basic groups of arginine and of lysine (pK 12.48 and 10.53) are strongly basic, their ionization being more pronounced than that of the amino groups of monoamino acids; the imidazole group of histidine is but weakly basic. Table VI-2 shows that the ionization constants of amino acids are only slightly affected by remote molecular groups. One can expect, therefore, that ionization constants similar to amino acids will be found for proteins. The isoelectric pH values recorded in Table VI-2 are the arithmetic means of the two principal pK values. They are equal to (pK$_1$ + pK$_2$)/2 for the first four amino acids in the table, and equal to (pK$_2$ + pK$_3$)/2 for the last three amino acids.

TABLE VI-2

| Amino acid | pK$_1$ | pK$_2$ | pK$_3$ | Isoelectric point |
|---|---|---|---|---|
| Aspartic acid | 2.09 (COOH) | 3.87 (COOH) | 9.82 (NH$_3{}^+$) | 3.0 |
| Glutamic acid | 2.19 (COOH) | 4.28 (COOH) | 9.66 (NH$_3{}^+$) | 3.2 |
| Tyrosine | 2.20 (COOH) | 9.11 (NH$_3{}^+$) | 10.1 (OH) | 5.7 |
| Cysteine | 1.96 (COOH) | 8.18 (NH$_3{}^+$) | 10.28 (SH) | 5.07 |
| Arginine | 2.02 (COOH) | 9.04 (NH$_3{}^+$) | 12.48 (guanido) | 10.8 |
| Lysine | 2.18 (COOH) | 8.95 ($\alpha$-NH$_3{}^+$) | 10.53 ($\epsilon$-NH$_3{}^+$) | 9.7 |
| Histidine | 1.77 (COOH) | 6.10 (imidazole) | 9.18 (NH$_3{}^+$) | 7.6 |

## B. Ionization of Proteins

### Combination of Proteins with Hydrogen Ions

In electrophoresis, proteins display essentially the same behavior as amino acids. They are transferred to the cathode in acid solutions, to the anode in alkaline solutions, and exhibit an isoelectric point, at which no migration occurs (B-1). It is generally assumed that proteins are multivalent *zwitterions* at their isoelectric point, and that they differ from the simpler amino acids mainly by the multiplicity of their anionic and cationic groups.

The view that the isoelectric protein has numerous ionic groups was accepted somewhat reluctantly because proteins have a minimum solubility at their isoelectric point (B-2) whereas it is a well-established fact that the solubility of organic acids or bases increases with their ionization. The precipitation of proteins at their isoelectric point is attributed to the mutual attraction by electrostatic forces of positive and negative groups of adjacent molecules. This interpretation gains support from a consideration of the "salting-in" phenomenon, i.e., the prevention of precipitation by neutral salts. The ions of a neutral salt neutralize ionic groups of the opposite charge in the surface of the protein and thereby prevent the mutual attraction between groups of opposite charge and aggregation of protein particles.

The amphoteric nature of the ionized isoelectric protein particles is also proved by the magnitude of $\Delta$H, the *heat of neutralization* by strong acids and bases. The latter is either determined calorimetrically (B-3), or is calculated from the temperature coefficient of the ionization constant

(B-4). Both methods furnish values of approximately $-10{,}000$ cal per equivalent of hydrogen ion when acids are added to the slightly alkaline solution of the protein (pH about 8–9). This value is of the same order of magnitude as values obtained by the addition of strong acids (hydrogen ions) to organic bases (see above, Section A). It is therefore evident that the reaction taking place at pH 8–9 corresponds to the equilibrium $R \cdot NH_2 + H^+ \rightleftharpoons R \cdot NH_3^+$.

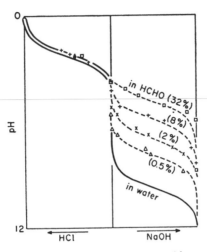

Fig. VI-2. Titration curve of glycine in the presence of increasing concentrations of formaldehyde (B-5).

The most convincing proof for the *zwitterion structure of the isoelectric protein* is furnished by a comparison of the electrometric titration in the presence and absence of formaldehyde. Formaldehyde combines with the amino groups of proteins so as to eliminate their basicity: $R \cdot NH_2 + HCHO \rightarrow R \cdot NH \cdot CH_2OH$ (see Chapter III, B). If amino acids or proteins are titrated with sodium hydroxide in the presence of formaldehyde, the typical inflection of the titration curve at pH 9 which corresponds to the conversion of $N^+H_3$ into $NH_2$, is shifted to lower pH (see Fig. VI-2), thereby showing conclusively that the buffering action of proteins in this pH region is due to their amino groups and not to the ionization of carboxyl groups (B-5). The course of the electrometric titration curve is also changed by the addition of ethanol, which considerably reduces the dissociation constant of the carboxyl group, while the dissociation constant of the ammonium groups is only slightly reduced (B-6) as shown by Fig. VI-3.

Although the curves obtained by the *electrometric titration of proteins* are similar to those of amino acids, several important differences are to be noted. The ionization of monoamino acids is due to the electrolytic dissociation of the grouping:

$$\begin{matrix} R & & COO^- \\ & \diagdown \; \diagup & \\ & C & \\ & \diagup \; \diagdown & \\ H & & N^+H_3 \end{matrix}$$

i.e., to the ionization of an $\alpha$-amino group and of the adjacent carboxyl group. Owing to the proximity of these two groups there will be electro-

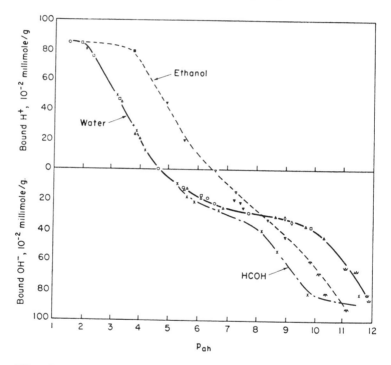

FIG. VI-3. Electrometric titration of ovalbumin in water, 80% ethanol, and 1% formaldehyde (B-7).

static interaction between them and also between nearby hydrogen and hydroxyl ions. The positive hydrogen ion, which is attracted by the negative carboxyl group, will be repelled from the neighborhood of this group by the positively charged ammonium group. The electrostatic field of the negatively charged carboxyl group will be weakened by the closely adjacent positive ammonium group. Evidently a very complicated interaction of electrostatic forces will take place, the result of which cannot be predicted. From Table VI-1, however, it is apparent that the $\alpha$-carboxyl groups of monoamino acids (pK about 2.2) are more strongly acidic than those of the corresponding fatty acids (pK about 4.8) and that the amino groups (pK about 9.8) are less basic than those of organic amines (pK about 10.6).

The *acidic groups of proteins* are mainly the free carboxyl groups of aspartic and glutamic acids, whose ionizations, according to Table VI-2, respectively correspond to pK 3.87 and 4.28. The basic groups of proteins are the guanidine groups of arginine (pK 12.48) and the ε-amino groups of lysine (pK 10.53). The hydroxyl groups of tyrosine and the sulfhydryl groups of cysteine lose their protons in the same pH range (pK about 10), whereas the imidazole groups of histidine are titrated at about pH 6 (Table VI-2). The great importance of the OH, SH, and imidazole groups for the biological activity of enzymes will be discussed in Chapter XII.

The electrometric titration curve of proteins distinctly shows the inflections due to the buffering action of the carboxyl and amino groups at pH 3–4 and 10–12, respectively. The pK values of the ionizing groups depend to a great extent on their environment inside the globular protein molecules, particularly on adjacent ionic groups. Thus, the pK values of the carboxyl groups of pepsin seem to be shifted to a higher pH by hydrogen bonding (B-8). Anomalous pK values were also found for the carboxyl and amino groups of serum albumin (B-9,10), and for the phenolic hydroxyl group of tyrosine residues (B-11). Evidently, the "microscopic ionization constants" (B-11) of the amino acid residues inside the globular protein molecules are frequently quite different from those of the same residues in free amino acids or small peptides. This allows us to draw certain conclusions concerning the accessibility of the titrated groups and the nature of adjacent amino acid residues (see Chapter VII, F).

In bovine serum albumin (BSA) the pK values of the carboxyl groups change on acidification near pH 4 abruptly from 4.4 to 3.7. This reversible change is attributed to the combination of the native N-form of BSA with 3 protons to yield the isomeric F-form in which there is less stabilizing interaction between the peptide chains (B-9,31). This causes expansion of the BSA molecules and an increase in the viscosity of the solution (see Chapter V).

The electrometric titration of proteins is carried out by adding strong acids such as hydrochloric acid or strong bases such as sodium hydroxide to the isoelectric solution of the protein. If we add hydrochloric acid to an isoelectric protein solution, one portion of the hydrogen ions (protons) will be bound by the $COO^-$ groups of the proteins and these will be converted into carboxyl groups (see Eq. VI-3). Moreover, some of the few $NH_2$ groups present in the isoelectric protein will be converted into positively charged ammonium groups, $NH_3^+$. The maximum acid binding is represented in the titration curve by the distance between the horizontal line which corresponds to the isoelectric point and the horizontal part of the titration curve at pH 1–2. Similarly the maximum base used is determined by electrometric titration of the isoelectric protein with sodium hydroxide. It is difficult, however, to determine precisely the amount of base used, because the end point is less sharp here than in the titration of acid-binding groups. In general the results of the electrometric titration agree satisfactorily with the results of amino acid analyses.

The *isoelectric protein* contains positively and negatively charged groups in equal amounts, its resulting *free charge* being zero, so that no unilateral migration occurs if an electrical field is applied. However, the isoelectric protein is not a truly homogeneous substance, but a mixture of neutral molecules with others containing a slight excess of positive and negative charges. Thus, it has been estimated that at the isoelectric point of hemoglobin, pH 6.9, only 22.4% of the hemoglobin molecules are strictly isoelectric (free charge, 0), while the percentages of molecules carrying 1, 2, 3, or more positive charges are, respectively 17, 9.4, 3.9, and 1.5%. The percentages of groups carrying 1, 2, 3, or more negative charges are 21.2, 14.2, 3.0, and 2.5% respectively (B-12). According to this view the protons would be distributed statistically among a population of protein molecules. Similarly, an investigation of the light scattering of bovine serum albumin has led to the view that the molecule contains on an average 3.5–4.0 protonic units which fluctuate between various proton-binding groups (B-13,14). This fluctuation of charges seems to stabilize to some extent the conformation of the protein molecules. Most of the fluctuating protons are contributed by the COOH and $N^+H_3$ groups, very few by the phenolic OH groups (B-10). Since the electrostatic forces are inversely proportional to the dielectric constant of the medium, and since the dielectric constant of water is much higher than that of the typical organic substances, one would expect that the concentration of protons in the surface of a globular protein molecule would be different from their concentration inside the molecule. Theoretical considerations have led to the view that the protons have a tendency to concentrate near the surface of the protein molecule (B-15).

The dependence of the isoelectric point on the ionization constants of the ionic groups is also evident from determinations of the isoelectric point in the presence of various concentrations of ethanol. Since ethanol reduces mainly the dissociation constants of the carboxyl groups (see Fig. VI-3), it shifts the isoelectric point to higher pH values. In gelatin solutions containing 80% ethanol the isoelectric pH value is 6.0, while the isoelectric point in aqueous solutions is 4.9 (B-7).

It may be clear from the preceding discussion that it is very difficult to draw definite conclusions from the electrometric titration of proteins. The titrations are further complicated by the *lability* of many proteins, by their combination with ions other than hydrogen ions, and by the mutual interaction of protein molecules with each other. Many of the proteins are *denatured* at pH values of less than 2 or more than 11, so that accurate determinations of the ionization of the native protein in these pH ranges are impossible. The rate of denaturation can be minimized by working at lower temperatures and by the use of very rapid measuring techniques. Thus, upon rapid electrometric titration (within 3 sec) carbon monoxide hemoglobin binds only one-third of the hydrogen ions which are bound in the customary electrometric titration; approximately 35 proton-binding groups are unavailable to hydrogen ions in the native protein, but become accessible during denaturation (B-16).

*Combination of Proteins with Ions Other Than Hydrogen Ions*

Considerable difficulties are encountered in the ready *combination of proteins with ions other than hydrogen ions.* Proteins combine particularly with multivalent ions such as calcium, magnesium, the heavy metal ions, and ions of the multivalent acids. Among the latter phosphate must be mentioned particularly since phosphate buffers are widely used in bio-chemical laboratories. In order to avoid complications caused by complex formation, it is preferable to use univalent acids or bases for the electro-metric titration. Although sodium, potassium, and chloride ions are much less firmly bound to the proteins, their binding is not negligible. Thus, serum albumin has 11 groups which bind chloride ions with an association constant of $10^3$ and about 30 other groups which have a much lower affinity to chloride ions (B-17). Binding of sodium ions has been observed in casein (B-18), and binding of sodium and potassium ions occurs in pepsin, zein, lactoglobulin, $\alpha$-, and $\beta$-globulins. Serum albumin, ovalbumin, and $\gamma$-globulins do not appear to bind these cations (B-19). The reason for the different behavior of these proteins is not yet known. It may be caused by particular structures, such as an agglomeration of numerous negative charges within a small portion of the molecule. Obviously the binding of alkali or chloride ions interferes seriously with electrometric titrations when alkali hydroxides are used as proton acceptors or HCl as a proton donor. In accurate determinations of the binding of protons to proteins it will be necessary to apply corrections for bound alkali or chloride ions. The extent of alkali or chloride binding can be determined by equilibrium dialysis or electrophoresis (B-19,20).

Since most of the physiological buffers are multivalent, non-physio-logical *univalent buffer systems* are preferred in protein chemistry. Very few of them are usable in the most important pH range, pH 6–8. One of these is the veronal buffer; it contains diethylbarbituric acid (veronal, barbital) and its sodium salt (B-21). Another useful buffer is the "Tris" buffer which consists of tris(hydroxymethyl)aminomethane, $H_2N \cdot C(CH_2OH)_3$ and its hydrochloride (B-22). Electrophoretic measurements indicate, however, that serum albumin binds the veronal anion and migrates as a complex (B-23).

Since the binding of extraneous ions considerably alters the value of the *isoelectric point* of a protein, this point is *not a constant*. The term *isoionic point* is used to designate the pH of the pure protein in salt-free water (B-24). The direct determination of this constant is difficult and frequently impossible because many proteins are insoluble in the absence of salts and because the conductance of salt-free solutions is extremely low. The isoionic point is usually determined indirectly, that is, by measuring the isoelectric points at different concentrations of neutral salts and extrapolating to zero concentration. The value of the isoionic point may differ from the isoelectric point by more than a pH unit.

Thus the isoionic point of carbon monoxide hemoglobin is at 7.6 whereas the isoelectric point in phosphate buffer solution varies from 6.70 to 7.16 (B-25). The isoionic point of gelatin is 9.3 whereas the isoelectric point at 0.1 ionic strength is 5.1 (B-26). Similarly, the isoelectric point of the enzyme enolase is 5.1 although its molecule has more basic than acidic groups (B-27). Evidently, enolase and gelatin combine preferentially with anions in their aqueous solutions, so that the isoelectric molecule is not the pure protein but its salt with the bound anions.

Many proteins are less soluble at the isoelectric or isoionic point than at higher or lower pH values. This indicates *protein-protein interaction* and may involve changes in the charge of the protein molecules; very little is known about this. Finally, it should be mentioned that the hydrogen ion concentration measured by means of the usual electrodes is not the true concentration of the hydrogen ions but merely a measure of their electromotive activity, $a_{H^+}$, which is equal to $[H^+] \times \gamma$, where $\gamma$ is the activity coefficient. The activity of the hydrogen ions depends on the nature and concentration of other ions present. Although the true concentration of the hydrogen ions is not determinable (B-28), the values measured are usually considered to be close to the actual concentration (B-29,30).

## C. *Electrophoresis of Proteins (C-1 to 5)*

Electrophoresis is used for determining the isoelectric point of proteins and also for the fractionation of protein mixtures. The devices used originally for electrophoresis were simple U-tubes containing the solution of the protein in a buffer of the desired pH (C-6,7). Protein-free buffer solution of the same concentration and pH was carefully layered over the protein solution. By means of nonpolarizable electrodes, a direct current of about 1–3 v/cm was passed through the solution and the motion of the protein boundary was observed. Since frequently only very small amounts of protein are available, micromodifications of the simple U-tube have been developed.

The motion of the boundary at a pH near the isoelectric point is very slow and difficult to measure. It is customary therefore to make measurements in buffer solutions of different pH values at both sides of the isoelectric point. If the motility of the boundary in both directions is plotted against the pH of the buffered protein solution, a curve is obtained. The pH at which the curve intersects the line of zero motility is the isoelectric point of the protein.

In order to avoid differences in the buffer concentrations, the protein-buffer mixture is dialyzed against the buffer solution. Difficulties in obtaining a sharp boundary are overcome by a U-tube consisting of three separate cells which can be slid over one another along two horizontal planes. It is customary to place the protein-buffer mixture in the bottom compartment and the cathodic middle compartment while the anodic compartment and the top cells contain the buffer solution (C-8). This has the advantage of permitting one to observe a descending and an ascending boundary. The methods used for the measurement of the displacement of the boundary are similar to those used for the measurements of sedimentation during ultracentrifugation (Chapter V, Section E).

Although it is possible to see the boundary when the protein is colored (e.g., hemoglobin) or to determine it by photography in ultraviolet light, this method is rarely used. Most of the measurements are based on the high refractivity of proteins and on the "schlieren phenomenon," that is, the appearance of a shadow on a ground-glass screen or a photographic plate indicating the position of the boundary (C-9). The shadow is brought about by the *difference* in refractivity between the buffer solution and the protein-buffer mixture. Since the latter forms the lower layer, an incident beam of light which arrives with a slight angle will be bent downward by both layers, but much more by the protein-buffer mixture than by the buffer solution. The refractivity and the deflection of the incident light increases with increasing concentrations of protein.

The refractive indices of the solutions of various proteins are very similar. The increase in refractive index amounts to 0.00185 ($\pm 0.00005$) per gram of protein in 100 ml of the solution (C-10). Therefore the height of the peak $H_m$ in Fig. VI-4 is very nearly proportional to the concentration gradient and the area under the peak to the total gradient through the boundary. It must be kept in mind, however, that the boundary between the protein-buffer mixture and the protein-free buffer solution is not a sharp geometrical plane because diffusion of protein into the adjacent protein-free solution takes place almost immediately. What we designate as the boundary is a rather narrow region of refractive index, varying between that of the homogeneous protein mixture and that of the homogeneous buffer solution. If we designate the height of the arm of the U-tube by $h$ (in millimeters) and the refractive index of the solution by

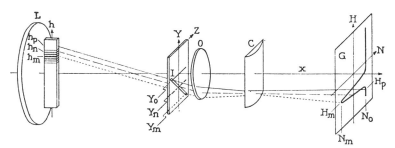

FIG. VI-4. The astigmatic schlieren camera. From L. G. Longsworth (C-28), p. 145. Light passes through a horizontal slit at the left of the schlieren lens ($L$), then through the solution at $h$, an oblique slit ($Y$), a spherical lens ($O$), and a cylindrical lens ($C$), to form a picture on the screen ($G$). The boundary between the protein solution and the protein-free buffer solution is shown by $h_m$–$h_p$. At $H_p$ the line $N_o$ corresponds to the protein-free buffer solution $h_p$. The upper portion of the vertical line $N_o$ is the image of the homogeneous protein solution in the lower part of the cuvette. The pencil of light which comes from the protein-free buffer solution passes the oblique slit at its highest point at $Y_o$ and is deflected by $C$ to the right, thereby causing a vertical line ($N_o$) on the right part of the screen. Light passing through the boundary at $h_m$ is deflected downward to $Y_n$ and passes through the lowest portion to the oblique slit where it is deflected by $C$ to the left to $N_m$ and appears on the screen at $H_m$. The peak at $H_m$ is an image of the concentration gradient at $h_m$.

$n$, the gradient of refraction, i.e., the increase or decrease in refraction per unit height is represented by the differential quotient $dn/dh$. The gradient in the two homogeneous solutions is obviously zero and has its maximum value in the boundary. It increases when the concentration of protein in the solution increases. The intensity of the dark bands depends on the magnitude of the gradient of refraction, so that the maximum of the shadow will correspond to the maximum of the gradient of refraction and of concentration.

While a single dark band is produced by solutions containing only one protein, two or more bands are obtained when the solution contains two or more proteins of different mobility. In such a case only at the time $t = 0$ is a single boundary present. Under the influence of the direct current the different proteins will migrate at different rates so that a multiplicity of regions of varying refraction is produced. The single dark band will be replaced by a number of horizontal dark bands.

The protein mixture which has been most frequently examined is that of normal and pathological blood sera. The diagrams of its ascending and descending boundaries are commonly presented horizontally, i.e., turning them through 90°. The ascending and the descending diagrams are mirror images of each other but show small differences. Usually only the descending boundary is shown.

At pH 7.35 (the normal pH of blood serum) and at higher pH values, the serum proteins are present as anions; hence, all of them migrate in the same direction, to the anode. It is customary to examine blood sera in veronal buffer solution at pH 8.6, at which the anodic migration is more pronounced. Serum albumin, whose isoelectric point is near pH 4.6, moves with the greatest velocity, the α-, β-, and γ-globulins with decreasing velocities. The maxima of the curves shown in Fig. VI-5 indicate the position of the respective boundaries at the time the picture was taken. The height of the curves indicates the magnitude of $dn/dh$, the gradient of refraction, while the concentration of the protein is indicated by the area included under each of the curves. If the narrow oblique slit (Fig. VI-4) is replaced by a metal wire, a black curve on a white background is obtained.

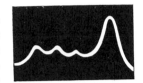

Fig. VI-5. Electrophoresis diagram of blood serum obtained by Svensson's method. The direction of migration is from left to right. The high peak at the right is caused by serum albumin, the three lower peaks by the three globulins.

Electrophoresis has not only revealed the presence of three different types of globulins in the human blood serum (C-11), but has also demonstrated the heterogeneity of β-lactoglobulin (C-12), crystalline serum albumin (C-13), and crystalline ovalbumin (C-14). Many proteins which appear to be homogeneous at one pH value are found to be heterogeneous at other pH values (C-15).

While electrophoresis is an excellent tool for determining isoelectric points, mobilities, and the uniformity of proteins, difficulties arise when an attempt is made to interpret the results quantitatively and to calculate the number of charged groups per protein molecule. These difficulties are partly due to the fact that the protein boundary is subject to *diffusion*, even when no current passes through the solution. As a result of diffusion and of *thermal convection*, the boundary widens with time. This can be reduced by working at low temperatures; actually, electrophoresis is carried out at 4°C, where water has its maximum density. Another more serious difficulty is that migration of the protein molecules might lead to *chemical reactions in the boundary*, for instance to the formation of salt-like compounds between proteins of different acidities or between proteins and the buffer ions (C-16). If such reactions take place, the observed electrophoretic behavior will not be that of the protein, to which we ascribe it, but that of a protein complex.

If all these complications are taken into consideration, the valence of a spherical protein ion can be calculated from its migration velocity. Since protein ions are not spherical, their valence cannot be determined reliably. Approximate values have been obtained by using simplifying assumptions. Thus it has been claimed that the ovalbumin anion at pH 7 carries an excess of 10–12 negative charges and that serum albumin under the same conditions has an excess of 16–17 anionic groups (C-1). The net charge (valence) at the isoelectric point is obviously zero.

Although free-boundary electrophoresis, described in the preceding paragraphs, is the best method for the determination of isoelectric and isoionic points and is also very valuable in the preparation of electrophoresis diagrams of the type shown in Fig. VI-5, its application for the separation and isolation of the components of a protein mixture is very limited. The reason for this limitation is that only those two proteins which have the highest and the lowest migration velocity are separated from the other components during their migration, since these two proteins form the top fractions of the ascending and descending protein columns. All other components are contaminated by at least one of the other components.

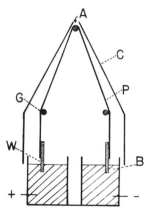

FIG. VI-6. Paper electrophoresis. The protein solution is fed to the filter paper strips (P) through the slot (A) in the lucite cover (C). The paper strips are supported by three glass rods (G) or by a plastic rack. The buffer solution (B) rises to the paper strips through two wicks (W) which consist of heavy filter paper. Evaporation of water is prevented by the cover.

If the purpose of electrophoresis is the separation and isolation of pure proteins from their mixtures, better results may be obtained by *zone electrophoresis*. In this modification of the classical electrophoresis small amounts of protein solution are applied to a solid carrier which is impregnated with the buffer solution of the desired pH. The carrier can be filter paper, starch, cellulose, acrylamide (C-29), or other material. Particularly excellent separations have been obtained by starch gel electrophoresis (C-4,17). Colorless proteins are made visible on the paper or on the starch gel by fixation and subsequent staining with bromphenol blue, amido black (C-18), or other dyes. Amido black, which gives very dark staining, is unstable in alkaline media, whereas bromphenol blue is more stable (C-19). Attempts have been made to determine the amount of each of the protein fractions by measuring the intensity of the color of the bound dye. Careful determinations have shown, however, that the amount of dye bound differs for different proteins, and that albumin binds more dye than the globulins of the blood serum (C-20).

The high resolving power of zone electrophoresis is due to the fact that a very small amount of protein is applied to a very narrow zone and that the length of the path of migration is about 100 times larger than that of the zone of application. In free-boundary electrophoresis the protein solution migrates only 1–2 times the length of its own column. Therefore the resolution cannot be high. One of the possible arrangements of paper electrophoresis is shown in Fig. VI-6.

Migration in paper electrophoresis is slow and usually requires 16–20 hours. During this time considerable diffusion takes place. In high voltage electrophoresis migration takes place much faster. However, evolution of heat caused by the high current intensity prevents application of this method to the heat-labile proteins, although it gives excellent results with peptides. The increase in temperature can be avoided by cooling devices and by using solvent mixtures of low conductivity, for instance, pyridine-acetic acid mixtures. These are good solvents for peptides but not for proteins.

The qualitative results obtained with zone electrophoresis agree well with those in free-boundary electrophoresis. Attempts have been made to calculate the free mobility from the mobility in paper (C-21). Whereas in free-boundary electrophoresis all fractions of the normal human blood serum migrate anodically at pH 8.6 in barbital buffer, the γ-globulins do not migrate on paper under these conditions but remain at the origin or are carried to some extent towards the cathode (C-22). This may be caused by the osmotic flow of the solvent towards the cathode. One of the principal advantages of zone electrophoresis is the simplicity of the equipment and of the procedure. Usually several determinations can be made at the same time. The conditions are much less critical than those required for free-boundary electrophoresis. It is not surprising, therefore, that the method is widely used.

Zone electrophoresis can be combined with other methods, such as chromatography or capillary flow due to gravity. In the combination of zone electrophoresis with chromatography, the protein sample is applied to the corner of a rectangular filter paper and is exposed in one direction to chromatography and then in a perpendicular direction to high voltage electrophoresis (C-18,23). The method is particularly valuable for the separation of peptides and large protein fragments (see Chapter II, B, and Chapter IV, D). In an apparatus developed for continuous flow (C-24) the mixture of proteins or peptides flows slowly, due to gravity, down along the central portion of a rectangular sheet of paper while an electrical potential is applied in horizontal direction (Fig. VI-7). The electrical field causes deviations from the vertical downward transport to the anode or cathode and leads thus to the separation of the components of the mixture.

The highest degree of resolution has been attained by immunoelectrophoresis, i.e., by the combination of gel electrophoresis with the serological precipitin test (C-25). In this method (Fig. VI-8) the protein mixture is first exposed to agar gel electrophoresis. Subsequently, an antiserum against the protein mixture is allowed to diffuse from one side into the agar strip which now contains a series of proteins differing from each other by their electrophoretical mobility. Proteins which are homogeneous give only one precipitation band with the immune serum when it diffuses into the agar gel. Heterogeneity of a protein fraction is revealed by the appearance of more than one precipitation band.   By means of

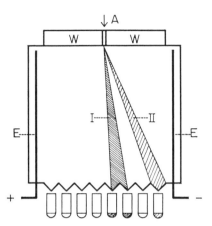

FIG. VI-7. Combination of continuous downward flow with electrophoresis of a protein mixture. *E*: electrodes; *A*: point of entry of protein solution; *I* and *II*: components separated as a result of their different migration velocities toward the anode. They are collected in test tubes; *W*: wicks for the buffer solution.

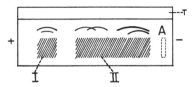

FIG. VI-8. Immunoelectrophoresis. The protein mixture is applied to the agar or starch gel or to gelatin at *A*. After migration toward the anode, two zones (*I* and *II*) are formed. If after completion of the electrophoresis the trough (*T*) is filled with an immune serum against the protein mixture, diffusion of the antibodies into the gel causes the formation of several precipitin zones. Each of these corresponds to a definite antigen-antibody (C-28; see also Chapter XV, Section B).

this method more than 20 different proteins have been discovered in the normal blood serum. Still higher resolution can be obtained when the principal fractions of the serum (albumin, $\alpha$-, $\beta$-, and $\gamma$-globulin) are first isolated, then exposed to immunoelectrophoresis.

Although we know that the electrophoretic motion of protein particles is due to their electrical charges, we may raise the question whether only *ionic groups on the surface* of globular protein particles are responsible for the migration, or whether ionic groups which are hidden in the interior of the protein particle also contribute. Since quartz particles coated with protein behave electrophoretically in the same way as the protein from which the coating layer was formed (C-26), it is evident that the mobility of the protein particles is determined by the potential of their surfaces. Because this potential manifests itself only during the motion of the particle or of the surrounding solution in an electrical field, it is called *electrokinetic potential* or ζ-potential.

The ζ-potential, according to Gouy's theory, is attributable to a diffuse ionic atmosphere surrounding the protein molecule, each ionic group of the protein attracting one or more ions of the opposite sign. The thickness of the double layer depends on the ionic strength of the solution according to the following equation (C-27):

$$d = 3.05 \times 10^{-8} \times \frac{1}{\sqrt{\mu}} \text{ cm}$$

where $d$ is the thickness of the double layer and $\mu$ the ionic strength of the solution. According to this equation, the thickness of the double layer at an ionic strength of 0.01 would be approximately 30 Å.

## D. Hydration and the Specific Volume of Solutes (D-1)

The phenomenon of hydration is due to the polar properties of water molecules. The electronic formula of water shows that the center of charge of the negatively charged electrons is nearer to the oxygen atom than to the positively charged hydrogen nuclei (Fig. VI-9). On the other hand, the center of the positive charges is nearer the two hydrogen atoms. We call such a molecule, in which the centers of the positive and the negative charges do not coincide, a polar molecule or a *dipole*. Figure VI-10 shows that dipoles are *always* attracted by ions. In a first phase, the ion attracts the opposite pole and repels the pole of the same sign; in a second phase, attraction is stronger than repulsion because the attracted pole is nearer the ion than the repelled pole. For similar reasons attraction takes place between two dipoles, as shown by the following diagram (Fig. VI-11).

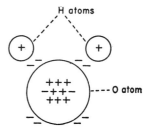

Fig. VI-9. Electron formula of water.

The dipoles represented in Fig. VI-11 are *permanent dipoles*. Actually, however, all molecules, including even typical nonpolar ones, become dipoles when placed in an external electrical field. This is due to the shift of their electrons toward the external positive pole and the analogous shift of their positive nuclei toward the negative pole. We call the dipoles which arise in an external electrical field *induced dipoles.* The total dipole

Phase I          Phase II

FIG. VI-10. Mutual attraction of a dipole and a positively charged ionic group.

moment of proteins, as well as that of other molecules, is due in part to dipole induction.

The forces operating between dipoles are less effective over large distances than those operating between ion and dipole. They are, however, responsible for H bonding and the mutual association of water molecules, and for the high boiling temperature of water as compared with that of analogous molecules such as $H_2S$ or $H_2Se$.

*Hydration* consists of the binding of water dipoles to ions or ionic groups, to dipoles, or to polar groups. Hydration takes place in solid substances as well as in solution. In some cases, hydrates of a definite

PHASE I      PHASE II      PHASE III

FIG. VI-11. Dipole association.

stoichiometric composition are formed. A well-known example is the hydration of copper sulfate. The dry substance, $CuSO_4$, attracts water from the air to form the hydrate, $CuSO_4 \cdot 5H_2O$.

Since the components of a compound are linked to each other in such a way that they have lost some of their free translational mobility, the volume of a hydrated molecule is always smaller than the sum of the volumes of its components. In other words, *hydration is accompanied by a decrease of the total volume.* This decrease can be determined by measuring directly the volume before and after the reaction or, more precisely, by measuring the density, $d$, i.e., the weight in grams per milliliter. The reciprocal value of the density, which is the volume occupied by 1 g of the substance, is called the specific volume, $V_{sp} = 1/d$.

*Proteins are the principal water-binding substances* which occur in animal organisms and, since water is the universal medium of biological reactions, the great importance of the phenomenon of protein hydration is obvious. The literature dealing with protein hydration is very confusing, because the hydration of dry proteins and hydration of proteins in solution have been investigated by different authors who used varying methods. Drying, moreover, even if the mild freeze-drying method is used, causes irreversible changes in the native structure of some of the labile proteins.

The amount of water bound to proteins depends primarily on the ratio of water to protein in the investigated system. The two extreme cases are (a) the dry protein (water content = 0) and (b) a highly diluted aqueous solution of the same protein. The dry protein undergoes hydration if it is exposed to water vapor of increasing vapor pressure. The extent of hydration can be determined by measuring the increase in weight. It is much more difficult to determine the extent of hydration in aqueous solutions of the proteins. Although hydration is accompanied by a volume contraction of the solute (protein) and the solvent (water), this change in volume is very small and difficult to measure directly. It is customary to measure the density of protein solutions by weighing them in a pycnometer of an accurately known volume, $V$. If the weight of the solution is $P$ g and the volume of the pycnometer is $V$ ml, the density is $d = P/V$.

Although it is easy to measure the density of protein solutions, great difficulties arise in the determination of the *density of dry proteins* or of wet protein preparations. Many proteins are denatured by drying, so that we are not able to use the dry preparations for experiments investigating hydration. This complication can be overcome by first determining the density of the wet protein crystal or of the protein solution and subsequently drying the protein. The principal problem, however, that of determining the density of dry or wet protein preparations, still remains. One of the methods used to solve it consists of placing the protein in a pycnometer, weighing it, and then filling the pycnometer with a liquid of known density. Obviously water cannot be used because it would increase the extent of hydration. Ethanol, acetone, or diethyl ether, on the other hand, would attract water and, consequently, reduce the extent of hydration of the protein. Benzene, bromobenzene, and other hydrophobic liquids are used to fill the pycnometer. Even so, the results obtained are not always very satisfactory, since proteins are denatured at the water-solvent interface and because small amounts of water are dissolved by these solvents. Difficulties also arise from the formation of small air bubbles in the protein powder when it is suspended in the organic liquid. Therefore, it is necessary to remove air bubbles by evacuation.

Another method applied to determine the density consists of measuring the rate of sedimentation of the protein particles in aqueous salt solutions of different densities (D-2). The interpretation of results obtained in this way is complicated by the fact that the protein combines not only with water but also with salts (see Chapter VI, Section B). The densities found are therefore the densities of the hydrated protein-salt compound and not those of the hydrated protein.

It is evident from these few remarks that we are confronted with great experimental obstacles when we attempt to determine the density of dry or hydrated proteins.   Moreover, the validity of the interpretation of the results is very uncertain. Since we are not able to indicate the amount of hydrated protein and of free water in the protein-water system, the thermodynamic notion of *partial specific volume* has been introduced and is frequently determined (D-3).

Let us consider a system consisting of $n_P$ moles of protein and $n_W$ moles of water. If the total volume of the system is $V$, we can consider this volume as the sum of the volumes of the two components forming the system. Let us further assume that the volume contribution per mole of protein is $\bar{v}_P$ and that per mole of water $\bar{v}_W$; then the total volume will be $V = n_P\bar{v}_P + n_W\bar{v}_W$. The values $\bar{v}_P$ and $\bar{v}_W$ are called the *partial molar volumes*. Since the molecular weight of proteins is not always known, the partial molar volumes $\bar{v}_P$ and $\bar{v}_W$ are frequently replaced by the *partial specific volumes*, $\bar{v}_p$ and $\bar{v}_w$. Their relation to $V_{sp}$, the specific volume, is shown by the equation:

$$V_{sp} = g_p\bar{v}_p + g_w\bar{v}_w \tag{VI-9}$$

where $g_p$ and $g_w$ are the amounts of protein and water, respectively, in 1 g of the mixture. Obviously, $g_p + g_w = 1$ g. Since $V_{sp}$, $g_p$, and $g_w$ can be determined, Eq. (VI-9) contains only two unknowns, $\bar{v}_p$ and $\bar{v}_w$. Their magnitude can be determined by varying the protein-water ratio $(g_p/g_w)$ and plotting $V_{sp}$ against $g_p$. An approximate value for $\bar{v}_p$ is obtained by drawing the best smooth curve through the points of the plot and determining the slope of the line.

Another less rigorous procedure is based on the arbitrary assumption that the partial specific volume of water remains unchanged and is always $v_w = 1.0$. Then Eq. (VI-9) is reduced to the simpler form:

$$V_{sp} = g_p(v_p)_{app} + g_w$$

The term $(v_p)_{app}$ is called the *apparent specific volume* of the protein.

It must be emphasized again that neither the partial specific volume, $\bar{v}_p$, nor the apparent specific volume, $(v_p)_{app}$ indicates the true volume occupied by the protein in the protein-water mixture. We have obtained these values by assuming that the volume of the protein-water mixture is the sum of the volume of a water phase and a protein phase. In this assumption the presence of hydrated protein molecules was not taken into account. Therefore, we are not able to calculate the extent of hydration from $\bar{v}_p$ or $(v_p)_{app}$. Nevertheless these values are important for the evaluation of results obtained by other methods. Their determination is also necessary for the calculation of molecular weights from the sedimentation velocity in the ultracentrifuge (see Chapter V, Sections D and E).

## E. Hydration of Dry Protein

It is a well known fact that the length of keratin fibers depends on their water content. Because of this, hair is used in hygrometers. Keratin, however, is not unique since all proteins bind water when exposed to an

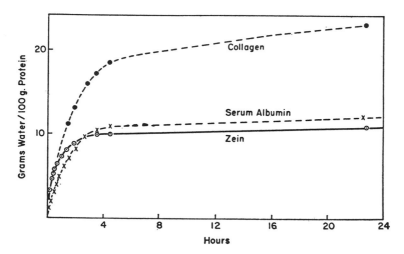

FIG. VI-12. Adsorption of water vapor by dry proteins at 20°C; relative humidity: 50% (E-1a).

atmosphere containing water vapor. Figure VI-12 shows the adsorption of water by powdered proteins which had been dried at room temperature in a vacuum over phosphorus pentoxide. It is evident from the figure that the adsorption of water takes place at a rather rapid rate until a certain saturation is reached (E-1a,1b).

The amount of water bound depends on the temperature and the water vapor pressure (see Fig. VI-13) (E-2). The figure shows a sigmoid curve. Similar curves have been obtained for all proteins hitherto investigated (E-3). The curves reveal that a certain amount of water, *a*, is bound very firmly and is given off only at low pressures of less than 0.2 saturation. A second amount, *b*, approximately equal to *a*, is bound if the water vapor pressure is increased further. At water vapor pressures of more than 0.7 saturation a sharp rise in the amount of water bound to the protein is

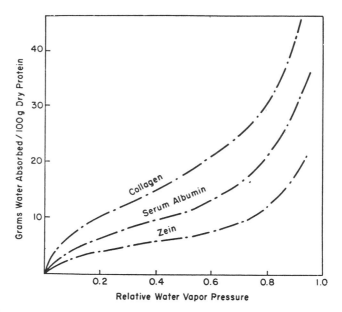

FIG. VI-13. Adsorption of water vapor by dry proteins at 25°C (E-2).

observed, until about 40 g of water are bound to 100 g serum albumin. The amount *a* varies from 4 to 10 g, and the maximum amount of water bound from 20 to 60 g per 100 g protein (E-2).

The particular shape of the curve of hydration cannot be accounted for by a simple adsorption of water or by the formation of stoichiometric hydrates. The amount, *a*, corresponding to the first steep part of the curve, is about one-fifth of the amount of water required to cover the whole protein molecule with a monomolecular layer. It is assumed, therefore, that this water is bound to certain hydrophilic groups, forming a layer of water one molecule in thickness between the peptide chains (D-3, E-2). Water molecules which are bound in this manner between adjacent peptide chains have less freedom of rotation than water molecules bound to only a single polar group of the protein molecule. They have been designated as irrotationally bound water molecules (E-3). The amount of this

irrotationally bound water has been calculated from measurements of the
dielectric constant (see Section H); it forms less than 50% of the total
water of hydration. The adsorption of the firmly bound water takes place
immeasurably fast, but is inhibited in the presence of air (E-4). Not only
water vapor, but also the gaseous molecules of HCl, $BF_3$, or $NH_3$ are
bound to dry protein molecules (E-4). This is not surprising since the
molecules of all these gases, like those of $H_2O$, are strongly polar.

The question arises as to whether or not the second water layer, and also additional layers of water molecules, could possibly be bound by dipole induction to
the primarily oriented layer of water molecules. This matter is of significance in
our understanding of the forces operating in biological systems. Actually the
occurrence of multiple layers of oriented water molecules has been assumed by
several investigators (E-5) and denied by others (E-6). The force by which the
$m$th water layer is bound to the $(m - 1)$ layer has been calculated by Brunauer
*et al.* (E-7) and it has been found far too weak to fix more than the first layer of
water molecules in a definite position. Similar results were obtained by Harkins
(E-8) who found that the first water layer adsorbed to $TiO_2$ was bound with an
energy of 6550 cal per mole water, but that this energy decreased to 1380, 220,
and 70 cal for the second, third, and fourth water layers, respectively. Since the
latter values are lower than the energies involved in the thermal movements of
the water molecules, oriented layers of water dipoles can hardly be formed.

A similar calculation of the bond energy of water molecules bound to
dry protein preparations furnishes values of 3000–6000 cal per mole for
the water bound in the steep first part (*a*) of the curve shown in Fig. VI-13.
The bond energy of additionally bound water (*b*) is much lower (E-9).
The steep third part of the curve (*c*) is probably due to the condensation
of randomly oriented water molecules on the hydrated surfaces and has
been compared with the formation of water clusters in gases (E-7). While
the first process (*a*) consists of the adsorption of an oriented layer of polar
water molecules and results in a higher degree of order in the system, i.e.,
in a decrease of entropy, the last part of the curve is due to the tendency
of the peptide chains to dissolve in water as a solvent; it results in a less
ordered system, i.e., in an increase in entropy (E-10).

The great strength by which the first portions of water are bound to a
protein is evident from the appreciable contraction of the volume attend-
ing this process. The density of dry ovalbumin, $d = 1.2655$, increases to
1.2855 when 6.15% water is bound. The specific volume of the protein
decreases accordingly from $1/1.2655 = 0.792$ ml/g to $1/1.2855 = 0.777$
ml/g. On further binding of water the density decreases; ovalbumin containing 56.26 g water/100 g has a density of 1.1280 ($V_{sp} = 0.887$) (E-11).
Important information on the hydration of protein crystals has been
obtained by X-ray analysis of the crystals. It has been found that the
crystals maintain their shape when they are exposed to water vapor of

different pressures or to aqueous salt solutions of a different salt content, but that stepwise swelling or shrinking of the crystals occurs. The conclusion has been drawn that the water bound to the crystals is bound mainly at the surface of the protein molecules and that it does not penetrate their interiors (E-12). The water content found in the crystals of proteins varied within wide limits; thus 32% water was found in insulin crystals, but more than 90% in crystals of tropomyosin (E-12).

There is no general agreement at present concerning the chemical grouping in proteins to which water is bound. While X-ray analyses, as mentioned above, suggest that this water covers the surface of the molecule, some authors attribute the water of hydration to the ionic groups of the protein or to other polar groups such as the peptide bonds. The total number of water molecules bound corresponds approximately to the number of positively and negatively charged ionic groups (E-13). The extent of hydration decreases on benzoylation of the protein (E-14), but is hardly affected by the reduction of dithio bonds (E-14) or by methylation; methylated zein, fibroin, and hemoglobin bind approximately the same amount of water as the untreated proteins (E-1a).

Not only the ionic groups, but also the peptide bonds, $—CO \cdot NH—$, are to be considered points of attachment for water molecules, since it has been shown that peptides synthesized from nonhygroscopic amino acids are hygroscopic, i.e., able to bind water molecules (E-9). Moreover, nylon, which can be considered an artificial polypeptide of the formula

$$\ldots \ —NH \cdot R \cdot NH—CO \cdot R \cdot CO—NH \cdot R \cdot NH—CO \cdot R \cdot CO— \ \ldots$$

also binds water in accordance with the typical sigmoid curve (Fig. VI-13) (E-2). Since nylon has no ionic side chains, there is no doubt but that most of the water is bound to the peptide bonds; possibly the water molecules are interlinked between two peptide bonds of adjacent peptide chains.

## F. Hydration of Proteins in Solution (F-9)

Attempts to determine the hydration of proteins in solution were made before the behavior of dry proteins, described in the preceding section, was known. It was assumed at that time that a definite part of the water in protein solutions was free and another definite part bound to the protein.

One of the methods used to determine these two portions of water consisted of measuring the expansion which occurred when the solution was frozen. Since the density of ice is considerably lower than that of water, freezing of "free" water is accompanied by a distinct increase in volume, which can be measured dilatometrically. In this way it was found that a certain part of the water of protein solutions did not freeze. This was thought to be the water of hydration. However, this conclusion is not very convincing, first, because water can be supercooled below its freezing point without solidifying; second, because the water of hydration might consist of "ice" molecules which would not undergo any volume change on cooling down below freezing temperatures (F-1). The presence of "icebergs," i.e., clusters of ice molecules (F-2), is indicated by the volume contraction which proteins undergo when they are digested by proteolytic enzymes (F-1). This process is probably accompanied by an increase in hydration and by the conversion of ice molecules into water molcules (F-1). The volume contraction during proteolysis might also be caused by electrostriction due to the release of numerous charged groups.

In a second method, the solubility of urea, glucose, or other nonelectrolytes in a protein solution was determined. It was assumed that the water of hydration, being bound to the proteins, would not be available to function as a solvent for other substances. The solubility of the nonelectrolytes added was determined either chemically or by measuring the lowering of the freezing point or the vapor pressure. The objection raised against this method is that the added anelectrolyte may reduce the extent of hydration by combining with water molecules or with the protein molecules, thus displacing bound water from the protein molecules (F-3). Indeed, the hydration of lactoglobulin is higher in water than in concentrated sucrose solutions (F-4) or in salt solutions (F-4). It is remarkable that the water-binding capacity of proteins decreases only slightly upon denaturation and heat coagulation of the proteins (F-5).

Since the phenomenon of hydration is accompanied by a decrease in volume, information concerning the extent of hydration can be obtained by determining the density of dry proteins and of protein solutions using one of the methods discussed in Section D, above. The decrease in volume is considerable, amounting to 5–8 ml/100 g of dry protein (F-6). Thus the density of dry ovalbumin suspended in benzene is 1.2715, corresponding to a specific volume of 0.786 ml/g, whereas the apparent specific volume of ovalbumin in aqueous solution is 0.726 ml/g (F-12) corresponding to a density of 1.377 g/ml. The densities of most dry proteins are close to 1.27, the "apparent densities" of proteins in solution close to 1.34. A straight line is obtained if the density of the solution is plotted against the concentration of the protein dissolved; the value $\dot{I} = (d_s - d_w)/c_p$ has been called *density increment*; $d_s$ and $d_w$ are the densities of the protein solution and of its dialyzate, respectively, and $c_p$ is the protein concentration in grams per milliliter. The density increment depends very little on the concentration of the protein and of pH, but is highly dependent on the salt concentration (D-2). Thus the density increment, $\dot{I}$, of carboxyhemoglobin is 0.251 in water, but 0.21 in 1% sodium chloride solution.

The amount of water bound to the protein in solution has been determined by Adair and Adair (D-2, F-7) who measured the density of protein solutions in equilibrium with their dialyzates. If the density increments of a protein in solutions containing different amounts of a buffer are $\dot{I}_1$ and $\dot{I}_2$ and the densities of the dialyzates are $d_1$ and $d_2$, the nonsolvent volume $V_{ns} = (\dot{I}_1 - \dot{I}_2)/(d_2 - d_1)$. The nonsolvent volume is the sum of the *"bound water"* and of the "apparent specific volume" of the protein, that is, the volume occupied by the dissolved hydrated protein in 1 ml of the solution. The bound water is obtained by subtracting the apparent specific volume from the nonsolvent volume. The values of bound water vary between 20 and 50 g of water per 100 g protein. This is close to the maximum amount of water bound to 1 g of dry protein (see Fig. VI-13). Values of the same order have been obtained for the small serum albumin molecule (D-2) as for the macromolecules of tobacco mosaic virus (F-8).

From the foregoing discussion on the hydration of protein crystals and protein solutions it is evident that the main volume contraction occurs in the region of low water pressure (Fig. VI–13). The apparent specific volume of proteins in their crystals is hardly different from the apparent specific volume found in protein solutions. Both however, are much lower than the specific volume of dry proteins. It is also clear that we are not able to determine precisely the extent of hydration because it depends on our definition of "bound water" and on the method used for its determination (F-9).

Although the amount of water bound by hydration amounts to only 20–50 g/100 g protein, much higher amounts of water are bound by living matter. The average water content of muscle and of parenchymatous organs is approximately 70–80%. Since their protein content is 20–30%, this indicates that 200–300 g water are found in a tissue containing 100 g protein. In gelatin gels and in the tissues of jellyfish the water:protein ratio is still higher. It was believed for a time that these large amounts of water are bound by proteins. Examination of organs and of gelatin gels by means of the methods described in the preceding sections have proved, however, that most of this water is "free." It freezes at the same temperatures as pure water and has the same solvent power. Investigation of the water of agar gels by means of nuclear magnetic resonance has revealed that the water molecules in these gels behave like intermediates between free water and ice (F-10). The same may be true for the water molecules of gelatine gels, which are able to immobilize about 30 g of water per gram of dry gelatin.

If ethanol is added carefully to gelatin solutions, a two-phase system is obtained consisting of two liquid layers. The designation *coacervate* has been used for such a system. Its lower layer contains coiled up threads of gelatin molecules which include a large amount of immobilized water (F-11). All these observations indicate that most of the water present in the living matter consists of free water molecules which are immobilized by the network of cellular protein filaments.

## G. Solubility of Proteins in Water

The solubility of proteins in water varies within wide limits. While some proteins dissolve easily in salt-free water, others dissolve only in the presence of certain concentrations of salts; a third group is insoluble in water, but dissolves in mixtures of water and ethanol, while a fourth group, the scleroproteins, does not dissolve in any solvent.

This differing behavior of proteins has been chosen as a basis for a classification of proteins into albumins, globulinlike proteins, prolamins, and scleroproteins. Classifications of this kind are more or less artificial and are of restricted value. Thus, some hemoglobins are readily soluble in salt-free water, while other hemoglobins are almost insoluble in the absence of salts but promptly dissolve upon the addition of salts.

One would expect that those proteins which have a high affinity for water and are strongly hydrated would be the most readily soluble in water. Solubility and hydration, however, do not run parallel. Thus, collagen has a much higher affinity for water than serum albumin and binds more water when its dry powder is exposed to water vapor (see Fig. VI-13). Nevertheless, it is insoluble in water, while serum albumin dissolves easily.

There is no doubt that the solubility of a protein in water is determined by its chemical structure, i.e., by the kind and number of the amino acids forming the molecule, and by the folding of the peptide chains in the molecule. One would expect that a loose peptide structure and a high content of the positively or negatively charged ionic groups would increase both the affinity of the protein for water and its solubility. While this is certainly true, ionic groups at the same time exert the opposite effect by combining readily with ionic groups of the opposite sign, thus forming saltlike bonds within a protein molecule as well as between adjacent protein molecules.

It is well known that the solubility of proteins depends to a great extent on pH and on the concentration of salts present in the solution. The minimum solubility is found at the isoelectric point. The solubility is increased by the addition of acids or bases to the isoelectric protein. Use is made of the low solubility of isoelectric proteins when it is desired to isolate them from mixtures with other proteins. The precipitation of the isoelectric protein is probably due to the mutual formation of saltlike links between anionic and cationic groups of adjacent protein molecules.

Neutral salts have a two-fold effect on protein solutions. Low concentrations of the salts increase the solubility of proteins, while high concentrations of neutral salts reduce the solubility and give rise to the formation of precipitates. The first of these two effects is known as the *salting-in* effect while the precipitation caused by high concentrations of the salts has been designated as *salting-out* (Chapter II, A).

The *salting-in* effect is particularly striking in those cases in which the protein is insoluble in salt-free water. Thus the euglobulins of the blood serum or the vegetable globulins are insoluble in water, but are readily dissolved upon the addition of sodium chloride to their suspension in water. The solubility-promoting action of the neutral salts is due to the electrostatic interaction between their ions and the charged groups of the protein. Each of the ionic groups of the protein, because of its electrostatic action, is probably surrounded by an atmosphere of salt ions of the opposite charge. In some cases this leads to the formation of permanent bonds between protein and inorganic ions (see Chapter X, B).

The solubility of proteins is also increased by glycine and other dipolar molecules which raise the dielectric constant of water (G-1); on the other hand, the solubility of proteins in water is reduced by organic solvents, such as ethanol, which lower the dielectric constant of water. Use has been made of these actions of glycine and of organic solvents in the fractionation of plasma proteins (see Chapter VIII, B).

Soluble proteins are precipitated by high concentrations of neutral salts (see Chapter II, A). The *salting-out* effect of high concentrations of salts is evidently due to a competition between the salt and the protein for molecules of the solvent. So much water is bound by the salt ions that not enough water is available for the dissolution of the proteins.

The salting-in effect of low salt concentrations and the salting-out effect of higher concentrations of salts are shown in the following table, where $\mu$ is the ionic strength of a potassium phosphate buffer solution at pH 7.4 and 0°C, and $G$ the amount of muscle globulin dissolved, measured in milligrams of nitrogen per liter of salt solution (G-2).

| $\mu$ | 0.1 | 0.20 | 0.30 | 0.50 | 3.0 | 3.3 | 3.4 | 3.5 |
|---|---|---|---|---|---|---|---|---|
| $G$ | 1.8 | 3.2 | 149.0 | 1275 | 822 | 204 | 41 | 26 |

In general, the solubility of proteins is decreased when the temperature is lowered. If, however, sodium sulfate is used for salting-out, lowered temperature results in heightened solubility, because solid sodium sulfate crystallizes and the sodium sulfate concentration of the solvent decreases. For this reason salting-out by sodium sulfate is carried out at 20°C or higher.

We are not yet able to predict the solubility of a protein in water from its amino acid composition, nor can we explain the solubility of the vegetable prolamins in ethanol. It has frequently been attributed to their high content of proline, which is soluble in alcohol. Despite its much higher content of proline, however, collagen is insoluble in alcohol.

From the observations noted in this section it is clear that we cannot yet correlate solubility with the occurrence and content of certain amino acids in the protein molecule. The solubility of a solute in a solvent depends quite generally on the intensity of the mutual interaction between solvent and solute molecules and for similar reasons on the lattice energy of the solid phase, i.e., on the forces effective between molecules of the solute. If the intensity of the attraction between molecules of the solute and the solvent exceeds the mutual attraction of solute molecules, dissolution will result. Since the diameter of globular protein molecules is very large, only the superficial molecular groups will be available for solute-solute interaction. We have to conclude, therefore, that the solubility of proteins will depend mainly on the nature of those groups which form the *surface* of the large particles, and particularly on the distribution of ionic and nonpolar groups beteeen the surface and the interior of the protein molecule (G-1,3). Unfortunately our knowledge about this arrangement of polar and nonpolar groups is very limited.

## H. *The Dielectric Properties of Protein Solutions (H-1 to 3)*

Two electrical charges of opposite sign attract each other with the force $F = q_1q_2/Dr^2$ where $q_1$ and $q_2$ are the charges (in electrostatic units), $r$ the distance between the charges, and $D$ the dielectric constant. The constant $D$ is determined by measuring the electrostatic force between two charged plates of a condenser first *in vacuo* and then in the substance to be tested. If the force of attraction (or repulsion) in a vacuum is $F_0$ and that in the examined medium $F_s$, the latter value is always smaller than $F_0$ because of dipole induction in the medium (see below). The ratio $F_0/F_s$ is called the dielectric constant. The dielectric constant depends primarily on the polarity of the medium as shown by the following values of $D$ at 20°C: water 80, ethanol 24, diethyl ether 4.3, paraffin 2. Quite generally, high dielectric constants are found in polar molecules, low constants in nonpolar molecules.

Because they are highly polar substances, one would expect zwitterions to have high dielectric constants. Unfortunately it is usually not possible to determine the dielectric constants of amino acids or of proteins directly, because they cannot be liquefied; their melting points are so high that decomposition occurs at the temperatures required to melt them. However, significant results have been obtained by measuring the dielectric constant of their aqueous solutions. The high dielectric constant of water is due to its polarity, discussed in Section D, this chapter. If organic molecules are dissolved in water, its dielectric constant is lowered in general. However, if amino acids or proteins are dissolved in water, the dielectric constant increases. This behavior, which is somewhat atypical, is a consequence of the highly polar character of amino acids and proteins. The increase in the dielectric constant, while different for different amino acids, peptides, and proteins, is proportional to the concentration of the dissolved substance. If $D_w$ and $D_s$ are the dielectric constants of water and of the solution respectively, and $C$ the molar concentration of the dissolved substance, the molar *dielectric increment* is $i_M = (D_s - D_w)/C$.

The dielectric increments of $\alpha$-amino acids vary from 22 to 28, while those of the peptides are much higher; for glycine polymers containing 2, 3, 4, 5, and 6 glycine molecules the following values were found: 70, 113, 159, 215, and 239 (H-4). One has to bear in mind, however, that molar solutions of the higher peptides contain larger amounts of substance per unit volume than the solutions of the peptides with low molecular weight. If we divide the values reported by the number of glycyl residues per molecule we obtain the following values for the dielectric increment per equivalent of glycyl residue: 35, 38, 40, 43, and 39. Evidently the increase in the dielectric constant of water is the same, whether we add a certain amount of triglycine or the same weight of hexaglycine.

The dielectric increments in protein solutions vary widely. Since the concentration of proteins is not usually expressed in moles, the molar dielectric increment, $i_M$, is replaced by dielectric increment per gram, $i_G = (D_s - D_w)/c$, where $c$ is the concentration in grams per liter. The $i_M$ and $i_G$ values found are shown in Table VI-3 (H-5).

Calculating the dielectric increments of amino acids per gram of amino acid, we find values between 0.18 and 0.36, i.e., values of the same order of magnitude as those of some of the proteins. For a long time it was difficult to interpret the fact that the dielectric constant of water was increased to approximately the same extent by 1 g of protein as by 1 g of an amino acid. In order to understand this observation it may be useful to discuss the dipole moment of amino acids since this is a measure of their polarity.

The magnitude of the dipole moment of $\alpha$-amino acids can be estimated from the distance between the positively and the negatively charged groups. This distance is approximately 3 Å ($3 \times 10^{-8}$ cm). The charge of an electron is $4.8 \times 10^{-10}$ electrostatic unit. Hence, the dipole moment will be roughly $15 \times 10^{-18}$ electrostatic unit, i.e., 15 debye. Since the distance between positive and negative charge is the same in all $\alpha$-amino acids, it is not surprising that dielectric increments of the same order of magnitude have been found in most of the amino acids.

As they occur in protein molecules, only a few of the amino acids contain free positive or negative groups. To a first approximation it can be assumed that, of 10 amino acids, 1 has a positively charged group and 1 a negatively charged group. Hence, the number of charges in 1 g protein is approximately one-tenth of those present in 1 g of an amino acid mixture and one would expect less polarity in proteins. The dielectric constant of protein solutions depends not only on that of the solvent and solute, but also on the frequency of the applied electrical field. The principle underlying this method will be discussed since it has been used to obtain some

TABLE VI-3

| Substance | Dielectric increments | | Substance | Dielectric increments |
| | $i_M$ (per mole) | $i_G$ (per gram) | | $i_G$ (per gram) |
|---|---|---|---|---|
| Glycine | 22.6 | 0.30 | Ovalbumin | 0.1 |
| Alanine | 23.2 | 0.26 | Gliadin | 0.1 |
| Valine | 25 | 0.21 | Serum albumin | 0.17 |
| Leucine | 25 | 0.19 | Insulin | 0.3 |
| Diglycine | 70.6 | 0.54 | Carboxyhemoglobin | 0.33 |
| Triglycine | 113 | 0.60 | Zein | 0.4 |
| Tetraglycine | 159 | 0.65 | Edestin | 0.7 |
| Pentaglycine | 215 | 0.71 | Serum pseudoglobulin | 1.1 |
| Hexaglycine | 239 | 0.65 | Lactoglobulin | 1.5 |

information on the shape of the protein molecules. If a polar molecule is placed in an electrical field, the molecule will become oriented, its positive pole shifting in the direction of the external negative pole, and the negative pole of the molcule shifting toward the external positive pole. If the polar molecule is spherical and if the distance of the poles from the center is equal, a revolution of the molecule around its center will result, as shown in Fig. VI-14 by 1a and 1b. If however, the shape of the polar molecule is nonspherical, the molecule as a whole will yield to the electrostatic forces exerted on its two poles and it will be oriented as shown in Fig. VI-14 by IIa and IIb.

The relaxation time, i.e., the time that will be necessary for the completion of the movements shown in Fig. VI-14 can be determined by increasing the frequency of the external alternating electrical field. When the frequency exceeds a certain limiting value, the molecule can no longer alter its position rapidly enough. Above this frequency the molecule does not contribute to the dielectric increment of the solution. The *relaxation time* is short for small molecules and long for large molecules. For the small water dipoles the relaxation time is approximately $10^{-11}$ sec; for amino acids it is $10^{-10}$ sec; for proteins, $10^{-6}$ to $10^{-8}$ sec. If the frequency of the external alternating field is $10^3$ cycles/sec (wave length $3 \times 10^7$ cm), amino acids and proteins give the dielectric increments reported above. At frequencies of $10^7$ cycles/sec. (wave length $3 \times 10^3$ cm) and above, however, proteins do not increase the dielectric constant of water.

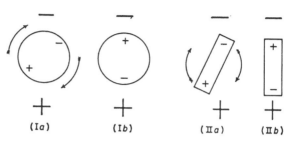

(Ia)     (Ib)     (IIa)     (IIb)

FIG. VI-14. Orientation of dipoles in an electric field.

A slight decrease is observed because that portion of the water which is bound by the protein is unable to follow the oscillations of the alternating field (H-14).

By varying the frequencies from high to low, the critical frequency corresponding to the relaxation time of the molecule can be determined. If the critical frequency is $n$, the relaxation time $t = 1/(2\pi n)$. Particles which are not spherical but ellipsoidal give two different relaxation times, one of them possibly corresponding to rotation around the short axis, the other to rotation around the long axis. By determining these two values the axial ratio, $a/b$, of proteins in their aqueous solutions has been estimated. The values obtained for $a/b$, neglecting hydration, are much higher than those found by other methods (H-3,5,15). This indicates strongly that the dielectric increment of proteins is not caused by high permanent dipole moments of the protein molecules.

If the positive and negative charges in a protein molecule were invariably localized at certain of the free amino or carboxyl groups, the dipole moment of proteins in electrostatic units per gram protein should be much lower than that of amino acids and close to that of the weak organic acids or bases. However, estimates of the dielectric constant in the surface of globular protein molecules give values close to 50, and for a depth of one Ångstrøm unit the value 20 (H-6,7). Moreover, the dielectric increment caused by elongated, fibrous proteins in their aqueous solutions should depend on their orientation in the solution and should be higher in a streaming solution where all molecules have the same orientation (see Fig. V-6) than in a resting solution where they are randomly oriented. It seems, however, that the dielectric increment is the same in both the flowing and the resting solution (H-8). This has given rise to a new interpretation of the dielectric increment. It has been attributed to the ice-like structure of bound water molecules (H-8). In agreement with this view, the dielectric constant of water is increased to higher values not only by proteins but also by filter paper, although the dielectric constant of the paper is very low, close to 7 (H-8). A similar low value has been found for dry keratin (H-10a). If dry proteins are exposed to increased water vapor pressures, the dielectric constant increases with increasing amounts of bound water (H-9). All these observations seem to indicate that the increase in the dielectric constant of water caused by proteins is due to the *rigid orientation of the bound water molecules* rather than to charges on the protein molecules (H-10).

Another interpretation of the high dielectric increment of proteins in water is based on the *theory of fluctuating charges* discussed in Section B of this chapter. It will be recalled that the protons, according to this theory, fluctuate between various proton acceptors ($-NH_2$ or $-COO^-$ groups) inside the protein molecule. If these fluctuations take place randomly, the mean permanent dipole moment must be zero. However, the fluctuations of charges induce nonvanishing electrical dipole moments. The average root of the mean-square dipole moments is responsible for long-range forces, proportional to $1/r^2$ ($r$ = distance) and also responsible for hydration of the proteins and for protein-protein interaction (H-11, 12). Attempts have been made to determine the dipole moment of fibrinogen, an elongated protein molecule, in its aqueous solutions by measuring the electrical birefringence during the application of rectangular electrical impulses (H-13). Assuming a molecular weight of $3 \times 10^5$, the dipole moment parallel to the long axis of the fibrinogen molecule was zero at pH 6–7, increased to 500 debye at pH 8.5, and decreased again with increasing pH.

## I. The Role of Proteins in Energy Transmission (I-10)

The presence of positively and negatively charged groups in proteins is suggestive of a role of these charges in the transmission of electrical stimuli. Similarly, the groups which absorb visible or ultraviolet light might be responsible for the transmission of light energy. Indeed, it has been suspected that the peptide chains of proteins might share with semiconductors the properties of electron conductivity and photoconductivity. This would be of great importance for our understanding of the physiological effects of electricity and light.

Careful investigations have shown that proteins are not semiconductors

(I-1,2). In spite of the occurrence of charge transfers in biological systems, the values found for electron spin resonance are very low (I-2). Most of the energy transfer in or between protein molecules seems to take place through the transfer of ions without any significant amount of free radicals. As mentioned earlier (Section B), many of the protons bound to carboxyl or amino groups of the protein molecules fluctuate between these groups without any permanent localization. One can imagine the secretion of acid by the gastric mucosa as a unidirectional flow of the "fluctuating protons." Obviously, energy is required for such flow against the concentration gradient. Different "pump" mechanisms have been suggested to explain it. We know from investigations of the intermediary metabolism that energy is frequently supplied by oxidative phosphorylation, which involves the transfer of phosphate ions to enzyme proteins and from these to the ultimate phosphate acceptor (I-3). Another type of transfer reaction between proteins may occur by means of the —S—S— interchange with —SH as described in Chapter VII, E (I-4). In this reaction the —SH group is oxidized (loses electrons), and the SS bond is reduced (gains electrons). Since free radicals are not detected, the reaction may take place by the heterolytic cleavage of the dithio group (see Chapter III, D). The common factor in all these transfer reactions is the presence of a carrier ion which is transferred together with the electrons (I-3).

We are dealing with a different mechanism of energy transfer when the energy is supplied by *light*. It is a prerequisite for this process that the light of the used wave length be absorbed by the protein. In the colored proteins, absorption takes place in the prosthetic groups, for instance in the heme residues of hemoglobin or in the flavin moiety of the flavoproteins. In the colorless proteins and apoproteins light is absorbed predominantly by the tryptophan and tyrosine residues (Chapter VII, C). The absorbing groups are thereby excited. The excited state of tryptophan has a lifetime of approximately $4 \times 10^{-9}$ sec (I-5); that of tyrosine is about $10^{-8}$ sec. Part of the absorbed energy can be re-emitted as fluorescent light of slightly longer wave length (I-6,7); part of it can be utilized for chemical reactions, for instance for the cleavage of CO-myoglobin (or CO-hemoglobin) into CO and free myoglobin (or hemoglobin) (I-3). Since the CO residue in CO-myoglobin is bound to heme, one might have expected that in this reaction only visible light, which is absorbed by heme, would be effective. However, the release of CO is also brought about by ultraviolet light of the wave length 280–290 m$\mu$ which is absorbed by tryptophan (I-3). Evidently, part of the light energy absorbed by tryptophan is transmitted to the CO-heme complex. The energy of light absorbed by tryptophan or tyrosine residues causes the emission of blue phosphorescence by the proteins (I-8). The emitted light energy can be transmitted to other similar residues even if these are separated from the absorbing residue by a distance of about 20 Å (I-5). However, when fluorescent dyes or nucleic acids are bound to proteins and irradiated with light absorbed by the dye or the nucleic acids, no fluorescence is produced in the protein. Evidently, the light energy is not transferred through the chains of the nucleic acids or proteins (I-9). Transfer of light energy takes place without a carrier ion and differs essentially in this respect from the transfer reactions described in the preceding paragraph. In ion transfer reactions, the protein molecule supplies the carrier ions for the transmission of energy; in the excitation process it provides groups which are able to absorb and transmit light energy. A third type of energy transfer, in which electrical or chemical energy is converted into mechanical energy, is mediated by the muscle proteins (see Chapter IX).

## REFERENCES

### Section A

**A-1.** E. Q. Adams, *JACS* 38: 1503(1916). **A-2.** N. Bjerrum, *Z. physik. Chem.* (*Leipzig*) 104: 147(1923). **A-3.** L. Ebert, *Z. physik. Chem.* (*Leipzig*) 121: 385(1926). **A-4.** J. T. Edsall, *J. Chem. Phys.* 4: 1(1936). **A-5.** F. T. Gucker, Jr., *et al.*, *J. Phys. Chem.* 43: 153(1939); 45: 309(1941). **A-6.** S. Miyamoto and C. L. A. Schmidt, *JBC* 90: 165(1931). **A-7.** S. Sørensen, *Biochem. Z.* 71: 174(1909).

### Section B

**B-1.** W. Hardy, *J. Physiol.* 24: 288(1899). **B-2.** L. Michaelis and P. Rona, *Biochem. Z.* 28: 193(1910). **B-3.** O. Meyerhof, *Pflüger's Arch. ges. Physiol.* 195: 53(1922). **B-4.** W. C. Stadie and K. Martin, *JBC* 60: 191(1924). **B-5.** L. J. Harris, *BJ* 24: 1080(1930). **B-6.** H. A. Abramson, "Electrokinetic Phenomena." Chem. Catalogue Co., New York, 1939. **B-7.** I. Lichtenstein, *Biochem. Z.* 303: 26(1939). **B-8.** H. Edelhoch, *JACS* 80: 6640(1958). **B-9.** J. F. Foster and K. Aoki, *J. Phys. Chem.* 61: 1369(1957). **B-10.** C. Tanford *et al.*, *JACS* 77: 6414(1955). **B-11.** J. T. Edsall *et al.*, *Proc. Natl. Acad. Sci. U.S.* 44: 505(1958). **B-12.** E. J. Cohn and J. T. Edsall, "Proteins, Amino Acids and Peptides," p. 467. Reinhold, New York, 1943. **B-13.** C. G. Kirkwood, *Discussions Faraday Soc.* 20: 78(1956). **B-14.** S. N. Timasheff *et al.*, *JACS* 79: 782(1957). **B-15.** T. L. Hill, *JACS* 78: 1577(1956). **B-16.** J. Steinhardt and E. M. Zaiser, *JBC* 190: 197(1951). **B-17.** G. Scatchard *et al.*, *JACS* 72: 535, 540(1950). **B-18.** C. W. Carr and L. Topol, *J. Phys. Chem.* 54: 176(1950). **B-19.** C. W. Carr, *ABB* 62: 476(1956). **B-20.** R. A. Alberty and H. H. Marvin, *JACS* 73: 3220(1950). **B-21.** L. Michaelis, *JBC* 87: 33(1930). **B-22.** G. Gomori, *Proc. Soc. Exptl. Biol. Med.* 62: 33(1946); see also on Amine Buffers, *Ann. N.Y. Acad. Sci.* 92: pp. 333–812(1961). **B-23.** W. Bockemüller, *Z. Naturforsch.* 13b: 772(1958). **B-24.** S. Sørensen *et al.*, *J. Gen. Physiol.* 8: 543(1927). **B-25.** S. G. Adair and M. E. Adair, *BJ* 28: 1230(1934). **B-26.** D. S. Jackson and A. Neuberger, *BBA* 26: 638(1957). **B-27.** B. G. Malmström *et al.*, *JBC* 234: 1108(1959). **B-28.** R. Bates, *Chem. Revs.* 42: 1(1948). **B-29.** J. Sendroy, Jr., *Ann. Rev. Biochem.* 7: 231(1939). **B-30.** D. A. McInnes, *Science* 108: 693(1948). **B-31.** J. F. Foster and P. Clark, *JBC* 237: 3162 (1962).

### Section C

**C-1.** H. A. Abramson, L. S. Moyer, and M. H. Gorin, "Electrophoresis of Proteins," p. 159. Reinhold, New York, 1942. **C-2.** L. G. Longsworth, *Chem. Revs.* 24: 271 (1939); 30: 323(1942). **C-3.** A. Tiselius and P. Flodin, *Adv. in Protein Chem.* 8: 461(1953). **C-4.** O. Smithies, *Adv. in Protein Chem.* 14: 65(1959). **C-5.** H. Svensson, *Adv. in Protein Chem.* 4: 251(1948). **C-6.** K. Landsteiner and W. Pauli, cited by W. Pauli and E. Valko, "Kolloidchemie der Eiweisskörper," p. 17. Steinkoff, Dresden, 1933. **C-7.** L. Michaelis and H. Davidsohn, *Biochem. Z.* 41: 102(1912). **C-8.** A. Tiselius, *Kolloid-Z.* 85: 129(1938). **C-9.** J. E. L. Philpot, *Nature* 141: 283(1938); H. Svensson, *Kolloid-Z.* 87: 181(1939). **C-10.** G. S. Adair and M. Robinson, *BJ* 24: 993(1930). **C-11.** A. Tiselius, *BJ* 31: 1464(1937). **C-12.** T. L. McMeekin *et al.*, *JACS* 70: 881(1948); J. A. Bain and H. F. Deutsch, *ABB* 16: 223(1948). **C-13.** D. G. Sharp *et al.*, *JBC* 144: 139(1942). **C-14.** A. Tiselius and B. Erickson-Quensel, *BJ* 33: 1752(1939); L. G. Longsworth *et al.*, *JACS* 62: 2580(1940). **C-15.** R. Alberty *et al.*, *J. Phys. Chem.* 52: 217(1948). **C-16.** R. Alberty and J. Nichols, *JACS* 70: 1675, 2297(1948). **C-17.** O. Smithies, *BJ* 61: 629(1955). **C-18.** W. Grassmann and K. Hannig, *Naturwiss.* 37: 496(1951). **C-19.** C. Moncke, *Klin. Wochschr.* 34: 100(1956). **C-20.** F. W. Sunderman, Jr., and F. W. Sunderman, *Clin. Chem.* 5: 171(1959). **C-21.** W. Pohlit and H. Schittko, *Kolloid-Z.* 156: 73(1958). **C-22.** R. J. Slater, *ABB* 59: 33(1955). **C-23.** K. Hannig, *Z. physiol. Chem.* 311: 63(1958). **C-24.** J. Brattsten and A. Nilsson, *Arkiv Kemi* 3: 337(1951). **C-25.** P. Grabar, *Methods of Biochem. Anal.* 7: 1(1959). **C-26.** L. S. Moyer and H. A. Abramson, *J. Gen. Physiol.* 19: 727 (1936). **C-27.** H. B. Bull, "Physical Biochemistry," p. 187. Wiley, New York, 1951. **C-28.** M. Bier, ed., "Electrophoresis." Academic Press, New York, 1959. **C-29.** S. Raymond *et al.*, *Nature* 195: 697(1962).

### Section D

**D-1.** A. D. McLaren, *J. Polymer Sci.* 7: 289(1951). **D-2.** G. S. Adair and M. E. Adair, *Proc. Roy. Soc.* A190: 341(1947). **D-3.** H. B. Bull, "Physical Biochemistry," p. 320. Wiley, New York, 1951.

### Section E

**E-1a.** F. Haurowitz and F. Bursa, unpublished experiments. **E-1b.** S. W. Benson, *in* (E-4). **E-2.** H. B. Bull, *JACS* 66: 1499(1944). **E-3.** T. J. Buchanan *et al.*, *Proc. Roy. Soc.* A213: 379(1952). **E-4.** S. W. Benson *et al.*, *JACS* 72: 2102(1950); 75: 3925 (1953). **E-5.** H. R. Kruyt, *Natuurw. Tijdschr.* 18: 38(1936). **E-6.** F. Haurowitz, *Kolloid-Z.* 74: 208(1936); F. C. Tompkins, *Ann. Repts. on Progr. Chem.* 47: 58(1951). **E-7.** S. Brunauer, P. H. Emmett, and E. Teller, *JACS* 60: 309(1938). **E-8.** W. Harkins, *Science* 102: 294(1945). **E-9.** M. Dole and A. D. McLaren, *JACS* 69: 651 (1947); 70: 3040(1948). **E-10.** S. Davis and A. D. McLaren, *J. Polymer Sci.* 3: 16(1948). **E-11.** H. Neurath and H. B. Bull, *JBC* 115: 519(1936). **E-12.** M. F. Perutz *et al.*, *Trans. Faraday Soc.* 42: B, 187(1946); *Proc. Roy. Soc.* A191: 83(1947); *Research (London)* 2: 52(1949). **E-13.** L. Pauling, *JACS* 67: 555(1945). **E-14.** E. F. Mellon *et al.*, *JACS* 69: 827(1947); 71: 2761(1949).

### Section F

**F-1.** K. Linderström-Lang, *Cold Spring Harbor Symposia Quant. Biol.* 14: 117(1950). **F-2.** I. M. Klotz *et al.*, *JACS* 80: 2132(1958); 81: 5119(1959). **F-3.** H. B. Bull, *J. Gen. Physiol.* 17: 83(1933). **F-4.** T. L. McMeekin *et al.*, *JACS* 72: 3662(1950); 76:

407(1952). **F-5.** F. Haurowitz, *Kolloid-Z.* 71: 198(1935). **F-6.** H. Chick and J. Martin, *BJ* 7: 92(1913). **F-7.** G. S. Adair and M. E. Adair, *Proc. Roy. Soc.* B120: 422 (1936). **F-8.** D. C. Douglas *et al.*, *BBA* 44: 401(1960). **F-9.** A. G. Ogston, *Ann. Rev. Biochem.* 24: 181(1955). **F-10.** O. Hechter *et al.*, *Proc. Natl. Acad. Sci. U.S.* 46: 783(1960). **F-11.** S. Basu and G. Battacharya, *Science* 115: 544(1952). **F-12.** F. Haurowitz, unpublished experiments.

*Section G*

**G-1.** E. J. Cohn and J. T. Edsall, "Proteins, Amino Acids and Peptides," p. 615. Reinhold, New York, 1943. **G-2.** J. T. Edsall, *JBC* 89: 289(1930). **G-3.** F. Haurowitz, *Kolloid-Z.* 74: 208(1936); 71: 198(1935); 77: 65(1936).

*Section H*

**H-1.** J. L. Oncley, *Chem. Revs.* 30: 433(1942). **H-2.** J. T. Edsall, *in* "Proteins, Amino Acids and Peptides" (E. J. Cohn and J. T. Edsall, eds.), p. 140. Reinhold, New York, 1943. **H-3.** J. L. Oncley, *Ibid.*, p. 543. **H-4.** J. Wyman and T. L. McMeekin, *JACS* 55: 908(1933). **H-5.** J. L. Oncley, *J. Phys. Chem.* 44: 1103(1940). **H-6.** C. Tanford and J. G. Kirkwood, *JACS* 79: 5333(1957). **H-7.** J. A. Schellman, *J. Phys. Chem.* 57: 472(1953). **H-8.** B. Jacobsen and M. Wenner, *BBA* 13: 577(1954); B. Jacobsen, *JACS* 77: 2919(1955). **H-9.** S. T. Bailey, *Trans. Faraday Soc.* 47: 509(1951). **H-10.** T. J. Buchanan *et al.*, *Proc. Roy. Soc.* A213: 379(1952); G. H. Haggis *et al.*, *Nature* 167: 607(1950). **H-10a.** G. King, *Trans. Faraday Soc.* 43: 601(1947). **H-11.** J. G. Kirkwood *et al.*, *Proc. Natl. Acad. Sci. U.S.* 38: 855, 863(1952); J. G. Kirkwood, *in* "Mechanisms of Enzyme Action" (W. D. McElroy and B. Glass, eds.), p. 4. Johns Hopkins, Baltimore, 1954. **H-12.** S. Timasheff *et al.*, *Proc. Natl. Acad. Sci. U.S.* 41: 710(1955). **H-13.** I. Tinocco, *JACS* 77: 3476(1955). **H-14.** J. Wyman, Jr., *JBC* 90: 443(1931); J. Wyman, Jr., and E. N. Ingalls, *Ibid.* 147: 297(1943). **H-15.** J. T. Edsall, *Fortschr. chem. Forsch.* 1: 119(1949).

*Section I*

**I-1.** K. Wirtz, *Z. Elektrochem.* 54: 47(1950). **I-2.** A. Szent-Györgyi, *Discussions Faraday Soc.* 27: 111(1959); I. Isenberg and A. Szent-Györgyi, *Proc. Natl. Acad. Sci. U.S.* 45: 1229, 1232(1959). **I-3.** T. Bücher, *Angew. Chemie* 62: 256(1950); N. Riehl, *Naturwiss.* 43: 145(1957). **I-4.** I. M. Klotz *et al.*, *JACS* 80: 2132(1958). **I-5.** G. Weber and F. W. J. Teale, *BJ* 72: 15(1959). **I-6.** G. Weber, *BJ* 75: 335, 345(1960). **I-7.** F. W. J. Teale, *BJ* 76: 381(1960). **I-8.** P. Debye and T. O. Edwards, *Science* 116: 143(1952). **I-9.** V. G. Shore and A. B. Pardee, *ABB* 62: 355(1956). **I-10.** A. Pullman and B. Pullman, *in* "Horizons in Biochemistry" (M. Kasha and B. Pullman, eds.), p. 568. Academic Press, New York, 1962.

# Chapter VII

## The Internal Structure of Globular Proteins*

### A. General Considerations

When we speak of globular proteins on the one hand and of fibrous proteins on the other, we must keep in mind that this is an artificial classification based on the oversimplifying assumption that the former group consists of spherelike or ellipsoidal molecules, the latter of elongated, threadlike molecules. The inadequacy of this classification is best exemplified by the fact that some proteins, such as actin, can exist in both a globular and a fibrous state. Although the protein molecules vary in their molecular weights from about $10^4$ to several millions, and although their shape also may vary widely, many, if not most of the proteins of the body fluids have molecular weights of the order of $10^5$. Their shape does not deviate very much from that of a sphere or an ellipsoid of rotation whose diameters in the three dimensions of space vary from about 15 to 60 Å. Since a protein of the molecular weight $10^5$ contains approximately 800 amino acid residues and since the length of each residue in the extended peptide chain is approximately 3.6 Å, the peptide chains of such a globular molecule must be somehow folded or coiled in order to conform with the aforementioned dimensions. The marked species specificity of proteins further indicates that this folding occurs in a definite manner (Chapter XV, B and D).

The species specificity of proteins is not affected by dissolution of the protein in water or in dilute salt solutions, nor by salting-out with neutral ammonium or sodium sulfate. Evidently the *specific internal structure* of the globular proteins is maintained during these procedures. Proteins differ in this respect from the macromolecules of rubber or gum arabic which, in suitable solvents, undergo continuous folding and unfolding and have no definite three-dimensional shape. The *maintenance of the three-dimensional specific structure* of the proteins is of fundamental importance because this structure is essential for the biological activity of those proteins which act as enzymes, hormones, antigens, or antibodies.

* See references (A-3,7).

In the present chapter, methods will be discussed by means of which some insight into details of the internal structure of the globular protein molecules can be gained. Until a few years ago it seemed quite hopeless that the precise arrangement of the peptide chains within the globular protein macromolecule could ever be elucidated and described. At present there is no doubt that we are very close to a solution of this problem. This is partly due to ingenious experimental work, especially the brilliant application of X-ray diffraction, which will be described in subsequent sections of this chapter. The rapid progress was facilitated by the fact that most globular protein molecules consist of a single peptide chain or of a very small number of such chains, and that these seem to be unbranched chains free of ramifications or rings. Work in this field has been greatly facilitated by the sequential analysis of amino acids described in Chapter IV.

Different methods have been used in order to get more information on the internal structure of the globular proteins. Some of these are *chemical methods*, based on the reactivity of various functional groups of the amino acid residues. If these functional groups (e.g., the ε-amino groups of lysyl residues or the phenolic hydroxyl groups of tyrosyl residues) are located on or near the surface of the globular protein molecule, they can react with specific reagents whereas groups of the same type which are buried inside the globular protein molecule react very slowly or do not react at all. Further information on the internal structure can be obtained by *optical methods*, particularly by determinations of the ultraviolet and infrared absorption and by measurements of the optical rotation and dispersion. The third and most powerful method is based on the *X-ray diffraction* of single crystals of the protein.

The specific folding of the peptide chains and the maintenance of their unique arrangement inside the globular protein macromolecule involves cross-links and mutual attraction between the folded chains. The following types of intramolecular bonding occur in protein molecules: (1) dithio bonds of the cystine residues, (2) salt-bridges, formed by the electrostatic attraction between anionic and cationic groups, (3) hydrogen bonds, and (4) mutual attraction of nonpolar groups. Our views on the importance of each of these types of bonding have changed considerably during the last few years. The evidence for the presence of each of these bonds and the significance of their contribution to the stability of the specific protein structure will be discussed in Section E of this chapter, where the problems of branching and cyclization of peptide chains will also have to be taken into consideration.

Before going further, it may be necessary to comment on the nomenclature used in this area of protein chemistry. As long as we had no knowledge on the three-dimensional arrangement of the peptide chains inside the globular protein molecules, the terms *chemical constitution* and *configuration* were adequate, the former describing the sequence of the amino acids in the peptide chain, the latter their sterochemical configuration. When it became clear that an elongated peptide chain of a definite amino acid sequence can exist in different geometrical arrangements, the terms *constellation* (A-1) and *conformation* (A-6) were introduced into protein chemistry. The latter is at present commonly used. Both terms referred to the folding of the peptide chains and also to the cross-links responsible for the resulting folds. It was then necessary to differentiate the primary from the secondary and tertiary structures of the protein molecule; the term *primary structure* was used for the unique sequence of amino acids characteristic of the respective protein. The folds and coils caused by cross-links were designated as *secondary structure*, and modifications brought about by the side chains of the amino acids as *tertiary structure*. Unfortunately, we are frequently unable to decide whether a specific loop in the peptide chain is caused by cross-links or by the interference of the side chains of amino acid residues. In view of these difficulties it would be best (A-2) to abandon the terms primary, secondary, and tertiary structure and to speak merely of *chain sequence* and *chain conformation*. The term *internal structure* used in this chapter comprises covalent cross-links in addition to sequence and conformation of the peptide chain.

According to some authors the conformation of the *native* proteins is uniquely determined by the amino acid sequence. This view is based on the observation that the enzyme activity of ribonuclease, which is lost after reductive cleavage of the dithio bonds, is restored, to a great extent, upon reoxidation (A-4). On the other hand, it has been claimed that pork and whale insulin, which have identical amino acid sequences, can be differentiated by immunochemical methods and therefore have different conformations (A-5). The phenomenon of denaturation, in which the protein molecule undergoes merely a change in its conformation, demonstrates clearly that a long peptide chain of unique amino acid sequence can exist in many different conformations although one of these may be favored by a particularly low energy content and high stability.

## B. Examination of Globular Proteins by X-Ray Diffraction (B-1,2,3,13)

Although examination of protein powders by means of X-ray diffraction has been used for many years, very little information is gained from the resulting Debye-Scherrer diagrams since the molecules in a protein powder are randomly oriented. Consequently, scattering occurs in all directions and causes blurred circles in the diagram rather than definite localized reflections. Deep insight into the structure of protein molecules can be gained, however, by the investigation of single crystals of the protein, since only in such crystals do all molecules have identical orientation in space. Great difficulties in work of this type arise in the preparation of protein crystals of a size (0.3–0.5 mm) sufficient for exposure to a beam of monochromatic X-rays. Many proteins lose their crystalline structure when they are dried. Therefore wet crystals in their saturated mother liquor are frequently investigated.

Exposure of a single crystal to monochromatic X-rays produces a scattering in the same manner as visible light is scattered by a regular lattice of a suitable width. Indeed, the crystal can be considered as a three-dimensional diffraction grating. The scattering of the beam of X-rays is recorded on a photographic film which is exposed to the rays emerging from the protein crystal in different directions. The film displays an array of spots (Fig. VII-1), each of them corresponding to the diffraction of the monochromatic X-rays from the structural elements of the protein. If the wave length of the X-rays is $\lambda$, and the angle of incidence is $\alpha$, the X-ray beam will be reinforced by diffraction when $2d \sin \alpha = n\lambda$, where $d$ is the distance between parallel planes of identical structural arrays in the investigated protein and $n$ is an integer. The wave length of the X-rays used in crystallographic analyses is of the same order as the interatomic distances in organic compounds, close to 1.5 Å. Since the dimensions of small organic molecules are usually within the range of 5–25 Å, it is possible to determine the value of $d$ from the X-ray diffraction diagram. Further analysis of diagrams obtained at different angles of incidence may lead to a complete resolution of the position of all carbon atoms and atoms of higher atomic weight.

In the X-ray analysis of proteins, difficulties arise from the fact that each protein molecule consists of thousands of atoms of the elements C, N, O, and H and a few sulfur atoms, and that the average electron density is not very different in various parts of the molecule. Moreover, each exposure gives several hundreds of spots, each of them corresponding to diffraction from different planes of repeating units. The evaluation of the intensity and position of each of these spots involves complicated calculations which require the use of electronic computers.

An important improvement in the analytical procedure has been brought about by the introduction of heavy metal substituents into the protein molecule (isotope replacement method) (B-4,5). Thus silver or mercury atoms can be bound specifically to the sulfhydryl groups of proteins. These heavy atoms, owing to their high electron density, cause interference patterns which are easily recognizable. For the same reason, the presence of iron-containing heme groups in hemoglobin or myoglobin facilitate the examination of these proteins. Further improvement is possible by using X-rays of very short wave length. Whereas the first X-ray analyses gave resolution to approximately 6 Å, recent investigations of myoglobin reveal details down to about 2.0 Å (B-6).

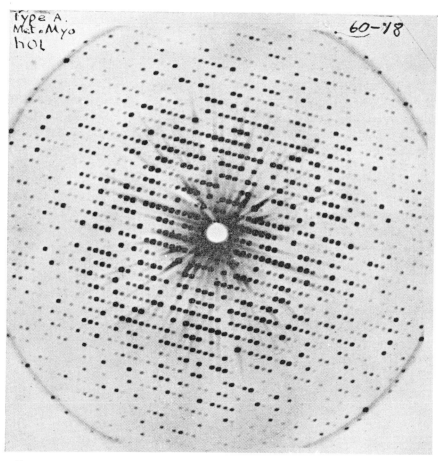

Fig. VII-1. X-ray diffraction pattern of metmyoglobin. (Courtesy of Dr. J. C. Kendrew, Cambridge, England.)

The idea of a regular type of folding of the peptide chains had first been advanced by Astbury (B-7), who investigated fibrous proteins, in which all molecules are oriented parallel to the long axis and thus form a suitable object for X-ray analysis. Since the peptide chain as such does not contain any double bonds, it was assumed at that time that it can rotate freely around its long axis. In 1951, Pauling and Corey (B-8), who examined the

X-ray diffraction of amino acids and simple peptides, showed, however, that the group

$$\begin{array}{cc} O & H \\ \| & | \\ -C & -N- \end{array}$$

which contains the peptide bond is *coplanar*, owing to resonance with the form

$$\begin{array}{cc} ^{(-)}O & H \\ | & |^{(+)} \\ -C & =N- \end{array}$$

Since the natural amino acids are asymmetric L-amino acids, it was reasonable to assume that they might be present in the peptide chains as helices (cylindrical spirals) held in position by intramolecular hydrogen bonds. However, attempts to build models with an integral number of amino acid residues per turn of the helix failed. A solution to this difficulty was proposed by Pauling *et al.* (B-9) who concluded from their X-ray analyses that proteins contain a helix with a nonintegral number of 3.7 amino acid residues per turn. This helix has been designated as the α-helix. Each turn of the helix corresponds to a length of 5.4 Å in the direction of the long axis of the helix. Accordingly, the dimension of each amino acid residue parallel to the long axis of the helix is $5.4/3.7 = 1.5$ Å. Indeed, a reflection of 1.5 Å spacing along the axis was discovered shortly thereafter by Perutz (B-10) in hemoglobin and some other proteins.

The α-helix is at present considered an important structural element in some of the proteins. It consists of a rigid, straight structure which is stabilized by intramolecular hydrogen bonds between adjacent turns of the helix (see Fig. VII-2). The α-helix cannot be the only structural element in globular proteins since the peptide chains of these proteins are folded and coiled whereas the α-helix is straight. Obviously, the length of the α-helix in a globular protein cannot be longer than the largest diameter of the molecule. Another reason for the absence of α-helices, at least in some parts of the molecule, is the presence of prolyl residues. These, owing to the particular structure of proline, cause kinks in the peptide chain which are incompatible with the straight α-helix structure (see Chapter IX).

At present it seems that some other regular types of peptide chains exist, e.g., a right-handed and a left-handed α-helix, two types of folded α-keratins, and an extended β-keratin (B-11; see also Chapter IX). It can be seen from Fig. VII-2 that the peptide bonds are shown in the *trans*-configuration

$$\begin{array}{c} H \\ | \\ -N-C- \\ \| \\ O \end{array}$$

whereas the *cis*-configuration

$$\begin{array}{cc} H & O \\ | & \| \\ -N & -C- \end{array}$$

may be more stable in aqueous solutions (B-12). The configuration of the peptide bond and the conformation of the peptide chain most probably depend on the nature of the amino acids involved.

The most advanced and complete X-ray analysis of a protein is at present the analysis of myoglobin, the pigment of red muscles. Its molecule

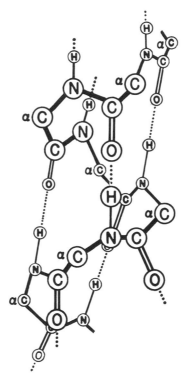

FIG. VII-2. The α-helix.

contains 1 heme residue per molecule (mol. wt. 17,000). In this molecule a resolution of 2 Å has been attained (B-6). This is close to the interatomic distances and makes it possible to describe not only the conformation of the peptide chain within the globular molecule but also to identify many of the side chains of the amino acid residues. The myoglobin molecule is a globule of 40 × 35 × 25 Å size. Its peptide chain is so closely folded that there is not much space for water molecules inside the globular parcel.

The course of the peptide chain in myoglobin is shown in the diagram of Fig. VII-3. It can be seen that only certain parts of the chain are straight. Only these can consist of α-helices. They form about 70% of the total

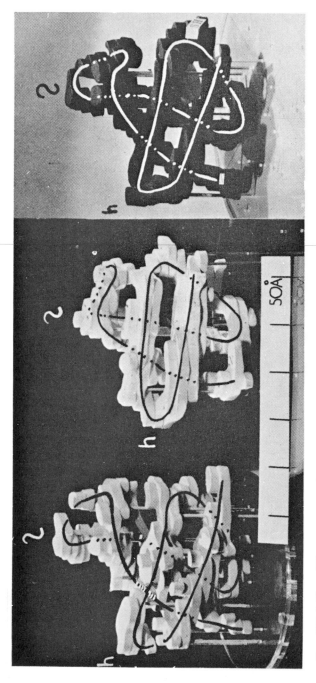

Fig. VII-3. Two different polypeptide chains in the asymmetric unit of hemoglobin compared with myoglobin (left). The heme groups are at the back of the chains. From Perutz et al. (B-4) and Kendrew et al. (B-6).

peptide chain. Some of the corners of the chain are formed by prolyl residues. The molecular weight of hemoglobin, which contains 4 heme residues, is approximately four times higher than that of myoglobin. The X-ray analysis reveals the presence of 4 subunits inside the hemoglobin molecule (B-4). Although the amino acid composition of hemoglobin is different from that of myoglobin, the conformation of each of the 4 subunits resembles strikingly the conformation of the peptide chain of myoglobin (B-4,6) as shown in Fig. VII-3. It must be pointed out, however, that hemoglobin and myoglobin differ from most of the other globular proteins by their high helix content. The helix content of most of the globular proteins is very low.

The brilliant results of the X-ray analysis of the heme proteins reported in the preceding paragraphs are of the greatest importance for protein chemistry. They show us a new way of obtaining direct information not only on the conformation of the peptide chains inside the globular protein molecules but also on the chain sequence. In spite of the availability of high-speed computers, analyses of this type still present formidable difficulties. However, there is no doubt that this method will enable us to gain deep insight into the internal structure of various protein molecules.

## C. Refractivity and Absorbancy of Protein Solutions (C-1)

In discussing electrophoresis it has been pointed out that aqueous solutions of proteins have a higher *refractive index* than pure water and that use is made of this fact to determine the position of the moving boundary. The refractive index of protein solutions increases linearly with the concentration of the protein. The difference between the refractive indices of a 1% protein solution and of water is called the *specific refractive increment*. It varies slightly from protein to protein; the following typical values have been found: bovine serum albumin, 0.001901, human serum albumin, 0.001887; ovalbumin, 0.001876; and human γ-globulin, 0.001875 (C-2). Since the temperature and the presence of salts do not greatly alter the refractive increment, the concentration of proteins can be determined very rapidly by measuring the refractive index of their solution and subtracting the refractive index of their dialyzate. It must be borne in mind, however, that the refractive increment of lipoproteins, 0.00171, is lower than that of fat-free proteins (C-3).

The molecular refraction of organic substances is an additive property and is equal to the sum of the atomic refractions. Similarly, the molecular refraction of proteins can be calculated as the sum of the refractions of the amino acid residues; a correction must be made for the specific volume or density of the protein (C-26).

Some information on the structure of proteins is obtained by measuring the *absorption of light* of different wave lengths. Visible light is absorbed only by colored proteins such as hemoglobin, the yellow enzyme, visual purple, and other chromoproteins. While very important information concerning the colored prosthetic group is obtained by measuring the visible absorption spectrum between 450–650 m$\mu$, the absorption curves of visible light do not permit conclusions concerning the structure of the colorless protein carrier.

All proteins strongly absorb ultraviolet light; the maximum of the absorption is near 280 m$\mu$. If the specific absorbancy of a protein at these wave lengths is known, its concentration can be determined by photometry. Nucleic acids have their absorption maximum at shorter wave lengths, but may interfere with protein determinations if their concentration is comparable with that of the proteins.

The absorption at 280 m$\mu$ is caused by the aromatic residues of tryptophan, tyrosine, and phenylalanine. Their absorption maxima are at 280, 275, and 258 m$\mu$, respectively. The absorption coefficients at these wave lengths have the ratio 27:7:1 (C-1). Hence phenylalanine contributes very little to the absorption of proteins. The ratio tyrosine:tryptophan can be estimated from the ratio of the absorbancies at 280.0 and 294.4 m$\mu$ (C-5). The absorption maximum of tyrosine undergoes a significant shift towards longer wave lengths at alkaline reactions when the phenol residue dissociates to form a phenolate anion. Although the pK of free tyrosine is 10.1, it usually is increased when tyrosine is protein-bound. This has been attributed to the formation of H-bonds between the phenol group and neighboring carboxyl groups (C-6). However O-methyl-tyrosine, which cannot form a phenolate, behaves in a manner similar to that of tyrosine (C-7). The abnormal behavior of the protein-bound tyrosyl residues may therefore be caused by other factors, for instance, by their embedding in a hydrophobic environment or by polarization rather than by hydrogen bonding (C-8; A-3). The same environmental factors may be responsible for the shift of the absorption maximum at 275–280 m$\mu$ towards longer wave length in proteins as compared to mixtures of amino acids or protein hydrolyzates (Fig. VII-4). The figure also demonstrates a steep increase of the absorption in the far ultraviolet. This absorption is attributed to the peptide bonds and has its maximum at 185–240 m$\mu$ (C-9,24). The absorbancy of amino acids is much lower in this range than that of peptides, so that peptides can be determined spectrophotometrically in the presence of amino acids (C-10). It is not yet quite clear whether the high absorbancy of peptides in the near ultraviolet is caused only by the intrinsic amide absorption (C-22,25), which may involve an $n$-$\pi^*$ transition, i.e., excitation of a nonbonding lone electron pair of the nitrogen atom (C-27). Some of the lower transmittance of peptides may be due to higher light scattering (C-23). Formation of peptenols —C(O⁻):N— is less probable (C-11).

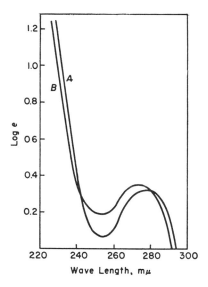

Fig. VII-4. Spectral absorption of (A) serum globulin and (B) its tryptic hydrolyzate (C-4).

The *infrared absorption spectra* (C-12) of proteins give information on the vibrations of certain groups characteristic of the proteins, particularly the CO and NH groups of the peptide bonds which can undergo stretching or deformation out of their planes. Similar information can be obtained by the analysis of the Raman spectra (C-21). The frequency of the atomic or molecular vibrations depends to a certain extent on the environment of the respective group, on hydrogen-bond formation between NH and CO or other groups, and on the folded (helical) or unfolded state of the peptide chain.

It is customary to express the spectral range in wave numbers per centimeter. The wave length in microns is easily obtained when 10,000, i.e., the number of microns per centimeter, is divided by the wave number. For instance, the frequency 3000 cm⁻¹ corresponds to the wave length of $10,000/3,000 = 3.3\mu$. Since water absorbs in the near infrared, the infrared absorption of proteins cannot be determined in aqueous solutions. This difficulty is avoided by the preparation of homogeneous pellets of dry protein or of a mixture of the protein with dry KBr which does not absorb infrared light; if these pellets are prepared under high pressure, they are transparent and give good absorption spectra. Alternatively, solutions of the proteins in $D_2O$ can be analyzed, since $D_2O$ does not absorb in some of the regions in which $H_2O$ is opaque (C-12,13).

By means of these methods, several typical absorption bands have been discovered in proteins. One of them, in the near infrared at about 3300 $cm^{-1}$, is caused by the stretching of the N—H bonds; another band at about 1650 $cm^{-1}$ is due to C=O stretching, and a third band at 1550 $cm^{-1}$ is due to the deformation of the N—H bond (C-14). The wave numbers of these bands undergo changes of 1–2% of their value, depending on the state of the peptide chain. The frequency of the vibrations is, in general, lowered by folding of the peptide chains (C-15) or by the transition from random coils to helices (C-16). In prolylpeptides, which can occur only in a folded conformation, the N—H stretching band is shifted from 3350 to 3330 $cm^{-1}$ (C-17). Hydrogen bonding between CO and NH lowers the frequency of the CO stretching and increases that of the N—H deformation. If proteins with elongated parallel peptide chains are investigated, or if the peptide chains are made parallel by stretching the protein, the C=O and N—H groups absorb incident polarized light more strongly when the plane of polarization is parallel to the direction of stretching, whereas the N—H deformation frequency is absorbed more strongly perpendicular to the direction of stretching; this causes *dichroism* either parallel or perpendicular to the orientation of the peptide chains (C-18; see Chapter IX, C). The dichroism is designated as *parallel* when the optical density parallel to the fiber axis is higher than that perpendicular to the fiber axis; if the absorbancy perpendicular to the fiber axis is higher, the dichroism is called *perpendicular*. The infrared absorption maxima and the dichroism allow us to draw certain conclusions concerning the presence or absence of helical structures in proteins (C-19). Indeed, it has been concluded from such investigations that most of the globular proteins contain only small amounts of $\alpha$-helixes (C-20).

The *scattering of light* by protein solutions and the *depolarization of incident fluorescent light* have been discussed in Chapter V, Sections H and I, since both phenomena are used for the determination of the molecular weight of proteins.

## D. *The Optical Rotation and Rotatory Dispersion* (D-1 to 5,27,29)

Very interesting results have been obtained in recent years by investigating the *optical rotation of polarized light* in protein solutions and by measuring the rotatory dispersion. As is well known, all amino acids with the exception of glycine have an asymmetric $\alpha$-carbon atom. Some of them are dextro- and some levorotatory. All of them, however, belong to the L-series of amino acids and have the same stereochemical configuration of the 4 substituents of the α-carbon atom: the $NH_2$ group, the COOH group, the α-hydrogen atom, and the side chain which is characteristic of each of the amino acids. For a long time it was believed that the optical rotation of polypeptides and proteins would be an additive property and that it would depend merely on the contributions of each of the amino acid residues to the molecular rotation. Comparison of the rotatory power of synthetic peptides prepared from L-amino acids with the rotation of the component amino acids seemed to confirm this view (D-6). However, two well-known phenomena indicated that the optical rotation of proteins depends to a great extent on the conformation of the peptide chain. One of these is the increase in levorotation of proteins when they undergo denaturation. While the specific rotation of native proteins varies from about $-30°$ to $-70°$, the specific rotation of the denatured proteins is usually $-100(\pm 50)°$. A very strong increase in levorotation is also observed when a solution of gelatin ($[\alpha]_D$ = about $-100°$) is cooled down to form a gel ($[\alpha]_D$ = about $-300°$). This reaction is reversible and can be produced repeatedly. Since neither gelification nor mild denaturation involve changes in the configuration of the 4 substituents at the α-carbon atom of the amino acid residues, these changes in rotation must be attributed to conformational changes of the peptide chains (D-7).

It is customary to express optical rotation by the specific rotation which indicates the angle of rotation of polarized light in 1 dm (10 cm) length of a 100% solution of the analyzed material. The wave length of the incident light is indicated by a subscript. The subscript $D$ is an old symbol for the wave length of the yellow light of a sodium lamp, $\lambda$ = 5895.9 and 5889.9 Å. If the angle of rotation is $\alpha$, the depth of the solution $b$ decimeter, and the concentration of the optically active substance $c$ g in 100 ml of the solution, the specific rotation is defined by: $[\alpha]$ = $100/bc$. The temperature is indicated by a superscript. It is clear from the definition of $[\alpha]$ that the specific rotation is that of a solution containing 100 g in 100 ml, a condition which cannot be met experimentally when solid substances like proteins or amino acids are investigated. Although it is assumed that $\alpha$ is proportional to $c$, the value of $[\alpha]$ depends to a certain extent on the concentration. The deviations from linearity are small and need not concern us here.

The dependence of the optical rotation of polarized light on the wave length of the incident light becomes very striking when the wave length of the light approaches the absorption band of the investigated material. The specific rotation then increases to very high positive or negative values and changes its sign when the wave length of the absorption band is passed (Fig. VII-5). The dependence of the rotation on the wave length is designated as *rotatory dispersion* and the increase in rotation and change of its sign near the absorption band as the Cotton effect. If the range of measurement is at wave lengths that are longer than those of the absorption bands of proteins in the ultraviolet, the rotatory dispersion can be expressed by the equation

$$[\alpha]_\lambda = \frac{|K|}{{}_i\lambda^2 - \lambda_0^2}$$  (VII-1)

which is based on a more general equation for rotatory dispersion (Drude dispersion equation). As will be shown below, the rotatory dispersion of

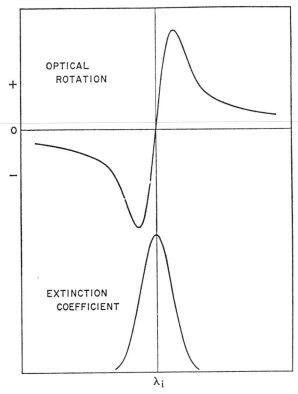

Fig. VII-5. Cotton effect. The maximum of the absorption band (shown below) is at $\lambda_i$ (D-32). Optical rotation (shown above) plotted versus wave length.

proteins sometimes cannot be expressed by the simple single-term equation (VII-1), but follows the two-term equation:

$$[\alpha]_\lambda = \frac{K_1}{\lambda^2 - \lambda_1^2} + \frac{K_2}{\lambda^2 - \lambda_2^2}$$  (VII-2)

We then speak of complex dispersion. In both equations, $[\alpha]$ signifies the specific optical rotation, $\lambda$ the wave length of the incident light; $K$, $K_1$, $K_2$, $\lambda_0$, $\lambda_1$, and $\lambda_2$ are constants. It can be seen from Eq. (VII-1) that $[\alpha]_\lambda$ approaches $+\infty$ as $\lambda$ approaches $\lambda_0$ from higher wave lengths, passes then through zero to $-\infty$, and changes its sign when $\lambda < \lambda_0$. The $\alpha$-helix seems to cause a trough in the rotatory dispersion curve (Fig. VII-5, upper portion) at 204 m$\mu$ and a peak at 190 m$\mu$. They are attributed to a $\pi$-$\pi^*$ transition of the amide bond (D-39).

When it became probable that some proteins contain helices, attempts were made to correlate the conformation of the peptide chains with the optical rotation and dispersion. It was reasonable to assume that the increase in levorotation on denaturation was caused by the uncoiling of the helices and randomization of the peptide chains (D-8). However, other factors are also involved. Very valuable information was obtained from investigations of synthetic polyamino acids.

Polymers of numerous amino acids have been prepared by Katchalsky and his co-workers (D-9). Since all of the natural amino acids have the L-configuration, it was expected that all of them would form either right-handed or left-handed helices. The right-handed helix seems to be more stable and has been found in poly-$\alpha$-L-glutamic acid and some of the other poly-L-amino acids (D-10). However, at high humidities the conformation of poly-$\alpha$-L-glutamate films is of the $\beta$-type; in solution, randomization occurs (D-11). In the $\beta$-conformation the peptide chains are extended and interlinked by interchain bonds; the infrared absorption of the $\beta$-form is similar to that of fibrous proteins (D-12). In contrast to polybenzyl-$\alpha$-L-glutamate, which in the dry state forms a left-handed $\alpha$-helix, dry polybenzyl-$\alpha$-L-aspartic acid forms a right-handed helix (D-10). Poly-L-proline occurs in two different modifications, a right-handed helix with a specific rotation of $+50°$ and a left-handed helix with an $[\alpha]$ of $-540°$ (D-13). In neither of these proline helices are hydrogen bonds involved. The two modifications are interconvertible and show mutarotation in their solutions. They are not $\alpha$-helices.

If the polyamino acids are dissolved in solvents with which they can form hydrogen bonds, these hydrogen bonds compete with those which maintain the helices in their spiral form (Fig. VII-2) and uncoiling of the helices may take place (D-14a). Poly-L-glutamic acid is helical at pH below 5.5, poly-L-lysine at pH above 10.5; at pH 7, when their side chains are charged, the helices are converted into random coils (D-14a). Changes in the rotation are also observed when polyamino acids are transferred from a highly polar solvent to a less polar solvent (D-14a). The rotatory dispersion of the polyamino acids follows the *two-term equation* (VII-2). If a dye is bound to the polyamino acid, the rotatory dispersion is deeply affected. The Cotton effect also appears at longer wave lengths, in the region of maximum absorption of the dye. It must be concluded that the bound dye acquires optical rotatory power although it does not contain the asymmetric center in free solution (D-12). If the length of the polyamino acid molecule decreases, the helices become less stable. This is easily understandable since the terminal amino acids of the helix are bound by a smaller number of H-bonds than the amino acid residues inside the helix. The minimum number of amino acids for the formation of a stable helix seems to be about 15, but depends on the side chains and the solvent (D-14b).

The data given in the preceding paragraph reveal that the simple polymers of L-amino acids may occur in different forms: left-handed or right-handed helixes, extended $\beta$-chains, and randomly folded peptide chains. In a protein, which is formed by a large number of different amino acids, all of these different conformations may occur. In addition, complications by dithio bonds and other factors can be expected. The extremely high levorotation of cystine, $[\alpha]_D$ = about $-300°$, has been mentioned in Chapter III, D. Cleavage of the S—S bond of cystine by oxidation with performic acid, by reduction with thioglycol, or by sulfitolysis results in the conversion of cystine into cysteic acid, cysteine, or cysteine-S-sulfonic acid, respectively. All three reactions are accompanied by a drastic decrease of the levorotation, the two first-mentioned reactions resulting in dextrorotation. Similar changes have been observed in the author's laboratory when proteins with a high cystine content are treated with performic acid, thioglycol, or sulfite (D-15). The decrease in levorotation is of the same order as that expected from the cleavage of S—S bonds. The rotatory power of insulin, lysozyme, ribonuclease, or serum albumin (D-15,16) all of which are rich in cystine, is deeply affected by the cleavage of S—S bonds, whereas the rotation of ovalbumin and of other proteins with low cystine content is not significantly changed. However, the cleavage of the S—S bonds does not account completely for the differences in the rotatory dispersion of the native and the oxidized or sulfitolyzed proteins (D-31). Evidently, the cleavage of S—S bonds is accompanied by conformational changes which are reflected by changes in the rotatory dispersion (D-26). The optical rotation of insulin is also altered by its combination with zinc ions (D-31). Chelation of metal-containing proteins with phenanthroline causes anomalous rotation (D-17). Similar changes are brought about by sodium dodecyl sulfate (D-18,35) or by dyes (D-12).

In spite of all these complicating factors, the optical rotation and rotatory dispersion of the proteins is surprisingly uniform (D-19). Most of the proteins follow the one-term Drude equation (VII-1) (D-1,19,22), which can more conveniently be written in the form (D-32):

$$[m']_\lambda = \frac{a_0\lambda_0^2}{\lambda^2 - \lambda_0^2} \qquad \text{(VII-3)}$$

where $[m']$ is the effective residue rotation which is obtained by multiplication of $[\alpha]$ by the product $(M_0/100) \times [3/(n^2 + 2)]$. $M_0$ is the mean equivalent weight of the amino acid residues and $n$ the refractive index of the medium. In most proteins $M_0$ is approximately 125. If the value of $\lambda_0$ is higher than 240, the protein does not follow Eq. (VII-3), but follows the two-term equation which in a form proposed on theoretical grounds by Moffit (D-27) can be written as (D-32):

$$[m']_\lambda = \frac{a_0\lambda_0^2}{\lambda^2 - \lambda_0^2} + \frac{b_0\lambda_0^4}{(\lambda^2 - \lambda_0^2)^2} \tag{VII-4}$$

The symbols $m'$ and $\lambda$ have here the same meaning as in Eq. (VII-3); $a_0$, $b_0$, and $\lambda_0$ are constants characteristic of the conformation of the peptide chain and its helix content. Since polypeptides which contain the $\alpha$-helix follow the two-term Eq. (VII-4), it would be logical to assume also that those proteins which follow the two-term equation contain a significant amount of $\alpha$-helices. The constant $b_0$ in Eq. (VII-4) has a value of $-600$ to $-650$ in the helical polyamino acids, but decreases to $-200$ or less in randomly coiled peptides. In poly-L-tyrosine, $b_0$ has a positive value (D-30). Since the value of $b_0$ in Eq. (VII-4) is lowered on treatment of proteins with cationic or anionic detergents or with urea, it has been suggested that $b_0$ might be a measure of the helix content of the proteins. This interpretation has been contested. It is more probable that $b_0$ is lowered by processes which weaken the hydrophobic bonds in proteins (see Section E, and D-38).

The levorotation of proteins increases like that of polyamino acids if the polarity of their solvents is increased, for instance when the less polar solvent chloroethanol is replaced by the highly polar trifluoroacetic acid (D-14a). Heating with chloroethanol must be avoided in such experiments since it causes cleavage and esterification (D-20). Low levorotation has been observed when proteins are dissolved in concentrated solutions of lithium bromide (D-21a). This salt seems to interact with the water of hydration of the proteins (D-21b,33). The rotatory dispersion of numerous proteins has been measured by Jirgensons (D-22). The results of these analyses indicate that the proteins can be divided into four groups, three of them following the one-term Drude equation (VII-3), one following the two-term equation (VII-4) (D-1,4).

The latter group of proteins consists chiefly of the muscle proteins myosin, meromyosin, and tropomyosin; their $b_0$ values vary from 300° to 600° (D-23). The proteins which show simple rotatory dispersion belong to three classes. The largest of these consists of many of the crystalline enzymes and has $\lambda_0$ values between 231 and 293 m$\mu$. These proteins may contain both $\alpha$-helices and non-helical regions. The term "random coils" is used to indicate the absence of helical or other periodicity but does not mean complete randomization. Even after denaturation, proteins retain their specificity to a certain extent. Another smaller number of proteins has $\lambda_0$ values between 210 and 230 m$\mu$; they may consist of random coils or $\beta$-chains. The molecules of the third class, with $\lambda_0$ values of less than 210 m$\mu$, probably consist only of $\beta$-chains; this class contains the serum $\gamma$-globulins, macroglobulins, myeloma proteins, and Bence-Jones proteins. The Cotton effect at $\lambda = 225$ m$\mu$ is an approximate measure of the $\alpha$-helix content; the trough adjacent to the peak is lost when random coils are formed (D-36).

It can be seen from this survey that the optical rotation data do not yet allow us to draw unambiguous conclusions concerning the chain conformation of proteins. Although an $\alpha$-helix has been found in myoglobin by X-ray diffraction, most of the proteins, in contrast to the synthetic poly-L-amino acids, follow the one-term Drude equation. This does not support the claim that their principal conformation is that of the $\alpha$-helix. Neither in lysozyme (D-24) nor in $\beta$-lactoglobulin (D-25) do helices seem to be present in significant amounts. The low levorotation of some of the globular proteins may be due to the low polarity of their interiors, which according to X-ray diffraction analyses contain very littler water. The increase in levorotation during denaturation could then be caused by the transfer of the unfolding peptide chains from the almost anhydrous state to the polar aqueous medium (D-25,34,37). The earlier assumption that each degree change in specific rotation of a protein corresponds to a change of 1% in the number of residues in the helical conformation is an oversimplification and is certainly not justified. The most important result of the analysis of optical rotation and rotatory dispersion is the recognition of the fact that the optical rotation of proteins is a complex phenomenon which depends not only on the rotation of the amino acid residues but also on the chain conformation, cross-links such as the S—S bonds, hydration (see Section E), and other factors.

Since all experimental data indicate that the polyamino acids form either right- or left-handed helices, but not mixed right- and left-handed helices, it seems that the first turn of a newly formed helix acts as a primer and determines the screw sense of the growing helix (D-28). The difficult problem of the significance of helices in globular proteins will be taken up again after a discussion of the nature of hydrogen bonds, which are responsible for the rigidity of the helix structure.

## E. Cross-Links in Globular Proteins (E-1)

In the preceding sections of this chapter it has been mentioned that the $\alpha$-helices found in some of the globular proteins owe their stability to *hydrogen bonds* between NH and CO groups of adjacent turns of the peptide spirals which form the helix (see Fig. VII-2). The structure of the hydrogen bonds is shown by the electronic formula:

$$: \overset{|}{\underset{|}{N}}-H : \overset{..}{O}=\overset{|}{\underset{|}{C}} \quad \rightleftharpoons \quad : \overset{|{(-)}}{\underset{|}{N}} \overset{(+)}{:} H : \overset{..}{O}=\overset{|}{\underset{|}{C}}$$

Hydrogen bond

in which the unshared electron pairs of the electron octets are represented by double dots, while each of the shared electron pairs is represented by the usual dash. It is evident from the formula that attraction between the proton and the unshared electron pair of the oxygen atom causes a shift of the proton from N to O. Nevertheless, the hydrogen nucleus which forms the hydrogen bond remains nearer to the nitrogen than to the oxygen nucleus. Analogous hydrogen bonds may be formed between phenolic OH groups and the carbonyl oxygen atom of the carboxyl groups (E-2), or between two carboxyl groups (E-3). In the latter case a double H-bond

$$-C \begin{matrix} O \cdots HO \\ \\ OH \cdots O \end{matrix} C-$$

is formed (E-4). In all hydrogen bonds the H atom forms a bridge between 2 atoms which have unshared electron pairs. The formation of a hydrogen bond causes shortening of the distance between the 2 adjacent atoms. Thus the distance between the 2 oxygen atoms in the bond —OH · · · O= is reduced from 3.0 to 2.5 Å. Estimates of the bond energy of hydrogen bonds in dry proteins vary from 2 to 10 kcal/mole. In aqueous solutions, the water molecules compete with the proton donors and acceptors and form hydrogen bonds according to reaction (VII-5):

$$-C=O \cdots H-N \diagup + 2 H_2O \rightleftharpoons -CO \cdots HOH + \begin{matrix} H \\ O \\ H \end{matrix} \cdots H-N \diagup$$

$$\text{(VII-5)}$$

Consequently, a large portion of the energy required for the cleavage of the H-bonds is liberated when H-bonds are formed between water molecules and the nucleophilic residue. The over-all energy requirement for the cleavage of intramolecular H-bonds in proteins in their aqueous solution is therefore much lower than assumed earlier and may even be negative. Estimates of $\Delta F$ vary from $-0.095$ kcal/mole (E-5) to $+1.3$ (E-6) or $+3.5$ kcal/mole (E-7), depending on the nature of the hydrogen bond formed.

In aqueous solutions the hydrogen atoms of OH, $NH_2$, or NH groups are readily exchanged with the hydrogen atoms of water molecules. If $D_2O$ is used as a solvent, the rate and extent of the *hydrogen-deuterium exchange* (HD-exchange) can be measured. Experiments of this type show that the HD-exchange of acids, alcohols, or amines is almost instantaneous and too fast for accurate measurements. In proteins, however, only part of the O- or N-bound hydrogen atoms are rapidly exchanged, whereas the exchange of other O- or N-linked H-atoms may require several hours for completion (A-3; E-8).

Thus 24 of the labile H-atoms of whale myoglobin are only slowly exchanged at neutrality, but are exchanged rapidly at pH 3.6 (E-9). Similarly, only 175 of the O-, N-, or S-bound H-atoms of ribonuclease are exchanged rapidly at 0°C. Of the remaining O-, N-, and S-bound H-atoms, 50 are exchanged slowly and 20 not at all (E-10). Since the HD-exchange of poly-α-glutamic acid proceeds much faster in the random-chain conformation of the molecule than in the form which is considered as α-helical, the slow HD-exchange has been attributed to the H-bonds of the helices (E-11). For the same reason the slow HD-exchange of ribonuclease in concentrated solutions of LiBr has been ascribed to tighter folding of the peptide chains in this solvent (E-12). However, alanyl-alanine and the tripeptide alanyl-alanyl-alanine show the same slow HD-exchange as poly-DL-alanine, in which the exchange of 33 of the labile H-atoms takes place with a half-life of 1 hour (E-13).

In view of these contradictory results it is doubtful whether hydrogen exchange is a suitable method for studying the conformation of proteins. Changes in the viscosity, optical rotation, and other physical properties of the proteins have frequently been attributed to changes in the hydrogen bonding and to the conversion of helices into random coils or vice versa. This interpretation may not be justified (E-14 to 16). It indeed appears doubtful whether the low bond energy of hydrogen bonds in aqueous solutions is sufficient to maintain the rigid chain conformation of proteins (E-5,17,32).

Whereas polyamino acids which are formed by a single type of amino acid usually have a low solubility in water, copolymers formed from different amino acids are much more soluble (E-18). This is attributed to the greater ease of H-bond formation between the chains of the homogeneous polymers in which the side chains have identical shape and length. In proteins, H-bond formation between the side chains of amino acid residues should be just as difficult as in the heterogeneous copolymers.

The arguments which have been raised against the significance of intramolecular H-bonds in proteins apply also to the *salt-bridges* which are formed by the mutual attraction of positively and negatively charged groups of the peptide chains. The addition of acids or bases causes rupture of the salt-bridges by the conversion of $COO^-$ into $COOH$ or of $^+NH_3$ into $NH_2$ (see Chapter III, D). However, the expansion of protein molecules at low pH values (E-19) is reversible on neutralization if excessive pH values are avoided during expansion (E-20) and, therefore, can hardly involve drastic changes in the chain conformation.

The absence of water inside the globular protein molecules, proved by X-ray diffraction (see Chapter VII, B), leads to the conclusion that the maintenance of the specific chain conformation must be attributed to attraction between the side chains of the amino acid residues in adjacent peptide chains. This attraction has been attributed to *dipole induction* of polar side chains of certain amino acids, and also to short-range *van der Waals forces* between the nonpolar side chains of the amino acids valine, leucine, isoleucine, and phenylalanine. In aqueous solutions of the proteins these nonpolar side chains are surrounded by water molecules which tend to combine with other water molecules by means of hydrogen bonds. The mutual attraction of water molecules and the tendency of the nonpolar groups to coalesce with groups of the same kind, result in repulsion of the nonpolar groups from the aqueous medium and intensify their affinity for each other. The bonds formed in this manner have been designated as *hydrophobic bonds*. They are, to a great extent, responsible for the maintenance of the unique conformation of native proteins (E-31,32,35). On denaturation the hydrophobic bonds are cleaved by penetration of water molecules into spaces between the unfolded peptide chains. The assumption of the presence of hydrophobic bonds has been supported by changes observed in the ultraviolet absorption spectra of the aromatic amino acids after alkalinization, and also by the denaturing action of urea (see Section H). The hydrophobic bonds between nonpolar groups in protein molecules are of the same type as the bonds formed between the paraffin chains in the micels of soaps or detergents (E-31,35). The bond energy corresponds to $\Delta F$ values varying from $-0.2$ to $-1.5$ kcal/mole (E-34).

In proteins which contain cystine residues, the *dithio bonds* have an important role in the cross-linking of peptide chains (E-21). It has been mentioned earlier (Chapter III, D) that the equilibrium in the reaction (VII-6)

$$2RSH + \tfrac{1}{2}O_2 \rightleftharpoons RSSR + H_2O \qquad (VII-6)$$

is shifted far to the right in neutral and alkaline solutions and that the sulfhydryl groups are stable only in acid solutions. Consequently proteins in the neutral biological medium contain predominantly cystine residues. During the biosynthesis of proteins the situation may be different; the newly formed protein molecules may contain more cysteine residues whose SH groups, during aging and also during the isolation procedures, may later be converted into S—S cross-links (E-22). The number of SH groups in proteins can be determined by amperometric titration with silver, mercury, or copper ions (E-23), or by polarography (E-24). It has been shown in this manner that serum albumin, in addition to 17 S—S bonds, contains 0.5–0.7 SH groups per protein molecule (E-23). This was at first attributed to heterogeneity caused by the presence of molecules with and without SH groups. It seems more probable that the solution contains dimers formed according to reaction (VII-6); one of the S—S bonds would then form an *intermolecular* S—S bridge between the 2 monomer units of serum albumin. *Intramolecular S—S bonds* between 2 separate subunits of a protein molecule have been found in insulin where these bonds form cross-links between the A- and the B-chain of the hormone molecule (see Chapter XIII). In most of the globular protein molecules the S—S bonds are intramolecular bonds which form cross-links between different parts of a single peptide chain. This has been found in serum albumin (E-25a), lysozyme (E-26), serum globulin, ovalbumin, edestin (E-25a), and pepsin (E-27). The molecular weights of these pro-teins remain unchanged when they are reduced by thioglycol or similar reducing reagents. If the S—S bonds were interchain bonds, their cleavage would be accompanied by a decrease in the molecular weight.

Some of the S—S bonds of the examined pro-teins seem to be buried inside the globular molecules and are accessible to the reducing reagents only in the presence of sodium dodecyl sulfate (E-28a), urea, guanidinium chloride (E-28b), or other denaturing agents. A very mild method of cleavage of the S—S bonds is their exposure to sulfite (E-28c) and a weak oxidizing agent such as tetrathionate or iodosobenzoate (see Chapter III, D). As mentioned in Section D of this chapter, the cleavage of the S—S bonds of free or protein-bound cystine is accompanied by a drastic decrease in the levorotation at $\lambda = 589$ m$\mu$ (D-15). Since homocystine ($-S \cdot CH_2CH_2CHNH_2COOH)_2$ does not show this phenomenon (E-29), the anomalous optical rotation of cystine has been attributed to mutual sterical interference of the two halves of the molecule which are attached to the S—S bond and which in cystine cause a distortion of the dihedral angle about the —S—S— linkage as shown in Fig. VII-6 (E-30).

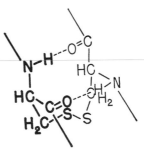

Fig. VII-6. Conforma-tion of cystine residue in peptide chain (E-30).

Although the numerous S—S bonds in the serum albumin molecule are of great importance for the maintenance of the specific chain conforma-tion, they do not prevent flexibility of the peptide chains as shown by the reversible expansion of the molecule on acidification (E-19). Indeed, the Drude constant $\lambda_0$ (Eq. VII-3) of oxidized or reduced serum albumin is the same as that of the native serum albumin (E-33). Evidently, the con-formation of serum albumin does not change significantly when the numer-ous dithio bonds are cleaved. If, however, protein molecules are exposed for considerable periods of time to the action of denaturing agents, interchange reactions of the type:

$$\text{RSSR}' + \text{R}''\text{SSR}^* \rightleftharpoons \text{RSSR}'' + \text{R}'\text{SSR}^*$$

(see Chapter III, D) may take place and cause irreversible denaturation.

## F. Chemical Reactivity of Proteins (F-1a,1b)

Many of the side chains of the aminoacyl residues of globular proteins react easily with specific reagents and can thus be detected or determined quantitatively. Higher values are usually obtained for the denatured than for the native proteins. Apparently, some of the side chains are inaccessible in the tightly folded structure of the native globular proteins but become accessible after unfolding in the denatured state. Comparison of the results obtained with native and denatured proteins can therefore give some information on the accessibility of the analyzed side chains.

The reactivity of the terminal α-amino groups and their substitution by alkylating and acylating reagents have been discussed in Chapter IV, B. The same reagents also combine with the ε-*amino groups of lysine* residues. These groups can therefore be *alkylated* by dinitrofluorobenzene or substituted with phenylisothiocyanate.

Divalent alkylating agents such as 1,5-difluoro-2,4-dinitrobenzene (F-2a), or $p,p'$-difluoro-$m,m'$-dinitrophenylsulfone (F-2a), form bridges between distant lysine groups. Similar cross-linking is accomplished by diisocyanate (F-2b). The highly cross-linked products are more resistant to denaturation than the untreated protein. *Alkylation of the amino groups* is accomplished by a large number of alkyl halides and also by mustard gas, $S(CH_2CH_2Cl)_2$, and its derivatives. In this reaction, thioethers of the structure $R \cdot NH \cdot CH_2CH_2SCH_2—CH_2OH$ or their anhydrides (thiazan derivatives) of the structure $R \cdot N \cdot CH_2CH_2SCH_2CH_2$ are formed (F-3a,b,c). *N*-Acetylation of the amino groups is achieved by treatment of the proteins with ketene (F-4) or with acetic anhydride and sodium acetate (F-35), sulfonation by the sulfonyl chlorides of toluene, naphthalene, or iodobenzene (F-5c). Fluorescent proteins can be produced by acylation with 1-dimethylamino-naphthalene-5-sulfochloride (F-6) or by coupling with the isocyanate (F-7a) or isothiocyanate (F-7b) of fluorescein. The fluorescent proteins are widely used in immunochemical research.

Interesting derivatives in which polyamino acids are bound to the ε-amino groups of proteins are obtained when the proteins are treated with the anhydrides of N-carboxylpolyamino acids. In this manner polyglycyl protein and other polyaminoacyl proteins have been produced (F-8). *N*-Phosphoproteins can be prepared by the action of phosphorus oxychloride on proteins. The alkylating and acylating substituents are not bound exclusively to the α- and ε-amino groups of the proteins. Many of them react also with the imidazole groups of histidine, the guanidino groups of arginine, and the hydroxyl groups of the hydroxyamino acids. The O-acyl and O-alkyl derivatives are more easily hydrolyzed by acids and can thus be differentiated from the analogous N-acyl or *N*-alkyl derivatives. *O*-Methylisourea converts the amino groups of lysine into guanidino groups, thus providing amidinated proteins in whose molecules lysine is replaced by homoarginine (F-9).

On treatment with *formaldehyde* the amino groups of the proteins are converted into N-hydroxymethyl (N-methylol) derivatives (F-10):

$$RNH_2 + HCHO \rightarrow RNH \cdot CH_2OH; + HCHO \rightarrow RN(CH_2OH)_2$$

and not into methyleneimines $RN{:}CH_2$ as assumed earlier. Some of the formaldehyde residues form methylene bridges of the structure $R \cdot NH \cdot CH_2 \cdot NH \cdot R'$ between two adjacent amino groups or between one amino group and the imidazole ring of histidine, the indole ring of tryptophan (F-10), the amide groups of asparagine or glutamine, or the phenol residue of tyrosine (F-11a,b,c). The formation of these methylene cross-links renders the proteins insoluble and converts their solutions into gels which can swell to a limited extent by taking up water (F-11a). In these reactions formaldehyde can be replaced by other aldehydes. It is worthy of note that the ε-amino groups of lysine combine even with glucose at 37°C and pH 8.5 (F-12).

By treatment with nitrous acid in dilute HCl or acetic acid, the proteins are deaminated. The primary amino groups are converted into hydroxyl groups. Since the hydrolytic cleavage of the deaminated proteins does not yield any α-hydroxyglutaric acid nor its lactone (F-13), neither glutamyl nor glutaminyl residues with free amino groups occur in the proteins, contrary to earlier suggestions (F-37).

The *free carboxyl groups* of proteins are readily esterified by treatment of the proteins with ethanol in the presence of small amounts of mineral acids (F-14) or by exposure to other alkylating agents. Since the ester bonds formed are easily hydrolyzed by water, the protein esters are unstable in aqueous solutions. The terminal α-carboxyl groups can be differentiated from the β- and γ-carboxyl groups of aspartic and glutamic acid residues by virtue of the fact that only the former undergo condensation with thiocyanates to yield thiohydantoin derivatives. The mechanism of this reaction has been discussed in Chapter IV, C. Poly-γ-glutamic acid (I) combines with several thiocyanate residues since it contains numerous α-carboxyl groups whereas poly-α-glutamic acid (II) does not react with thiocyanate (F-15a).

$$
\begin{array}{cc}
COOH & CH_2 \cdot CH_2 \cdot COOH \\
| & | \\
(-NH \cdot CH \cdot CH_2 \cdot CH_2 \cdot CO-)_n & (-NH \cdot CH \cdot CO-)_n \\
(I) & (II)
\end{array}
$$

Since some of the proteins, particularly γ-globulin from blood serum, combine with more than 1 equivalent of thiocyanate per mole of protein, it is possible that these proteins contain $\gamma$-glutamyl or $\beta$-aspartyl residues, although their presence has not yet been proved (F-15b). The carboxyl groups of proteins can be reduced to alcohol groups by treatment of the proteins with $LiAlH_4$, lithium aluminum hydride, or by treatment of the esterified proteins with $LiBH_4$, lithium borohydride (F-16a,b,c) (see also Chapter IV, C). Unfortunately, some of the carbonyl groups of the peptide bonds are reduced to methylene groups under the same conditions (F-17a,b).

The hydroxyl groups of the *aliphatic hydroxyamino acids* are easily esterified by treatment of the proteins with ice-cold sulfuric acid, chlorosulfonic acid, or phosphorus pentoxide (F-18a). The biological activity of insulin is hardly affected by this procedure (F-18b). Whereas treatment with acetic anhydride or with arylsulfonylchlorides, as mentioned above, results in acylation not only of hydroxyl groups but also of amino groups, selective acylation of the hydroxyl groups has been reported when trifluoroacetic acid, or acetic anhydride and perchloric acid (F-19), or phenylphosphodichloride (F-20) were used as acylating agents. Selective acylation of a definite serine residue in the catalytically active group of enzyme proteins has been accomplished by DFP (diisopropylfluorophosphate, diisopropylphosphofluoridate) (F-21):

$$R \cdot OH + [(CH_3)_2CH \cdot O]_2PO \cdot F \rightarrow [(CH_3)_2CH \cdot O]_2PO \cdot OR + HF$$

and by other similar nerve poisons. The great importance of this reaction in enzyme chemistry will be discussed in Chapter XII,B. Since the N-terminal serine or threonine residues have an $NH_2$ and an OH group at adjacent carbon atoms, they are easily oxidized by periodate at alkaline reaction and can thus be differentiated from seryl and threonyl residues inside the peptide chains. The latter are resistant to periodate.

The phenolic hydroxyl groups of *tyrosine* give a large number of typical *color reactions* (see Chapter III, D). When native proteins are treated with Folin's reagent (phosphomolybdic acid) some of the hydroxyl groups of tyrosine are oxidized and the reagent is reduced to molybdenum blue. Higher color yields are obtained when the proteins are denatured prior to exposure to the reagent (F-22). *Electrometric titration* and *difference spectrophotometry* in the presence of glycerol or other bulky solvents (F-39) also show that some of the phenolic OH groups are inaccessible "endo" groups in the native state and become reactive and titratable "exo" groups at higher pH values or after the denaturation. The phenol groups of protein-bound tyrosine react readily with concentrated nitric acid to give the intensely colored yellow *xanthoproteins*. These contain 3-mononitro and 3,5-dinitro derivatives of tyrosine (F-24a,b). Similar yellow derivatives are obtained when the proteins are exposed to nitrous acid for deamination. It seems that the yellow products are proteins in which the phenol residues were converted into their nitro, nitroso, and diazo derivatives (F-1a). The phenol group of one of the tyrosine residues of fibrinogen occurs in this protein as phenol sulfate (F-23). The absorption of ultraviolet light by tyrosine and tryptophan has been described in Section C of this chapter. Some of the energy of the incident ultraviolet radiation is reemitted as luminescence. The luminescence spectra extend from 282 to 530 m$\mu$ (F-38).

*Iodination of proteins* yields products in which the iodine atoms are bound in *ortho* position to the phenolic OH group of tyrosine (F-24b). Enzymatic partial hydrolysis of iodinated native ribonuclease has revealed that the 3 tyrosine residues which are not iodinated in the native state are those in the positions number 92, 97, and 115 (F-36). Evidently, these residues are buried inside the native RNase molecule. Iodination with $I^{131}$ is frequently used for the preparation of radioactive proteins (F-25a).

Another reaction of great importance in immunochemistry is the coupling of the tyrosyl residues of proteins with *diazo compounds* in slightly alkaline solution at room temperature or at 0°C. The diazo compounds react with the carbon atoms which are in *ortho* position to the phenolic OH group and convert the tyrosyl residues into azo derivatives. If sulfanilic acid is diazotized and the resulting diazobenzenesulfonic acid coupled to the tyrosine residues, these are substituted by azophenyl-sulfonate groups (F-26):

$$\text{—}\langle\bigcirc\rangle\text{—OH} + HO_3S \cdot C_6H_4 \cdot \overset{+}{N}\equiv N \longrightarrow \text{—}\langle\bigcirc\rangle\overset{\text{—N}=N \cdot C_6H_4 \cdot SO_3H}{\underset{\text{—OH}}{}}$$

The reaction has been used by Landsteiner (F-27) and other investigators (F-25b) for the preparation of numerous antigens. Sugars, amino acids, and other substituents can be coupled to proteins by converting them first into their nitranilinides, then reducing the nitro group, and finally diazotizing the amino group and coupling the diazo derivative to the protein. The reaction is not specific for tyrosine residues since the diazo derivatives are also bound to the imidazole rings of histidine and to the $\epsilon$-amino groups of lysyl residues (F-28a,b,c). The *azo proteins*, like other azo derivatives of phenols, are red or brown azo dyes. The absorption maxima of the lysine derivatives at 363 m$\mu$ are different from those of the tyrosine or histidine derivatives (F-28a) at 330 and 380 m$\mu$, respectively. The azo proteins act as indicators; on acidification their color changes to light yellow. If isotopically labeled amines such as $S^{35}$-sulfanilic acid or $C^{14}$-anthranilic acid are diazotized and coupled to proteins, radioactive proteins are obtained (F-25a).

The indole rings of *tryptophan* are responsible for the color reactions of proteins with aldehydes and with Folin's reagent. As mentioned earlier, the peptide bonds formed by the carboxyl groups of tryptophan residues are split selectively by *N*-bromosuccinimide (Chapter III, D-60).

The imidazole rings of *histidine* react with diazo compounds like the phenol rings of tyrosine (see above). The histidyl residues of proteins are destroyed by photo-oxidation in the presence of methylene blue (Chapter III, D-54).

The guanidino groups of *arginine* residues give an intensely red color with sodium hypochloride and α-naphthol or 8-hydroxyquinoline (Chapter III, D-49). Nitric acid converts arginine into nitroarginine (F-29):

$$R \cdot NH \cdot C(NH)NH_2 + HNO_3 \rightarrow R \cdot NH \cdot C(NH) \cdot NHNO_2$$

The reactivity of the *dithio groups* of proteins has been discussed extensively in Section E of this chapter. All of these groups seem to form cross-links between peptide chains. If some of these groups were bound by only one-half of the cystine residues and had free amino and carboxyl groups in the other half, oxidation with performic acid would split the S—S bond and yield free cysteic acid. In none of a number of examined proteins was free cysteic acid formed by this treatment (Section E, E-30). The *sulfhydryl groups* of cysteine residues can be oxidized to dithio groups according to reaction (VII-6). In some of the proteins a yellow intermediate has been observed in this reaction if iodine is used as the oxidizing reagent. The yellow product has been identified as a sulfenyl iodide (F-30; see Chapter III, D). Sulfhydryl groups also react with alkylating reagents, particularly with monoiodoacetic acid or its amide, or with 2-bromoethylamine (see Chapter III, D). In contrast to cysteine, the reaction products are not destroyed by the customary hydrolysis with concentrated hydrochloric acid. Inhibition of biologically active proteins by alkyl halides is usually considered an indication that the biological activity is connected with the presence of sulfhydryl groups (F-31). Other reagents frequently used for the same purpose are PCMB (*p*-chloromercuribenzoate) and other mercuric compounds (F-32).

An evaluation of the different substitution reactions discussed in the preceding paragraphs leads to the conclusion that some of the side chains of the protein-bound amino acids are reactive, whereas others react only after denaturation of the protein. It is reasonable to assume that the former are *endo groups*, buried in the interior of the globular protein molecules, the latter *exo groups*, present on or near the surface of the molecules. Clearly, we cannot draw a sharp borderline between these two types of groups since the reactivity of a group depends not only on its position in the protein molecule but also on the ability of the applied reagent to penetrate between the peptide chains. One can expect that alkylation by reagents with small alkyl groups will proceed more rapidly and to a higher extent than alkylation with large and bulky alkyl groups. The latter may react only with exo groups which project from the surface of the protein molecule.

Attempts have been made to determine the number of easily accessible *acidic and basic exo groups* of proteins by means of large counter ions. The large anion of Congo red, an azo dye with sulfonate groups, was used to determine the number of cationic exo groups (F-33a); the protamine salmine was used for the determination of acidic exo groups (F-33b). Denaturation increases the number of both acidic and basic exo groups of native serum albumin and other proteins.

The reactivity of the different side chains depends not only on the steric factors mentioned in the preceding paragraph but also on their environment inside the protein globule. This is demonstrated by the anomalous pK values of carboxyl, amino, and other titratable groups of some of the amino acid side chains inside the globular protein molecules (see Chapter VI, B), and also by the resistance of some proteins or peptides to carboxypeptidase or amino peptidases (F-34) in spite of the presence of terminal COOH or $NH_2$ groups.

## G. The Problem of Branched or Cyclic Peptide Chains

As mentioned in Section E of this chapter, cystine residues can form cross-links between peptide chains and thus can give rise to branching or cyclization. Both phenomena occur in the molecules of the hormones oxytocin and vasopressin which contain a pentapeptide ring formed by 1 cystine residue and 4 other amino acids (G-1; see Chapter XIII, D).

True ramifications of the peptide chains, which have been found in some of the toxic bacterial peptides (see Chapter XIV), are more stable than the S—S bonds. In these toxins, points of ramification are formed by aspartyl and lysyl residues. It would not be surprising to find in proteins, a number of γ-glutamyl residues, $—NH \cdot CH(COOH)CH_2 \cdot CH_2 \cdot CO—$, since these occur in glutathione, in folic acid conjugates, and in the capsular poly-γ-glutamic acid of *Bacillus anthracis* and other bacteria (G-2). Although the presence of γ-glutamyl residues in proteins has not yet been proved directly, their presence is suggested by the ability of some of the typical proteins to undergo condensation with thiocyanate (G-3). Another indication for γ-glutamyl bonds is the formation of α,γ-diaminobutyric acid when dinitrophenylhydroxamates of gelatin in alkaline solutions undergo a Lossen rearrangement to isocyanates and are then hydrolyzed (G-4). However, the number of such γ-glutamyl residues is very small (G-5). A β-aspartyl residue seems to occur in the yellow enzyme (G-6).

True points of ramification require the presence of trifunctional aminoacyl residues. Branching of this type seems to occur in collagen by means of lysyl residues. In these, both amino groups form peptide bonds with adjacent amino acids (G-7). Lysyl residues in which the ε-amino group is methylated have been discovered in the flagellar protein of *Proteus* and *Salmonella typhimurium* (G-8).

In evaluating the role of the aminodicarboxylic acids and of lysine as points of ramification, it is important to exclude the artificial formation of these unusual bonds during the isolation procedures. It is well known that glutamine in aqueous solution at neutral reaction is slowly converted to the cyclic pyrrolidonecarboxylic acid if the temperature is raised above 40°C, and that this conversion occurs more rapidly at higher temperatures (Chapter III, D). Glutamic acid undergoes the same cyclization at higher temperatures in acid solutions. It is not surprising, therefore, that pyrrole derivatives are obtained when proteins are exposed to dehydrating agents (G-9). Cyclic anhydrides are also formed from amino acid esters in anhydrous media, for instance, in ethanol. These anhydrides contain the piperazine ring and are designated as diketopiperazines. Their formation is shown below:

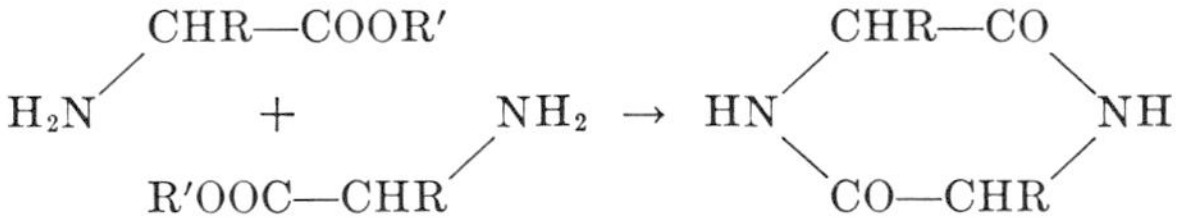

$$\text{H}_2\text{N} \diagup \begin{matrix} \text{CHR—COOR'} \\[4pt] + \\[4pt] \text{R'OOC—CHR} \end{matrix} \diagdown \text{NH}_2 \rightarrow \text{HN} \diagup \begin{matrix} \text{CHR—CO} \\[4pt] \\[4pt] \text{CO—CHR} \end{matrix} \diagdown \text{NH}$$

Several authors have claimed that diketopiperazines found in protein hydrolyzates occurred as such in the protein molecules and were not artifacts (G-10,11). It is difficult to reconcile the presence of diketopiperazines in proteins with the fact that proteins are completely hydrolyzed by proteolytic enzymes without the formation of diketopiperazines (G-12,18). For the same reason the occurrence of oxazole, thiazole, or azlactone rings in proteins is extremely improbable.

The X-ray diagrams of protein crystals have been interpreted to indicate the presence of trihydroxytriazine rings formed by the condensation of 3 enolized peptide linkages (G-13):

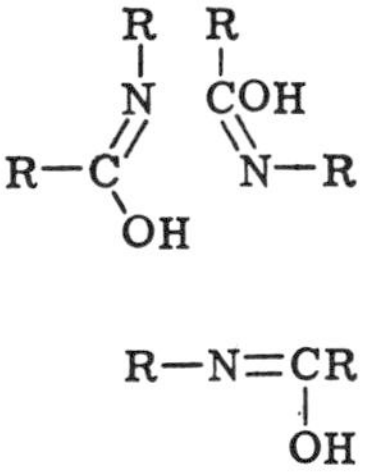

Enolized peptides 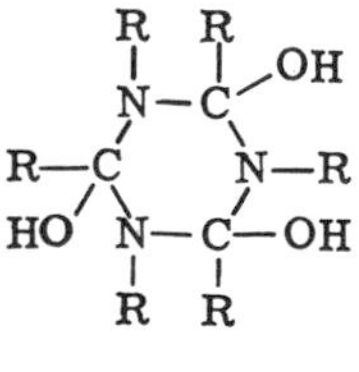        Trihydroxytriazine (cyclol)

This interpretation has been criticized by the X-ray analyst (G-14). The assumption of trihydroxytriazine rings is invalidated by the fact that proteins do not possess the large number of hydroxyl groups required for the postulated structure (G-15). Moreover, the lattice of polymeric trihydroxytriazine rings is so restricted in size that it could not contain amino acids other than glycine and alanine; there would be no space for the side chains of the other amino acids (G-16). Although similar ring systems occur in ergot alkaloids these are not hydrolyzed by proteolytic enzymes. For all these reasons the occurrence of the postulated ring systems in proteins is extremely improbable.

Some proteins reveal no N-terminal amino acid when they are treated with dinitrofluorobenzene or with phenylisothiocyanate. It was first suspected that the peptide chains of these proteins form large ring systems similar to those found in numerous bacterial toxins (see Chapter XIV). In the meantime it has been found that the nonreactivity of the N-terminal amino acid in some of these proteins is due to the presence of N-acetyl groups (G-17). Similarly, the absence of a C-terminal α-carboxyl group in oxytocin and vasopressin is due to amidation of the C-terminal glycine residue (see Chapter XIII, D). The absence of terminal α-amino or α-carboxyl groups in some of the other proteins is also caused by substitution of these groups by acyl or amide groups, respectively.

## H. Denaturation of Proteins (H-1,2)

The irreversible coagulation of egg white on heating is a phenomenon familar to all. Similar changes can be brought about not only by other physical means, such as vigorous shaking or stirring, and irradiation with ultraviolet light or ultrasonic waves, but also by the action of acids, bases, organic solvents, salts of heavy metals, by urea, guanidine, salicylates, detergents, and other compounds. In all these reactions the proteins lose their original solubility; in most instances they become insoluble at their isoelectric range. Collagen, however, becomes soluble if heated with water. We call the changed protein "denatured" in contrast to the original "native" protein. Denaturation is frequently accompanied by a loss of the biological activity of the protein; enzymes lose their catalytic activity, hormones their physiological action, and antibodies their ability to combine with antigens.

It was first proposed that heat causes a dehydration of the protein molecule or an establishment of peptide linkages between some of the free amino and carboxyl groups. In addition, the reverse reaction, the cleavage of peptide bonds by heat, was also considered possible. Dilatometric measurements show, however, that denaturation by heat is not accompained by any noticeable change in volume (H-3,4). The hydration of denatured proteins in humid air is only slightly lower than that of native proteins; their water-binding power is of the same order of magnitude (H-5).

The first reasonable *theory of denaturation* was advanced by Wu (H-6). In this theory it was proposed that denaturation consists of a rearrangement of the peptide chains in the protein molecule due to rupture by the denaturing agent of the weak bonds which hold these chains together. Applying these concepts of Wu to the picture outlined in the preceding sections we may safely say that *denaturation consists of an alteration of the chain conformation.* The closely folded peptide chains are unfolded and/or refolded (H-7). The particular mode of denaturation will determine whether the disrupted peptide chains remain in the unfolded state, whether they are refolded to give the original specific pattern, or whether there will result some other pattern different from the original internal structure. Obviously, the extent of denaturation can vary from slight structural changes to complete rearrangement of the peptide chains.

That denaturation is accompanied by unfolding of the peptide chains is indicated by the more intensive *color reactions* given by denatured protein than by the same protein in the native state. Some examples have been given in Section F of this chapter where the chemical reactivity of native proteins was discussed. The higher reactivity of the denatured protein shows that some of the reacting groups are either buried inside the native globular protein or screened off in another manner, e.g., by neighboring groups which repel the molecules of the reagent used or bind them without giving a color reaction.

The higher reactivity of the denatured proteins was first shown for the sulfhydryl groups of cysteine and the dithio groups of cystine (H-8). By the nitroprusside test, by titration with ferricyanide (H-9), iodine, or $p$-Cl-mercuribenzoate (H-10), and by polarography (H-11), more sulfhydryl and disulfide groups are found in the denatured than in the native protein. Similarly, the denatured protein gives more intensive color reactions for tyrosine with phosphomolybdic acid (F-22) and with the diazo reagent (H-12), for arginine with Sakaguchi's reagent (H-13), and combines with larger amounts of iodine (H-14). Whereas in the native $\beta$-lactoglobulin, 12 $\epsilon$-amino groups of lysine react with dinitrofluorobenzene, and 3 in native ovalbumin, in the denatured proteins react 31 and 9 groups, respectively (H-15). Similarly, the number of dinitrophenyl groups bound to histidine increases after denaturation (H-16). The number of Congo red molecules bound per ovalbumin or serum albumin molecule increases after denaturation by 4 and 2, respectively (H-17). The strongly basic protamine salmine combines readily with denatured serum albumin and causes precipitation; it does not react with the native serum albumin (H-18). The resistance of many native proteins (e.g., serum albumin, hemoglobin) to trypsin and other proteolytic enzymes as compared with the rapid proteolytic degradation of the same proteins in the denatured state (H-19) may be explained on a similar basis. The hydrolysis of other native proteins by the same enzymes may be due to the cleavage of traces of denatured protein, so that the equilibrium "native protein $\rightleftharpoons$ denatured protein" is continually disturbed and shifted to the right (H-20).

All these observations support the view that some of the reactive groups of the native protein molecule are inaccessible to the different reagents, but become accessible through the unfolding of the peptide chains. A rearrangement of the peptide chains upon denaturation is also indicated by an increase in flow birefringence (H-21,28) and viscosity (H-3,22). As mentioned earlier (Section D), denaturation is usually accompanied by an increase in levorotation which is correlated with the change in chain conformation. In contrast to the polyamino acids, the denatured proteins more than the native ones, follow the simple Drude equation [Eq. (VII-1)] in their rotatory dispersion. However, the value of $\lambda_0$ is smaller for the denatured than for the native proteins (H-23). Neither will obey the one-term equation near 233 m$\mu$, i.e., at the absorption maximum of peptides when a Cotton effect is observed. Since changes in conformation involve numerous hydrogen bonds, it is not surprising that denaturation is accompanied by a shift of the infrared absorption maximum of the CO band (H-24), and also by a shift of the isoelectric point towards higher pH values (H-25). For the same reasons the rate of denaturation depends, to a high extent, on pH and temperature. The rate of denaturation is low at the isoelectric point of the protein and increases in acid or alkaline solutions (H-26,27).

The complicated specific arrangement of the peptide chains in the native protein molecule can be disturbed by agents of different types. Almost any physical or chemical agent will alter the labile structure of the native protein. *Mineral acids* convert the negative —COO$^-$ groups into —COOH groups, whereas the positively charged ammonium groups remain unchanged. The mutual electrostatic repulsion of these $NH_3^+$ groups causes subsequent unfolding, expansion of the molecules, and changes in the specific rotation (H-29,55). Similarly, treatment of the protein with *alkali* causes unfolding due to the mutual repulsion of the negatively charged groups. If the added acid or alkali is neutralized, the protein is reconverted to its amphoteric state, although some of the original chain conformation may be changed (H-53).

If denaturation is brought about by *heat*, the protein remains in the zwitterionic state. Hydrogen bonds between the peptide chains are cleaved by the thermal motion of the peptide chains, and bonds between hydrophobic groups may "melt." The insolubility of the heat-treated protein is probably caused by the S—S interchange reaction described in Section E of this chapter, and by the resulting formation of new intermolecular S—S bonds (H-30,31). This view is based on the inhibition of coagulation by iodoacetate (H-32) and *p*-CMB (H-22). It is also in agreement with the fact that collagen, which is free of cystine residues, is converted into soluble gelatin after heating. As long as the S—S bonds of the native protein are intact, denaturation seems to be reversible (H-33).

The most interesting denaturing agents are those substances which are neutral and apparently indifferent such as *urea*, some acid amides, guanidinium salts, and detergents. Urea is neither acidic nor basic; it is neither toxic nor surface active. Yet high concentrations of urea, 6 to 8 $M$, denature proteins at room temperature. This has been attributed to the presence in urea of the structure —CO·NH— which enables it to form hydrogen bonds with the peptide linkages and thus to compete with the intrachain hydrogen bonds which maintain the native protein in its specific structure. It is imaginable that urea molecules would elbow in between the closely folded peptide chains and would thus rupture bonds responsible for their mutual attraction (H-34). Recent investigations have revealed, however, the low energy of hydrogen bonds in aqueous solutions (H-52; see also E-5 to 7). It has been suggested, therefore, that urea breaks hydrophobic rather than hydrogen bonds (H-35,52), and that the high concentrations of urea required for denaturation disrupt the hydration lattice of proteins (H-36). Changes in hydration may also explain the fact that denaturation of some proteins by urea takes place more rapidly at 6°C than at higher temperatures (H-37). Whatever the mechanism of the action of urea may be, it also gives rise to S—S interchange and to the formation of new, intermolecular, S—S bonds (H-38). This phase of denaturation is preceded by an increase in the number of detectable sulfhydryl groups (H-39). If the nuclear magnetic resonance spectra of proteins are investigated in the native state and in the presence of urea, the peaks produced by the different types of hydrogen atoms appear much sharper in the denatured state (H-51). This is attributed to the disruption of the structure by urea and to an increase in the configurational freedom of the hydrogen-containing groups.

Several proteins are dissociated by urea into smaller *subunits*. Thus osmometric measurements have shown that hemoglobin (mol. wt. 68,000) and edestin (mol. wt. 212,000) are split into subunits of molecular weights 34,000 and 49,500, respectively (H-40). Serum albumin, globulin, and ovalbumin have the same molecular weights in urea solutions as in water. It is not yet quite clear why some protein molecules dissociate into smaller units in urea solutions whereas others form aggregates held together by intermolecular dithio bonds.

The peptide chains of the closely folded native protein cannot unfold unless *water* flows into the space between the chains. Therefore, dry proteins are much more resistant to heat denaturation than proteins in solution. For the same reason concentrated protein solutions are more stable than dilute solutions. Water molecules are essential for the native structure of proteins because they form hydrogen bonds with carboxyl, amino, and other polar groups (H-41). Urea may interfere with the formation of these bonds. Little attention has been paid, hitherto, to the fact that concentrated solutions of urea always contain small amounts of cyanate owing to the equilibrium:

$$NH_2 \cdot CO \cdot NH_2 \rightleftharpoons NH_4^+ + NCO^-$$

The cyanate ion forms carbamylates with amino or sulfhydryl groups (H-42a) and thus may cause irreversible changes.

Denaturation by vigorous *shaking* or *stirring* is caused by the formation of a foam and consists essentially of surface denaturation in the protein films which extend over the air bubbles constituting the foam (see Section J). Although many proteins remain in the native state when their solutions are frozen and thawed repeatedly, the solubility of lipoproteins (H-42b) and the biological properties of some antibody proteins are affected by such treatment. Ultrasonic waves cause denaturation by mechanical agitation, thermal effects, production of $O_2$ from water, and oxidative degradation, particularly of the aromatic amino acids (H-47).

As mentioned above (Section C), *ultraviolet radiation* is absorbed mainly by the aromatic rings of tyrosine and tryptophan. Actually it has been found that peptide bonds adjacent to the aromatic rings are the ones split by photolysis (H-43). The effective wave lengths are those absorbed by the aromatic amino acids. The quantum yield found in such experiments is very low; it was only 0.026 for the denaturation of ribonuclease by ultraviolet light, and 0.000043 for the denaturation of tobacco mosaic virus (H-44). Since the size of the virus is about 3000 times larger than that of ribonuclease, the "sensitive volume" of both is of the same order of magnitude. Exposure of proteins to X-rays or to $\gamma$-rays results in the transient formation of free radicals which can be detected by electron-spin resonance (H-45). The radiation-induced rupture of the peptide chains occurs predominantly between the $\alpha$-carbon atoms and the side chains of the amino acid residues (H-45), but also in the S—S bonds thus giving rise to S—S interchange reactions (H-46). Although radicals can have a long lifetime in the dry state, they are usually unstable in solutions. It is therefore doubtful whether protein radicals are normal intermediates in metabolic reactions (see H-54, and Chapter XII, H).

Denaturation and coagulation of certain proteins are inhibited by concentrated solutions of glucose and of other sugars (H-48). Heat coagulation of serum albumin is prevented by small amounts of alkali salts of fatty acids (H-49), anionic dyes such as Congo red (F-33a), urea, and certain other polar substances. The action of these substances may be due to their adsorption to the globular protein particles, and to the formation of large hydrophilic complexes in which the protein is coated by the adsorbed molecules and thus prevented to form aggregates with other protein molecules. It may finally be mentioned that denaturation can be suppressed by high pressures (H-50).

## I. Kinetics and Thermodynamics of Denaturation

If drastic methods of denaturation are used and if the molecular weight of the native protein is very high, denaturation is usually an irreversible process. Complete restoration of every minute detail of the original structure of the large protein macromolecule is very improbable. When mild methods of denaturation are used, almost complete *renaturation,* i.e., reversion to the native state, is observed in some of the smaller proteins, for instance, in a trypsin inhibitor from soy beans (I-1) and in chymotrypsinogen (I-2). Reversible denaturation also takes place when an aqueous solution of Bence-Jones protein is heated. The solutions of this protein are transparent at room temperature and remain clear up to 90°C but become turbid upon further heating; after cooling, the turbidity disappears again. Similarly, the denaturation of ribonuclease by sulfhydryl compounds is reversible (see Chapter XII, G). It seems that in all cases of renaturation the chain conformation of the native protein is so stable that almost complete reversion to the native state takes place at low temperatures. When we represent the irreversible denaturation by the symbols: $N \rightarrow D$, the reversible denaturation is represented by the equilibrium $N \rightleftharpoons D$. In the simplest case we have only native and completely denatured molecules, represented by the symbols $N$ and $D$. Denaturation, in this case, would be an "all-or-none" reaction. It is more probable that denaturation takes place stepwise, through many intermediates (I-3,4). In many instances protein denaturation proceeds according to first-order kinetics. However, the order of the reaction sometimes depends on the concentration not only of the protein but also on that of the denaturing agent (I-4,5).

In most instances denaturation is an *irreversible* process. This makes it impossible to determine the thermodynamic data for the forward and the reverse reaction. We can, however, investigate the rate of denaturation at different temperatures and compute the activation energy for this process from the classical Arrhenius equation:

$$k = A e^{-E/RT} \qquad\qquad \text{(VII-7)}$$

where $k$ is the velocity constant, $A$ a constant, $e$ the base of the natural logarithm, $E$ the activation energy, $R$ the gas constant (1.98 cal per mole per degree), and $T$ the absolute temperature. By determining the velocity constants $k_m$ and $k_n$ at two different temperatures, $T_m$ and $T_n$, and dividing $k_m$ by $k_n$, the term $A$ can be eliminated:

$$k_m/k_n = \frac{e^{-E/RT_m}}{e^{-E/RT_n}}$$

Taking the natural logarithms ($2.3 \times \log_{10}$) we obtain:

$$\ln (k_m/k_n) = -E/RT_m - (-E/RT_n) = \frac{E}{R}\left(\frac{1}{T_n} - \frac{1}{T_m}\right) = \frac{E}{R}\left(\frac{T_m - T_n}{T_m \times T_n}\right)$$

and

$$E = R \ln (k_m/k_n) \left(\frac{T_m \times T_n}{T_m - T_n}\right) = 2.3R \log (k_m/k_n) \frac{T_m \times T_n}{T_m - T_n} \quad \text{(VII-8)}$$

In most organic or biological reactions the velocity increases by a factor of 2 to 3 when the temperature is raised by 10°C. At the body temperature, which is about 310°K, this gives an activation energy of

$$E = 4.6 \log 2[(310 \times 320)/10] = 4.6 \times 0.3 \times 9920$$
$$= \text{approximately } 13{,}000 \text{ cal or } 13 \text{ kcal/mole}$$

when the factor is 2, and of about 21 kcal when the factor $k_m/k_n$ is 3. Much higher activation energies have been found for the denaturation of proteins since the velocity of denaturation increases extremely rapidly in certain temperature ranges. The increase at this critical temperature is so steep that it was originally considered as the "coagulation temperature" and as a constant, characteristic of the protein. Later it was found that coagulation takes place slowly at slightly lower temperatures but that an increase by even 1°C increases the rate of denaturation considerably. This leads to very high activation energies, close to 100 kcal per mole. The interpretation of these high values remained problematic for a long time since it was not possible to determine experimentally the entropy change taking place during denaturation nor the free energy of the activation process.

Determinations of these values were carried out for the first time by Kunitz (I-1) who investigated the reversible denaturation of the trypsin inhibitor from soy beans by dilute HCl. In the temperature range of 35–50°C, the solution of this protein contains a mixture of native and denatured protein; the ratio, *native protein:denatured protein*, depends on the temperature, and decreases when the system is heated. The denatured protein differs from the native protein by its insolubility at pH 4.5; since the mutual conversion of native and denatured proteins takes place very slowly at this pH, they can be separated by precipitation.

The energetics of denaturation can be treated quantitatively by applying the principles of *thermodynamics*. According to the second law of thermodynamics, for each reaction we can write the equation:

$$\Delta H = \Delta F + T\Delta S \tag{VII-9}$$

where $\Delta H$ is the heat of reaction, i.e., the difference between the heat content of the system before and after the reaction. If we call the heat content of the native protein $H_N$, and that of the denatured protein $H_D$, the heat of denaturation is $\Delta H = H_D - H_N$. It is customary to represent $\Delta H$ by a positive value when energy is taken up by the reacting system, and to write negative $\Delta H$ values when heat is given off during the reaction. A positive $\Delta H$ value, accordingly, indicates an endothermic reaction, whereas negative $\Delta H$ values represent exothermic reactions. It must be kept in mind, however, that the $\Delta H$ values refer to the entire system including the solvent (water) and that they must not be attributed to the protein only.

*The heat of denaturation, $\Delta H$, can be computed from the equilibrium constants $K_1$ and $K_2$ measured at two different temperatures, $T_1$ and $T_2$.* If the concentrations of native and denatured protein at equilibrium are $[N]$ and $[D]$, respectively, the equilibrium constant of the denaturation is $K = [D]/[N]$. If, at the two temperatures $T_1$ and $T_2$, the equilibrium constants are $K_1$ and $K_2$, $\Delta H$ can be calculated from van't Hoff's equation:

$$\Delta H = R \ln (K_1/K_2) \left( \frac{T_1 \times T_2}{T_1 - T_2} \right) = 2.3R \log (K_1/K_2) \left( \frac{T_1 \times T_2}{T_1 - T_2} \right) \tag{VII-10}$$

Such a procedure is possible only if a reversible equilibrium between native and denatured protein is established. Kunitz (I-1), applying this method to the reversible denaturation of the trypsin inhibitor, found that $\Delta H$ is $+57,000$ cal per mole of protein.

Similar high values of $\Delta H$ were found by direct *calorimetric measurements* of the heat of denaturation produced by acids or bases (I-6). Although this method is simple in principle, it requires extremely sensitive differential calorimeters and suffers from difficulties in the interpretation of the results since addition of the denaturing agent to a protein-free control solution or to a solution of the denatured protein also causes heat changes. The principle of the method is exemplified by a measurement of the heat of the reactions $A$ and $B$ (I-7):

Native protein + KOH → denatured alkali proteinate     (A)
Denatured protein + KOH → denatured alkali proteinate     (B)

If the heat produced by the first reaction is $\Delta H_A$, and that produced by the reaction $B$ is $\Delta H_B$, the difference, $\Delta H_B - \Delta H_A$, is equal to the heat of denaturation. The value found for the denaturation of crystalline pepsin was $+15$ cal per gram nitrogen; this corresponds to $+85,000$ cal per mole pepsin, when a molecular weight of $35,000$ is assumed for pepsin. The heat of denaturation of methemoglobin was of the same order of magnitude (I-7). Investigations carried out at different acidities have shown that $\Delta H$ depends to a great extent on pH. Thus the heat of denaturation of pepsin at 15°C has its maximum at pH 7.2 where it is 24 kcal per mole (I-6). At 35°C $\Delta H$ was about 70 kcal/mole. However, the loss of enzyme activity did not parallel the heat absorption.

The important result of these investigations is that the *denaturation* of proteins is an *endothermic reaction*, and that it is necessary to increase the energy content of the protein in the native state in order to cause denaturation. This is somewhat surprising since one should expect that the oriented labile state of the peptide chains in the native protein corresponds to a higher energy content than the disoriented stable state of the same peptide chains in the denatured protein. In order to understand this discrepancy, it is necessary to discuss the changes of the two other thermodynamic constants, $\Delta F$ and $\Delta S$, indicated in Eq. (VII-9). In analogy to the procedure applied for $\Delta H$, we can write $\Delta F = F_D - F_N$ and $\Delta S = S_D - S_N$, where $\Delta F$ is the "free energy" and $\Delta S$ the entropy of denaturation, respectively.

It is well known that the molecules of each substance are continuously in motion and that the intensity of this motion depends on the temperature. At a given temperature $T$, the energy represented by this motion is equal to $T \times S$. If the total energy content of the system is $H$, the difference $H - TS$ is designated as $F$ and is called the *free energy* of the system. The free energy can be converted into electrical, mechanical, or other forms of energy, whereas the energy represented by $TS$ cannot be used to produce work. It is the energy required for the thermal motion of the molecules at the given temperature. If the energy $TS$ decreases by $T\Delta S$, we speak of a decrease in *entropy*. This occurs, for instance, at the freezing point, when a liquid passes into the solid state. Conversely, energy must be supplied to a crystalline solid in order to convert it into a liquid at the same temperature. Thus, 79 cal are necessary for the conversion of 1 g of ice, at 0°C, into water at the same temperature. Since the free energy, $F$, is the same in both ice and water at 0°C, all the energy is consumed in increasing the entropy of the system.

The change in free energy, $\Delta F$, can be calculated from the equilibrium constant, $K$, according to Eq. (VII-11):

$$\Delta F° = -RT \ln K = -2.3 \, RT \log K \qquad \text{(VII-11)}$$

where $\Delta F°$ is the standard free energy change, i.e., the difference in free energy between a system containing all reactants in molar concentration and the same system after it has been allowed to reach equilibrium.

It is customary to consider the reaction as proceeding from the left to the right; accordingly, $\Delta F°$ for the reaction $N \rightleftharpoons D$ is the free energy for the conversion of native into denatured protein at the standard state, where the concentrations of $N$ and $D$ are equal. If the conversion of $N$ into $D$ brings the system closer to equilibrium, where $[D]/[N] = K$, the constant $K$ must be larger than 1.0; in this case log $K$ is a positive and $\Delta F°$ a negative figure. Conversely, if $K$ is smaller than 1.0, the reaction proceeds in a direction away from the equilibrium state; in this case log $K$ is negative and $\Delta F°$ a positive value. Evidently, only reactions with a negative $\Delta F°$ value can proceed spontaneously, without energy supply. We call them *exergonic* reactions.

Since $\Delta H$, according to Eq. (VII-9) is equal to the sum of $\Delta F$ and $T\Delta S$, we can also write:

$$\Delta F = \Delta H - T\Delta S \qquad (VII\text{-}12)$$

$\Delta F$ will be negative if either $\Delta H$ is negative and $\Delta S$ positive or slightly negative, or if $\Delta H$ is positive and $\Delta S$ has a high positive value. The latter case is of particular significance in connection with the denaturation of proteins. It was shown above that denaturation is an endothermic reaction, requiring a supply of approximately $10^5$ cal per mole protein. On the other hand, we know that denaturation is brought about very easily, even at room temperature, by a variety of agents such as urea or the detergents. We have to assume that $\Delta F$ in these cases is negative because of an increase in entropy and a positive value of the term $T\Delta S$ in Eq. (VII-12). The process of denaturation would then be of the same type as the melting of ice, namely, an energy-consuming endothermic process which takes place spontaneously because of the increase in entropy. If $\Delta F$ and $\Delta H$ are known, $\Delta S$ can be computed from Eq. (VII-12). The complete thermodynamic analysis of the reversible denaturation of the crystalline trypsin inhibitor gave for $\Delta H$ the value $+57,300$ cal per mole. The values for $K$, $\Delta F$ (cal/mole), and $\Delta S$ (cal/mole/degree) (I-1) are listed in the following tabulation.

| Thermodynamic values | Temperature (°C) | | | | | |
|---|---|---|---|---|---|---|
| | 30 | 35 | 40 | 45 | 47 | 50 |
| $K = [D]/[N]$ | 0.010 | 0.042 | 0.220 | 0.870 | 2.03 | 4.35 |
| $\Delta F = -4.58T \log K$ | 2780 | 1920 | 950 | 87.5 | −450 | −950 |
| $\Delta S = (\Delta H - \Delta F)/T$ | 180 | 180 | 180 | 180 | 180 | 180 |

The table shows that native and denatured protein, in this case, are present in equal concentrations ($K = 1$ and $\Delta F = 0$) at a temperature between 45 and 47°C. At lower temperatures the equilibrium is shifted toward the native state, at higher temperatures toward the denatured state. The table also shows that the reaction "native protein → denatured protein" is accompained by an increase in entropy amounting to 180 cal per mole per degree. This large increase in entropy is, apparently, the driving force in denaturation; it renders $\Delta F$ at higher temperatures negative to such an extent that denaturation takes place at an increased rate.

If an attempt is made to interpret these results of thermodynamic analysis in terms of molecular kinetics, the process of denaturation can again be compared with the melting of ice. Upon an increase of the temperature the various cross-linking bonds between peptide chains "melt" so that the released peptide chains acquire a higher thermal motility.

The finding of high values of $\Delta H$, which vary from $10^4$ to $10^5$ kcal per mole in protein denaturation came as a surprise to the first observers because this large amount of energy corresponds to that required for the splitting of strong covalent bonds, and is higher than the heat of reaction observed in many chemical reactions. $\Delta H$ for the formation of a peptide bond or an ester bond is approximately 2000–5000 cal per mole; values of a higher order of magnitude are observed when stable compounds are split into their components; thus, 94,380 cal are required for the cleavage of 1 mole of carbon dioxide into carbon and oxygen, and 115,660 cal for the cleavage of water into hydrogen and oxygen. In order to understand the high values of $\Delta H$ in protein denaturation, one has to bear in mind that the structure of the native protein molecule is due to a large number of cross-links formed by hydrogen bonds, salt bridges, dithio groups, and bonds between hydrophobic groups. A molecule of the molecular weight 25,000, such as chymotrypsinogen, contains approximately 200 amino acid residues. Since the $\Delta H$ value is 143,000 cal per mole at pH 3.0 (I-2), it is approximately 700 cal or 0.7 kcal per 1 g equivalent of amino acid residue. This value is of the same order of magnitude as that computed for the cleavage of a hydrogen bond in an aqueous solution of the protein (see Section E of this chapter; refs. E-5, 6, 7) and is therefore not exceptional.

From the standpoint of kinetics, the equilibrium between native and denatured protein can be considered the over-all result of the forward reaction "$N \rightarrow D$" and the reverse reaction "$D \rightarrow N$." If the velocity constant of the first reaction is $k_1$, and that of the second reaction, $k_2$, the equilibrium is represented by the equation $k_1[N] = k_2[D]$, where $[N]$ and $[D]$ are the concentrations of native and denatured protein, respectively, at equilibrium. The equilibrium constant is $K = k_1/k_2$. The ratio $[D]/[N]$ and the magnitude of $K$ depend on the temperature of the protein solution; at higher temperatures $k_1$ increases greatly, whereas $k_2$ increases only slightly; the result is an increase in the ratio $k_1/k_2$ and a shift of the equilibrium "$N \to D$" to the right side.

The great influence of the temperature on the velocity of slow reactions is ascribed to the fact that only "activated" molecules are able to react. The "activation energy" can be computed from the classical Arrhenius equation (VII-7) as was shown above, or according to the theory of absolute reaction rates (I-8). The latter is based on the assumption that chemical reactions take place in a manner represented diagrammatically in Fig. VII-7, where the energy content of the initial system is $H_i$, that of

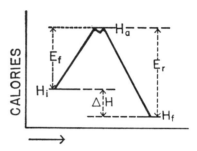

Fig. VII-7. Energy path during an exothermic reaction. $H_i$, $H_a$, $H_f$, are the energy levels of the initial, the activated, and the final state, respectively. $E_f$ and $E_r$ are the activation energies for the forward and the reverse reactions.

the final system $H_f$, and that of the activated state $H_a$. The heat of activation for the forward reaction, $\Delta H_1^{\ddagger}$, is equal to $H_a - H_i$; the heat of activation for the reverse reaction, $\Delta H_2^{\ddagger}$, is $H_a - H_f$. Since $H_a$ is larger than both $H_i$ and $H_f$, the heat of activation will be positive in both cases. In other words, heat must be supplied for both the forward and the reverse reaction. Thus, while we have to heat an ester in order to split it into acid and alcohol, we also have to heat the mixture of acid and alcohol in order to produce an ester. The net heat of the reaction, $\Delta H$, is $H_f - H_i$.

In protein denaturation we are then dealing with the conversion of the native protein into activated protein, $N \rightleftharpoons P^*$, and subsequent conversion of the latter into denatured protein, $P^* \rightleftharpoons D$. If this process is reversible, the velocity of denaturation and renaturation can be measured at different temperatures. The measured velocity is identical with the velocity of the activation processes $N \to P^*$ or $D \to P^*$ since the unstable activated molecule $P^*$ has a very short half-life and is immediately converted into either native or denatured protein. If the velocities of the reactions $N \to D$ and $D \to N$ are measured at two different temperatures, the activation energy of each of these reactions can be calculated from Eq. (VII-7). In this manner the activation energies $E_f$ and $E_r$ for the forward and the reverse reaction (see Fig. VII-7) have been determined. The heat of activation in solutions is $\Delta H^{\ddagger} = E - RT$, where $RT$ is approximately 0.6 kcal per mole. Having computed, in this way, the heat of activation for denaturation, $\Delta H_f^{\ddagger}$, and renaturation, $\Delta H_r^{\ddagger}$, the fundamental equation of thermodynamics, Eq. (VII-9) can be applied to the two activation processes $N \rightarrow P^*$ and $D \rightarrow P^*$. This gives the two equations:

$$\Delta H_f^{\ddagger} = \Delta F_f^{\ddagger} + T\Delta S_f^{\ddagger} \text{ and } \Delta H_r^{\ddagger} = \Delta F_r^{\ddagger} + T\Delta S_r^{\ddagger}$$

where the subscripts $f$ and $r$ indicate forward reaction (denaturation) and reverse reaction (renaturation), and the daggers indicate that the symbols $\Delta H^{\ddagger}$, $\Delta F^{\ddagger}$, and $\Delta S^{\ddagger}$ refer to the activation process and not to the over-all reaction. It is also evident that the thermodynamic values for the over-all reaction $N \rightleftharpoons D$ are the differences between the analogous values for activation, as shown in Fig. VII-7 for the heat of the reaction, $\Delta H$. Hence, $\Delta H = \Delta H_f^{\ddagger} - \Delta H_r^{\ddagger}$; $\Delta F = \Delta F_f^{\ddagger} - \Delta F_r^{\ddagger}$; and $\Delta S = \Delta S_f^{\ddagger} - \Delta S_r^{\ddagger}$. In reactions of the type shown in Fig. VII-7, activation does not occur spontaneously, but requires energy supply in both directions. Accordingly, the values of $\Delta H^{\ddagger}$ and $\Delta F^{\ddagger}$ usually have a positive sign. $\Delta S^{\ddagger}$ can be positive or negative. In the reversible denaturation of proteins we are dealing with an unusual reaction in so far as renaturation does not require any activation. The denatured protein is reconverted to native protein spontaneously, without energy supply. This is reflected in a negative value of $\Delta H_r^{\ddagger}$ as shown by the following data obtained for the trypsin inhibitor from soy beans: $\Delta H_f^{\ddagger} = +55,000$ and $\Delta H_r^{\ddagger} = -1900$ cal per mole, $\Delta S_f^{\ddagger} = +95$ and $\Delta S_r^{\ddagger} = -84$ entropy units. The two values of $\Delta H^{\ddagger}$ were obtained by measuring the rate of denaturation and renaturation at two different temperatures, computing $E$ for each of these according to Eq. (VII-10) and deducting $RT$ in order to convert $E$ into $\Delta H^{\ddagger}$. $\Delta F^{\ddagger}$ was evaluated from Eq. (VII-13) (I-8).

$$\Delta F^{\ddagger} = RT \left( \ln \frac{\kappa}{h} + \ln T - \ln k \right) \qquad \text{(VII-13)}$$

where $\kappa$ is the Boltzmann constant ($1.38 \times 10^{-16}$ erg per degree), $h$ the Planck constant ($6.62 \times 10^{-27}$ erg sec), and $k$ the velocity constant. If these figures are inserted in Eq. (VII-13), and natural logarithms are converted into decadic logarithms, we obtain $\Delta F^{\ddagger} = 4.58T(10.318 + \log T - \log k)$. The entropy of activation, $\Delta S^{\ddagger}$, is then determined from the equation $\Delta S^{\ddagger} = (\Delta H^{\ddagger} - \Delta F^{\ddagger})/T$.

The values of the heat, and the entropy of activation for the denaturation and renaturation of chymotrypsinogen at pH 3.0 are: $\Delta H_f^{\ddagger} = 80.2$ and $\Delta H_r^{\ddagger} = -63.7$ kcal/mole; $\Delta S_f^{\ddagger} = +178$ and $\Delta S_r^{\ddagger} = -258$ entropy units. Hence, $\Delta H$ (for the over-all reaction $N \to D$) is

$$80.2 - (-63.7) = +143.9 \text{ kcal/mole}$$

and $\Delta S = 178 - (-258) = +436$ entropy units. The $\Delta F^{\ddagger}$ values depend on pH and the temperature. As shown below, $\Delta F_f^{\ddagger}$ decreases and $\Delta F_r^{\ddagger}$ increases when the temperature is raised (I-2). $\Delta F$ is obtained as the difference between $\Delta F_f^{\ddagger}$ and $\Delta F_r^{\ddagger}$.

| $T(°C) =$ | 56.7 | 57.4 | 58.2 | 59.2 | 60.1 | |
|---|---|---|---|---|---|---|
| $\Delta F_f^{\ddagger}$ | $= +21.7$ | $+21.5$ | $+21.4$ | $+21.2$ | $+21.1$ | kcal/mole |
| $\Delta F_r^{\ddagger}$ | $= +21.3$ | $+21.3$ | $+21.6$ | $+21.8$ | $+22.1$ | kcal/mole |
| $\Delta F$ | $= + 0.4$ | $+ 0.2$ | $- 0.2$ | $- 0.6$ | $- 1.0$ | kcal/mole |

The change in sign of $\Delta F$ reflects merely the fact that the equilibrium $N \rightleftharpoons D$ is shifted toward the left at low temperatures and toward the right at high temperatures. At a temperature between 57.4 and 58.2°C the product $k_f [N]$ is equal to $k_r [D]$ so that equilibrium is maintained and $\Delta F = 0$. The positive value of $\Delta S_f$ is a measure of the "melting" of the peptide chains on denaturation; the negative value of $\Delta S_r$ is a measure of the "freezing" of the peptide chains on renaturation. These values include also the increased or reduced freedom of the solvent molecules on denaturation and renaturation, respectively.

The high values of $\Delta H$ and $\Delta F$ must not give the impression that the heat or the free energy of denaturation are comparable with the $\Delta H$ values of oxidation or similar organic reactions. Let us assume, for the sake of simplification, that the molecular weight of a protein is 100,000 and that the heat of denaturation is 100,000 cal per mole. Then the difference between the energy content of 1 g of the native protein and 1 g of the denatured protein will be 1 cal. On the other hand, the complete oxidation of 1 g protein *in vivo* furnishes ~4000 cal. Evidently the heat of denaturation is so small that it does not play any role in the energy balances of organisms.

## J. Proteins in Interfaces (J-1 to 3)

Proteins are capillary-active substances. Their concentration at the interfaces between their aqueous solutions on the one hand, and air, organic solvents, or solid phases on the other is higher than in the bulk of the solution. The formation of the interfacial film requires some time which usually does not excede several minutes, but may amount to several hours. This time may be necessary for the migration of the protein molecules into the film and for the changes which their conformation undergoes in the film. Since the reactivity and other properties of molecules in interfacial films are different from those in their solutions, it would be interesting to know more about the state of protein molecules in biological interfaces, for instance at the interface between cell sap and cell membrane. Unfortunately, these systems are extremely complicated and not suitable for quantitative measurements. Most of the investigations on the behavior of proteins in interfaces deal with the air-water interface, a smaller number of analyses with the interface between water and organic solvents such as benzene or xylene. The term "oil-water" interface is frequently used for these systems since it is assumed that results obtained with them may also be valid for the interface between the aqueous and the lipid phase in biological systems.

Soluble proteins such as serum albumin or hemoglobin readily form surface films on water or on salt solutions. The protein is spread from small drops delivered by a capillary pipette. Stable films are obtained at the pH of the isoelectric point. Surprisingly, the area covered by 1 mg of most proteins at the point of minimum compressibility is about the same, close to 0.7 to 0.85 m² (J-4). Since the volume of 1 mg of dry protein is approximately 0.75 mm³, the average height of such a compact, homogeneous protein layer is 9–10 Å. This is much less than the diameter of the globular protein molecules (see Chapter V). It has been concluded, therefore, that the peptide chains of the globular protein molecules unfold in the film and that the film consists of a monomolecular layer of peptide chains, their long axes parallel to the surface of the solution.

The lateral pressure exerted by protein films can be measured by means of a film balance in which the film is compressed by a barrier. In the region of low film pressure, the protein molecules behave like those in a dilute solution. Their osmotic pressure is the same as that of a dilute solution of the same concentration. As shown in Chapter V (Section B) this pressure results from the different activities of water inside and outside the osmometer membrane. In the same way the pressure of the film can be attributed to the different activities of water in the bulk of the solution, $a_1$, and in the protein film, $a_2$. The free energy required for the transfer of 1 mole of water from the solution into the film is $\Delta F = RT \ln (a_1/a_2)$. On the other hand, it will be recalled that the molecules of a solute behave like gas molecules for which the equation $pv = RT$ expresses the dependence of volume ($v$) and pressure ($p$) on the temperature ($T$) and the gas constant ($R$). Since the molecules in a film move only in two dimensions, the volume ($v$) is replaced by the area $A$. The product $FA$ (force × area) of a film consisting of $n$ molecules is $FA = nSF + nRT$, where $S$ is the area per protein molecule, $T$ the absolute temperature, and $R$ the gas constant ($8.31 \times 10^7$ erg per degree per mole). If the molecular weight of the protein does not change during expansion and compression of the film, an almost straight line is obtained when $FA$ is plotted versus $F$ (see Fig. VII-8). The intercept at $F = 0$ is $nRT$ and the slope of the line is equal to

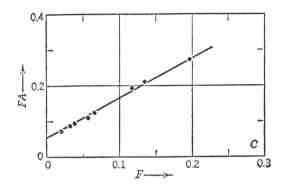

FIG. VII-8. Plot of $FA$ versus $F$ of a film of pepsin spread on 5% ammonium sulfate. Area is expressed in square meters per milligram and pressure in dynes per centimeter. From H. B. Bull, "Physical Biochemistry," p. 229. Wiley, New York, 1951.

$nS$. Hence $n$, the number of protein molecules, is obtained when the intercept is divided by $RT$ (J-1).

Since $F$ is usually expressed in dynes/cm, and the area $A$ in cm², the dimensions of $FA$ are dynes × cm. The area covered by the protein is expressed in m²/mg of protein. At 25°C (which = 298°K) the product $RT$ equals

$$8.31 \times 298 \times 10^7 = 2476 \times 10^7 \text{ erg per mole}$$

Converting gram into milligram, and m² into cm², we obtain

$$RT = 2476 \times 10^7/(10^3 \times 10^4) = 2476$$

Example: If extrapolation of the intercept for $F = 0$ gives $FA = 0.06$, the molecular weight is 2476/0.06 = 41,300.

The method furnishes values of 17,000 and 20,100 for the molecular weights of β-lactoglobulin and zein, respectively (J-1) and 34,400 for pepsin (J-5). The value recorded for β-lactoglobulin indicates dissociation of the molecule into halves when it is spread on the water surface (see Chapter VIII, D). A similar dissociation of hemoglobin into halves has been observed at the water-xylene interface when the pH was lowered to 3 (J-6). Edestin and glycinin had, in the air-water interface, the molecular weight 240,000, ribonuclease the molecular weight 18,000; these values are of the same order as those obtained in their homogenous solutions (J-7).

Although the results of the measurements of the interfacial pressure exerted by films allow us to calculate the number of osmotically active protein molecules in the interface and to determine their molecular weights they do not give any information on the state of the protein molecules in the interface. It is particularly doubtful whether the proteins in the interface are indeed fully unfolded. This would involve denaturation and loss of the biological activity. Since many proteins maintain their serological specificity and biological activity in the interfacial films we must conclude that they either do not unfold completely or that their denaturation is reversible (J-8).

Calculations of film heights are based on the simplifying assumption that the films are homogeneous and that their protein molecules occupy the same volume as in wet crystals or in solution. It is quite possible, however, that the protein molecules expand in the film and that their volume is larger than in wet crystals. There may also be holes in the films; or the film may consist of clusters of protein molecules. Such inhomogeneities can be expected, particularly in films of those proteins which easily undergo denaturation and form insoluble aggregates in the film. For all these reasons, statements on the height of the films are uncertain. The only measurable parameter is the interfacial pressure of the film which causes a lowering of the interfacial tension of the solvent, and can be determined by measuring the interfacial tension before and after spreading of the protein film.

In the low pressure region of less than 1 dyne/cm, the protein molecules are diluted in the interfacial boundary and are separated from each other by protein-free spaces. Since they can freely move in both directions of the plane of the boundary, they can also exert the same osmotic pressure in these two dimensions as molecules of a homogenous dilute protein solution. In the high-pressure region, the protein molecules form condensed films. The height of these films varies from 10 to 50 Å or more. Such films contain globular protein molecules (J-9). Globular protein molecules are also present in the $p$-xylene-water interface (J-6), and in films formed on the surface of glass beads (J-10), or on layers of stearate or cholesterol (J-11). In the intermediate range between the high and the low pressure region, the films consist of a mixture of globular and unfolded protein molecules (J-12). Trypsin is enzymatically active in its condensed films but inactive in the expanded films (J-12). In the benzene-water interface or in other similar "oil-water" boundaries, the proteins occupy a larger area than in the air-water interface (J-13). This is attributed to the formation of mixed protein-oil films. Mixed protein-alcohol films are formed when protein solutions containing ethanol or propanol are spread on the water-air interface (J-14).

If condensed films are loaded with protein, they reach a maximum pressure which may be the pressure exerted by the condensed film of globular molecules (J-15). At this point, films of serum albumin at the xylene-water interface have a height of at least 30 Å, those of the bovine γ-globulin a height of at least 40 Å (J-6). The maximum interfacial pressure of bovine hemoglobin in its films is approximately 13 dynes/cm; that of the bovine globin is about twice as high, indicating the cleavage of the hemoglobin molecules into halves (J-6,15) after the loss of the heme moiety (see Chapter XI, E,F,G). This reaction is reversible as shown by the decrease of the interfacial pressure of globin after the addition of hemin (J-16).

Our knowledge about the factors causing film formation of proteins is still very limited. It seems that the hydrophilic ionic groups of the proteins are directed toward the aqueous solution, while the nonpolar groups cover the layer of peptide chains and are directed toward the air (J-17). Of particular biological interest are the properties of mixed lipid-protein films since similar films may be present in the organism. Such protein-lipid films are permeable for water-soluble and also for fat-soluble substances; the permeation of the water-soluble molecules occurs, apparently, through the protein patches of the film, the permeation of fat-soluble substances through the lipid phases. Mixed molecules diffuse, in all probability, along the water-oil interface, the polar head in the protein phase, the nonpolar paraffin group in the lipid phase (J-18). Penetration has also been observed when a protein film is spread on the surface of a solution of another protein (J-19).

Since proteins carry positively and negatively charged groups, their films give rise to the formation of interfacial potentials. The origin of these potentials was for a long time the subject of a controversy. Since very little is known on relations between these potentials and the structure and conformation of proteins, they will not be discussed here. The problems of interfacial potential and film viscosity are treated in books on surface chemistry (J-20).

## REFERENCES

### Section A

**A-1.** W. Kuhn and H. Kuhn, *Helv. Chim. Acta* 28: 1539(1945); *J. Colloid Sci.* 3: 11(1948). **A-2.** D. B. Wetlaufer, *Nature* 190: 1113(1961). **A-3.** H. A. Scheraga, "Protein Structure," Academic Press, New York, 1961. **A-4.** F. H. White, *JBC* 235: 383(1960). **A-5.** S. A. Berson and R. S. Yalow, *Nature* 191: 1392(1961). **A-6.** R. Lumry and H. Eyring, *J. Phys. Chem.* 58: 110(1954). **A-7.** C. Tanford, "Physical Chemistry of Macromolecules." Wiley, New York, 1961.

### Section B

**B-1.** J. C. Kendrew and M. F. Perutz, *Ann. Rev. Biochem.* 26: 327(1957). **B-2.** B. W. Low, *J. Polymer Sci.* 99: 153(1961); *in* "Proteins" (H. Neurath and K. Bailey, eds.), Vol. 1A, p. 235. Academic Press, New York, 1953. **B-3.** J. T. Edsall and J. Wyman,

"Biophysical Chemistry," Vol. 1, p. 47. Academic Press, New York, 1958.    **B-4.** M. F. Perutz *et al., Proc. Roy. Soc.* A225: 287(1954); A265: 15(1961); *Nature* 185: 416 (1960).    **B-5.** F. C. Crick and J. C. Kendrew, *Adv. in Protein Chem.* 12: 134(1957). **B-6.** J. C. Kendrew *et al., Nature* 185: 422(1960).    **B-7.** W. T. Astbury *et al., Proc. Roy. Soc.* A230: 72; A232: 333(1935).    **B-8.** L. Pauling and R. B. Corey, *Proc. Natl. Acad. Sci. U.S.* 37: 235, 241, 282(1951); R. B. Corey and J. Donohue, *JACS* 72: 2899(1950).    **B-9.** L. Pauling, R. B. Corey, and H. R. Bransson, *Proc. Natl. Acad. Sci. U.S.* 37: 205(1951).    **B-10.** M. F. Perutz *et al., Nature* 167: 1053, 1054(1951). **B-11.** S. Mizushima and T. Shimanouchi, *Adv. in Enzyme Chem.* 23: 1(1961).    **B12.** D. T. Warner, *Nature* 190: 120(1961).    **B-13.** A. Rich and D. W. Green, *Ann. Rev. Biochem.* 30: 93(1961).

### Section C

**C-1.** P. Doty and E. P. Gaiduschek, *in* "Proteins" (H. Neurath and K. Bailey, eds.), Vol. 1A, 433. Academic Press, New York, 1953.    **C-2.** G. E. Perlmann and L. G. Longsworth, *JACS* 70: 2719(1948).    **C-3.** S. H. Armstrong *et al., JACS* 69: 1747 (1947).    **C-4.** F. Haurowitz and T. Astrup, *Nature* 143: 118(1939).    **C-5.** E. R. Holiday, *BJ* 30: 1795(1936); G. H. Beaven and E. R. Holiday, *Adv. in Protein Chem.* 7: 320(1952).    **C-6.** J. A. Crammer and A. Neuberger, *BJ* 37: 302(1943).    **C-7.** D. B. Wetlaufer *et al., JBC* 233: 1421(1958).    **C-8.** E. J. Williams and J. F. Foster, *JACS* 81: 865(1959).    **C-9.** A. R. Goldfarb *et al., JBC* 193: 397(1951); L. J. Saidel and H. Lieberman, *ABB* 76: 401(1958).    **C-10.** M. A. Mitz and R. J. Schlueter, *BBA* 27: 168(1958); A. Schmitt and G. Siebert, *Biochem. Z.* 334: 96(1961).    **C-11.** I. M. Klotz *et al., JACS* 71: 1615(1949); E. Schauenstein and G. Perko, *Z. Elektrochem.* 57: 529 (1953).    **C-12.** G. B. B. M. Sutherland, *Adv. in Protein Chem.* 7: 291(1952); G. Ehrlich and G. B. B. M. Sutherland, *Nature* 172: 671(1953).    **C-13.** E. R. Blout and H. Lenormant, *J. Opt. Soc. Am.* 43: 1093(1954).    **C-14.** S. E. Darmon and G. B. B. M. Sutherland, *JACS* 69: 2074(1947).    **C-15.** S. Mizushima *et al., JACS* 73: 1330(1951). **C-16.** T. Miyazawa and E. R. Blout, *JACS* 83: 712(1960).    **C-17.** S. Mizushima *et al., Nature* 169: 1058(1952).    **C-18.** A. Elliott, *Proc. Roy. Soc.* A211: 490(1952); R. D. B. Fraser and W. C. Price, *Nature* 170: 491(1952).    **C-19.** E. J. Ambrose and A. Elliott, *Proc. Roy. Soc.* A206: 206(1951); S. Krimm, *J. Phys. Chem.* 23: 1371(1955); F. M. Perutz *et al., Discussions Faraday Soc.* 9: 423(1951).    **C-20.** A. Elliott, *Proc. Intern. Congr. Biochem., 3rd Congr., Brussels (1955)* p. 106(1956); S. J. Leach, *Revs. Pure and Appl. Chem. (Australia)* 9: 33, pp. 37–38 (1959).    **C-21.** D. Garfinkel and J. T. Edsall, *JACS* 80: 3818(1958).    **C-22.** A. N. Glazer and E. L. Smith, *JBC* 236: 2942(1961).    **C-23.** J. F. Wootton and G. P. Hess, *JACS* 83: 4234(1961).    **C-24.** K. Rosenheck and P. Doty, *Proc. Natl. Acad. Sci. U.S.* 47: 1775(1961).    **C-25.** D. L. Peterson and W. T. Simpson, *JACS* 79: 2375(1957).    **C-26.** T. L. McMeekin *et al., Biochem. Biophys. Research Communs.* 7: 151(1962).    **C-27.** H. Holzwarth *et al., JACS* 84: 3194(1962); A. N. Glazer and K. Rosenheck, *JBC* 237: 3674(1962).

### Section D

**D-1.** B. Jirgensons, *Tetrahedron* 13: 166(1961).    **D-2.** S. J. Leach, *Revs. Pure and Appl. Chem. (Australia)* 9: 33(1959).    **D-3.** G. E. Perlmann and R. Diringer, *Ann. Rev. Biochem.* 29: 151(1960).    **D-4.** J. A. Schellman and C. G. Schellman, *J. Polymer Sci.* 49: 129(1961).    **D-5.** W. Kauzmann, *Ann. Rev. Phys. Chem.* 8: 413(1957).    **D-6.** B. F. Erlanger and E. Brand, *JACS* 73: 4025, 4027(1951).    **D-7.** C. Robinson and M. J. Bott, *Nature* 168: 325(1951); W. Kauzmann *et al., JACS* 75: 5139(1953).    **D-8.** E. R. Blout *et al., JACS* 79: 749(1957); J. T. Yang and P. Doty, *Ibid.,* 761(1957).

**D-9.** M. Sela and E. Katchalsky, *Adv. in Protein Chem.* 14: 392(1959).  **D-10.** A. Elliott *et al.*, *Nature* 178: 1170(1956); R. H. Karlson *et al.*, *JACS* 82: 2268(1960). **D-11.** H. Lenormant *et al.*, *JACS* 80: 6191(1958).  **D-12.** E. R. Blout *et al.*, *JACS* 82: 3787(1960); 83: 1411(1961).  **D-13.** I. Z. Steinberg *et al.*, *JACS* 82: 5263(1960). **D-14a.** P. Doty *et al.*, *JACS* 78: 947(1956); *Proc. Natl. Acad. Sci. U.S.* 44: 424(1958). **D-14b.** M. Goodman *et al.*, *JACS* 84: 1283(1962).  **D-15.** J. E. Turner *et al.*, *JACS* 80: 4117(1958); H. Würz and F. Haurowitz, *JACS* 83: 280(1961).  **D-16.** G. Markus and F. Karush, *JACS* 80: 89(1958).  **D-17.** D. D. Ulmer and B. L. Vallee, *J. Biol. Chem.* 236: 730(1961).  **D-18.** I. M. Klotz and R. E. Heiney, *BBA* 25: 205(1957). **D-19.** K. Linderström-Lang and J. A. Schellman, *BBA* 15: 156(1954).  **D-20.** A. F. Richter and J. Duchon, *Acta Univ. Carolinae Med. (Prague)* 1958: 413.  **D-21a.** W. F. Harrington and J. A. Schellman, *Compt. rend. trav. lab. Carlsberg, Sér. chim.* 30: 21 (1956); **D-21b.** L. Mandelkern and D. E. Roberts, *JACS* 83: 4292(1961).  **D-22.** B. Jirgensons, *ABB* 74: 57, 70(1958); 78: 235(1958); 85: 89(1959); 93: 172(1961). **D-23.** C. Cohen and A. Szent-Györgyi, *JACS* 79: 248(1957).  **D-24.** A. Elliott *et al.*, *Discussions Faraday Soc.* 25: 167(1958).  **D-25.** C. Tanford *et al.*, *JACS* 81: 3255, 4032(1959); 82: 6082(1960); *JBC* 236: 1711(1961).  **D-26.** B. Jirgensons and T. Ikenaka, *Makromol. Chem.* 31: 112(1959).  **D-27.** W. Moffitt, *J. Chem. Phys.* 25: 467 (1956).  **D-28.** R. D. Lundberg and P. Doty, *JACS* 79: 3961(1957).  **D-29.** W. Kuhn, *Tetrahedron* 13: 1(1961).  **D-30.** G. D. Fasman, *Nature* 193: 681(1961).  **D-31.** M. M. Marsh, *JACS* 84: 1896(1962).  **D-32.** P. Urnes and P. Doty, *Adv. in Protein Chem.* 16: 401(1961).  **D-33.** C. C. Bigelow and I. I. Geschwind, *Compt. rend. trav. lab. Carlsberg, Sér. chim.* 32: 89(1961).  **D-34.** C. Tanford, *JACS* 84: 4240(1962).  **D-35.** B. Jirgensons, *ABB* 94: 59(1961).  **D-36.** N. S. Simmons *et al.*, *JACS* 83: 4766(1961). **D-37.** V. P. Kushner, *Biokhimiya* 26: 1077(1961).  **D-38.** M. L. Meyer and W. Kauzmann, *ABB* 99: 348(1962).  **D-39.** E. R. Blout *et al.*, *JACS* 84: 3193 (1962).

*Section E*

**E-1.** J. Bjorksten, *Adv. in Protein Chem.* 6: 343(1951).  **E-2.** J. L. Crammer and A. Neuberger, *BJ* 37: 302(1943).  **E-3.** M. Laskowski, Jr., and H. Scheraga, *JACS* 76: 6305(1954); 78: 5793(1956).  **E-4.** H. Scheraga, "Protein Structure," Academic Press, New York, 1961.  **E-5.** J. A Schellman, *Compt. rend. trav. lab. Carlsberg, Sér. chim.* 29: 230(1955).  **E-6.** D. B. Wetlaufer, *Compt. rend. trav. lab. Carlsberg, Sér. chim.* 30: 135(1956).  **E-7.** I. M. Klotz and J. S. Franzen, *JACS* 82: 5241(1960).  **E-8.** K. Linderström-Lang, *BBA* 18: 308(1955); *Chem. Soc. (London) Spec. Publ.* No. 2: 1955. **E-9.** E. S. Benson, *Compt. rend. trav. lab. Carlsberg, Sér. chim.* 31: 235; 32: 579(1960). **E-10.** C. L. Schildkraut and H. Scheraga, *JACS* 82: 58 (1960).  **E-11.** E. R. Blout *et al.*, *JACS* 83: 1895(1961).  **E-12.** A. Stracher, *Compt. rend. trav. lab. Carlsberg, Sér. chim.* 31: 468(1960).  **E-13.** A. Berger and K. Linderström-Lang, *ABB* 69: 106 (1957); W. P. Bryan and S. O. Nielsen, *BBA* 42: 552(1960).  **E-14.** W. Kauzmann, *BBA* 28: 87(1958).  **E-15.** C. Tanford *et al.*, *JACS* 81: 4032(1959).  **E-16.** G. H. Haggis, *BBA* 23: 494(1957).  **E-17.** D. T. Warner, *Nature* 190: 120(1961).  **E-18.** R. R. Becker and M. A. Stahmann, *JACS* 76: 3707(1954).  **E-19.** J. T. Yang and J. F. Foster, *JACS* 76: 1588(1954); C. Tanford *et al.*, *JACS* 77: 6421(1955).  **E-20.** A. G. Pasynskii, *J. Polymer Sci.* 29: 61(1958).  **E-21.** R. Cecil and J. R. McPhee, *Adv. in Protein Chem.* 14: 256(1959); P. D. Boyer, *Brookhaven Symposia in Biol.* 13: 1(1960).  **E-22.** D. Mazia, *in* "Sulfur in Proteins" (R. Benesch, ed.), p. 367. Academic Press, New York, 1959; F. Karush, "Immunochemical Approaches to Problems in Microbiology," p. 368. Rutgers Univ. Press, New Brunswick, New Jersey, 1961. **E-23.** R. E. Benesch *et al.*, *JBC* 216: 663(1953); I. M. Kolthoff and B. R. Willeford, Jr., *JACS* 80: 5673(1958); O. Pihar, *Chem. listy* 47: 1647(1953).  **E-24.** G. J. Millar,

*BJ* 53: 393(1953). **E-25a.** F. Haurowitz and M. Kennedy, *FP* 11: 227(1952); **E-25b.** M. J. Hunter and F. C. McDuffie, *JACS* 81: 1400(1959). **E-26.** H. Fraenkel-Conrat *et al.*, *AS* 73: 625(1951). **E-27.** H. L. Kern and R. M. Herriott, *FP* 12: 229(1953). **E-28a.** G. Markus and F. Karush, *AS* 79: 134(1957); **E-28b.** I. M. Kolthoff *et al.*, *AS* 82: 4147(1960); **E-28c.** J. L. Bailey and R. D. Cole, *JBC* 234, 1733(1959). **E-29.** L. E. Fieser, *Rec. trav. chim.* 69: 410(1950). **E-30.** F. Haurowitz, M. B. Kennedy, and J. E. Turner, *in* "Sulfur in Proteins" (R. Benesch, ed.), p. 25. Academic Press, New York, 1959. **E-31.** S. Yanari and F. A. Bovey, *JBC* 235: 2818(1960). **E-32.** I. M. Klotz, *Brookhaven Symposia in Biol.* 13: 25(1960). **E-33.** J. Stauff and R. Jaenicke, *Kolloid-Z.* 175: 1(1961). **E-34.** H. A. Scheraga *et al. JBC* 237: 2506(1962); *J. Phys. Chem.* 66: 1773(1962). **E-35.** W. Kauzmann, *Adv. in Phys. Chem.* 14: 1(1959).

*Section F*

**F-1a.** F. Haurowitz, *in* "Hoppe-Seyler/Thierfelder, Handbuch der physiologisch-und pathologisch-chemischen Analyse," Vol. 4: p. 125. Springer, Berlin, 1960. **F-1b.** R. M. Herriott, *Adv. in Protein Chem.* 3: 170(1947). **F-2a.** H. Zahn, *Angew. Chem.* 67: 561(1955); H. Zahn *et al., Ber.* 89: 407 (1956). **F-2b.** A. F. Schick and S. J. Singer, *J. Biochem. Biophys. Cytol.* 9: 519(1961). **F-3a.** A. Wormall *et al., BJ* 40: 730(1946). **F-3b.** S. Moore *et al., J. Org. Chem.* 11: 675(1946). **F-3c.** V. du Vigneaud *et al., AS* 70: 2547(1948). **F-4.** R. M. Herriott and J. H. Northrop, *J. Gen. Physiol.* 18: 35 (1934); 19: 383(1936). **F-5a.** H. N. Christensen, *JBC* 160: 75(1945). **F-5b.** S. Edlbacher and B. Fuchs, *Z. physiol. Chem.* 144: 133 (1921). **F-5c.** S. Udenfriend and S. F. Velick, *JBC* 190: 733(1951). **F-6.** G. Weber, *BJ* 51: 145, 155(1952). **F-7a.** A. H. Coons and M. H. Kaplan, *J. Exptl. Med.* 91: 1(1944). **F-7b.** J. D. Marshall *et al., Proc. Soc. Exptl. Biol. Med.* 98: 898(1958). **F-8.** M. Sela and E. Katchalski, *Adv. in Protein Chem.* 14: 392(1959); R. R. Becker and M. A. Stahmann, *JBC* 204: 737, 745(1953). **F-9.** W. L. Hughes *et al., AS* 71: 2476(1949). **F-10.** D. French and J. T. Edsall, *Adv. in Protein Chem.* 2: 277(1945). **F-11a.** K. H. Gustavson, *Adv. in Protein Chem.* 5: 354(1949). **F-11b.** K. H. Gustavson, *Kolloid-Z.* 103: 43(1943). **F-11c.** P. Alexander *et al., BJ* 48: 435(1950). **F-12.** R. S. Hannan and C. H. Lea, *BBA* 9: 293 (1952). **F-13.** F. Haurowitz and M. Tunca, *BJ* 39: 443(1945). **F-14.** H. Fraenkel-Conrat and H. S. Olcott, *JBC* 161: 259(1945). **F-15a.** F. Haurowitz and J. Horowitz, *BBA* 20: 574(1956). **F-15b.** F. Haurowitz *et al., JBC* 224: 827(1957). **F-16a.** G. Fromageot *et al., BBA* 6: 283(1950/51). **F-16b.** A. C. Chibnall and M. W. Rees, *BJ* 48: XLVII(1951). **F-16c.** W. Grassmann *et al., Z. physiol. Chem.* 296: 208(1954). **F-17a.** J. C. Crawhall and D. E. Eliott, *Nature* 175: 299(1955). **F-17b.** M. Jutisz *et al., Bull. chim. France* 1954: 1087. **F-18a.** H. S. Olcott and F. Fraenkel-Conrat, *AS* 70: 2101(1948); *JBC* 161: 259(1945). **F-18b.** R. E. Ferrel *et al., AS* 70: 2101 (1948). **F-19.** J. Bello and J. R. Vinograd, *AS* 78: 1369(1956). **F-20.** P. Grabar *et al., Bull. soc. chim. biol.* 33: 690(1951). **F-21.** A. K. Balls and E. F. Jansen, *Adv. in Enzymol.* 13: 321(1952). **F-22.** M. L. Anson, *J. Gen. Physiol.* 24: 399(1941). **F-23.** F. R. Bettelheim, *AS* 76: 2838(1954). **F-24a.** O. Kratky *et al., Z. Naturforsch.* 10b: 68(1955). **F-24b.** A. Wormall, *J. Exptl. Med.* 51: 295(1930). **F-25a.** F. Haurowitz, *Ergeb. Mikrobiol.* 34: 1(1961). **F-25b.** F. Haurowitz, *Biol. Revs.* 27: 247(1952). **F-26.** H. Pauly, *Z. physiol. Chem.* 94: 284(1915). **F-27.** K. Landsteiner, "The Specificity of Serological Reactions," 2nd ed. Harvard Univ. Press, Cambridge, Massachusetts, 1946. **F-28a.** H. G. Higgins and K. J. Harrington, *ABB* 85: 409(1959). **F-28b.** E. W. Gelewitz *et al., ABB* 53: 411(1954). **F-28c.** M. Tabatschnick and H. Sobotka, *JBC* 234: 1726(1959). **F-29.** A. Kossel and F. Weiss, *Z. physiol. Chem.* 84: 1(1913). **F-30.** H. Fraenkel-Conrat, *JBC* 217: 373(1955); L. W. Cunningham and B. J. Nuenke, *JBC* 236: 1716(1961). **F-31.** E. S. G. Barron and T. P. Singer, *Science* 97:

356(1943). **F-32.** L. Hellerman, *Physiol. Revs.* 17: 454(1937). **F-33a.** F. Haurowitz *et al., AS* 74: 2265(1952). **F-33b.** F. Haurowitz, *Z. physiol. Chem.* 315: 127(1959). **F-34.** R. L. Hill *et al., Ann. Rev. Biochem.* 28: 106(1959). **F-35.** H. S. Olcott and H. Fraenkel-Conrat, *Chem. Revs.* 41: 151(1947). **F-36.** C. Y. Cha, and H. A. Scheraga, *Biochem. Biophys. Research Communs.* 6: 369(1961). **F-37.** A. C. Chibnall, *Proc. Roy. Soc.* B131: 136(1942). **F-38.** Yu. A. Vladimirov, *Biofizika* 7: 270(1962). **F-39.** T. Herskovits and M. Laskowski, Jr., *JBC* 237: 2481(1962).

### Section G

**G-1.** V. du Vigneaud *et al., JACS* 74: 3713(1952); *JBC* 199: 929(1952). **G-2.** V. Bruckner *et al., Experientia* 9: 63(1953). **G-3.** F. Haurowitz *et al., JBC* 224: 827 (1957). **G-4.** P. M. Gallop *et al., JBC* 235: 2619(1960). **G-5.** J. Bello, *Nature* 185: 241(1960). **G-6.** F. Weygand and R. Junk, *Z. physiol. Chem.* 300: 27(1955). **G-7.** G. Mechanic and M. Levy, *JACS* 81: 1889(1959). **G-8.** R. P. Ambler and M. W. Rees, *Nature* 184: 56(1959). **G-9.** N. Troensegaard, "The Structure of the Protein Molecule." Munksgaard, Copenhagen, 1944. **G-10.** E. Abderhalden, *Z. physiol. Chem.* 277: 248(1943). **G-11.** N. I. Gavrilov and L. N. Akimova, *Zhur Obshchei Khim.* 24: 563(1954); K. S. Makarov and N. I. Gavrilov, *Zhur. Obshchei Khim.* 29: 2143(1959). **G-12.** J. P. Greenstein, *JBC* 112: 517(1936); E. Waldschmidt-Leitz and G. v. Schuckmann, *Ber. deut. chem. Ges.* 62: 1891(1929). **G-13.** D. Wrinch, *Science* 107: 445(1948); "Chemical Aspects of the Structure of Small Peptides." Munksgaard, Copenhagen, 1960. **G-14.** D. Crowfoot, *Nature* 135: 591, 891(1935). **G-15.** F. Haurowitz, *Z. physiol. Chem.* 256: 28(1938). **G-16.** H. Neurath, *J. Phys. Chem.* 44: 296(1940). **G-17.** K. Narita, *BBA* 28: 184(1958). **G-18.** R. L. Hill and W. R. Schmidt, *JBC* 237: 389(1962).

### Section H

**H-1.** F. Haurowitz, *in* "Hoppe-Seyler/Thierfelder, Handbuch der physiologisch-und pathologisch-chemischen Analyse," Vol. 4, p. 116. Springer, Berlin, 1960. **H-2.** W. Kauzmann, *Adv. in Protein Chem.* 14: 1(1959). **H-3.** W. D. Loughlin and W. C. M. Lewis, *BJ* 26: 476(1932); 27: 99, 106(1933). **H-4.** F. Haurowitz, *Kolloid-Z.* 71: 198 (1935). **H-5.** H. Neurath and H. B. Bull, *JBC* 115: 519(1936). **H-6.** H. Wu, *Chinese J. Physiol.* 5: 321(1931). **H-7.** R. Lumry and H. Eyring, *J. Phys. Chem.* 58: 110(1954). **H-8.** A. E. Mirsky and M. L. Anson, *J. Gen. Physiol.* 19: 427(1936). **H-9.** R. M. Herriott, *Adv. in Protein Chem.* 3: 170(1947). **H-10.** R. L. McDonell *et al., ABB* 32: 288(1951). **H-11.** R. Brdička and J. Klumpar, *Casopis Ceskoslov. lékárnictva* 17: 243(1937). **H-12.** F. Haurowitz and S. Tekman, *BBA* 1: 484(1947). **H-13.** J. Roche and M. Mourgue, *Bull. soc. chim. biol.* 30: 322(1948). **H-14.** C. H. Li, *JACS* 67: 1065(1945). **H-15.** R. R. Porter, *BBA* 2: 105(1948); F. S. Steven and G. R. Tristram *BJ* 70: 179(1958). **H-16.** R. R. Porter, *BJ* 46: 304(1950). **H-17.** F. Haurowitz *et al., JACS* 74: 2265(1952). **H-18.** F. Haurowitz *et al., Z. physiol. Chem.* 315: 127(1959). **H-19.** M. L. Anson and A. E. Mirsky, *J. Gen. Physiol.* 17: 399(1934); F. Haurowitz *et al., JBC* 157: 621(1945). **H-20.** K. Linderström-Lang *et al., Nature* 142: 996(1938). **H-21.** E. Fredericq, *Bull. soc. chim. Belges* 56: 223(1947); M. Joly and E. Barbu, *Bull. soc. chim. biol.* 31: 908(1950). **H-22.** W. Kauzmann, *JACS* 75: 5139(1953). **H-23.** W. Kauzmann, *Ann. Rev. Phys. Chem.* 8: 513(1957). **H-24.** A. Elliott *et al., Nature* 166: 194(1950). **H-25.** K. Bailey, *BJ* 36: 140(1942). **H-26.** M. Levy and A. E. Benaglia, *JBC* 186: 829(1950). **H-27.** J. Steinhardt and E. M. Zaiser, *Adv. in Protein Chem.* 10: 152(1955). **H-28.** J. Foster and E. G. Samsa, *JACS* 73: 5388(1951). **H-29.** J. Foster *et al., JACS* 76: 6044(1954); *JBC* 236: 2662(1961). **H-30.** M. Halwer, *JACS* 76: 183(1954); V. D. Hospelhorn *et al., JACS* 76: 2827(1954).

**H-31.** R. C. Warner and M. Levy *JACS* 80: 5735(1958). **H-32.** E. V. Jensen *et al.*, *JBC* 185: 411(1950). **H-33.** L. B. Gorbacheva, *Biokhimiya* 22: 70(1957). **H-34.** F. Haurowitz, "Chemistry and Biology of Proteins," 1st ed., p. 130–131. Academic Press, New York, 1950; K. Hamaguchi, *J. Biochem. (Tokyo)* 43: 83(1956). **H-35.** M. Levy and J. Magoulas, *J. Am. Chem. Soc.* 84: 1345(1962). **H-36.** I. M. Klotz and V. H. Stryker, *JACS* 82: 5169(1960). **H-37.** C. F. Jacobsen and L. Koorsgaard, *Nature* 161: 30(1948). **H-38.** C. Huggins *et al.*, *Nature* 167: 592(1951). **H-39.** M. L. Groves *et al.*, *JACS* 73: 2790(1951); R. Benesch and R. E. Benesch, *JACS* 75: 4367(1953). **H-40.** N. Burk, *JACS* 87: 197(1930); 98: 353(1932); 131: 373(1937). **H-41.** C. Tanford and P. K. De, *JBC* 236: 1711(1961). **H-42a.** G. R. Stark *et al.*, *JBC* 235: 3177(1960). **H-42b.** R. D. Cole, *JBC* 236: 2670(1961); J. T. Edsall, *Fortschr. chem. Forsch.* 1: 119(1949). **H-43.** J. S. Mitchell and E. K. Rideal, *Proc. Roy. Soc.* A167: 342(1939). **H-44.** A. D. McLaren, *Science* 113: 716(1951). **H-45.** W. Gordy and H. Shields, *Radiation Research* 9: 611(1958); *Proc. Natl. Acad. Sci. U.S.* 46: 1124(1960). **H-46.** D. Cavallini *et al.*, *Science* 131: 1441(1960). **H-47.** P. Grabar *et al.*, *J. chim. physique* 44: 145(1948); *Bull soc. chim. biol.* 32: 620(1950). **H-48.** T. Brosteaux and B. Erickson-Quensel, *Arch. phys. biol.* 12: 209(1935). **H-49.** P. D. Boyer *et al.*, *JBC* 162: 181(1946). **H-50.** F. H. Johnson and D. H. Campbell, *JBC* 163: 689(1946). **H-51.** A. Kowalsky, *JBC* 237: 1807(1962). **H-52.** P. L. Whitney and C. Tanford, *JBC* 237: PC 1735(1963). **H-53.** C. Tanford *et al.*, *JACS* 81: 4032(1959). **H-54.** J. F. Kirby-Smith and M. L. Randolph, *J. Cellular Comp. Physiol.* 58: Suppl. 1., p. 1(1961). **H-55.** J. F. Foster *et al.*, *JBC* 237: 2509, 2514(1962).

*Section I*

**I-1.** M. Kunitz, *J. Gen. Physiol.* 32: 241(1948). **I-2.** M. A. Eisenberg and G. W. Schwert, *J. Gen. Physiol.* 34: 583(1951). **I-3.** A. S. Tsiperovich, *Ukraïn. Biokhim. Zhur.* 20: 108(1948). **I-4.** W. Kauzmann, *JACS* 75: 5139(1953). **I-5.** E. J. Casey and K. J. Laidler, *Science* 111: 110(1950). **I-6.** J. M. Sturtevant *et al.*, *JACS* 74: 1983(1952); *J. Phys. Chem.* 58: 97(1954). **I-7.** G. Kistiakowsky *et al.*, *JACS* 62: 1895 (1940); 63: 2080(1941). **I-8.** H. Eyring and A. E. Stearn, *Chem. Revs.* 29: 253(1939).

*Section J*

**J-1.** H. B. Bull, *Adv. in Protein Chem.* 3: 95(1947). **J-2.** A. E. Alexander, "Surface Chemistry," Longmans, Green, New York, 1951. **J-3.** D. F. Cheesman and J. T. Davies, *Adv. in Protein Chem.* 9: 440(1954). **J-4.** E. Gorter and F. Grendel, *Proc. Koninkl. Ned. Akad. Wetenschap.* 29: 1262(1926); 32: 770(1929). **J-5.** H. A. Dieu and H. B. Bull, *JACS* 71: 450(1949). **J-6.** F. Haurowitz *et al.*, *ABB* 59: 52(1955). **J-7.** G. A. Deborin *et al.*, *Doklady Akad. Nauk S.S.S.R.* 114: 602(1957). **J-8.** I. G. Kaplan and M. J. Frazer, *JBC* 210: 57(1954). **J-9.** R. R. Ray and L. G. Augenstine, *J. Phys. Chem.* 60: 1193(1956). **J-10.** H. B. Bull, *JACS* 80: 1901(1958). **J-11.** J. B. Bateman and E. D. Adams, *J. Phys. Chem.* 61: 1039(1957); D. D. Eley and D. G. Hedge, *Trans. Faraday Soc.* 21: 221(1956). **J-12.** L. G. Augenstine *et al.*, *J. Phys. Chem.* 62: 1231(1958). **J-13.** F. A. Askew and J. T. Danielli, *Trans. Faraday Soc.* 36: 785(1940); J. T. Davies *et al.*, *BBA* 11: 165(1953); *Proc. Roy. Soc.* A227: 537(1955). **J-14.** T. Tachibana *et al.*, *BBA* 24: 174(1957). **J-15.** N. Benhamou, *J. chim. phys.* 1956: 32. **J-16.** F. Haurowitz *et al.*, *Nature* 180: 437(1957). **J-17.** H. Neurath and H. B. Bull, *Chem. Revs.* 23: 391(1938); D. G. Dervichian, *J. Chem. Phys.* 11: 236 (1943). **J-18.** T. H. Schulman, *Proc. Roy. Soc.* A155: 701(1936); F. Sebba and E. K. Rideal, *Trans. Faraday Soc.* 35: 1200(1939). **J-19.** J. D. Arnold and C. Y. C. Pak, *J. Colloid Sci.* 17: 348(1962). **J-20.** J. T. Davies and E. K. Rideal, "Interfacial Phenomena," pp. 56; 251. Academic Press, New York, 1961.

# Chapter VIII

## Albumins, Globulins, and Other Soluble Proteins

### A. Classification of Soluble Proteins

As mentioned in Chapter I, the classification of soluble proteins according to their solubility in different solvents is not very satisfactory.

It will be recalled that the fraction salted out by 50% saturation with ammonium sulfate was called *globulin* and that precipitated by higher saturation with the same salt *albumin*. Some of the globulins, called *euglobulins*, are insoluble in salt-free water while others, the *pseudoglobulins*, are soluble. Globulin-like proteins from the cereals, which are soluble in 80% ethanol, have been designated as *prolamins*. We are not yet able to correlate the solubility of proteins with a definite composition or chain conformation.

The soluble proteins do not form a uniform group. They have very different functions in living organisms. Some of them may act merely as nutrients for the growing tissues. Many of them are active as enzymes, hormones, or antibodies. Fibrous soluble proteins have important functions in the clotting of blood and in muscle contraction. Although many of these proteins will be mentioned in this chapter, more detailed information on enzymes will be found in Chapter XII, on hormones in Chapter XIII, and on antibodies in Chapter XV. The muscle proteins will be discussed in Chapter IX.

### B. Proteins of the Blood Serum (B-1 to 3)

The presence of more than one type of protein in human or animal blood serum was first demonstrated by salting-out methods. The neutral salts used are ammonium sulfate (B-4), magnesium sulfate (B-5), sodium sulfate (B-6), and mixtures of primary and secondary alkali phosphates (B-7). If the amount of protein precipitated is plotted against the amount of salt added, a curve is obtained which shows a steep increase in precipitated protein at certain salt concentrations (Fig. VIII-1). When blood plasma is subjected to the salting-out procedure, the first protein precipitated is fibrinogen (F). Since this protein has a particular function in clotting, it will be discussed separately in Section C of this chapter. The next part of the curve corresponds to the precipitation of the globulins (G), the last part to that of albumin (A). If magnesium sulfate is used for

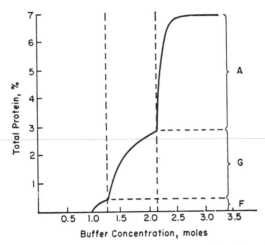

FIG. VIII-1. Salting-out of human blood plasma by phosphate buffer solutions, pH 6.5. The ordinate shows the amount of protein precipitated (in grams per 100 ml blood plasma).

salting-out, the globulins require saturation for their complete precipitation; albumin is precipitated by adjustment to pH 4.6, its isoelectric point.

In discussing electrophoresis, it was mentioned (Chapter VI, C) that the serum globulin fraction is a mixture of at least three electrophoretic components called α-, β-, and γ-globulin. If ammonium sulfate is added cautiously to horse blood serum, the γ-globulins are precipitated at a concentration of 1.34 moles of ammonium sulfate per liter of solution; β- and α-globulins are precipitated by 1.64 and 2.05 moles, respectively, and albumins by 2.57 moles (B-8). Although this method is more convenient than electrophoresis for the preparation of large amounts of the globulins, the fractions obtained are somewhat impure; each of them is contaminated with small amounts of the other neighboring fractions. Ammonium sulfate has the disadvantage of containing nitrogen, so that direct Kjeldahl analyses of the protein fractions are rendered impossible. It has been replaced by sodium sulfate in the clinical determination of albumins and globulins (B-6). Globulins are precipitated by 21.5% sodium sulfate, while albumins are determined in the filtrate; the precipitation and filtration are carried out at 37°C, because the solubility of sodium sulfate at room temperature is not sufficiently high.

*Serum albumin*, which forms more than half of the serum proteins, can be prepared conveniently by precipitation of the serum proteins with

trichloroacetic acid and extraction of the insoluble protein trichloroace-
tates with ethanol or acetone. The extract contains serum albumin free of
globulins. Trichloroacetic acid and the organic solvent are removed by
dialysis against water. Surprisingly, the serum albumin prepared by this
method is in the native state as demonstrated by chemical, serological,
and metabolic methods (B-9 to 13).

The salting-out methods, although convenient for the preparation of
protein fractions in the laboratory, are not suited for large-scale preparations since it is difficult to remove salts from large volumes of solutions by
dialysis. A more suitable method was worked out by E. J. Cohn and his
co-workers when large amounts of plasma albumin and globulin were required during World War II. The plasma proteins were precipitated in
this method by the fractional addition of ethanol at low temperatures;
ionic strength, pH, and the dielectric constant of the solutions were varied
in a systematic manner. The original method of the Harvard laboratories
has been repeatedly modified (B-14a,b).

In the modification designated as method 10, the blood plasma is precipitated
at 0°C by 8–10 % of chilled ethanol. *Fraction I* obtained in this way contains
predominantly fibrinogen (see Fig. VIII-2). *Fractions II* and *III* are precipitated

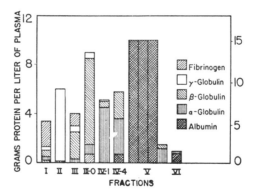

Fig. VIII-2. Distribution of electrophoretic components in fractions of human
plasma. The scale at the right side indicates the amount of the fractions in per cent of
the total plasma protein (B-3).

by raising the ethanol concentration to 25 % (at −5°C). The proteins soluble in
20 % ethanol (*Fraction III-0*) consist of the $\beta_1$-globulins, whereas the insoluble
precipitate contains γ-globulins, prothrombin, and proteolytic enzymes. *Fraction
III* is insoluble in 17 % ethanol at pH 5.2, while *Fraction II* remains dissolved;
it is precipitated by raising the ionic strength to 0.05. The precipitate contains
most of the antibodies. *Fraction IV-1* is precipitated from the supernatant of
Fractions II and III by lowering the ethanol content to 18 % and adjusting pH to
5.2. It contains chiefly α-globulins. Raising the ethanol concentration to 40 %

and pH to 5.8 causes precipitation of *Fraction IV-4*, which contains some of the serum enzymes and several glycoproteins. *Fraction V*, which consists mainly of albumin, is precipitated by lowering pH to 4.8 while the ethanol content is maintained at 40%. The supernatant *Fraction VI* contains only small amounts of proteins. The composition of these fractions is shown in Fig. VIII-2.

Fractionation of some of the serum proteins has also been accomplished by precipitation with 0.02 $M$ zinc sulfate (method 12 of the Harvard Laboratories, B-14a). Clotting can be prevented by passing the blood plasma through a cation exchange resin (B-15). If it is desired to isolate merely the γ-globulins, all other plasma proteins can be precipitated by adding 3 volumes of a 0.4% solution of the acridine dye rivanol at pH 7.6–7.8 (B-16). The supernatant solution, after removal of the excess of dye by means of Norit SX 30, contains almost pure γ-globulins. Fractionation of the plasma proteins can also be accomplished by treatment with ether at low temperatures at decreasing pH values (B-17), by electrophoresis or ultracentrifugation (B-18), and by chromatography (B-19).

The total amount of protein in human blood plasma is about 7%. About one-half of the proteins consist of serum albumin, 15%, 19%, and 11% of α-, β-, and γ-globulins, respectively, and about 5% of fibrinogen. The physicochemical properties of the best-investigated plasma proteins are shown in Table VIII-1 (B-20). It is difficult at present to interpret the results of fractionations of the plasma and serum proteins. Very few of them have been obtained in the crystalline state. Some of them may be saltlike complexes formed by the mutual attraction of acidic and basic proteins, or complexes of proteins with lipids, mucopolysaccharides, or other macromolecules. Differences in the solubility may also be caused by dimerization or polymerization, by the binding of calcium or magnesium ions or of phosphate. The properties of some of the labile serum proteins are distinctly altered by the action of ethanol even at low temperatures.

*Pre-albumin* is the designation used for a fast moving α-glycoprotein which, in zone electrophoresis, moves in front of the intense albumin band. Pre-albumin has been precipitated from the supernatant of crystalline serum albumin by acidification to pH 3.9 (B-21). It is very rich in tryptophan and therefore has a very intense absorbancy at 280 m$\mu$ (B-21). The protein-bound thyroxine of the blood serum is bound in part to pre-albumin (B-22).

*Serum albumin* is obtained in crystalline form by salting-out with sodium or ammonium sulfate at its isoelectric point. Nevertheless, the purified protein is heterogeneous when examined by electrophoresis at low ionic strength (B-23), by chromatography (B-26b), by interferometry (B-24), or serologically (B-25). Part of the heterogeneity can be attributed to the ease of SS—SH interchange in this protein (B-26a), and to polymerization (B-26b). Serum albumin has 17 cystine residues and 1 cysteine residue per molecule; if the latter reacts with the cysteine residue of another albumin molecule to form an intermolecular S—S bond, a dimer is formed (B-27). Dimers can also be produced by the combination of serum albumin molecules with mercuric salts; they contain an intermolecular S—Hg—S bridge (B-28). If the dithio bonds of serum albumin are

TABLE VIII-1

PHYSICOCHEMICAL PROPERTIES OF SOME OF THE HUMAN PLASMA PROTEINS

| Protein | Isoelectric point | Mol. wt. | $s_{20}$ | mg/ml[j] | $f/f_0$[a] |
|---|---|---|---|---|---|
| Prealbumin[j] | — | 61,000 | 4.1 | 0.1–0.5 | — |
| Serum albumin[b] | 4.7 | 69,000 | 4.6 | 50–65 | 1.28 |
| $\alpha_1$-acid glycoprotein (orosomucoid[c]) | 2.7 | 44,100 | 3.1 | 0.7–1.0 | 1.78 |
| $\alpha_1$-lipoproteins[d] | — | 435,000 | 5.5 | — | — |
|  | — | 195,000 | 5.0 | 0.5–1.5 | — |
| Haptoglobin 1-1[e] | 4.1 | 85,000 | 4.2 | 1[e] | — |
| $\alpha_2$-glycoprotein[f] | 3.8 | — | 2.6 | — | — |
| Plasminogen[g] | 5.6 | 143,000 | 4.3 | — | 2.09 |
| Ceruloplasmin[h] | 4.4 | 151,000 | 7.1 | 0.2–0.3 | — |
| $\beta$-Globulin[i] | — | 93,000 | 4.6[i] | 4–14 | — |
| $\beta$-Lipoprotein[d] | — | $3.2 \times 10^6$ | — | — | — |
| Transferrin[k] | 4.4 | 88,000 | 5.8 | 3–6.5 | 1.25 |
| $\gamma$-Globulins[b] | 7.3 | 160,000 | 6.6 | 13–22 | 1.38 |
| Fibrinogen[l] | 5.8 | 330,000 | 7.8 | — | — |

[a] $f/f_0$: frictional ratio, usually calculated from diffusion and sedimentation data.

[b] J. L. Oncley et al., J. Phys. Chem. 51: 184(1947).

[c] E. L. Smith et al., JBC 185: 569(1950).

[d] B. Shore, ABB 71: 1(1957).

[e] M. Nyman, Scand. J. Clin. & Lab. Invest. 11, Suppl. 39(1959).

[f] K. Schmid, BBA 21: 399(1956).

[g] S. Shulman et al., JBC 233: 91(1958).

[h] C. G. Holmberg and C. B. Laurell, Acta Chem. Scand. 2: 550(1948).

[i] S. Wallenius et al., JBC 225: 253(1957).

[j] H. E. Schultze and G. Schwick, Clin. Chim. Acta 4: 15(1959); H. E. Schultze, Verhandl deut. Ges. inn. Med. 66: 225(1960).

[k] C. B. Laurell and B. Ingleman, Acta Chem. Scand. 1: 770(1947).

[l] S. Shulman, JACS 75: 5846(1953).

cleaved by reduction with thioglycolate in the presence of urea and if reoxidation to S—S is prevented by subsequent treatment with iodoacetamide which converts —SH into —S·CH₂·CONH₂, the molecular weight of the acetamide-substituted serum albumin is the same as that of native serum albumin, 68,000 (B-29). Also, the molecular weight of serum albumin does not decrease when the protein is treated with an excess of thioglycolate or sulfite; this and the presence of a single terminal $\alpha$-amino group indicate that the albumin molecule consists of a single peptide chain and that all S—S bonds are intrachain bonds (B-30,31a).

Serum albumin, in contrast to the serum globulins, has the ability to bind numerous anions, for instance, acidic azo dyes, fatty acids, chloride ions, or alkyl sulfates (see Chapter X). Cations such as dextranamines are not bound by native serum albumin although dextran sulfate anions are bound (B-31b). It has been concluded from such experiments that the surface of the serum albumin molecule contains positively, but not negatively, charged groups (B-31b,33b). The high reactivity of serum albumin with a variety of anions has been attributed to a particular configurational "adaptability" of the molecule (B-32). On acidification to pH 4 the serum albumin molecule undergoes changes in the viscosity, the optical rotation, and the polarization of fluorescence of complexes formed with fluorescent dyes. All these changes are ascribed to expansion of the molecule brought about by the mutual repulsion of the positively charged ammonium groups (B-33a). Whereas the radius of gyration of serum albumin at pH 5.3 is 30.6 Å, it increases to 68.5 Å at pH 3.6 (B-39b). Evidently, the conformation of the serum albumin molecule is less rigid than that of other proteins (B-34). This is surprising in view of the numerous dithio bonds present in serum albumin. At neutral pH the serum albumin molecule is only sparingly hydrated; nevertheless its intrinsic viscosity is higher than that calculated for a sphere; its shape may be ellipsoidal with an axial ratio of approximately 3:1 (B-35). The amino acid composition of bovine serum albumin is similar to that of the human serum albumin, as shown in Table VIII-2. However, the N-terminal dipeptide is aspartylthreonyl in the former, and aspartylalanyl in the latter protein (B-36).

In the frog, albumin is formed only after the metamorphosis; the blood serum of tadpoles is poor in serum albumin (B-37). In some human blood sera electrophoresis reveals the presence of two serum albumin peaks; the phenomenon seems to be genetically determined (B-38). It was a great surprise when another genetic abnormality, the almost complete absence of serum albumin, was discovered by Bennhold (B-39a) in members of a family. It had been assumed quite generally that serum albumin is essential for the maintenance of the oncotic (colloid-osmotic) pressure of the blood serum. Yet the members of the analbuminemic family appeared to be quite healthy. The oncotic pressure in the analbuminemic persons was about one-half the normal pressure. Evidently, the oncotic pressure has not the great importance attributed to it in earlier years.

The *α-globulin* fraction consists of several glycoproteins (mucoproteins) and lipoproteins (see Chapter XI, B and C). One of the α-glycoproteins, discovered by Jayle (B-40), has the ability to combine stoichiometrically with 1 molecule of hemoglobin or globin (B-41) and has been designated as *haptoglobin*. It can be detected when paper electrophoresis is carried out in the presence of hemoglobin since the haptoglobin-hemoglobin complex gives the peroxidase test (B-42,43). Starch gel electrophoresis reveals the occurrence of three different phenotypes of haptoglobin in human blood sera (B-44,45), a fast moving type 1-1, a slowly moving type 2-2, and an intermediate type 2-1. The distribution of these in families is determined by two alleles Hp[1] and Hp[2]. The haptoglobins contain

TABLE VIII-2
AMINO ACID COMPOSITION OF SOME PLASMA PROTEINS[a]

| Amino acid | Human albumin[b] (Mol. wt.: 69,000) | | Bovine albumin[b] (Mol. wt.: 69,000) | | Human γ-globulin[b] (Mol. wt.: 160,000) | | Human fibrinogen[c] | | Bovine pro- thrombin[d] | |
|---|---|---|---|---|---|---|---|---|---|---|
| | A | C | A | C | A | C | A | B | A | B |
| Glycine | 1.6 | 14.7 | 1.82 | 16.7 | 4.47 | 95.3 | 5.6 | 74.7 | 4.49 | 59.8 |
| Alanine | — | | 6.25 | 48.4 | 4.05 | 72.8 | 3.7 | 41.5 | 4.09 | 46.0 |
| Valine | 7.7 | 45.4 | 5.92 | 34.9 | 9.42 | 128. | 4.1 | 35.0 | 4.88 | 41.6 |
| Leucine | 11.9 | 62.6 | 12.3 | 64.6 | 8.57 | 104. | 7.1 | 54.2 | 7.60 | 58.0 |
| Isoleucine | 1.7 | 8.95 | 2.61 | 13.7 | 2.59 | 31.6 | 4.8 | 36.7 | 3.14 | 24.0 |
| Proline | 5.1 | 30.6 | 4.75 | 28.5 | 7.90 | 109. | 5.7 | 49.5 | 5.52 | 48.0 |
| Phenylalanine | 7.8 | 32.6 | 6.59 | 27.5 | 4.79 | 46.4 | 4.6 | 27.9 | 4.73 | 28.5 |
| Tyrosine | 4.66 | 17.7 | 5.06 | 19.3 | 6.75 | 59.6 | 5.5 | 30.4 | 4.38 | 24.2 |
| Tryptophan | 0.19 | 0.64 | 0.59 | 1.99 | 3.42 | 26.8 | 3.3 | 16.0 | 3.30 | 16.3 |
| Serine | 3.7 | 24.3 | 4.23 | 27.8 | 11.7 | 178. | 7.0 | 66.8 | 6.19 | 53.8 |
| Threonine | 5.0 | 29. | 5.83 | 33.8 | 8.9 | 119. | 6.1 | 51.7 | 4.65 | 39.2 |
| Cysteine | 0.70 | 3.99 | 5.91 | 33.9 | 2.26 | 30.1 | 0.4 | 3.3 | 3.14 | 26.3 |
| Cystine/2 | 5.58 | 32.1 | | | | | 2.3 | 19.0 | | |
| Methionine | 1.28 | 5.93 | 0.81 | 3.75 | 0.90 | 9.66 | 2.6 | 17.1 | 1.94 | 13.0 |
| Arginine | 6.15 | 24.3 | 5.90 | 23.4 | 4.45 | 40.9 | 7.8 | 44.8 | 8.19 | 47.0 |
| Histidine | 3.5 | 15.6 | 4.0 | 17.8 | 2.55 | 26.3 | 2.6 | 17.1 | 1.84 | 11.9 |
| Lysine | 12.3 | 58.0 | 12.8 | 60.5 | 8.01 | 87.7 | 9.2 | 63.0 | 5.73 | 39.3 |
| Aspartic acid | 10.4 | 53.9 | 10.9 | 56.5 | 9.05 | 108. | 13.1 | 98.5 | 10.1 | 75.3 |
| Glutamic acid | 17.4 | 81.6 | 16.5 | 77.4 | 12.5 | 135. | 14.5 | 98.7 | 12.85 | 87.4 |
| Ammonia | 1.07 | 43.5 | 0.95 | 38.6 | 1.75 | 165. | 1.49 | 106.4 | — | 68.2 |

[a] A: gram amino acid per 100 g protein; B: amino acid residues per $10^5$ g protein; C: amino acid residues per protein molecule.

[b] See (B-1).

[c] See (C-36).

[d] From (C-22).

about 20% carbohydrate. It is not yet known whether they differ from each other in the composition of their protein or their polysaccharide moiety. The blood serum also contains a heme-binding protein called *hemopexin;* in contrast to the other proteins it is not precipitated by perchloric acid (B-21).

The *β-globulin* fraction of the human serum contains glyco- and lipoproteins (Chapter XI, B and C) and also two or more metal-binding proteins (Chapter XI, D). One of these, called *transferrin* or siderophilin, is an iron containing $\beta$-globulin which occurs in three or more genetically determined types: BB, BC, and BD (B-46,47). A copper-containing protein of the $\beta$-globulin fraction has been called *ceruloplasmin;* it is found in Cohn's Fraction IV-1, 4 (see Fig. VIII-2) and has been purified chromatographically (B-48). Some of the metal-free $\beta$-globulins occur also in different, genetically determined, phenotypes; one of these, designated as type D, has been found only in the serum of Negroes (B-49). Various phenotypes of $\beta$-globulins have been found in sheep (B-50) and in cattle (B-51).

The $\gamma$-*globulin* fraction of the human or animal serum contains a very heterogeneous population of molecules. Their sedimentation constants are 6S and 9.5S (B-52). Heterogeneity of the $\gamma$-globulins is also indicated by the finding of different N-terminal amino acids; human $\gamma$-globulins contain approximately 1 N-terminal aspartyl and 1 N-terminal glutamyl residue per molecule of the molecular weight 160,000 (B-53). Evidently, the molecule consists of at least two peptide chains. Short hydrolysis with papain yields 3.5S fragments and a resistant 6.6S fraction (B-71). Since the molecular weight of $\gamma$-globulin decreases upon treatment with thioglycolate, mercaptoethylamine, or sulfite, it is evident that subunits of the $\gamma$-globulin molecule are held together by dithio bridges (B-54,55,69,70). Some of the $\gamma$-globulins are euglobulins, insoluble at low ionic strength, others are pseudoglobulins, soluble upon dilution with distilled water. The insolubility of the euglobulins has been attributed to a higher molecular weight and combination with lipoproteins (B-56). If the globulins are exposed to denaturing agents such as urea or guanidinium salts in the presence of reducing agents, they undergo dissociation into light (L) and heavy (H) subunits (B-74).

Newborn animals seem to be unable to produce $\gamma$-globulins, or they produce only small amounts of them. In the human fetus the $\gamma$-globulins form about 5% of the total plasma proteins whereas in the adult person, 11% or more of the plasma proteins are $\gamma$-globulins. During the first weeks after birth the $\gamma$-globulin content decreases to lower values. This seems to indicate that the fetal $\gamma$-globulins are produced in the maternal organs (B-57). Agammaglobulinemia, i.e., an almost complete absence of $\gamma$-globulins, has been observed as an inheritable disease in certain families. This inability to form $\gamma$-globulins presents a certain danger for persons affected by it since most of the antibodies are $\gamma$-globulins and cannot be produced by the persons suffering from agammaglobulinemia.

All of the plasma proteins contain small amounts of carbohydrate, as shown in Table VIII-3. The hexoses found in these proteins are galactose and mannose. Far higher carbohydrate percentages are found in the proteins which are designated as glyco- or mucoproteins or as mucoids (see Chapter XI, C). In contrast to the other plasma proteins, certain glycoproteins and mucoids are not precipitated by perchloric acid and can thus be separated from the bulk of albumin and the globulins (B-58). Many of the plasma proteins contain sialic acid in their carbohydrate moieties.

TABLE VIII-3

CARBOHYDRATE CONTENT (IN PER CENT) OF HUMAN PLASMA
PROTEIN FRACTIONS (B-59)

| Plasma fraction | Hexose | Hexosamine | Sialic acid | Fucose |
|---|---|---|---|---|
| Albumin | 0.2 | 0.06 | 0.1 | 0.01 |
| $\alpha_1$-Globulins | 7.5 | 6.3 | 4.1 | 0.55 |
| $\alpha_2$-Globulins | 5.9 | 4.2 | 3.0 | 0.4 |
| $\beta$-Globulins | 2.7 | 1.9 | 1.5 | 0.2 |
| $\gamma$-Globulins | 1.9 | 1.5 | 1.2 | 0.3 |
| Orosomucoid (in $\alpha_1$-globulins) | 15.0 | 12.0 | 12 | 1.0 |

Variations in the sialic acid content are characteristic of genetic variants (B-73). They cause differences in the electrophoretic mobilities (B-72).

Under pathological conditions the blood plasma may contain *abnormal proteins* which do not occur in the normal plasma (B-60 to 62). One of these is the *C-reactive protein* which is a crystallizable $\beta$-globulin endowed with the capacity for precipitating the polysaccharide of the pneumococcal type C (B-63). It is found in the blood serum of persons suffering from various infectious and inflammatory processes.

The globulin content of the blood serum increases to very high values in persons affected by myelomas, malignant tumors of the bones. The serum of these patients frequently contains globulins of very high molecular weight called *macroglobulins* (B-64). Similar macroglobulins occur also in other pathological conditions. The macroglobulins are a mixture of globulins having a sedimentation constant 19S with globulins of higher and lower sedimentation constants (B-65). Some of them become insoluble when the serum is cooled down; they are designated as *cryoglobulins*. The urine of patients suffering from multiple myelomas of the bones frequently contains *Bence-Jones protein* which becomes insoluble when the urine is heated to 45–55°C, but redissolves upon boiling. Its molecular weight is 45,000 or less (B-62). Determinations of the *N*-terminal amino acids have revealed that the Bence-Jones proteins which occur in different patients differ from each other in their chemical and physicochemical properties (B-66). The Bence-Jones proteins resemble the L subunits of the normal γ-globulins whereas the macroglobulins contain the large H subunits (B-74).

A specific globulin of high carbohydrate content called *fetuin*, has been discovered in the serum of newborn calves, where it constitutes about 90% of the globulins (B-67). Its molecular weight is 50,000 and it is salted out by 35–45% ammonium sulfate (see Chapter XI, C).

Deviations from the normal ratio of the plasma proteins are revealed by several simple *clinical tests* (B-61). One of them is the Takata-Ara reaction, in which the serum is tested for its protective action on the precipitation of mercuric chloride by sodium carbonate; the higher the albumin:globulin ratio, the higher the protective capacity of the serum. The test is positive in certain diseases of the liver. Another similar test is the thymol turbidity test (B-68), in which the serum is mixed with a saturated solution of thymol in barbiturate buffer at pH 7.8. While normal serum remains clear, serum of patients with liver diseases frequently gives a distinct turbidity, which is due to an increase in γ-globulins, phospholipids, and cholesterol and is inhibited in normal serum by the albumins.

## C. Fibrinogen and Its Conversion to Fibrin (C-1 to 6)

*Fibrinogen* is a globulin which is present in normal human blood plasma in amounts of 0.2–0.4%. It is precipitated by 5% ethanol, by half-saturation with sodium chloride, by repeated freezing and thawing of the plasma (C-7), or by treatment of the plasma with ether (C-8). Fibrinogen solutions are highly viscous and show flow birefringence (C-9). Both properties are indicative of an elongated shape of the fibrinogen molecules. Some of the other physical-chemical properties of fibrinogen are shown in Table VIII-4.

In electron micrographs, fibrinogen molecules appear as 3 nodules held together by threads or rods which link the first to the second, and the second to the third, nodule (C-11). The diameters of the three nodules are 65, 50, and 65 Å; that of the threads is 15 Å and the total length of the molecule 475 Å. This agrees well with earlier estimates of an elongated molecule of the dimensions $50 \times 600$ Å (C-12). Human fibrinogen contains 2.36% carbohydrates (1.32% hexoses and 1.04% aminosugars) (C-13) and one phosphorus atom per molecule (C-39).

The *conversion of fibrinogen into fibrin* is promoted by the proteolytic enzyme *thrombin* whose specificity is similar to that of trypsin; it catalyzes the hydrolysis of peptide bonds formed by the carboxyl group of the basic amino acids arginine or lysine (C-14). Intravascular normal blood does not clot because it is free of thrombin. The latter is formed from a precursor, *prothrombin*, by the action of thromboplastic substances. Prothrombin is a globulin present in the III-2 fraction of blood serum (see

Fig. VIII-2). It can be purified by asdorption to barium sulfate and subsequent elution. The best prothrombin preparations contain about 2000 thrombin units per milligram after activation (C-15). Normal human plasma contains about 15 mg of prothrombin/100 ml.

The mechanism of the conversion of prothrombin into thrombin is not yet clear. Numerous factors are involved in this process. Originally, it had been assumed that a single activator, *thromboplastin*, was responsible for

TABLE VIII-4

PROPERTIES OF SOME FACTORS CONCERNED IN CLOTTING[a]

| Factor | Electro-phoretic fraction | Molecular weight | Iso-electric point | Cohn fraction | Precipitation by ammonium sulfate (% saturation) | Stability at 56°C |
|---|---|---|---|---|---|---|
| Fibrinogen | F | 340,000[c] | 5.5 | I-2 | 25 | labile |
| Thrombin | Alb. | 62,000 | 4.1 | Alb. | 50 | labile |
| Prothrombin | $\alpha_2$ | 62,000 | 4.2 | III-2 | — | stable |
| V[b] | — | — | 5.3 | III | 45 | labile |
| VII[b] | $\beta$ | — | — | III-2 | — | stable |
| VIII[b] | $\beta_2$ | — | — | I | 33 | labile |
| IX[b] | $\beta$ | — | — | IV | 33–50 | stable |

[a] From R. G. Macfarlane (C-2).

[b] Factor V = Ac. globulin; Factor VII = prothrombin conversion factor = proconvertin = serum prothrombin conversion accelerator (SPCA); Factor VIII = antihemolytic factor; Factor IX = Christmas factor (PTC, plasma thromboplastin component).

[c] 407,000 from light scattering (C-10).

the activation of prothrombin. Since clotting occurs rapidly in wounds, this was attributed to the release of thromboplastin (C-16) from damaged cells and blood platelets. The term *thrombokinase* has been used by authors who attribute the activation of prothrombin to an enzymatic reaction. Fractionation procedures reveal that several factors are involved in the activation process. Some of them are listed in Table VIII-4. In addition to these factors, calcium ions are required for the formation of thrombin. Clotting can be prevented by the addition of oxalate, citrate, or fluoride, all of which form complexes with calcium ions. Decalcification can also be accomplished by ion exchange resins.

Factor V or accelerator globulin (Ac-globulin) (C-17) is a labile globulin fraction of the blood serum. It is involved in the conversion of Factor VIII ($\alpha$-prothromboplastin or thromboplastinogen A) and Factor IX ($\beta$-prothromboplastin or plasma thromboplastin component, PTC) into thromboplastin. An intermediate product formed from the Factors VIII and IX is converted into active thromboplastin by the action of Factor VII from tissue extracts (or by a platelet factor) and subsequent action of Factor V (C-1,2). The Factors V and VIII are labile and are consumed during the clotting process. The Factors VII and IX are stable and are not consumed. It is difficult to decide whether any of these factors is an enzyme since the clotting of blood is catalyzed by many nonspecific substances such as the surface of glass vessels. When glass vessels are coated with paraffin or silicone, the clotting process is slowed down. Complications also arise from the fact that traces of thrombin, formed by the action of the thromboplastic factors on prothrombin, catalyze the further conversion of prothrombin into thrombin (C-18). The failure of blood to clot in persons suffering from hereditary hemophilia is attributed to deficiency of the Factor VIII (C-19).

Under biological conditions the conversion of prothrombin to *thrombin* takes place very rapidly, usually in less than 12 sec. This and the presence of the different thromboplastic factors render an investigation of this reaction difficult. The conversion of prothrombin into thrombin takes place very slowly when prothrombin is dissolved in 25% sodium citrate, in the absence of thromboplastic factors. Under these conditions, prothrombin is degraded to smaller fragments (C-20), some of which form aggregates secondarily. It is doubtful whether thrombin formed by this process is identical with the natural thrombin formed during the clotting of whole blood.

The activity of thrombin preparations is expressed in thrombin units. As defined by the U.S. National Institutes of Health, one unit is the amount of thrombin that clots one ml of standard fibrinogen in 15 sec at 28°C (C-2). If a unit of thrombin is allowed to catalyze the hydrolysis of TAME (tosyl-arginine methyl ester) at pH 8 and 30°C, it hydrolyzes 0.1 $\mu$mole of TAME/min (C-21). The amino acid composition of bovine prothrombin (C-22) is shown in Table VIII-2. The active group of thrombin seems to be very similar to that of trypsin. Both enzymes, after inhibition by DFP and subsequent partial hydrolysis, yield the peptide Gly.Asp.Ser.Gly in which the seryl residue is acylated by the diisopropylphosphoryl group (C-23).

The *conversion of fibrinogen into fibrin* had originally been considered as consisting merely of the aggregation of fibrinogen molecules to form large aggregates of fibrin. In 1952, Lorand (C-24) found that the number of N-terminal residues increased during the conversion of fibrinogen into fibrin and that during this process new peptides, called fibrinopeptides A and B, were formed. Their respective molecular weights were approximately 1900 and 2400 (C-25). The composition of the bovine fibrinopeptide A is Glu.Arg.Gly.Ser.Asp.Pro.Pro.Ser.Gly.Asp.Phe.Leu.Thr.Glu.Gly.-Gly.Gly.Val.Arg—; that of the fibrinopeptide B is N-acetyl-Thr.Glu.-

Phe.Pro.Asp.Tyr(OSO$_3$⁻).Asp.Glu.Gly.Glu.Asp.Asp.Arg.Pro.Lys.Val.-Gly.Leu.Gly.Ala.Arg— (C-26). In the fibrinogen molecule the arginyl residues are bound to glycine which has been found as the $N$-terminal amino acid of fibrin (C-24). Since the loss of the 2 fibrinopeptides involves a loss of 12 acidic and only 5 basic amino acid residues, fibrinogen migrates anodically faster than fibrin (C-27). Tyrosine-$O$-sulfate has been discovered among the acidic residues lost from the fibrinogen molecule (C-32). This residue has not yet been found in any other protein. Human fibrinopeptide contains a phosphoserine residue; its amino acid sequence is: Ala.-Asp.P-Ser.Gly.Glu.Gly.Asp.Phe.Leu.Ala.Glu.Gly.Gly.Gly.Val.Arg.OH (C-38). The conversion of fibrinogen into the fibrin monomer does not involve any change in the optical rotation (C-33), nor in the rotatory dispersion (C-34).

The immediate product of the action of thrombin on fibrinogen is an unstable *monomer of fibrin* which remains soluble provided the solution is acidified to pH 4 (C-28,29). The soluble fibrin monomer is also obtained by the disaggregating action of urea on the insoluble fibrin polymer (C-28). In the presence of hexamethylene glycol the polymerization of the fibrin monomer is slowed down and proceeds only to the formation of a pentadecamer in which 15 fibrin monomer molecules are aggregated laterally and also end-to-end (C-30a).

The first steps of the polymerization are reversible; if the process proceeds beyond the hepta- or octamer, it becomes irreversible (C-30b). Since polymerization does not take place at pH of less than 6.1, it has been concluded that the imidazole residues of histidine, which are positively charged at lower pH values, prevent the mutual lateral approach of fibrin monomers and their aggregation (C-31). The conversion of the soluble fibrin monomer into a stabilized form which then polymerizes has been attributed to the action of an enzyme called fibrinase (C-37). The polymerization is brought about by the formation of SS bonds (C-37), but also involves intermolecular interaction of ionic groups, subsequent interlocking of nonpolar groups (C-30a), and the formation of intermolecular hydrogen bonds (C-30). In the electron microscope fibrin threads show a cross-striation with intense bands at intervals of approximately 249 Å and finer intermediate bands (C-10). The intense bands may correspond to double rows of laterally aggregated terminal nodules of the elongated parallel fibrin molecules, while the finer bands may correspond to rows of the single central nodules (C-10). The fibrin scaffold of the clot is extremely loose. Upon stirring with a glass rod, it collapses, winds around the rod, and can thus be isolated almost quantitatively. A fibrin clot which originally occupied a volume of 100 ml is reduced in this way to a fibrin particle whose volume is less than 1 ml.

## D. Milk Proteins (D-1)

The principal protein of milk is *casein*, a phosphoprotein which is precipitated by acidification of the milk. It contains about 0.9% phosphorus. Most of the phosphoric acid in casein is bound to the hydroxyl groups of serine. This has been proved by the isolation of phosphoserine and phosphoserylglutamic acid from enzymic hydrolyzates of casein (D-2). In addition to primary phosphate ester bonds, casein may also contain $N$-phosphoryl, pyrophosphate, or secondary phosphoryl groups (D-3). However, most of the phosphate can be liberated by the action of phosphomonoesterases (D-4). By fractional precipitation with aqueous or alcoholic acids casein has been resolved into several fractions designated as $\alpha$-, $\beta$-, and $\gamma$-casein (D-5). $\alpha$-Casein contains 0.85% of P; it is the principal component of casein, forming about 75% of the total casein (D-6). $\gamma$-Casein, which forms only 3% of the total casein, has the lowest P content; its isoelectric point is at pH 5.8–6.0. Since the casein components are insoluble in water, the molecular weight has been determined from their solutions in $6M$ urea; at pH 7 the molecular weight of $\alpha$-casein was 27,600, that of $\beta$-casein 19,800 (D-7). The amino acid composition of the three casein components, as shown in Table VIII-5, is quite different (D-8). None of the three casein components seems to be a uniform pure substance. They may be mixtures of genetically determined variants (D-30). Further purification of $\alpha$-casein yielded preparations designated as $\alpha_1$-casein (D-9) and $\alpha_3$-casein (D-10). From the tryptic digest of $\beta$-casein a phosphopeptone of the molecular weight 3130 has been isolated (D-11). Its composition is: $Asp_3,Thr,Ser_3,Glu_8,Val,Leu,Ileu_2,Ala,Met,P_3,Fe$ (D-11).

Casein is precipitated from the milk at neutral reaction by the action of the enzyme rennin (rennet) which is produced in the gastric mucosa (O. Hammarsten, 1897). The *clotting* has been described as a conversion of casein into paracasein, or of caseinogen into casein. Very little is known about the mechanism of this process. Like the clotting of fibrinogen, it depends on the presence of calcium ions and seems to be caused by a proteolytic action of rennin (D-12). According to another view (D-13), the enzyme acts as a phosphoamidase which hydrolyzes N—P bonds in N-P-O ester bridges between amino and hydroxyl groups. The insolubility of the product formed from casein is attributed to the formation of an insoluble calcium complex.

Casein is easily soluble in dilute solutions of alkali hydroxides. If such solutions are extruded through fine jets into an acid bath, threads are obtained which can be spun and hardened by treatment with formaldehyde. Textile fibers are manufactured in this manner. $\kappa$-Casein is the designation for a casein fraction which has been prepared by converting casein into its calcium complex, centrifuging off the insoluble calcium micelles, removing calcium by means of oxalate, and removing the latter by dialysis (D-23). $\kappa$-Casein seems to be the fraction which stabilizes the casein micelles. The molecular weight of purified $\kappa$-casein is 26,000 (D-24).

TABLE VIII-5

AMINO ACID COMPOSITION OF SOME PROTEINS FROM MILK

| Amino acid | $\alpha$-Casein[a] | | $\beta$-Casein[a] | $\gamma$-Casein[b] | Lactalbumin[c] (Mol. wt.: 15,554) | | $\beta$-Lactoglobulin[d] (Mol. wt.: 37,300) | |
|---|---|---|---|---|---|---|---|---|
| | A[f] | B[f] | A | A | A | C[f] | A | C |
| Glycine | 2.8 | 37 | 2.4 | 1.5 | 3.21 | 6.6 | 3.3 | 16.4 |
| Alanine | 3.7 | 42 | 1.7 | 2.3 | 2.14 | 3.7 | 1.9 | 7.9 |
| Valine | 6.3 | 54 | 10.2 | 10.5 | 4.66 | 6.2 | 4.4 | 14.0 |
| Leucine | 7.9 | 60 | 11.6 | 12.0 | 11.52 | 13.7 | 9.8 | 27.9 |
| Isoleucine | 6.4 | 49 | 5.5 | 4.4 | 6.80 | 8.1 | 6.8 | 19.3 |
| Proline | 8.2 | 98 | 16.0 | 17.0 | 1.52 | 2.1 | 1.7 | 5.5 |
| Phenylalanine | 4.6 | 28 | 5.8 | 5.8 | 4.47 | 4.2 | 5.1 | 11.5 |
| Tyrosine | 8.1 | 45 | 3.2 | 3.7 | 5.37 | 4.6 | 4.8 | 9.9 |
| Tryptophan | 1.6 | 8 | 0.65 | 1.2 | 7.0 | 5.3 | 7.0 | 12.8 |
| Serine | 6.3 | 92[e] | 6.8 | 5.5 | 4.76 | 7.0 | 4.4 | 15.6 |
| Threonine | 4.9 | 41 | 5.1 | 4.4 | 5.50 | 7.2 | 5.2 | 16.3 |
| Cysteine | 0.43 | 4 | 0 | 0 | 6.4 | 4.1 | 6.1 | 18.9 |
| Cystine/2 | | | 0 | 0 | | | | |
| Methionine | 2.5 | 17 | 3.4 | 4.1 | 0.95 | 1.0 | 0.9 | 2.2 |
| Arginine | 4.3 | 25 | 3.4 | 1.9 | 1.15 | 1.0 | 1.5 | 3.2 |
| Histidine | 2.9 | 19 | 3.1 | 3.7 | 2.85 | 2.9 | 3.2 | 7.7 |
| Lysine | 8.9 | 61 | 6.5 | 6.2 | 11.47 | 12.2 | 10.2 | 26.0 |
| Aspartic acid | 8.4 | 63 | 4.9 | 4.0 | 18.65 | 21.8 | 18.3 | 51.1 |
| Glutamic acid | 22.5 | 153 | 23.2 | 22.9 | 12.85 | 13.6 | 12.2 | 30.9 |
| Ammonia | 1.6 | | 1.6 | 1.6 | 1.37 | 15.2 | | |

[a] From (D-25).　　[b] From (D-8).　　[c] From (D-15).　　[d] From (D-26).

[e] Sixty-four of these phosphorylated.

[f] A: gram amino acid per 100 g protein; B: amino acid residues in $10^5$ g protein; C: amino acid residues per protein molecule.

If casein is removed from milk by acidification and filtration, 2 additional proteins, *lactalbumin* and *β-lactoglobulin*, can be isolated from the filtrate, i.e., the whey, by precipitation with ammonium sulfate. $\beta$-Lactoglobulin is insoluble at low ionic strength and is obtained in crystalline form when the redissolved protein precipitate is dialyzed against dilute acid (D-14). *Lactalbumin* is salted out from the supernatant solution. It was first believed to be identical with serum albumin. Further purification, however, yields, a crystalline preparation, α-lactalbumin, whose amino acid composition is quite different from that of serum albumin (Table VIII-5) (D-15). The lactalbumin molecule consists of a single peptide chain of the molecular weight 16,300; it is resistant to heating for 10 min at temperatures of 100°C (D-16). The peptide chain of lactalbumin contains 4 intramolecular dithio bonds; the C-terminal sequence is Ileu.Val.-Thr.Lys.Leu.OH (D-17).

Since crystals of *β-lactoglobulin* can be obtained easily, the physicochemical behavior of this protein has been investigated very thoroughly (D-18). Its molecular weight is approximately 38,000 (see Table V-2). Although β-lactoglobulin for many years had been considered a homogeneous protein, electrophoresis reveals that it consists of two components called lactoglobulin A and B (D-19,20). The milk of cows contains either A or B or both β-lactoglobulins. Their occurrence is genetically determined by the genotypes $Lg^A$ and $Lg^B$ which determine the production of the phenotypes A/A, A/B, and B/B (D-19). Both β-lactoglobulin A and B are dimers of two monomers of the molecular weight 19,000; X-ray diffraction reveals the presence of a dyad axis in the dimer (D-27). In the monomer of β-lactoglobulin A, one glycine and one alanine residue of β-lactoglobulin B are replaced by one aspartyl and one valyl residue (D-28). Due to its higher aspartic acid content, β-lactoglobulin A migrates anodically faster than β-lactoglobulin B (D-21). The two lactoglobulins can be separated in this manner. In addition to lactalbumin and lactoglobulin, whey contains a *red protein* which has a molecular weight of 86,000 and contains 2 iron atoms per molecule (D-22). The iron can be removed at pH 2 by Dowex 50 resin; the red protein contains 7.2% carbohydrates and has its isoelectric point at pH 7.8 (D-22). It seems to be identical with lactotransferrin (D-29).

## E. Egg Proteins (E-1,2)

Ovalbumin, the principal protein of egg white in hens' eggs is one of the most thoroughly investigated proteins, since it is easily obtainable in crystals. The egg proteins were first isolated and described by Hofmeister (E-3) who removed globulins by half-saturation with ammonium sulfate; ovalbumin crystallized after acidification of the filtrate. If the use of ammonium salts is undesirable, a 36.7% solution of sodium sulfate may be used instead. The globulin-free filtrate is acidified with dilute sulfuric acid to pH 4.7 and anhydrous sodium sulfate is added to bring about slight opalescence. Upon standing, crystalline ovalbumin forms and settles out (E-4). Ovalbumin constitutes about 50% of the proteins of egg white. In spite of its crystalline appearance it is not quite homogeneous, but consists of two components, A and B, which can be separated by electrophoresis (E-5). Their occurrence seems to be determined genetically (E-6). The amino acid composition of ovalbumin is shown in Table VIII-6. Its carbohydrate content is discussed in Chapter XI, C. Ovalbumin contains

TABLE VIII-6
AMINO ACID COMPOSITION OF SOME PROTEINS OF HENS' EGGS[a]

| Amino acid | Ovalbumin[b] (Mol. wt.: 46,000) | Conalbumin[b] (Mol. wt.: 87,000) | Lysozyme[c] (Mol. wt.: 15,000) | Avidin[b] (Mol. wt.: 66,000) |
|---|---|---|---|---|
| Glycine | 19 | 66 | 12 | 40 |
| Alanine | 25 | 43 | 12 | — |
| Valine | 28 | 61 | 6 | 24 |
| Leucine | 32 | 58 | 8 | 25 |
| Isoleucine | 25 | 33 | 6 | 28 |
| Proline | 14 | 37 | 2 | 9 |
| Phenylalanine | 21 | 30 | 3 | 24 |
| Tyrosine | 9 | 22 | 3 | 3 |
| Tryptophan | 3 | 13 | 5–6 | 17 |
| Serine | 36 | 52 | 12 | 28 |
| Threonine | 16 | 43 | 7–8 | 58 |
| Cystine/2 | 7[d] | 14 | 8 | 1 |
| Methionine | 16 | 12 | 2 | 6 |
| Arginine | 15 | 38 | 10–11 | 25 |
| Histidine | 7 | 14 | 1 | 4 |
| Lysine | 20 | 60 | 6 | 28 |
| Aspartic acid | 32 | 87 | 22 ± 1 | 48 |
| Glutamic acid | 52 | 70 | 5 | 30 |
| Ammonia | 33 | 65 | 18 | 61 |

[a] Amino acid residues per protein molecule.
[b] From (E-23).
[c] P. Jolles, *in* "The Enzymes" (P. Boyer, H. Lardy, and K. Myrbäck, eds.), Vol. 4, p. 436. Academic Press, New York, 1959/61.
[d] Five SH.

two phosphoric acid groups which can be hydrolyzed by phosphatases (E-7). No *N*-terminal amino acid has been found in ovalbumin. The *C*-terminal amino acid sequence is —Val.Ser.Pro.OH (E-8). If ovalbumin is treated with subtilisin, a proteolytic enzyme from *Bacillus subtilis*, it is converted into an ovalbumin-like protein which crystallizes in platelets and has been called plakalbumin; a heptapeptide, Glu.Ala.Gly.Val.Asp.-Ala.Ala.OH, is split off from ovalbumin in this reaction (E-9). The native protein does not give any color reaction for sulfhydryl groups with sodium nitroprusside. Denatured ovalbumin gives an intense pink color with this reagent in ammoniacal solution.

The filtrate of the ovalbumin crystals contains *conalbumin* which forms about 15% of the proteins of the egg-white (E-10). It gives a pink complex with ferric ions and a yellow complex with cupric ions (E-11). Both conalbumin and its iron complex have been obtained in crystals (E-12). The molecular weight of conalbumin is 85,000 (E-13). Its amino acid composition is shown in Table VIII-6. Conalbumin is a glycoprotein whose protein moiety is identical with that of transferrin; the carbohydrate component is different from that of transferrin (E-25).

*Lysozyme*, which forms about 3% of the egg white, occurs also in nasal mucus, cartilage, and in tears, the secretions of the lacrymal glands. It is an enzyme which lyzes the capsular $N$-acetyl-chitosaccharides of certain bacteria (E-24). Lysozyme is rich in basic amino acids (see Table VIII-6); its isoelectric point is at pH 10.5–11.0; its molecular weight is approximately 17,000 (E-1). Partial hydrolysis has revealed some of the amino acid sequences in lysozyme (E-14,26). The lysozymes prepared from different sources differ from each other in their amino acid composition (E-15). X-ray diffraction has revealed part of the conformation of lysozyme (E-27).

A considerable portion of the egg-white protein is formed by *ovomucoid*, a glycoprotein (see Chapter XI, C). Very little is known of ovoglobulin, which is salted out by ammonium or sodium sulfate before the isolation of ovalbumin (E-16). Egg-white also contains *avidin*, a basic deoxyribonucleoprotein responsible for the "egg-white injury" in animals fed raw egg-white (see Chapter XI, I). Its toxic action is due to its combination with biotin and subsequent production of biotin deficiency. Avidin has been prepared in crystals (E-17a). It is a basic glycoprotein whose isoelectric point is close to pH 10 (E-17b). The egg-white proteins can be separated from each other by electrophoresis (E-18). Their amino acid composition is shown in Table VIII-6, their carbohydrate is discussed in Chapter XI, C.

The *yolk proteins* are quite different from those of the egg-white. Their mixture has been separated by the fractional addition of magnesium sulfate and centrifugation (E-19). The insoluble magnesium complex of *phosvitin* precipitates first. Phosvitin contains almost 10% of phosphorus and only 12% nitrogen; its molecular weight is 21,000 (E-20). This corresponds to 65 phosphate residues per molecule. Most of these occur in runs of 3 or 4 adjacent seryl-phosphate residues (E-21). Phosvitin is a heterogeneous protein which has been resolved into two fractions of different composition, one of them free of tyrosine (E-22). The filtrate, after precipitation of phosvitin with magnesium sulfate, contains the $\alpha$- and *β-lipovitellins*, which can be precipitated by raising the magnesium sulfate concentration (E-19). They are lipoproteins containing phospholipids. Another series of yolk proteins, called α-, β-, and γ-*livetin*, are salted out by ammonium sulfate or are coagulated by heat. They can be separated from each other by chromatography (E-19) and are similar to or identical with serum albumin, $α_2$-glycoprotein, and γ-globulin, respectively (E-25). Phosphoproteins, called *ichthulins* (Greek: *ichthys* = fish) have been found in fish eggs.

## F. Vegetable Proteins (F-1 to 6)

Vegetable proteins have been obtained principally from the seeds of cereals and leguminous plants. Osborne (F-4), in his pioneer work, removed lipids by extraction with ether and/or petrol ether and extracted the

TABLE VIII-7
AMINO ACID COMPOSITION OF SOME VEGETABLE PROTEINS[a]

| Amino acids | Edestin[b] | Pumpkin seed globulin[b] | Zein[c] | Gliadin[c] |
|---|---|---|---|---|
| Glycine | 5.01 | 4.59 | — | — |
| Alanine | 5.14 | 5.09 | 10.52 | 2.13 |
| Valine | 6.39 | 5.86 | 3.98 | 2.66 |
| Leucine | 7.86 | 8.63 | 21.1 | |
| Isoleucine | 5.49 | 4.91 | 5.0 | 11.9 |
| Proline | 3.78 | 4.62 | 10.53 | 13.55 |
| Phenylalanine | 6.33 | 5.89 | 7.3 | 6.44 |
| Tyrosine | 4.87 | 4.13 | 5.25 | 3.20 |
| Tryptophan | 1.23 | 1.74 | 0.16 | 0.66 |
| Serine | 5.85 | 6.57 | 7.05 | 4.90 |
| Threonine | 3.71 | 3.30 | 3.45 | 2.10 |
| Cystine[d] | 1.23 | 1.12 | 0.83 | 2.58 |
| Methionine | 2.44 | 2.48 | 2.41 | 1.69 |
| Arginine | 17.27 | 15.24 | 1.71 | 2.74 |
| Histidine | 2.88 | 2.70 | 1.32 | 1.82 |
| Lysine | 2.90 | 3.88 | 0 | 0.65 |
| Aspartic acid | 13.22 | 10.81 | 4.61 | 1.34 |
| Glutamic acid | 21.27 | 20.74 | 26.9 | 45.7 |
| Ammonia | 3.09 | 1.96 | 2.98 | 4.5 |
| $N$ as % of total $N$ | 106.45 | 98.21 | — | — |

[a] Amino acid composition in grams per 100 g protein.
[b] From (F-15).
[c] From (F-16).
[d] Including cysteine.

delipidated residue with aqueous solutions of salts or with 70 % ethanol. The protein content of the dry grains of cereals is approximately 8–15%, whereas the seeds of leguminous plants may contain up to 50 % protein. Most of the proteins extracted from seeds have the properties of globulins; only 0.1–0.5 % of albumins are present in the seeds. Those globulins which are soluble in aqueous ethanol have been called *prolamins*. Globulins which dissolve neither in salt solutions nor in alcohol, but are extracted by dilute acids or alkali hydroxides, have been called *glutelins*. Neither the prolamins nor the glutelins are uniform proteins; they are mixtures of various similar proteins, some of them denatured by the exposure to alkali hydroxide. Denaturation by ether and alcohol can be avoided by working at low temperatures.

Among the soluble globulins *edestin* from hemp seeds has received the most attention. Other similar proteins are excelsin from brazil nuts, amandin from almonds, legumin from peas and lentils, phaseolin from beans, glycinin from soybeans, canavalin and concanavalin from jack beans, and globulins from cotton seed, pumpkin seed, and other seeds. These globulins are extracted from the ground seeds by 2–10 % sodium chloride solutions. Not all of them correspond strictly to the definition of a globulin insofar as some of them are not precipitated by half-saturation with ammonium sulfate. The organic acids present in seeds are neutralized by adding baryta to the sodium chloride solution. Upon dialysis, edestin and some of the other globulins precipitate as crystals. Edestin can be recrystallized from 10 % sodium chloride solution. Its amino acid composition and that of a globulin from pumpkin seeds is shown in Table VIII-7. The following molecular weights have been found for some of the globulins (F-7,-8):

| | | | |
|---|---|---|---|
| Edestin | 310,000 | Concanavallin B | 42,000 |
| Excelsin | 295,000 | Concanavallin A | 96,000 |
| Amandin | 330,000 | Canavalin | 113,000 |

Many if not all of these high values are caused by the association of smaller subunits. Thus it has been shown that glycinin (F-9) and legumin (F-10) are hexamers, and also that arachin is a polymer (F-11). Dissociation into smaller units is accomplished by exposure to urea or dodecyl sulfate, or by increasing ionic strength or pH. The dissociation of arachin is a reversible reaction; in the nut, the protein occurs in the associated form (F-11). Electron micrographs of edestin show spherelike particles with a diameter of approximately 70 Å (F-12; see Fig. V-8). Their size agrees with the molecular weight of 310,000 found in the aqueous solutions of edestin.

The ethanol-soluble seed globulins have been called *prolamins* because of their high content of proline and amide nitrogen. Table VIII-7 shows that they contain only small amounts of basic amino acids but very large amounts of glutamic acid. Thus, more than 40% of the protein gliadin is formed by glutamic acid, most of it present as glutamine. Some of the prolamins are devoid of lysine, a fact which is important in nutritional problems.

The most extensively investigated prolamins are gliadin from wheat kernel, hordein from barley (*Hordeum vulgare*), and zein from corn (*Zea mays*). A convenient method for their preparation consists in kneading flour from the examined cereal with water to remove the starch grains from the sticky gluten. If the latter is extracted with 70–80% ethanol, the prolamins are dissolved, whereas the insoluble residue consisting of glutenin can be dissolved in dilute alkali. The insolubility and stickiness of glutenin is caused by dithio bonds between subunits of the molecular weight 46,000 (F-17). The cohesive properties are lost when the disulfide bonds are cleaved (F-18). Neither the prolamins nor the glutenins are homogeneous substances. Thus gliadin contains two fractions with isoelectric points of pH 5 and 7 (F-13). Since hordein is not attacked by aminopolypeptidase and since it yields $CO_2$ on treatment with ninhydrin, it seems to have an *N*-terminal γ-glutamyl residue (F-14).

## G. Protamines and Histones (G-1,2)

The *protamines* are strongly basic proteins of low molecular weight. They were discovered in the mature sperm of the salmon by Miescher (1874). Their chemical nature was elucidated by Kossel (G-1) who dissociated the nucleoprotamines by dilute sulfuric acid and precipitated protamine sulfate by alcohol. If an aqueous solution of the protamine sulfate is treated with sodium picrate, insoluble protamine picrate is obtained which by sulfuric acid can again be converted into the sulfate. Picric acid is removed by toluene. Better preparations are obtained by isolation of the sperm cell nuclei, washing with dilute citric acid, dissolution of the nucleoprotamine in 10% saline solution, and reprecipitation with water. If the dry nucleoprotamine is extracted with 0.2% HCl, protamine hydrochloride goes into solution; it is precipitated by acetone (G-3). The most thoroughly investigated protamines are salmine from the salmon and clupein from the herring (*Clupea harengus*). Other protamines occasionally investigated are sturine from the sturgeon, scombrin from the mackerel (*Scomber scomber*), and iridine from the trout. The amino acid composition of some of the protamines is shown in Table VIII-8. It can be seen that the bulk of the molecule is formed by arginine and that it contains, beside this strongly basic amino acid, only a few of the monoamino acids and proline. Protamines contain neither tryptophan nor tyrosine; they are free of cystine and methionine.

Because of their high arginine content, the protamines are strong bases. If their hydrochlorides or sulfates are neutralized with one equivalent of strong alkali, the free protamines are obtained as a heavy oily layer. Like

TABLE VIII-8
AMINO ACID COMPOSITION OF SOME PROTAMINES AND HISTONES

| Amino acid | Calf thymus histone Fraction I[a] (%) | Calf thymus histone Fraction II[a] (%) | Salmine[b] (Mol./Mol.) | Clupeine[b] (Mol./Mol.) |
|---|---|---|---|---|
| Glycine | 5.2 | 6.4 | — | — |
| Alanine | 19.1 | 7.5 | 4 | 7 |
| Valine | 4.3 | 5.4 | 4 | 4 |
| Leucine | 3.8 | 8.9 | — | — |
| Isoleucine | | | 1 | 1 |
| Proline | 6.4 | 2.5 | 7 | 9 |
| Phenylalanine | 0.55 | 1.3 | — | — |
| Tyrosine | 0.34 | 1.5 | — | — |
| Serine | 5.1 | 3.0 | 6 | 4 |
| Threonine | 4.5 | 4.4 | — | 2 |
| Cystine[c] | 0 | 0 | — | — |
| Methionine | < 0.05 | 0.4 | — | — |
| Arginine | 4.8 | 20.6 | 51 | 65 |
| Histidine | 0.15 | 5.9 | — | — |
| Lysine | 40.7 | 15.5 | — | — |
| Aspartic acid | 1.3 | 3.9 | — | — |
| Glutamic acid | 2.6 | 6.3 | — | — |
| Ammonia | 1.8 | 3.1 | — | — |

[a] From (G-32), recorded as amino acid nitrogen in per cent of total histone nitrogen.
[b] From (G-29), recorded in number of amino acid residues per residue of isoleucine.
[c] Including cysteine.

other macromolecular cations, the protamines form insoluble complexes with nucleic acids and other polyanions. They also precipitate solutions of denatured, but not of native, albumins and globulins (G-4a,4b). Since arginine forms approximately two-thirds of the amino acids, it has been suggested that the protamine molecules consist of units of the composition MAA where A is an arginine and M a monoamino acid residue. However, the results of partial hydrolysis do not always agree with this assumption (G-5,6). Determinations of the N-terminal amino acid in various protamine preparations reveals that many of them are heterogeneous.

Heterogeneity has also been found in electrophoresis (G-7), chromatography (G-8,9), and countercurrent distribution (G-10).

In view of this heterogeneity, the amino acid analyses and the determinations of molecular weight lose much of their value. However, it is clear from results obtained by different methods that the molecular weights of the protamines are very low; those of salmine and clupeine are close to 6000 (G-11). The $N$-terminal amino acid of salmine and clupeine is proline (G-11,12). The composition of clupein Z is: Ala.Arg$_4$.Ser.Arg$_2$.Ala.-Ser.Arg.Pro.Val.Arg$_4$.Pro.Arg$_2$.Val.Ser.Arg$_4$.Ala.Arg$_4$.OH (G-33). Heterogeneity of salmine is indicated by the finding of only 0.5 leucyl residues per molecule (G-13). The heterogeneity of the protamine molecules is surprising in view of their small molecular weights. It has been attributed to replacement of some of the amino acids by other amino acids during maturation of the sperm (G-14).

A strongly basic protein of low molecular weight, called gallin, has been discovered in the sperm of roosters (G-16). Its amino acid composition is very similar to that of typical protamines (G-16,17). Very little is know about basic proteins which have been found in the spores of lycopodium clavatum (G-18) and in the fruit of tung trees (G-19).

The *histones* are basic proteins which occur in the nuclei of the somatic cells bound to deoxyribonucleic acid (DNA). Their molecules are less basic than those of the protamines since they contain only 10–30% arginine. In contrast to the protamines they contain lysine and histidine. Since the histones are least soluble at their isoelectric point, near pH 8.5, they can be precipitated by the addition of ammonia. The histones are free of tryptophan and the sulfur-containing amino acids; they contain the aromatic amino acids tyrosine and phenylalanine. Thus their amino acid composition (see Table VIII-8) is intermediate between that of the protamines and the typical proteins. Their molecular weights vary from about 5000 to 37,000 (G-20). The larger complexes seem to be polymers since they dissociate into smaller units in concentrated urea solutions (G-21).

The best investigated histones are those of the thymus gland of calves. The glands are first extracted with isotonic saline solution to remove cytoplasmic proteins and other substances. The nuclei are then extracted with concentrated solutions of NaCl which dissociate the nucleohistones and render their two components, histones and DNA, soluble. The nucleohistone can be reprecipitated by dilution with water, the histone extracted with dilute HCl and reprecipitated with ammonia (G-22). Alternatively, the sodium salt of DNA can be precipitated by ethanol from its solution in 2.6 $M$ NaCl whereas the histone remains soluble (G-23). If the nuclei are extracted with 0.05 $M$ citric acid and 0.5 $M$ NaCl, only lysine-rich histones go into solution; the arginine-rich histones remain insoluble (G-24). Fractionation of the histones can also be accomplished by electrophoresis (G-25,31), alcohol precipitation (G-26), or ion exchange resins (G-27,30). In this manner six or more different histones have been obtained from calf thymus nuclei (G-28). Evidently the histones, like the protamines, consist of a heterogeneous population of molecules of varying amino acid composition (G-34).

## REFERENCES

### Section B

**B-1.** F. W. Putnam, ed., "The Plasma Proteins," Vol. 1. Academic Press, New York, 1960. **B-2.** R. A. Kekwick, *Adv. in Protein Chem.* 14: 231(1959). **B-3.** J. T. Edsall, *Adv. in Protein Chem.* 3: 384(1947). **B-4.** F. Pohl, *Arch. exptl. Pathol. u. Pharmakol.* 20: 426(1886). **B-5.** O. Hammarsten, *Z. physiol. Chem.* 8: 467(1897). **B-6.** P. E. Howe, *JBC* 49: 93, 109(1921). **B-7.** A. M. Butler and H. Montgomery, *JBC* 99: 173 (1933). **B-8.** E. J. Cohn *et al.*, *JACS* 62: 3386(1940). **B-9.** S. Levine, *ABB* 50: 515 (1954). **B-10.** E. Kallee *et al.*, *Z. Naturforsch.* 12b: 777(1957). **B-11.** G. W. Schwert, *JACS* 79: 139(1957). **B-12.** S. Fleischer and F. Haurowitz, *Arzneimittel-Forsch.* 10: 362(1960). **B-13.** J. Sri Ram and P. H. Maurer, *ABB* 76: 28(1958). **B-14a.** E. J. Cohn, *in* "Blood Cells and Plasma Proteins" (J. L. Tullis, ed.), p. 33. Academic Press, New York, 1953. **B-14b.** E. J. Cohn, *JACS* 71: 541, 1223(1949); 72: 465(1950). **B-15.** R. B. Pennell, *in* "The Plasma Proteins" (F. W. Putnam, ed.), Vol. 1, p. 9. Academic Press, New York, 1960. **B-16.** J. Horejsi and R. Smetana, *Collection Czechoslov. Chem. Communs.* 19: 1316(1954). **B-17.** R. A. Kekwick and M. E. Mackay, *Med. Research Council (Brit.) Spec. Rep. Ser.* 286(1954). **B-18.** G. R. Cooper, *in* "The Plasma Proteins," Vol. 1, p. 51. Academic Press, New York, 1960. **B-19.** E. A. Peterson and H. A. Sober, *in* "The Plasma Proteins" (F. W. Putnam, ed.), Vol. 1, p. 105. Academic Press, New York, 1960. **B-20.** R. A. Phelps and F. W. Putnam, *in* **B-1**, p. 158. **B-21.** H. E. Schultze *et al.*, *Naturwiss.* 48: 696(1961); *Biochem. Z.* 328: 267(1958). **B-22.** J. R. Tata, *Nature* 183: 877(1959). **B-23.** A. Saifer *et al.*, *Proc. Soc. Exptl. Biol. Med.* 86: 46(1954); *ABB* 92: 409(1961). **B-24.** H. Hoch, *ABB* 53: 387 (1954). **B-25.** J. Oudin, *J. Immunol.* 81: 376(1958). **B-26a.** P. Bro *et al.*, *JACS* 80: 389(1958). **B-26b.** R. W. Hartley *et al.*, *Biochem.* 1: 60(1962). **B-27.** P. King *et al.* *JACS* 82: 3350, 3355(1960). **B-28.** J. T. Edsall *et al.*, *JACS* 76: 3131(1954). **B-29** M. J. Hunter and F. C. McDuffy, *JACS* 81: 1400(1959). **B-30.** J. E. Turner *et al.*, *in* "Sulfur in Proteins" (R. Benesch, ed.), p. 25. Academic Press, New York, 1959. **B-31a.** G. Biserte, *BBA* 34: 558(1959). **B-31b.** T. E. Thompson and W. M. Mc-Kernan, *BJ* 81: 12(1961). **B-32.** F. Karush, *JACS* 72: 2705(1950). **B-33a.** J. T. Yang and J. F. Foster, *JACS* 76: 1588(1954); J. F. Foster *in* **B-1**, p. 179. **B-33b.** F. Haurowitz *et al.*, *JACS* 74: 2265(1952); *Z. physiol. Chem.* 315: 127(1959). **B-34.** I. M. Klotz and S. W. Luborsky, *JACS* 81: 5119(1959). **B-35.** C. Tanford and J. G. Buzzell, *J. Phys. Chem.* 60: 225(1956). **B-36.** E. O. Thompson, *JBC* 208: 565(1954). **B-37.** A. E. Herner and E. Frieden, *JBC* 235: 2845(1960). **B-38.** F. Wuhrmann, *Schweiz. Med. Wochschr.* 89: 150(1958). **B-39a.** H. Bennhold *et al.*, *Verhandl deut. Ges. inn. Med.* 60: 630(1954). **B-39b.** V. Luzzati *et al.*, *J. Mol. Biol.* 3: 367(1961). **B-40.** M. F. Jayle, *Expos. Ann. Biochim. Med.* 17: 157(1955). **B-41.** C. B. Laurell, *Clin. Chim. Acta* 4: 79(1959). **B-42.** E. A. Hommes, *Clin. Chim. Acta* 4: 707(1959). **B-43.** M. Nyman, *Scand. J. Clin. & Lab. Invest.* 11, Suppl. 39(1959). **B-44.** O. Smithies and N. F. Walker, *Nature* 178: 694(1957). **B-45.** A. C. Bearn and E. C.

Franklin, *J. Exptl. Med.* 109: 55(1959).   **B-46.** O. Smithies and O. Hiller, *Nature* 181: 1203(1958).   **B-47.** B. S. Blumberg, *Proc. Soc. Exptl. Biol. Med.* 104: 25(1960). **B-48.** B. E. Sanders *et al.*, *ABB* 85: 366(1959).   **B-49.** W. R. Horsfall and O. Smithies, *Science* 128: 35(1958).   **B-50.** G. C. Ashton, *Nature* 182: 1101(1960).   **B-51.** G. C. Ashton, *Nature* 180: 917(1957).   **B-52.** J. R. Cann, *JACS* 75: 4213, 4218(1953). **B-53.** F. W. Putnam, *JACS* 75: 2785(1953).   **B-54.** F. Haurowitz and M. Kennedy. *FP* 11: 227(1952).   **B-55.** G. M. Edelman, *JACS* 81: 3155(1959).   **B-56.** J. L. Oncley *et al.*, *J. Phys. Chem.* 56: 85(1952).   **B-57.** R. A. Kekwick, *Adv. in Protein Chem.* 14: 231(1959).   **B-58.** P. Grabar *et al.*, *Bull. soc. chim. biol.* 42: 853(1960).   **B-59.** R. J. Winzler, *in* "The Plasma Proteins" (F. W. Putnam, ed.), Vol. 1, p. 309. Academic Press, New York, 1960.   **B-60.** A. B. Gutman, *Adv. in Protein Chem.* 4: 156(1948). **B-61.** M. L. Peterman *in* **B-1**, Vol. 2, p. 310.   **B-62.** F. W. Putnam, *in* "The Plasma Proteins," (F. W. Putnam, ed.), Vol. 2, p. 346. Academic Press, New York, 1960. **B-63.** W. S. Tilett and T. Francis, Jr., *J. Exptl. Med.* 52: 561(1930).   **B-64.** H. G. Kunkel, *in* "The Plasma Proteins" (F. W. Putnam, ed.), Vol. 1, p. 279. Academic Press, New York, 1960.   **B-65.** H. F. Deutsch *et al.*, *Science* 125: 600(1957).   **B-66.** F. W. Putnam and A. Miyake, *JBC* 227: 1083(1957).   **B-67.** K. O. Pedersen, *Nature* 154: 575, 642(1944).   **B-68.** N. F. Maclagan and D. Bunn, *BJ* 41: 580(1947).   **B-69.** F. Franek, *Biochem. Biophys. Research Communs.* 4: 28(1961).   **B-70.** G. M. Edelman and M. D. Poulik, *J. Exptl. Med.* 113: 861(1961).   **B-71.** F. W. Putnam *et al.*, *JBC* 237: 717(1962); *BBA* 58: 279(1962).   **B-72.** H. E. Schultze, *ABB*, *Suppl.* 1: 290 (1962).   **B-73.** W. C. Parker and A. G. Bearn, *J. Exptl. Med.* 115: 83(1962).   **B-74.** G. M. Edelman and J. A. Gally, *J. Exptl. Med.* 116: 207(1962).

### Section C

**C-1.** R. G. Macfarlane, *Physiol. Revs.* 36: 479(1956).   **C-2.** R. G. Macfarlane, *in* "The Plasma Proteins" (F. W. Putnam, ed.), Vol. 2, p. 137. Academic Press, New York, 1960.   **C-3.** F. D. Mann, *Ann. Rev. Physiol.* 19: 205(1957).   **C-4.** H. A. Scheraga and M. Laskowski, Jr., *Adv. in Protein Chem.* 12: 1(1957).   **C-5.** T. Astrup, *Adv. in Enzymol.* 10: 1(1950).   **C-6.** W. H. Seegers, *Adv. in Enzymol.* 16: 23(1955).   **C-7.** A. G. Ware *et al.*, *JBC* 172: 699(1948).   **C-8.** R. A. Kekwick *et al.*, *BJ* 60: 671(1955). **C-9.** J. T. Edsall *et al.*, *JACS* 69: 2731(1947).   **C-10.** C. S. Hocking *et al.*, *JACS* 74: 775(1952).   **C-11.** C. E. Hall and H. S. Slayter, *J. Biophys. Biochem. Cytol.* 5: 11 (1959).   **C-12.** S. Schulman, *JACS* 75: 5846(1953).   **C-13.** Z. Stary *et al.*, *Z. physiol. Chem.* 295: 28(1953).   **C-14.** W. Troll *et al.*, *JBC* 208: 85, 95(1954).   **C-15.** W. H. Seegers *et al.*, *ABB* 6: 85(1945).   **C-16.** H. Jensen *et al.*, *Proc. Soc. Exptl. Biol. Med.* 86: 387(1954); 90: 679(1955).   **C-17.** A. G. Ware and W. H. Seegers, *JBC* 172: 699(1948). **C-18.** T. Astrup, *Enzymologia* 5: 119(1938).   **C-19.** L. Lorant and K. Laki, *BBA* 13: 448(1954).   **C-20.** F. Lamy and D. F. Waugh, *Physiol. Revs.* 34: 722(1954).   **C-21.** S. Sherry and W. Troll, *JBC* 208: 95(1954).   **C-22.** K. Laki *et al.*, *ABB* 49: 276(1954). **C-23.** J. A. Gladner and K. Laki, *JACS* 80: 1263(1958).   **C-24.** L. Lorand, *BJ* 52: 200 (1952).   **C-25.** J. A. Gladner *et al.*, *JBC* 234: 62, 67(1959).   **C-26.** J. E. Folk *et al.*, *JBC* 234: 2317(1959); *BBA* 44: 383(1960).   **C-27.** E. Mihaly, *Scand. Chim. Acta* 4: 351(1950).   **C-28.** S. Schulman *et al.*, *J. Phys. Chem.* 54: 66(1950); *JACS* 73: 1388 (1951).   **C-29.** K. Laki and W. F. H. M. Mommaerts, *Nature* 156: 664(1945).   **C-30a.** J. Ferry *et al.*, *JACS* 74: 5709(1952); *Physiol. Revs.* 34: 753(1954).   **C-30b.** H. A. Scheraga and J. K. Backus, *JACS* 74: 1979(1952).   **C-31.** S. Schulman *et al.*, *ABB* 42: 245(1953).   **C-32.** F. R. Bettelheim, *JACS* 76: 2838(1954).   **C-33.** V. A. Belitser and E. L. Khodorova, *Biokhimiya* 17: 672(1952).   **C-34.** C. M. Kay and M. M. Marsh, *Nature* 189: 307(1961).   **C-35.** A. G. Loewy and J. T. Edsall, *JBC* 211: 829

(1954). **C-36.** T. Donelly *et al.*, *ABB* 56: 369(1955). **C-37.** A. G. Lowey, *JBC* 236: 2625(1961). **C-38.** B. Blombäck *et al.*, *Nature* 193: 883(1962). **C-39.** P. Fantl and H. A. Ward, *BBA* 64: 568(1962).

## Section D

**D-1.** T. L. McMeekin, *Adv. in Protein Chem.* 5: 202(1949); *in* "Proteins" (H. Neurath and K. Bailey, eds.), Vol. 2A, p. 389. Academic Press, New York, 1954. **D-2.** F. Lipmann, *Biochem. Z.* 262: 9(1933); T. Posternak and H. Pollaczek, *Helv. Chim. Acta* 24: 921(1941). **D-3.** G. Perlmann, *Adv. in Protein Chem.* 10: 1(1955); *BBA* 13: 452 (1954); *Nature* 174: 273(1954). **D-4.** E. B. Kalan and M. Telka, *ABB* 79: 275(1959); T. Hofmann, *BJ* 69: 139(1958); T. A. Sundarajan and P. S. Sarma, *Enzymologia* 18: 234(1957). **D-5.** R. C. Warner, *JACS* 66: 1725(1944). **D-6.** T. L. McMeekin *et al.*, *ABB* 83: 35(1959). **D-7.** H. A. McKenzie and R. G. Wake, *Australian J. Chem.* 12: 734(1959). **D-8.** W. G. Gordon *et al.*, *JACS* 72: 4282(1950); 75: 1678(1953). **D-9.** T. L. McMeekin *et al.*, *ABB* 83: 35(1959). **D-10.** N. J. Hipp *et al.*, *ABB* 93: 245(1961). **D-11.** E. Pantlitschko and E. Gründig, *Monatsh. Chem.* 89: 489(1958). **D-12.** E. Cherbuliez and P. Baudet, *Helv. Chim. Acta* 33: 1673(1950); R. G. Wake, *Australian J. Biol. Sci.* 12: 479(1959). **D-13.** P. F. Dyachenko, *Doklady Akad. Nauk S.S.S.R.* 23: 38(1958). **D-14.** A. H. Palmer, *JBC* 104: 359(1934). **D-15.** W. G. Gordon and J. Ziegler, *ABB* 57: 80(1955). **D-16.** D. B. Wetlaufer, *Compt. rend. trav. lab. Carlsberg, Sér. chim.* 32: 125(1961). **D-17.** L. Weil and T. S. Seibles, *ABB* 93: 193(1961). **D-18.** M. V. Tracey, *Ann. Repts. Chem. Soc.* 46: 214(1950). **D-19.** R. Aschaffenburg and J. Drewry, *BJ* 65: 273(1957); *Nature* 180: 376(1957). **D-20.** M. P. Tombs, *BJ* 65: 517(1957). **D-21.** C. Tanford and Y. Nozaki, *JBC* 234: 2874(1959). **D-22.** M. L. Groves, *JACS* 82: 3345(1960). **D-23.** D. F. Waugh and P. H. von Hippel, *JACS* 78: 4576(1956). **D-24.** H. A. McKenzie and R. G. Wake, *BBA* 47: 240(1961). **D-25.** W. G. Gordon *et al.*, *JACS* 71: 3293(1949). **D-26.** R. J. Block and K. W. Weiss, *ABB* 55: 315(1955). **D-27.** D. W. Green and R. Aschaffenburg, *J. Mol. Biol.* 1: 54(1959). **D-28.** W. G. Gordon *et al.*, *JBC* 236: 2908(1961); K. A. Piez *et al.*, *Ibid.* 2912(1961). **D-29.** B. Blanc and H. Isliker, *Bull. soc. chim. biol.* 43: 929(1961). **D-30.** R. Aschaffenburg, *Nature* 192: 431(1961).

## Section E

**E-1.** H. L. Fevold, *Adv. in Protein Chem.* 6: 188(1951). **E-2.** R. C. Warner, *in* "Proteins" (H. Neurath and K. Bailey, eds.), Vol. 2A, p. 435. Academic Press, New York, 1954. **E-3.** F. Hofmeister, *Z. physiol. Chem.* 14: 165(1889). **E-4.** R. A. Kekwick and R. K. Cannan, *BJ* 30: 227(1936). **E-5.** A. Tiselius and B. Erickson-Quensel, *BJ* 33: 1752(1939); L. G. Longsworth *et al.*, *JACS* 62: 2580(1940). **E-6.** I. E. Lush, *Nature* 189: 981(1961). **E-7.** G. Perlmann, *Nature* 166: 870(1950). **E-8.** C. I. Niu and H. Fraenkel-Conrat, *JACS* 77: 5882 (1955). **E-9.** M. Ottesen, *Compt. rend. trav. lab. Carlsberg, Sér. chim.* 30: 212(1958). **E-10.** H. Wu, *Chinese J. Physiol.* 1: 131(1927). **E-11.** H. Fraenkel-Conrat and R. E. Feeney, *ABB* 29: 101(1950). **E-12.** R. C. Warner and I. Weber, *JBC* 191: 233(1950). **E-13.** R. A. Fuller and D. R. Briggs, *JACS* 78: 5253(1956). **E-14.** R. Acher *et al.*, *BBA* 14: 151(1954). **E-15.** P. Jolles and M. Ledieu, *BBA* 31: 100(1959). **E-16.** T. Osborne and G. Campbell, *JACS* 22: 422(1900). **E-17a.** D. Pennington *et al.*, *JACS* 64: 469(1942). **E-17b.** D. W. Woolley and L. G. Longsworth, *JBC* 142: 285(1942). **E-18.** R. H. Forsythe and J. F. Foster, *JBC* 181: 377, 385(1950). **E-19.** G. Bernardi and W. H. Cook, *BBA* 44: 86(1960). **E-20.** D. A. Mecham and H. S. Olcott, *JACS* 71: 3670(1949). **E-21.** J. Williams and F. Sanger, *BBA* 33: 294(1959). **E-22.** C. Connelly and G. Taborsky,

*JBC* 236: 1364(1961). **E-23.** G. R. Tristram, *in* "Proteins" (H. Neurath and K. Bailey, eds.), Vol. 1A, p. 181. Academic Press, New York, 1953. **E-24.** H. P. Lenk *et al.*, *Naturwiss.* 47: 516(1960). **E-25.** J. Williams, *BJ* 83: 346, 355(1962). **E-26.** J. Jolles and P. Jolles, *Compt. rend.* 253: 2773(1962). **E-27.** C. C. F. Blake *et al. Nature* 196: 1173 ff. (1962).

## Section F

**F-1.** J. W. H. Lugg, *Adv. in Protein Chem.* 5: 230(1949). **F-2.** S. Brohult and E. Sandegren, *in* "Proteins" (H. Neurath and K. Bailey, eds.), Vol. 2A, p. 487. Academic Press, New York, 1954. **F-3.** F. D. Steward and J. F. Thompson, *Ibid.*, p. 513. **F-4.** T. B. Osborne, "Vegetable Proteins." Longman Greens, New York, 1924. **F-5.** C. E. Danielson, *Ann. Rev. Plant Physiol.* 7: 215(1957). **F-6.** M. Pirie, *Ann. Rev. Plant Physiol.* 9: 33(1959). **F-7.** T. Svedberg, *Tabulae Biol.* 11: 231(1936). **F-8.** J. B. Sumner *et al.*, *Science* 87: 395(1938). **F-9.** V. L. Kretovich, *Biokhimya* 23: 547 (1958). **F-10.** B. P. Brand and P. Johnson, *Trans. Faraday Soc.* 52: 438(1956). **F-11.** P. Johnson and E. M. Shoter, *BBA* 5: 361(1950). **F-12.** C. E. Hall, *JBC* 185: 45(1950). **F-13.** D. Pennington *et al.*, *JACS* 64: 469(1942). **F-14.** E. Waldschmidt-Leitz and E. Reicheneder, *Z. physiol. Chem.* 323: 124(1961). **F-15.** J. R. Kimmel and E. L. Smith, *Bull. soc. chim. biol.* 40: 2049(1958). **F-16.** G. R. Tristram, *in* "Proteins" (H. Neurath and K. Bailey, eds.), Vol. 1A, p. 181. Academic Press, New York, 1953. **F-17.** E. Waldschmidt-Leitz *et al.*, *Z. physiol. Chem.* 323: 93(1961). **F-18.** H. C. Nielsen *et al.*, *ABB* 96: 252(1962).

## Section G

**G-1.** A. Kossel, "Protamines and Histones." Longmans, Green, New York, 1928. **G-2.** K. Felix, *Adv. in Protein Chem.* 15: 1(1960). **G-3.** K. Felix *et al.*, *Z. physiol. Chem.* 287: 224(1951). **G-4a.** V. Ross, *ABB*, 50: 34, 46(1954). **G-4b.** F. Haurowitz *et al.*, *Z. physiol. Chem.* 315: 127(1959). **G-5.** F. Šorm and Z. Šormova, *Collection Czechoslov. Chem. Communs.* 16: 207(1951). **G-6.** T. Ando *et al.*, *BBA* 34: 600(1959). **G-7.** K. Felix *et al.*, *Z. physiol. Chem.* 291: 228, 275(1953). **G-8.** T. Ando and F. Sawada, *J. Biochem.* (Tokyo), 46: 517(1959). **G-9.** F. S. Scanes and B. T. Tozer, *BJ* 63: 565 (1956). **G-10.** H. M. Rauen *et al.*, *Z. physiol. Chem.* 292: 101(1953). **G-11.** T. Ando *et al.*, *J. Biochem.* (Tokyo) 45: 27, 429(1958). **G-12.** K. Felix *et al.*, *Bull. soc. chim. biol.* 40: 1973(1958). **G-13.** M. Callanan *et al.*, *JBC* 228: 279(1957). **G-14.** E. Waldschmidt-Leitz and H. Gudernatsch, *Z. physiol. Chem.* 309: 266(1958). **G-15.** T. Ando *et al.*, *J. Biochem.* (Tokyo) 46: 933(1959). **G-16.** M. M. Daly *et al.*, *J. Gen. Physiol.* 34: 439(1955). **G-17.** H. Fischer and L. Kreuzer, *Z. physiol. Chem.* 293: 176(1953). **G-18.** G. S. d'Alcontres, *Giorn. biochim.* 4: 128(1955). **G-19.** A. P. Sadakova, *Trudy, Vsesoyuz. Nauch. Issledovatel. Inst. Zhivotnovodstva* 21: 260(1957); extract in *Chem. Abstr.* 53: 5358. **G-20.** N. Ui, *BBA* 22: 205(1956). **G-21.** R. Trautman and C. F. Crampton, *JACS* 81: 4036(1959). **G-22.** A. E. Mirsky and A. W. Pollister, *J. Gen. Physiol.* 30: 117(1947). **G-23.** C. F. Crampton *et al.*, *JBC* 225: 363(1957). **G-24.** M. M. Daly and A. E. Mirsky, *J. Gen. Physiol.* 38: 405(1955). **G-25.** P. Davison *et al.*, *Trans. Faraday Soc.* 50: 297(1954). **G-26.** J. Gregoire and M. Limozin, *Bull. soc. chim. biol.* 36: 15(1954). **G-27.** C. F. Crampton *et al.*, *JBC* 215: 787(1955). **G-28.** H. J. Cruft *et al.*, *Nature* 180: 1107(1957). **G-29.** R. J. Block *et al.*, *Proc. Soc. Exptl. Biol. Med.* 70: 494(1949). **G-30.** K. Satake *et al.*, *JBC* 235: 2801(1960). **G-31.** E. W. Johns *et al.*, *BJ* 77: 631(1960). **G-32.** J. M. Luck *et al.*, *JBC* 233: 1407(1958). **G-33.** T. Ando *et al.*, *BBA* 56: 628(1962). **G-34.** P. S. Rasmussen *et al.*, *Biochemistry* 1: 79(1962).

# Chapter IX

## Structural Proteins (Scleroproteins and the Muscle Proteins)

### A. General Remarks (A-1,2)

The scleroproteins were originally defined as the insoluble proteins of cells and tissues. Their structure is fibrous like that of the muscle proteins. The two most important classes of scleroproteins are the collagens and keratins. They differ from each other by their amino acid composition, their conformation, their wide-angle X-ray-diffraction pattern, and also by their origin. The collagens are the fibrous proteins of the connective tissue. They comprise the proteins which form tendons, cartilage, bone, the dentin of the teeth, ligaments, fasciae, and other parts of the connective tissue. The keratins are produced in the epidermal cells; to them belong the proteins of the hair (wool), of horns, hoofs, nails, and other products of ectodermal cells.

During the last few years it has been shown that a considerable portion of the collagen present in tendons, ligaments, skin, and other tissues can be solubilized by extraction with dilute acetic or citric acid. The amino acid composition and other properties of the solubilized material are typical of collagen. Hence, we have to abandon the classical definition of collagens as insoluble proteins. Indeed, most of the recent investigations of collagen have been performed with the soluble portion since this can be purified, recrystallized, and investigated like other soluble proteins (see Chapter V). The soluble scleroproteins are in many respects similar to the soluble muscle proteins and to fibrinogen, although these proteins are never designated as scleroproteins. This again demonstrates the inadequacy of the present classification of these proteins. In the present book fibrinogen, because of its occurrence in the blood, has been treated in the preceding chapter, whereas the muscle proteins, in view of their importance for the structure of the muscle, will be discussed in the present chapter.

Even though some of the scleroproteins are insoluble in water or neutral salt solutions, they can undergo certain typical reactions. One of these is the contraction which the wet fibers undergo at a certain critical temperature, and which can be considered as a process akin to the melting of solid materials. The entropy change in this process, $\Delta S$, is expressed by the equation $\Delta S = (\Delta H - f\Delta L)/T$, where $\Delta H$ is the heat change (enthalpy), $L$ the length of the fibers, $T$ the absolute temperature, and $f$ the retractive force exerted by the fiber during contraction (A-3). The "melting point" of the fibers is raised by increasing the number of cross links between the peptide chains. The nature of these cross links is the same as that of cross-links within the globular protein molecules (see Chapter VII, E) and will be discussed in detail below. Extensive use is made of covalent cross-linking in the industrial preparation of synthetic fibers from proteins (A-4).

In addition to collagen and keratin, numerous other insoluble proteins have been found in nature. It is sometimes difficult to draw a sharp line between these proteins and the slightly soluble proteins of the tissue cells. Their classification is at present unsatisfactory. The fact that we discuss them briefly in this chapter does not imply that they are related to collagen or keratin.

## B. Collagen and Gelatin (B-1 to 3,10,46,52,53)

Collagen, as mentioned above, is the structural protein of the connective tissues. It occurs as the principal constituent in tendons, bones, and ligaments; in the skin it is found in the dermis whereas keratin is present in the epidermal layer. Keratin can be solubilized by reduction of the dithio groups of its cystine residues and can thus be removed from insoluble collagen. In the preparation of leather the collagen of hide is purified in this manner and then tanned (B-3).

Depilation (removal of the hair from the hide) is accomplished by the action of sulfides, sulfite, thioglycolic acid, or other reducing agents in solutions which are made alkaline by calcium hydroxide. Concentrated solutions of urea also have a depilating effect (B-4). The first step in the cleavage of the dithio bonds is their hydrolytic breakdown and the formation of a thiol and a sulfenic acid (B-5): $RSSR \rightarrow RSH + HOSR$. The sulfenic acid is further reduced to a thiol. After removal of the solubilized keratin, the insoluble collagen is treated with various tanning agents. These are polyfunctional reactants which form cross-links between the peptide chains and thus increase the stability of the fibrils. The best-known of the tanning agents is tannic acid. Other substances used in the process of tanning are chromium salts, formaldehyde or other aldehydes, and quinones (B-3). A similar process seems to take place in the cuticle of the insects, where a protein of unknown nature is hardened by the action of phenols (B-6).

If connective tissues are extracted first with cold neutral salt solutions at physiological ionic strength some of the collagen is solubilized. Another portion is dissolved by dilute acetic acid or by citrate at pH 3.8. The acid-soluble protein has been designated as *procollagen* (B-7,8). Labeled amino acids are incorporated first into the collagen fraction which is extracted at neutral pH values by isotonic salt solutions (B-10,11). On maturation the collagen molecules are converted into the *acid-soluble collagen* and finally into *insoluble collagen* (B-12). The molecules of the salt-soluble collagen consist of three individual peptide chains (B-41) which have been called $\alpha$-chains. Two of the $\alpha$-chains, denoted as $\alpha_1$-chains, are identical, whereas the third, denoted as $\alpha_2$-chain, has a different amino acid composition. In the acid-extracted collagen two of the three peptide chains occur in pairs designated as $\beta$-components. They consist either of a pair of $\alpha_1$-chains or of one $\alpha_1$- and one $\alpha_2$-chain (B-12). The molecular weight of the $\alpha$-chains is approximately $1.25 \times 10^5$ (B-9), that of the $\beta$-components twice as high (B-49). The designations collastromin and metacollagen have been used for insoluble proteins of the connective tissue (B-13), whereas the soluble triple-chain unit has been called tropocollagen (B-49).

Collagen is particularly rich in glycine and proline and is the only protein which contains large amounts of hydroxyproline (see Table IX-1). In contrast to keratin it is free of cystine, cysteine, and tryptophan and contains only very small amounts of tyrosine and methionine. Attempts to determine the $N$-terminal amino acids of collagen failed; it may have a cyclic structure, or its $N$-terminal amino groups may be acetylated or may be substituted by other nonprotein residues (B-55). From the appearance of amino alcohols after treatment of collagen with $LiBH_4$, it has been concluded that collagen contains some ester groups (B-14). The presence of $\gamma$-glutamyl residues in gelatin has been suggested because of the formation of succinic semialdehyde on hydrolysis of alkali-treated dinitrophenylhydroxamates of gelatin (B-15,54).

In the reaction with hydroxylamine, ester groups are converted into hydroxamic acids: $R \cdot CO \cdot OR' + HONH_2 \rightarrow R \cdot CO \cdot NH \cdot OH + R'OH$. Treatment of the hydroxamates with dinitrofluorobenzene results in the formation of DNP-hydroxamates: $R \cdot CO \cdot NH \cdot O \cdot DNP$. With alkali and at elevated temperatures, the latter undergo the Lossen rearrangement to isocyanates and finally to amines:

$$R \cdot CO \cdot NH \cdot O \cdot DNP \rightarrow R \cdot N{=}C{=}O + \text{dinitrophenol} \rightarrow R \cdot NH_2 + CO_2$$

Glutamyl and aspartyl residues with free $\gamma$- or $\beta$-carboxyl groups, after esterification, are converted in this manner to $\alpha,\gamma$-diaminobutyric acid and $\alpha,\beta$-diaminopropionic acid respectively, whereas $\gamma$- or $\beta$-bound residues of these acids under the same conditions yield succinic semialdehyde and acetaldehyde, respectively. Gelatin, after esterification, treatment with hydroxylamine, dinitrophenylation, alkali, and hydrolysis yielded small amounts of succinic semialdehyde in addition to the expected large amounts of diaminopropionic and diaminobutyric acid. It is not yet quite clear whether the formation of succinic semialdehyde can be considered as a proof for the presence of esterified $\beta$- or $\gamma$-carboxyl groups, or whether it can be formed also from glutaminyl residues or other sources (B-48,54).

Another unusual feature of the collagen molecule is the presence of a branched chain in which lysine residues form the points of ramification. The presence of a branched chain, which originates from the $\epsilon$-amino group of lysine, has been proved by the isolation of $\epsilon$-glycyl-$\alpha$-glutamyllysine after partial hydrolysis with 7 $N$ HCl at 24°C (B-16). Since about one-third of all amino acid residues of collagen is formed by glycine and another third by proline and hydroxyproline, it had been claimed that the molecule has the periodic structure (Gly.Pro(or Hypro).X)$_n$ where X is any of the other amino acids. Although this may be true for large parts of the collagen molecule, it cannot be of general validity since peptides with adjacent Pro.Hypro residues have been isolated from partial hydrolyzates of gelatin (B-17). The principal sequences occurring in collagen are Gly.Pro.Gly.Gly.Pro.Ala and Gly.Pro.Hypro (B-44).

Collagen is a substrate for the bacterial enzyme collagenase (B-18). This enzyme requires for its action the presence of the sequence —Pro.-X.X'·Pro— (B-19,20) where Pro is a prolyl or hydroxyprolyl residue and X and X' are other amino acid residues. The synthetic peptide ester carbobenzoxy-Hypro.Gly.Gly·Pro.OCH$_3$ is hydrolyzed by collagenase (B-21). The action of collagenase from *Clostridum histolyticum* has very little initial effect on the molecular weight of the soluble collagen from the swim bladder of the carp, lowering it from 1.92 to 1.22 $\times$ 10$^6$; however, it lowers the viscosity considerably and causes a collapse of the rigid multistranded collagen structure (B-22). This initial phase is followed by breakdown to peptides of an average molecular weight close to 500 (B-47).

If collagen is heated for a long time with water, it is solubilized and converted to *gelatin* (B-46). Glue, obtained by the heating of hide with water, is impure gelatin. The conversion of collagen into gelatin seems to involve the hydrolysis of labile peptide bonds since gelatin, in contrast to collagen contains free $\alpha$-amino groups (detectable with dinitrofluorobenzene) and $C$-terminal amino acids (B-23). The average molecular weight, calculated from the number of $N$-terminal amino acid residues, is approximately 60,000 (B-24). Solutions of gelatin are highly viscous and form stiff gels at room temperature in concentrations above 2%. The formation of these gels depends on pH and the ionic strength (B-25), and is inhibited by cupric or cobaltous ions (B-26). The gelation is attributed to the formation of H-bonds between pairs of peptide bonds. It is accompanied by a drastic increase in the levorotation, which indicates that the peptide chains in the gelatin gel have a conformation similar to that of the left-handed poly-L-proline helix (B-46). When the gel is warmed up, it is solubilized at a certain temperature; the "melting point" increases with the proline and hydroxyproline content of the gelatin (B-1,46). Gelation takes place reversibly on cooling; it proceeds faster in $D_2O$ than in $H_2O$ (B-46).

3.4 Å

*Trans*-peptide chain                    Proline-containing peptide chain

Because of the absence of NH groups in the peptide bonds of poly-L-proline, this synthetic polypeptide cannot form the hydrogen bonds required for the production of the α-helix; poly-L-proline exists in its aqueous solutions either as a right-handed or a left-handed helix with $[\alpha]_D$ values of $+50$ and $-450°$ respectively (B-27). Interconversion of these two forms is accompanied by mutarotation. In the chilled gelatin gel triple-strand coils of the polyproline conformation may be present (B-28,46). If gelatin is treated with formaldehyde, $—CH_2—$ bridges are formed between amino or imino groups of adjacent peptide chains; these bonds do not rupture on adding more water or on cautious heating. Hence, formolized gelatin swells only to a limited extent and does not dissolve in an excess of water.

Collagen, unlike keratin, is only slightly extensible. X-ray analyses of collagen show that its structure is quite different from that of other proteins, including keratin and fibroin. Its spacings of 2.86 Å along the fiber axis correspond to the average translation of the amino acids parallel to the fiber axis (B-29,30). In addition to the meridional spacing of 2.86 Å, which is particularly typical of collagen, two equatorial spacings of 11.6 and 4.6 Å have been observed. Examination of collagen by low-angle X-ray diffraction reveals long spacing periods of approximately 640 Å in the direction of the longitudinal fiber axis (B-31,32). The same periodicity has been found in electron micrographs (Fig. IX-1) where it appears as cross-striation (B-33).

If collagen is cautiously heated with water, an irreversible contraction to about one-third of the original length of the fibers occurs at 62–63°C. It is probably caused by the "melting" of hydrophobic bonds or hydrogen bonds between the peptide chains. Higher melting points have been observed in those collagens which are particularly rich in hydroxyproline (B-34). The collagen fiber, whose diameter is approximately 20 μ and which is visible in the optical microscope as a single strand, contains in its cross-section several hundred fibrils. Each of these is formed by numerous filaments, and each of the latter by hundreds or thousands of laterally aggregated protofibrils (B-10,40,46,52). The diameter of each filament is approximately 200 Å.

In the collagen fiber the protein chains run predominantly longitudinally and parallel to each other. The smallest fibrous unit has been designated as *protofibril* (B-33). This protofibril is a polymer of elongated macromolecules which are linked to each other end-to-end (B-39). The

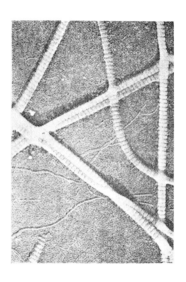

FIG. IX-1. Electron micrograph of collagen fiber. Prepared from fresh adult rat skin by cleaning with trypsin and fragmentation. Filaments in background are derived from elastic fibers. Chromium shadowed. × 16,400. (Courtesy of J. Gross and F. O. Schmitt, Massachusetts Institute of Technology.)

dimensions of this macromolecule called *tropocollagen* are approximately 15 × 2200 Å (B-35,42). In collagen solutions tropocollagen molecules are present as kinetic units. As mentioned earlier, the molecule consists of three peptide chains. In tropocollagen these chains are helically coiled and linked to each other by hydrogen bonds (B-33,34), possibly also by covalent bonds (B-49).

Although neither the tropocollagen macromolecule nor the protofibrils are banded, long spacings of 2000–3000 Å result when these *units aggregate laterally* either by hydrogen bonding or by means of chondroitin sulfate (see Chapter XI, C), hyaluronate, DNA, ATP, or similar fibrous polyanions (B-36 to 38,40). Most probably the acidic groups of the polyanions combine with the lysine side chains of the tropocollagen macromolecules (B-40). Since the tropocollagen units can be arranged in the protofibrils in parallel or antiparallel arrays, and since they can be either in register with respect to macromolecular ends or staggered by specific fractions of their length, different types of banding result under different conditions (B-40,50,51). The bands of the 640 Å spacing (Fig. IX-1) have been attributed to alternate zones of order and disorder, the latter caused by an aggregation of amino acids with bulky side chains (B-31 to 33,44,45). Alternatively, they may be caused by the staggered arrangement of the tropocollagen molecules within the collagen fibrils. Whereas the three peptide chains within the tropocollagen molecules form a helix with a threefold, left-handed screw axis, the tropocollagen molecules seem to coil gradually about one another in a "supercoil" with a right-handed twist (B-41,46).

## C. Keratin (C-1)

In contrast to collagen, which is a product of the connective tissue, keratin is formed exclusively by the epithelial cells, particularly in the outermost layers of the skin. Keratins are the structural proteins of hair, wool, feathers, horn, nails, and hoofs. Keratin is prepared from these products by pulverising them, extracting first with hot organic solvents, then with water, and finally digesting with proteolytic enzymes.

If the skin, nervous tissues, and certain other organs are treated in this manner, a small amount of insoluble material, frequently called pseudokeratin, is obtained. X-ray-diffraction analysis reveals that the structure of these substances is related to that of collagen rather than to that of keratin (C-22). Ovokeratin from egg shells (C-2) and other insoluble proteins belong to this group (C-23). In contrast to these pseudokeratins, the true keratins are sometimes designated as eukeratins.

The great stability of keratin and its insolubility are caused by the large number of dithio cross-links between the peptide chains (C-3). Amino acid analyses reveal that keratin has an unusually high cystine content. Thus keratin from human hair or from wool contains approximately 11–12% of cystine which corresponds to 3% of sulfur. Other eukeratins contain up to 5% S whereas pseudokeratins contain only 1–3% S (C-4). If the dithio cross-links of keratin are cleaved by oxidation, reduction, or hydrolysis (see Chapter VII, E; Chapter III, D), a product is formed which is soluble and digestible by proteolytic enzymes. The solubilizing action of inorganic sulfides has been known for many centuries. These substances, sulfides of calcium, barium, arsenic, and other metals, have been used for the removal of hair for cosmetic purposes and also for the depilation of animal hide mentioned in the preceding section of this chapter. The reduced keratin, which contains cysteine residues instead of cystine cross-links, has been called keratein (C-5). It can be reoxidized to keratin. Temporary reduction of hair by thioglycol or thioglycolate and reoxidation by the oxygen of the air is widely used in the cosmetic production of permanent waves.

The resistance of keratin to solvents and enzymes is lowered not only by exposure to chemical reagents which cleave the dithio bridges, but also by mechanical treatment. Finely ground wool or horn is partly soluble in water and digestible by proteolytic enzymes (C-6). When wool is boiled with a 2% solution of sodium carbonate, the dithio bonds are hydrolyzed to sulfide and sulfenate; owing to the lability of sulfenate, its sulfur atom is split off and the cystine residue is converted into lanthionine, which can be isolated from the acid hydrolyzate of the modified keratin (C-7) (see Chapter III, D). Like collagen, keratin is also a heterogeneous material. Wool consists of keratins of various sulfur content, some of them containing up to 6.5% of sulfur (C-8). Different amounts of amino acids have been found in keratins from animal sources, as shown in Table IX-1.

TABLE IX-1
AMINO ACID COMPOSITION OF SOME SCLEROPROTEINS
(g amino acid in 100 g protein)

| Amino acid | Collagen[a] (soluble) | Collagen[a] (Ox hide) | Elastin[b] | Wool[c] keratin | Feather[c] keratin | Silk[d] fibroin |
|---|---|---|---|---|---|---|
| Glycine | 26.1 | 26.6 | 26.7 | 5.2–6.5 | 7.2 | 42.8 |
| Alanine | 9.9 | 10.3 | 21.3 | 3.4–4.4 | 5.4 | 33.5 |
| Valine | 2.3 | 2.5 | 17.7 | 5.0–5.9 | 8.8 | 3.30 |
| Leucine | 3.2 | 3.7 | 9.0 | 7.6–8.1 | 8.0 | 0.890 |
| Isoleucine | 1.4 | 1.9 | 3.8 | 3.1–4.5 | 6.0 | 1.11 |
| Proline | 13.0 | 14.4 | 13.5 | 5.3–8.1 | 10.0 | 0.526 |
| Phenylalanine | 2.0 | 2.3 | 6.2 | 3.4–4.0 | 5.3 | 1.33 |
| Tyrosine | 0.50 | 1.0 | 1.5 | 4.0–6.4 | 2.2 | 11.9 |
| Tryptophan | 0 | 0 | 0 | 1.8–2.1 | 0.7 | 0.879 |
| Serine | 4.2 | 4.3 | 0.85 | 7.2–9.5 | 14 | 16.3 |
| Threonine | 2.2 | 2.3 | 1.12 | 6.6–6.7 | 4.8 | 1.38 |
| Cystine | 0 | 0 | 0.35 | 11.4–14.1 | 8.2 | 0 |
| Methionine | 0.78 | 0.97 | trace | 0.5–0.7 | 0.5 | 0 |
| Arginine | 8.3 | 8.2 | 1.3 | 9.2–10.6 | 7.5 | 1.04 |
| Histidine | 0.29 | 0.70 | — | 2.8–3.3 | 0.7 | 0.358 |
| Lysine | 3.6 | 4.0 | 0.50 | 0.7–1.1 | 1.7 | 0.601 |
| Aspartic acid | 6.0 | 6.9 | 1.1 | 6.4–7.3 | 7.5 | 2.22 |
| Glutamic acid | 11.0 | 11.2 | 2.4 | 13.1–16.0 | 9.7 | 1.92 |
| Ammonia | 0.60 | 0.56 | — | — | — | — |
| Hydroxyproline[e] | 13.6 | 12.8 | 1.6 | — | 0 | 0 |
| Hydroxylysine | 1.29 | 1.15 | — | 0.2 | 0 | 0 |

[a] From (B-43). The soluble collagen was extracted from calfskin by citrate.
[b] From (E-3).          [c] From (C-1).          [d] From (D-1).
[e] Tendon collagen contains 0.26% of the isomeric 3-hydroxyproline (B-56).

Fractionation of the solubilized wool keratin reveals that it consists of a matrix of very high sulfur content and no particular orientation of the fibers. In this matrix regularly oriented fibers of a lower sulfur content are embedded (C-25). The peptide chains of the matrix have an average molecular weight of 25,000–28,000. They contain approximately 24% cystine (C-26).

The length of keratin fibers depends on their water content; the hair hygrometer, used for the determination of the humidity of air, is based on the increased length of moist hair. Keratin fibers are extensible and possess elastic properties. The stretching of wool is accompanied by a decrease of the isoelectric point from pH 5.5 to 5.3 (C-9) and a typical change in the X-ray diffraction pattern. The extended keratin has been designated as β-keratin. Its diffraction pattern shows equatorial reflections of 4.65 and 9.7 Å and a meridional reflection of 3.33 Å, whereas the unextended keratin gives meridional reflections of 5.1 Å. Most probably, the extended β-keratin fibers consist of parallel peptide chains running in the same direction, and which are linked to each other in the extended state by numerous hydrogen bonds as shown below. This structure, designated as parallel pleated sheet structure (C-11), is less extended than the backbone of the

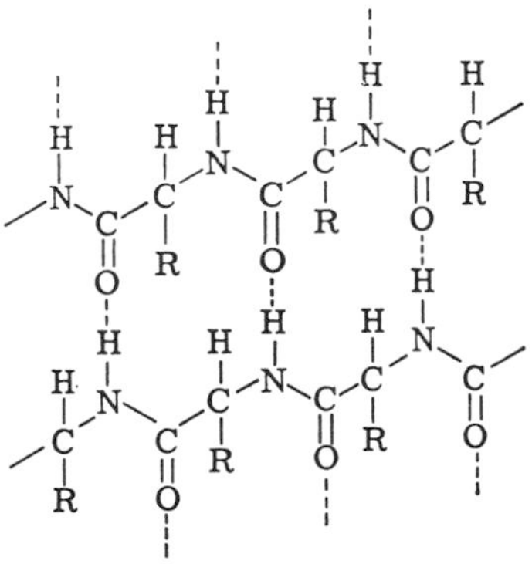

fully extended peptide chain. In his classic investigations on the structure of keratin, Astbury (C-10) had assumed meanderlike folding of the peptide chain in α-keratin. Later, a short spacing of 1.5 Å was discovered in the keratin of porcupine quills (C-12). It is indicative of the presence of α-helices in α-keratin. In the typical keratins, which are rich in sulfur, the numerous dithio linkages between peptide chains may interfere to a certain extent with the formation of α-helices. The X-ray-diffraction pattern of keratin is quite different from that of collagen, but resembles that of myosin and fibrinogen. Astbury denoted this group of proteins as the k-m-f group (k:keratin; m:myosin; f:fibrinogen) or k-m-e-f group (e:epidermin). The similar X-ray pattern indicates similar conformations of the peptide chains in keratin, myosin, and fibrinogen. However, in most other aspects these three classes of proteins are quite different from each other. Their development in ontogenesis, their function, and their amino acid composition have very little in common.

Some insight into the conformation of $\alpha$- and $\beta$-keratin has been obtained from the infrared dichroism caused by differences between the infrared absorption parallel and perpendicular to the fiber axis. The absorption maxima due to the stretching of C=O and N—H bonds and the deformation of N—H bonds are shown in the following tabulation.

| Bond | $\alpha$-Form (cm$^{-1}$) | $\beta$-Form (cm$^{-1}$) |
|---|---|---|
| C=O stretching | 1650/1660($\parallel$) | 1630($\perp$) |
| N—H deformation | 1540/1550($\perp$) | 1520/1525($\parallel$) |
| N—H stretching | 3290/3300($\parallel$) | 3280/3300($\perp$) |

The dichroism of the C=O and N—H stretching in the $\beta$-configuration (see above) indicates perpendicular orientation of these groups; whereas they are parallel to the fiber axis in the folded $\alpha$-configuration, in which parts of the peptide chain are folded in such a manner that the chains run perpendicular to the direction of the fiber axis (C-13,14). The low wave number of 3280/3300 cm$^{-1}$ for the N—H-stretching absorption indicates that these groups form hydrogen bonds. The absorption maximum of the unbonded NH groups is at 3460 cm$^{-1}$.

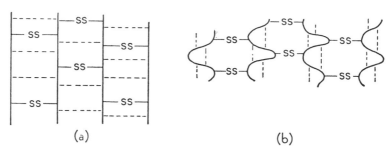

(a)                                (b)

Fig. IX-2. Diagram of keratin structure: (a) extended $\beta$-keratin structure; (b) the same supercontracted. The broken lines represent hydrogen bonds, bonds between hydrophobic groups, and salt-bridges.

Electron microscopy reveals a periodicity of about 110 Å in $\alpha$-keratin fibers (C-15). X-ray diffraction shows spacings of 198 Å (C-22) or 186 Å (C-17) in various keratins. It has been attributed to the formation of three-strand ropes of keratin with staggered peptide chains or to slightly distorted coiled coils (C-17). A long spacing of 190 Å in the feather keratin of birds has been attributed to a $\beta$-helix with a long pitch (C-16). Spacings of 95 Å (axial) and 34 Å (equatorial), discovered earlier in feather keratin, may be due to the presence of elongated keratin globules in the keratin fiber (C-24).

Since the nitroprusside test for SH groups is positive in the soft keratin of the skin epidermis, but becomes negative after keratinization (C-18), it is quite possible that a precursor of keratin exists in the SH- form and that coils are formed before the SH groups are oxidized to the dithio cross-links.

When wool is heated in water, it shrinks irreversibly at temperatures close to 90° C. Since the SS cross-links are not affected by heating, the reaction is attributed to the cleavage of other interchain bonds. These may be either hydrogen bonds, bonds between hydrophobic groups, or saltlike bridges between positively and negatively charged groups. As shown in Table IX-1, keratin is particularly rich in acidic and basic amino acids. It is quite possible that these also contribute to the mutual linkage of the peptide chains and that they are cleaved on heating, thereby causing supercontraction.

Keratin from feathers can be dissolved by boiling with detergents in the presence of sodium hydrogen sulfite (C-19). The average molecular weight of the solubilized keratin subunits is approximately 55,000. When dinitrofluorobenzene was used in these experiments, it was found that a mixture of subunits with various $N$-terminal amino acids was present. The ratio of the $N$-terminal components was 18 glycine to 8 threonine, 4 valine, 2 alanine, 2 serine, 2 glutamate, and 1 aspartate (C-20). This would indicate a molecular weight of about 2 millions. Feather keratin can also be solubilized by 10 $M$ urea in the presence of reducing agents or by sulfitolysis (C-21). The molecular weight of the soluble product is approximately 10,000. It has less than 1 equivalent of $N$-terminal amino acids and may be a cyclopeptide or may have its $N$-terminal groups substituted or masked.

## D. Fibroin (D-1)

Fibroin is the insoluble protein of silk which forms about two-thirds of the natural raw silk. The remaining third is composed of sericin, a protein which is soluble in hot water (D-2). The amino acid composition of fibroin is very simple (see Table IX-1) and quite different from that of keratin. The bulk of fibroin is formed by the 4 amino acids glycine, alanine, serine, and tyrosine. The tyrosine content is so high that fibroin is commonly used as a source for the preparation of this amino acid. Fibroin can be solubilized by exposure to concentrated solutions of lithium bromide (D-3) or lithium thiocyanate (D-4). It is also soluble in aqueous solutions of dichloroacetic acid (D-5) or of ethylenediamine copper (D-6). The molecular weight of the soluble particles is approximately 30,000 (D-6).

If fibroin is exposed to the action of trypsin (D-7) or chymotrypsin (D-8), part of it is digested whereas another part remains insoluble. The insoluble material obtained after digestion with trypsin consists of the amino acids glycine, alanine, and serine, whereas the soluble products contain tyrosine and the other amino acid with 4 or more carbon atoms (D-7). The soluble fraction consists of two octa-peptides, one of them of the structure Gly.Ala.Gly.Ala.Gly.Ala.Gly.Tyr, the other containing valine instead of one of the alanyl residues (D-16). The isolation of certain smaller peptides from the partial hydrolyzates of fibroin led several authors to the view that fibroin contains periodically repeated sequences of the type Ala.Gly.Ala.Gly.X.Gly., where X is serine or one of the larger amino acids (D-9 to 11). However, other investigations led to the isolation of the peptide Val.Ala.Gly.Asp.Gly.Tyr (D-12) and of numerous other peptides which do not reveal any recognizable periodicity (D-13). It is not yet clear whether fibroin consists of 2 different proteins, one of them consisting of repeated units of the smaller amino acids glycine, alanine, and serine, and the other formed by an aperiodic sequence of the larger amino acids.

The X-ray-diffraction pattern of fibroin is similar to that of $\beta$-keratin. Short range spacings of 6.94 Å have been interpreted to mean that the peptide chains of fibroin occur in the extended state (C-10, D-15). In contrast to $\beta$-keratin, in which parallel peptide chains run in the same direction, the "pleated sheet" of fibroin consists of an array of antiparallel peptide chains (D-14). The equatorial reflections differ for different types of fibroin and vary from 9.3 to 10.6 Å (D-15).

## E. Other Scleroproteins

*Elastin,* the structural protein of the elastic fibers, has usually been prepared from the cervical ligament of cattle. It resembles collagen in its high glycine and proline content (Table IX-1) and other properties, but is more resistant to the action of proteolytic enzymes. Small amounts of elastin occur together with collagen in many tissues. The elastin content of the lung increases with age from 0.05 to 15.5 % of the dry weight (E-1). Elastin is possibly formed from collagen by the loss of those parts of the molecule which contain most of the arginine and hydroxyproline (E-2). The higher resistance of elastin to boiling water, acids, and bases, has been attributed to its high content of nonpolar amino acid residues; it is hydrolyzed by elastase, an enzyme occurring in the pancreas and also in bacteria. Mild acid hydrolysis of elastin with 0.25 $M$ oxalic acid yields a mixture of 2 proteins called $\alpha$- and $\beta$-protein (E-3).

To the insoluble proteins, which are resistant to enzyme action, also belong *spongin, gorgonin,* and *antipathin,* the proteins of marine sponges and of corals, respectively. Gorgonin occurs in the octacorallaria, antipathin in hexacorallaria (E-4). Spongin contains approximately 1 % of iodine, whereas the iodine amounts of gorgonin and antipathin vary in different animal species from 0.15 to 9.3 % (E-4). The iodine atoms are bound to the 3- and/or 5-positions of tyrosine residues. In gorgonin small amounts of mono- and dibromotyrosine have also been found (E-4). Spongin and gorgonin resemble collagen in many properties (C-22). A structural protein, free of iodine, has been found in the flagella of thermophilic bacteria and has been designated as *flagellin;* in contrast to collagen it is free of proline, hydroxyproline, and histidine (E-5).

*Reticulin* is a collagenlike protein which occurs together with collagen and elastin in mammalian skin. It consists of a membranaceous or lamellar array of protein fibers. Reticulin, according to some authors, is a mucoprotein containing considerable amounts of polysaccharides (E-6,7, see also Chapter XI, C). Very little is also known about the *melanins*, black or dark brown pigments of the skin, the retina, and pathological tumors. They seem to be proteins in which the tyrosine residues underwent oxidation to quinoid derivatives. To the insoluble proteins belong numerous proteins of the invertebrates, such as conchiolin from the shells of mussels and the structural protein of Byssus (E-8, C-22).

The insoluble capsular substance of *Bacillus anthracis, Bacillus subtilis,* and a few other bacilli should be mentioned here, although it is not a typical protein; it is a polypeptide consisting exclusively of glutamyl residues (E-9). The polyglutamic acid is extracted from the bacteria by ethanol-HCl and is precipitated from the solution by neutralization with alkali hydroxide (E-10). The molecular weight is close to 50,000 (E-10). Degradation of the polyglutamic acid reveals that it consists predominantly or exclusively of $\gamma$-linked D-glutamyl residues (E-11,12).

## F. Biological Importance of Structural Proteins

The insolubility of some of the structural proteins might give the impression that they are lifeless substances with none other than passive, protective functions. This is certainly true for the keratinized products (hair, wool, silk fibroin), which are shed or otherwise removed from the organism. Much greater importance must be attributed to those proteins which form the insoluble framework of the cell and its membranes. The permeability and the metabolism of the cells depend to a great degree on the properties of these structural proteins and in particular on their electrical charges.

The amount of insoluble, structural protein in the cells is very considerable. Thus, an analysis of the particles of the electron transport system which contains the pyridine enzymes, flavoprotein, and various heme proteins, reveals that only 17% of the total proteins are contributed by these enzymes; 33% of the total proteins are insoluble and 55% soluble proteins, some of them present as lipoproteins (F-1). The enzyme proteins seem to be embedded in the nonenzymatic proteins; the influence of the latter on the electron transport is not yet known. It is evident from this example that we are not able to draw a sharp borderline between structural and other proteins. Thus cytochrome oxidase, which is insoluble in water and dilute salt solutions, could be called a structural protein, but is at the same time an enzyme protein and will, therefore, be discussed with other similar heme proteins in Chapter XII. For similar reasons, the lipoproteins, some of which are insoluble in water, will be treated with other conjugated proteins in Chapter XI. If a cell contains filamentous protein molecules which are arranged in a regular manner, birefringence is observed in polarized light. Thus the proteins of the nervous sheaths form concentric birefringent layers; the sense of their birefringence indicates that the protein fibers are oriented tangentially to the cross section of the cylinder (F-2). Elongated, filamentous protein molecules are found also in the highly viscous fluid which forms the liquid phase of the cytoplasm. These gelatinous masses are usually isotropic, free of birefringence, since their protein molecules are oriented randomly. They resemble in this respect the gelatin gels. If gels of this type are subjected to mechanical tension so that their protein threads are regularly oriented, they become birefringent. This phenomenon of "birefringence by stress" is analogous to flow birefringence and is explained in terms of a similar mechanism.

## G. *Muscle Proteins* (*Contractile Proteins*) (*G-1 to 7*)

In hardly any field of biochemistry have our views changed so frequently as in muscle chemistry, where we are dealing with a unique mechanochemical reaction, the direct conversion of chemical into mechanical energy. Although it has been established that contraction and relaxation involve changes in the conformation or arrangement of the proteins myosin and actin, very little is known about the immediate chemical reaction which necessarily must accompany this change in conformation and supply the required energy. Before describing the proteins involved, it may be useful to report briefly on the history of their discovery and isolation.

Muscle contains about 20% protein. Since the muscular tissue forms about 40% of the mammalian body, it is evident that most of the body proteins consist of muscle protein. When fresh muscle, immediately after its removal from the organism, is extracted with dilute saline solution, much of its protein is dissolved. If this solution is mixed with an equal volume of saturated ammonium sulfate, a large part of the proteins is precipitated. The protein in the supernatant albumin fraction has been called *myogen*. It is a mixture of numerous proteins and contains, among other substances, the enzymes phosphorylase, aldolase, phosphoglucomutase, and glyceraldehyde phosphate dehydrogenase (see Chapter XII). These enzymes form at least 20–25% of the myogen fraction. Whereas typical albumins are soluble in salt-free water, part of the myogen proteins becomes insoluble on dilution of the myogen solution with water. The insoluble protein has been called globulin X (G-8).

The muscle protein which is precipitated by half-saturation with ammonium sulfate has been called *myosin* (C-9). The designation myosin had been used first by Kühne (G-10) for a protein which was extracted from muscle by 10% saline solution and precipitated upon dilution with water. We now know that this is the fraction which contains the contractile proteins. The purification of myosin was improved considerably by extraction of the chilled muscle with a cold solution containing 0.5 $M$ KCl and 0.03 $M$ NaHCO$_3$. The highly viscous and birefringent extract gives a precipitate of myosin when diluted with water (G-11,13). When solutions of myosin in 0.5 $M$ KCl are extruded through a narrow orifice into salt-free water, fine threads of myosin are formed (G-12). The interest in myosin was further raised when Engelhardt and Ljubimova (G-14) discovered that myosin acts as an adenosinetriphosphatase and catalyzes the hydrolysis of ATP to ADP and inorganic phosphate. Since ATP is an energy donor in many biochemical reactions, it has been claimed that it is also the immediate energy donor for the muscle contraction and that its cleavage by myosin is an autoregulatory reaction.

When Szent-Györgyi and his co-workers (G-15) prepared myosin by repeated brief extractions of rabbit muscle with 0.6 $M$ KCl, they found that the first extracts were less viscous than those obtained after prolonged extraction. They assumed the presence of 2 different proteins and called them myosin A and B. Straub (G-16) discovered in the latter extracts another muscle protein which could also be extracted from acetone-precipitated, dry muscle powder with water. This protein was named *actin*. When its aqueous solution was added to a solution of myosin, the viscosity increased. It was then assumed that a complex of the 2 proteins, *actomyosin*, was formed, and that this was the true contractile matter of the muscle. Actin is a globular protein (G-actin) which, upon addition of 0.1 $M$ KCl and 0.0005 $M$ MgCl$_2$ at pH 7, is converted into a fibrous protein, F-actin (G-1). F-actin seems to be formed by the lengthwise association of G-actin molecules. Only F-actin combines with myosin to form actomyosin (G-17,18). During the last decade attempts have been made to unravel the complicated interrelations between myosin (ATPase), actin, and ATP. As will be seen, we are still far from a satisfactory interpretation of the processes taking place during the muscle contraction.

The best preparations of *myosin* approach homogeneity (G-19). The molecular weight of myosin is difficult to determine because of anomalies encountered in solutions of macromolecules of an elongated, threadlike shape. Values obtained by the light scattering technique or by the Archibald method vary from 400,000 to 600,000 (G-20,21). The length of the molecule is approximately 1600 Å, its diameter 26 Å (G-22,42). The isoelectric point is at pH 5.4, but is shifted to 7.0 by the addition of KCl, and

TABLE IX-2

AMINO ACID COMPOSITION OF SOME MUSCLE PROTEINS (G-29)

(The table gives the number of amino acid residues per $10^5$ g protein)

| Amino acid | Myosin | Tropo-myosin[b] | Actin | L-mero-myosin | H-mero-myosin |
|---|---|---|---|---|---|
| Glycine | 39 | 12.5 | 67 | 24 | 45 |
| Alanine | 78 | 110 | 71 | 76 | 73 |
| Valine | 42 | 38 | 42 | 39 | 45 |
| Leucine | 79 | 95 | 63 | 85 | 78 |
| Isoleucine | 42 | 29 | 57 | 35 | 42 |
| Proline | 22 | 0 | 44 | 8.5 | 29 |
| Phenylalanine | 27 | 3.5 | 29 | 9.5 | 40 |
| Tyrosine | 18 | 16 | 32 | 12 | 21 |
| Tryptophan | — | 0 | 10 | 6 | 3 |
| Serine | 41 | 40 | 56 | 37 | 43 |
| Threonine | 41 | 28 | 59 | 38 | 49 |
| Cystine[a] | 8.6 | 6.5 | 11.2 | 5.6 | 10.9 |
| Methionine | 22 | 16 | 30 | 14 | 19 |
| Arginine | 41 | 42 | 38 | 51 | 29 |
| Histidine | 15 | 5.5 | 19 | 19 | 11.5 |
| Lysine | 85 | 110 | 52 | 83 | 82 |
| Aspartic acid | 85 | 89 | 82 | 77 | 88 |
| Glutamic acid | 155 | 211 | 101 | 174 | 138 |
| Ammonia | — | 64 | 66 | — | — |

[a] Cystine and cysteine.

[b] Tropomyosin from smooth and striated muscle are different [D. R. Kominz et al., ABB 70: 16(1957)].

to 8.0 by magnesium salts (G-23). Evidently, the myosin molecule has a high affinity for cations, particularly the bivalent magnesium and calcium ions (G-24). This is in agreement with the unusually high content of amino acids with charged side chains (see Table IX-2). Approximately 18 % of the amino acid residues are formed by aspartic and glutamic acid, and 16 % by the basic amino acids. Hitherto it has not been possible to find an $N$-terminal amino acid residue in myosin or in tropomyosin (G-25). The $C$-terminal amino acid is isoleucine; the $C$-terminal sequence in tropomyosin is —Ala.Ileu.Met.Thr.Ser.Ileu.OH (G-26). *Tropomyosin* occurs in the smooth muscles of invertebrates in higher concentration than in mammalian muscles (G-27). Its molecular weight in salt solutions is close to 50,000, but increases at low ionic strength (G-28). In tropomyosin, 27% of the amino acid residues are formed by the aminodicarboxylic acids, and 18% by the basic amino acids (G-29). Since the amino acid composition of myosin is intermediate between that of actin and tropomyosin, it has been suggested that myosin may be an actin-tropomyosin complex (G-29).

If myosin is exposed to the action of trypsin, it is degraded to a heavy and a light fragment (G-29a) called *H-meromyosin* and *L-meromyosin*, respectively (G-30,22). It is not yet quite clear whether the meromyosins are products of hydrolytic cleavage of myosin (G-31), or whether they are linked to each other in the myosin molecule by hydrogen bonds or van der Waals' forces (G-32). The latter view is supported by the fact that the same *N*- and *C*-terminal amino acids have been found in meromyosins produced by the action of trypsin as in meromyosins formed by other proteolytic enzymes (G-33). In concentrated solutions of urea, the meromyosins are degraded to small fragments called protomyosins which have a molecular weight of 4600 (G-32). Apparently these are subunits which, in the meromyosins, are held together by association (H-bonds, van der Waals' bonds) rather than by covalent bonds (G-34).

*Actin* forms 12–15% of the muscle proteins; it is present in the muscle as F-actin (G-2). The molecular weight of actin is approximately $6 \times 10^4$ (G-35,36). The polymerized form of actin consists of elongated threads. The percentage of acidic amino acids is 13; that of basic amino acids 12% of the total amino acid residues (see Table IX-2) (G-29). No *N*-terminal amino acid has been found; the carboxyl end is formed by the sequence His.Ileu.Phe.OH (G-26).

*Actomyosin* contains under optimum conditions two monomers of actin for each molecule of myosin (G-20). The actomyosin complex dissociates into its components when potassium iodide or thiocyanate is added to its solutions (G-22). Sulfhydryl groups of actin and myosin seem to be involved in the combination of actin with myosin (G-40) and in the enzymatic action of actomyosin (G-48b). Actomyosin combines with 1 equivalent of ATP per myosin molecule (G-41,37,20). If an excess of ATP or of other nucleoside triphosphates is added to the highly viscous solutions of actomyosin, the viscosity decreases considerably (G-43,44). This is attributed (G-45) to dissociation of actomysin into its 2 components, actin and myosin; ATP and other nucleoside triphosphates apparently act as plasticizers and prevent the formation of the rigid actomyosin fibers (G-47,48a). The postmortal rigor of the muscles is attributed to the loss of the plasticizing action of ATP which is not regenerated in the muscles after the death of the organism (G-38).

As mentioned before, myosin acts as an *adenosinetriphosphatase* which catalyzes the hydrolytic cleavage of ATP to ADP and inorganic phosphate (G-49); its catalytic action is much lower than that of typical pyrophosphatases.

According to a recent report, myosin loses its ATPase activity upon dialysis, but regains it on addition of calcium and magnesium salts (G-50). Tracer experiments with $P^{32}$ or with water containing $O^{18}$ indicate that phosphorylmyosin is formed as an intermediate during the hydrolytic cleavage of ATP (G-51 to 53). It is not yet clear whether the physiological contraction of the muscle involves hydrolysis of ATP. Investigations with highly sensitive methods have demonstrated that neither during a prolonged tetanic contraction (G-54,55) nor during a single twitch (G-56,57) are significant amounts of ATP hydrolyzed. Moreover, there is no significant exchange of $P^{32}$ between ATP and inorganic phosphate during tetanic contraction (G-58). However, the small, apparently insignificant changes in the concentration of ADP or ATP discovered by spectrophotometry of the sarcoplasmic fluid may reflect much larger changes inside the mitochondria; moreover, the organic phosphates in the cell may be localized at different intracellular sites and need not be in metabolic equilibrium (G-59).

*Muscle Contraction*

It was assumed for a long period of time that muscle contraction involves merely shortening of the myosin molecules due to the formation of hydrogen bonds or other loose bonds between different groups of the same peptide chain. Cleavage of these bonds would result in elongation of the fiber and relaxation. Since the contraction takes place a few milliseconds after the electrical or nervous stimulus, a very rapid chemical reaction must be responsible for the deformation of the muscle fiber. Most probably it is an ionic reaction, since only these reactions proceed at a sufficiently rapid rate (G-60). Enzymatic reactions require a much longer time. One of the ionic reactions linked to muscle contraction is the increase in dialyzable $K^+$ ions. A portion of the potassium in the resting muscle is not dialyzable before, but becomes dialyzable immediately after, the contraction. This change is accompanied by a slight but measurable decrease in volume (G-61). The significance of these phenomena is not yet clear (G-72). In addition to $K^+$ ions, calcium and phosphate ions seem to be released during the muscular contraction. The shortening of muscle fibers during contraction might well be due to intra- or intermolecular neutralization of those positively and negatively charged groups of the peptide chains which have lost their inorganic counter ions. On the other hand one has to consider the possibility that bivalent ions such as $Ca^{++}$ or $Mg^{++}$ form bridges between carboxyl groups of adjacent peptide chain and that this might lead to contraction.

Several *contractile models* have been described during the past few years. Although these models may not directly reflect properties of muscle, they are of value in studies of the thermodynamics and chemistry of contractile fibers. One of these systems consists of a myosin-DNA complex which contracts upon acidification and is elongated when dilute alkali is added; the myosin-DNA fiber also contracts upon the addition of Ca or Mg salts; relaxation can then be brought about by ethylenediamine tetraacetate (G-73). Fibers of polymetacrylic acid react in a similar manner with HCl, NaOH, and with barium salts (G-62,63). Another contractile fiber consists of the copolymer of lysine, glutamic acid, and cystine; it undergoes contraction when acidified and reversion to the original length upon subsequent alkalinization (G-64).

Important information about the molecular rearrangement involved in muscle contraction has been gained by *electron micrography*. It must be

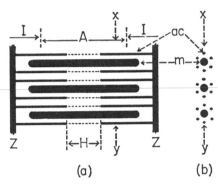

(a)                    (b)

Fig. IX-3. I: isotropic bands; A: anisotropic bands; Z: Z-zone, H: H-band; m: myosin fiber, ac: actin fiber. (a) is a longitudinal section through a typical striated muscle; (b) is a cross section through the plane x-y of Fig. (a). The S filaments are shown as dashed lines between the actin fibers.

remembered first that the typical cross-striated muscle is birefringent and that the birefringence decreases during the isotonic contraction. The cross striation is visible in the optical microscope and is caused by the presence of alternating isotropic (I) and anisotropic (A) bands. On high magnifica-tion a fine anisotropic layer is visible in the isotropic band, called the Z line. Similarly the anisotropic band is bisected by a thin, less anisotropic layer, the H band. Electron microscopy reveals that the thin actin fila-ments traverse the I bands and run into the A bands. Thick myosin fila-ments run through the entire length of the A bands (G-65,66). In the I bands, myosin is either absent or present in only small amounts (G-67, 68). If a cross section of the muscle is examined by electron microscopy, it is found that the myosin filaments form a hexagonal array in the A bands, that their diameter is approximately 100 Å, and that they are 455 Å apart from each other (G-69). The diameter of the actin filaments is approximately 40 Å; they run from the Z zone through the I band into the A band. On the cross section of the muscle, the actin filaments are seen surrounded by myosin filaments (G-69). The resulting arrangement is shown in Fig. IX-3, which represents, diagrammatically, a longitudinal section and a cross section of the striated muscle. It can be seen from the figure that the double filaments of actin on both sides of the narrow *H* zone are linked up by fine *S* filaments shown as dots (G-70). Contraction in this model is brought about by shortening of the *S* filaments and sliding of the actin filaments toward each other and past the myosin filaments. This model, which is strongly supported by beautiful electron micrographs, allows us to understand muscle contraction without the necessity to invoke coiling or other distortion of the actin or myosin filaments. In the electron micrographs the actin and myosin filaments are much closer to each other than in Fig. IX-3 where they are separated for the sake of clearer arrangement. The electron micrographs give no indication of the formation of actomyosin complexes.

Although the electron microscope has revealed a fascinating new picture of the mechanism of muscle contraction, we still do not know much about its essential process, the *conversion of chemical into mechanical energy*. It had first appeared obvious that the contraction of the muscle requires energy supply and that the relaxation is spontaneous. The contrary may be true. Contraction may be comparable to the shortening of an extended elastic spring; in other words, it may be a reaction which involves an increase in entropy and is therefore spontaneous (G-71). Relaxation and extension may be brought about by the adsorption of potassium ions or other cations to myosin and actin, and mutual repulsion of the filaments of these two proteins. This would allow the actin fibers to return to their expanded state.

## REFERENCES

### Section A

**A-1.** F. O. Schmitt, *Adv. in Protein Chem.* 1: 26(1944).    **A-2.** *Symposia Soc. Exptl. Biol.* 9(1955).    **A-3.** P. J. Flory, *JACS* 78: 5222(1956).    **A-4.** H. P. Lundgren, *Adv. in Protein Chem.* 5: 305(1949).

### Section B

**B-1.** K. H. Gustavson, "Chemistry and Reactivity of Collagen." Academic Press, New York, 1956.    **B-2.** R. S. Bear, *Adv. in Protein Chem.* 7: 69(1952).    **B-3.** K. M. Rudall, *Adv. in Protein Chem.* 7: 253(1952).    **B-4.** K. H. Gustavson, *J. Soc. Leather Chem.* 33: 162(1949).    **B-5.** J. B. Speakman, *Nature* 122: 930(1933).    **B-6.** R. Dennell and R. S. A. Malek, *Proc. Roy. Soc.* B143: 414(1955).    **B-7.** V. N. Oriekhovitch, *Communs. 2nd Intern. Congr. Biochem., Paris* (1952) p. 106.    **B-8.** R. D. Harkness *et al., BJ* 56: 558(1954).    **B-9.** V. N. Oriekhovitch *et al., Biokhimiya* 23: 285(1958). **B-10.** J. Gross, *in* "Connective Tissue, Thrombosis, and Atherosclerosis" (I. H. Page, ed.), p. 77. Academic Press, New York, 1959.    **B-11.** D. S. Jackson and J. P. Bentley, *J. Biophys. Biochem. Cytol.* 7: 37(1960).    **B-12.** K. A. Piez *et al., BBA* 53: 596(1961). **B-13.** A. V. Tustanovski *et al., Gerontologia* 4: 198(1960); J. Balo *et al., Acta Biol. Hung.* 7: 385(1958).    **B-14.** W. Grassmann *et al., Ber. deut. chem. Ges.* 88: 102(1955).

**B-15.** P. M. Gallop *et al.*, *JBC* 235: 2619(1960).   **B-16.** G. Mechanic and M. Levy, *JACS* 81: 1889(1959).   **B-17.** W. A. Schroeder *et al.*, *JACS* 76: 3556(1954).   **B-18.** I. Mandl, *Adv. in Enzymol.* 23: 163(1961).   **B-19.** O. A. Kazakova *et al.*, *Doklady Akad. Nauk S.S.S.R.* 122: 193(1959).   **B-20.** Y. Nagai and H. Noda, *BBA* 34: 298(1959); K. T. Poroshin *et al.*, *Biokhimiya* 26: 244(1961).   **B-21.** W. Grassmann *et al.*, *Z. physiol. Chem.* 316: 287(1959).   **B-22.** P. H. von Hippel *et al.*, *JACS* 82: 2774(1960). **B-23.** W. Grassmann and H. Hörmann, *Z. physiol. Chem.* 292: 24(1953).   **B-24.** A. Courts, *BJ* 58: 70(1954).   **B-25.** J. Bowes and R. Kenten, *BJ* 43: 358(1948).   **B-26.** J. Bello and J. Vinograd, *Nature* 181: 273(1958).   **B-27.** I. Z. Steinberg *et al.*, *JACS* 82: 5263(1960).   **B-28.** P. H. von Hippel and W. F. Harrington, *BBA* 36: 427(1959). **B-29.** W. T. Astbury, *Proc. Roy. Soc.* B134: 303(1947); *Adv. in Enzymol.* 3: 63(1943). **B-30.** W. Lotmar and L. E. R. Picken, *Helv. Chim. Acta* 25: 538(1942).   **B-31.** R. S. Baer, *JACS* 64: 727(1942).   **B-32.** O. Kratky, *J. Polymer Sci.* 3: 195(1948).   **B-33.** F. O. Schmitt *et al.*, *J. Cellular Comp. Physiol.* 20: 11(1942); *JACS* 64: 1234(1943).   **B-34.** K. H. Gustavson, *Acta Chem Scand.* 8: 1248(1954); B. J. Rigby and J. D. Spikes, *Nature* 187: 150(1960).   **B-35.** C. E. Hall, *J. Biophys. Biochem. Cytol.* 2: 625(1956). **B-36.** P. Vanamee and K. R. Porter, *J. Exptl. Med.* 94: 255(1951); H. Noda and R. W. G. Wyckoff, *BBA* 7: 494(1951).   **B-37.** J. H. Highberger *et al.*, *Proc. Natl. Acad. Sci. U.S.* 37: 286(1951).   **B-38.** F. O. Schmitt *et al.*, *Proc. Natl. Acad. Sci. U.S.* 39: 459(1953).   **B-39.** J. Gross *et al.*, *Proc. Natl. Acad. Sci. U.S.* 40: 679(1954); 41: 1(1955).   **B-40.** F. O. Schmitt, *J. Cellular Comp. Physiol.* 49: (Suppl.), p. 85(1957). **B-41.** G. N. Ramachandran, *Nature* 177: 710(1956).   **B-42.** H. Boedtker and P. Doty, *JACS* 78: 4267(1956).   **B-43.** J. H. Bowes *et al.*, *BJ* 61: 143(1955).   **B-44.** W. Grassmann *et al.*, *Z. physiol. Chem.* 323: 48(1961).   **B-45.** A. Nordwig *et al.*, *Z. physiol. Chem.* 325: 242(1961).   **B-46.** W. F. Harrington and P. H. von Hippel, *Adv. in Protein Chem.* 16: 1(1961).   **B47.** S. Seifter *et al.*, *JBC* 234: 285(1959).   **B-48.** M. Liefländer and O. Wacker, *Z. physiol. Chem.* 327: 195(1962).   **B-49.** K. Altgelt *et al.*, *Proc. Natl. Acad. Sci. U.S.* 47: 1914(1961).   **B-50.** K. Kuehn *et al.*, *Naturwiss.* 47: 258(1960); *Z.f. Naturforsch.* 16b, 648(1961).   **B-51.** H. B. Bensusan *et al.*, *Biochemistry* 1: 215(1962).   **B-52.** Z. Stary, *in* "Handbuch der Hautkrankheiten," Vol. 1, Part 3, p. 577. Springer, Berlin, 1962. (In english.)   **B-53.** G. R. Tristram and F. S. Stevens, *Ann. Repts. Progr. Chem.* 58: 367(1961).   **B-54.** O. O. Blumenfeld and P. M. Gallop, *Biochemistry* 1: 947(1962).   **B-55.** F. S. Stevens and G. R. Tristram, *BJ* 83: 245 (1962).   **B-56.** J. D. Ogle *et al.*, *JBC* 237: 3667(1962).

## Section C

**C-1.** W. H. Ward and H. P. Lundgren, *Adv. in Protein Chem.* 9: 244(1954).   **C-2.** Z. Stary, *Z. physiol. Chem.* 148: 83(1926).   **C-3.** Z. Stary, *Z. physiol. Chem.* 175: 178(1928).   **C-4.** R. J. Block and D. Bolling, *JBC* 127: 685(1939).   **C-5.** D. R. Goddard and L. Michaelis, *JBC* 106: 605(1934); 112: 361(1935).   **C-6.** H. Cohen, *ABB* 4: 145(1944).   **C-7.** M. J. Horn *et al.*, *JBC* 138: 141(1941).   **C-8.** J. M. Gillespie and D. H. Simmonds, *BBA* 39: 538(1960).   **C-9.** A. G. Pasynskii and V. Blokhina, *Doklady Akad. Nauk S.S.S.R.* 86: 1171(1952).   **C-10.** W. T. Astbury, *Adv. in Enzymol.* 3: 63(1943).   **C-11.** L. Pauling and R. B. Corey, *Proc. Natl. Acad. Sci. U.S.* 39: 247, 253(1953).   **C-12.** I. MacArthur, *Nature* 152: 38(1943).   **C-13.** T. Miyazawa and E. R. Blout, *JACS* 83: 712(1961).   **C-14.** C. H. Bamford *et al.*, "Synthetic Polypeptides." Academic Press, New York, 1956.   **C-15.** J. L. Farrant *et al.*, *Nature* 159: 535(1947).   **C-16.** R. Schor and S. Krimm, *Biophys. J.* 1: 489(1961).   **C-17.** R. D. B. Fraser and T. P. McRae, *J. Mol. Biol.* 3: 640(1961).   **C-18.** W. Montagna, *Intern. Rev. Cytol.* 1: 265(1952).   **C-19.** W. H. Ward, *J. Polymer Sci.* 1: 22(1946).   **C-20.**

W. R. Middlebrook, *BBA* 7: 547(1951). **C-21.** A. M. Woodin, *BJ* 63: 576(1956). **C-22.** R. S. Baer, *Adv. in Protein Chem.* 7: 69(1952). **C-23.** K. M. Rudall, *BBA* 1: 549(1947). **C-24.** R. S. Baer, *JACS* 66: 2043(1944). R. S. Baer and H. J. Rugo, *Ann. N.Y. Acad. Sci.* 53: 627(1951). **C-25.** R. D. B. Fraser *et al.*, *Nature* 193: 1052; 195: 1167(1962); H. P. Lundgren and W. H. Ward, *ABB, Suppl.* 1, 78(1962). **C-26.** J. M. Gillespie, *Austral. J. Biol. Sci.* 15: 564, 572(1962).

*Section D*

**D-1.** F. Lucas *et al.*, *Adv. in Protein Chem.* 13: 107(1958). **D-2.** K. Kodama, *BJ* 20: 1208(1926). **D-3.** P. P. von Weimarn, *Ind. Eng. Chem.* 19: 109(1927). **D-4.** E. Waldschmidt-Leitz and O. Zeiss, *Z. physiol. Chem.* 298: 239(1954). **D-5.** E. J. Ambrose *et al.*, *Nature* 167: 264(1951). **D-6.** D. Coleman and F. O. Howitt, *Proc. Roy. Soc.* A190: 145(1947). **D-7.** B. Drucker and S. G. Smith, *Nature* 165: 197(1950). **D-8.** E. Waldschmidt-Leitz and O. Zeiss, *Z. physiol. Chem.* 300: 49(1955). **D-9.** M. Levy and E. Slobodian, *JBC* 199: 563(1952). **D-10.** K. G. Joffe, *Biokhimiya* 19: 495(1954). **D-11.** F. Lucas, J. T. B. Shaw, and S. G. Smith, *BJ* 66: 468(1957). **D-12.** K. Ziegler and H. Spoor, *BBA* 33: 138(1959). **D-13.** I. M. Kay and W. A. Schroeder, *JACS* 76: 3564(1954). **D-14.** R. E. Marsh *et al.*, *BBA* 16: 1(1955). **D-15.** C. H. Bamford *et al.*, *Nature* 171: 1149(1953). **D-16.** F. Lucas *et al.*, *BJ* 83: 164(1962).

*Section E*

**E-1.** V. Scarselli and M. Repetto *Italian J. Biochem.* 8: 169(1959). **E-2.** D. Burton *et al.*, *Nature* 176: 996(1955). **E-3.** S. M. Partridge and H. F. Davies, *BJ* 61: 21 (1955). **E-4.** J. Roche *et al.*, *Bull. soc. chim. biol.* 33: 526, 1437(1951); *Progr. in Chem. Org. Nat. Prods.* 12: 349(1955). **E-5.** H. Koffler *et al.*, *ABB* 67: 246(1957); 84: 342 (1959). **E-6.** F. Bertalanffy *et al.*, *Can. Med. Assoc. J.* 70: 196(1954). **E-7.** G. M. Windrum *et al.*, *Brit. J. Exptl. Pathol.* 36: 49(1955). **E-8.** Z. Stary, *Z. physiol. Chem.* 148: 83(1926). **E-9.** G. Ivanovics and V. G. Bruckner, *Naturwiss.* 25: 250(1937). **E-10.** W. E. Hanby and H. N. Rydon, *BJ* 40: 297(1946). **E-11.** J. Kovacs and V. Bruckner, *JCS* 1952: 4255. **E-12.** F. Haurowitz and F. Bursa, *BJ* 44: 509(1949).

*Section F*

**F-1.** D. E. Green *et al.*, *Biochem. Biophys. Research Communs.* 5: 109(1961). **F-2.** F. O. Schmitt, *Adv. in Protein Chem.* 1: 25(1944).

*Section G*

**G-1.** A. Szent-Györgyi, "Chemistry of Muscle Contraction." Academic Press, New York, 1947. **G-2.** W. F. H. M. Mommaerts, "Muscular Contraction." Interscience, New York, 1950. **G-3.** W. F. H. M. Mommaerts *et al.*, *Ann. Rev. Physiol.* 23: 529 1961. **G-4.** H. H. Weber and H. Portzehl, *Adv. in Protein Chem.* 7: 162(1952). **G-5.** A. G. Szent-Györgyi, *Adv. in Enzymol.* 16: 313(1955). **G-6.** S. V. Perry, *Ann. Repts. on Progr. Chem.* 56: 343(1960). **G-7.** D. M. Needham, *in* "The Structure and Function of Muscle" (G. H. Bourne, ed.), Vol. 2, p. 55. Academic Press, New York, 1960. **G-8.** H. H. Weber and K. H. Meyer, *Biochem. Z.* 266: 137(1933). **G-9.** O. v. Fürth, *Z. physiol. Chem.* 31: 338(1900). **G-10.** W. Kühne, *Arch. Anat. Physiol.* 1859: 748; W. Kühne and R. H. Chittenden, *Z. Biol.* 7: 358(1889). **G-11.** A. v. Muralt and J. T. Edsall, *JBC* 89: 289, 315, 351(1930). **G-12.** H. H. Weber, *Ergeb. Physiol.* 36: 109 (1934). **G-13.** J. T. Edsall *et al.*, *JBC* 133: 397, 409(1940). **G-14.** V. A. Engelhardt and M. N. Ljubimova, *Nature* 144: 688(1939). **G-15.** A. Szent-Györgyi, *Science* 93: 158 (1941). **G-16.** F. B. Straub, *Studies Inst. Med. Chem. Szeged* 2: 3(1942). **G-17.**

A. Szent Györgyi, *Acta Physiol. Scand.* 9: Suppl. 25(1945).  **G-18.** S. Perry and R. Reed, *BBA* 1: 379(1947).  **G-19.** W. F. H. M. Mommaerts, *JBC* 188: 553(1951). **G-20.** W. F. H. M. Mommaerts and B. B. Aldrich, *BBA* 28: 627(1958).  **G-21.** W. W. Kielley and W. F. Harrington, *BBA* 41: 401(1960).  **G-22.** A. Holtzer and S. Lowey, *JACS* 81: 1370; *BBA* 34: 470(1959).  **G-23.** A. Szent-Györgyi, *Discussions Faraday Soc.* 11: 179(1952).  **G-24.** L. B. Nanninga and W. F. H. M. Mommaerts, *Proc. Natl. Acad. Sci. U.S.* 43: 540(1957).  **G-25.** K. Bailey, *BJ* 49: 23(1951).  **G-26.** R. H. Locker, *BBA* 14: 533(1954).  **G-27.** J. B. Rüegg, *BBA* 35: 279(1959).  **G-28.** C. M. Kay and K. Bailey, *BBA* 40: 149(1960).  **G-29.** D. R. Kominz *et al.*, *ABB* 50: 148(1954).  **G-29a.** S. V. Perry, *BJ* 48: 257(1951).  **G-30.** M. A. Lauffer and A. G. Szent Györgyi, *ABB* 56: 542(1955).  **G-31.** K. Laki, *Science* 128: 653(1958); K. Laki *et al.*, *BBA* 28: 656(1958).  **G-32.** A. G. Szent Györgyi and M. Borbiro, *ABB* 60: 180(1956).  **G-33.** W. R. Middlebrook, *Science* 130: 621(1960).  **G-34.** A. Szent-Györgyi, *Science* 124: 873(1956).  **G-35.** C. M. Kay, *BBA* 43: 259(1960).  **G-36.** W. F. H. M. Mommaerts, *JBC* 198: 445(1952).  **G-37.** L. B. Nanninga and W. F. H. M. Mommaerts, *Proc. Natl. Acad. Sci. U.S.* 46: 8(1960).  **G-38.** H. H. Weber, *Arzneimittel-Forsch.* 5: 404(1960).  **G-39.** A. Holtzer, *BBA* 42: 453(1960).  **G-40.** M. Barany and K. Barany, *BBA* 35: 293(1959).  **G-41.** A. Szent Györgyi, *Acta Physiol. Acad. Sci. Hung.* 1: 28(1946).  **G-42.** R. V. Rice, *BBA* 53: 29(1961); 52: 602(1961).  **G-43.** J. Needham *et al.*, *Nature* 150: 46(1942).  **G-44.** A. Szent Györgyi, *Studies Inst. Med. Chem. Szeged* 1: 1(1942).  **G-45.** A. Weber, *BBA* 19: 345(1954).  **G-46.** J. Gergely, *JBC* 220: 917(1956).  **G-47.** H. H. Weber, *BBA* 12: 150(1953).  **G-48a.** E. Boyler and J. T. Prince, *J. Gen. Physiol.* 37: 53, 63(1954).  **G-48b.** J. J. Blum, *ABB* 97: 309, 312(1962).  **G-49.** V. A. Engelhardt and G. A. Yarovaya, *Ukrain. Biokhim. Zhur.* 27: 312(1955).  **G-50.** Y. Tonomura and S. Kitagawa, *BBA* 40: 135(1960).  **G-51.** H. M. Levy and D. E. Koshland, *JBC* 234: 1102(1959).  **G-52.** I. Gruda *et al.*, *Bull. acad. polon. sci.* 8: 219(1960).  **G-53.** M. E. Dempsey and P. D. Boyer, *JBC* 236: PC6 (1961).  **G-54.** A. Fleckenstein *et al.*, *Nature* 174: 1081(1954).  **G-55.** W. F. H. M. Mommaerts, *Nature* 174: 1083(1954).  **G-56.** W. F. H. M. Mommaerts, *Am. J. Physiol.* 182: 585(1956).  **G-57.** B. Chance and C. M. Connelly, *Nature* 179: 1235 (1957).  **G-58.** J. Sacks and M. G. Cleland, *Am. J. Physiol.* 198: 300(1960).  **G-59.** S. V. Perry, *Ann. Rev. Biochem.* 30: 486(1961).  **G-60.** W. O. Fenn, *Physiol. Revs.* 16: 450(1936).  **G-61.** E. Ernst *et al.*, *Pflüger's Arch. ges. Physiol.* 239: 691(1938); E. Ernst, "Die Muskeltätigkeit." Budapest, 1958. (Cited in **G-3.**)  **G-62.** A. Katschalsky and M. Zwick, *J. Polymer Sci.* 16: 221(1955).  **G-63.** S. Basu and P. R. Chaudhury, *J. Colloid Sci.* 12: 19(1957).  **G-64.** H. Tani *et al.*, *JACS* 75: 3042(1953).  **G-65.** W. Hasselbach, *Z. Naturforsch.* 8b: 449(1953).  **G-66.** S. S. Spicer and G. Rozsa, *JBC* 201: 639(1953).  **G-67.** A. Corsi and S. V. Perry, *BJ* 68: 12(1958).  **G-68.** J. Hanson and H. E. Huxley, *BBA* 23: 250(1957).  **G-69.** J. Hanson and H. E. Huxley, *Symposia Soc. Exptl. Biol.* 9: 228(1954).  **G-70.** H. E. Huxley, *J. Biophys. Biochem. Cytol.* 3: 631(1957).  **G-71.** M. Morales *et al.*, *Discussions Faraday Soc.* 13: 125(1952); *Physiol. Revs.* 35: 475(1955).  **G-72.** L. B. Nanninga, *BBA* 36: 191(1959).  **G-73.** V. I. Vorobev, *Doklady Akad. Nauk S.S.S.R.* 137: 58(1962).

# Chapter X

## Combination of Proteins with Other Substances

### A. Intermolecular Forces

Proteins combine with nonprotein substances to form more or less stable complexes which, when they occur in Nature, have been designated as conjugated proteins. Other designations used are symplexes or cenapses. As a prelude to the description of different types of conjugated proteins in Chapter XI, this chapter will examine the formation of protein-nonprotein complexes *in vitro*.

The combination of proteins with nonproteins is brought about by intermolecular forces which operate between pairs of (a) ionic groups, (b) uncharged polar groups, and (c) nonpolar groups, or by certain combinations of such groups. The same forces are responsible for the formation of intramolecular cross-links in globular protein molecules (see Chapter VII, E). Bonds of the type (a) lead to the formation of *salt-bridges* between ionic groups of the protein on the one hand and organic or inorganic ions on the other. It is well known that proteins combine very firmly with certain organic acids such as picric, trichloroacetic, or sulfosalicylic acid, forming insoluble precipitates. Analyses of such precipitates produced by metaphosphoric acid reveal that the amount of metaphosphate ions bound to the precipitated protein is equivalent to the number of amino groups of the precipitated protein (A-1). Evidently the metaphosphate ions combine with the positively charged ammonium groups of the protein.

The mutual attraction of positively and negatively charged ionic groups is caused by the well-known *electrostatic forces*. If the charges of the two groups involved are $e_1$ and $e_2$, respectively, the force of attraction is $f = (e_1e_2)/(Dr^2)$, where $D$ is the dielectric constant and $r$ the distance between the positive and negative charge. It is evident that a bond of this type will be destroyed by any agent which causes the ionic groups to lose their electric charge. Since the positive and the negative groups of proteins lose their charges by the addition of strong bases or acids, respectively (see Chapter VI, A), the saltlike bonds can be cleaved by these reagents.

Complications may arise from the denaturing action of acids or alkalis which may increase the number of accessible anion or cation binding groups of the protein. Salt-bridges are highly localized bonds. The mutual attraction between protein and nonprotein is restricted to the two ionic groups which carry the positive and negative charge. Since the mutual force of attraction decreases only with the second power of the interionic distance $r$ (see above), salt-bridges are relatively stable to heat.

Example: Let the bond energy between two centers of charges of opposite sign be 5 kcal and their distance 3 Å. If this distance increases from 3 to 6 Å, the force of interionic attraction decreases from 5 kcal/mole to $5 \times (\frac{3}{6})^2 = 1.25$/kcal/mole. Hence, the two charges will still be held together by 25% of the original force of attraction and will therefore tend to resume their original position.

Bonds of the type (b) are caused by *dipole-dipole interaction* and have been discussed earlier in Chapter VI, H. In proteins, the most important dipolar groups are the proton donors OH, NH₂, NH, and SH, on the one hand, and the proton binding groups $\diagdown$CO and $\diagdown$N on the other. The interaction of these two types of groups results in the formation of *hydrogen bonds* which have been discussed extensively in preceding chapters (Chapter VI, D, E, F, and H; Chapter VII, D, E, and F). The forces responsible for the dipole-dipole interaction are the same electrostatic forces that cause salt formation. If the two poles are close to each other, the force of attraction is of the same order as that acting between two ionic groups. However, in the case of dipoles the force of attraction decreases in proportion to a much higher power of the mutual distance $r$, probably in proportion to $1/r^6$ (A-2). Hydrogen bonds, therefore, are much less stable at elevated temperatures and "melt" more easily than the salt-bridges. Although this is true for hydrogen bonds in nonpolar solvents, it does not apply to aqueous solutions where the water molecules as strong dipoles interact with proteins and other solutes (see Chapter VII, E).

Example: Let the intermolecular bond energy again be 5 kcal/mole and the distance between the dipoles 3 Å. An increase from 3 to 6 Å involves decrease of the intermolecular bond energy from 5 kcal/mole to $5 \times (\frac{3}{6})^6 = <0.1$ kcal/mole. The difference between the more stable salt-links and the less stable dipole-dipole attraction is exemplified by the differences in the melting points of the zwitterion glycine $N^+H_3CH_2COO^-$, mp 232°C, and the isomeric polar, but uncharged, glycolamide $CH_2OH \cdot CONH_2$, mp 117°C (see Chapter VI, A).

In spite of their sensitivity to heat, the dipole-dipole bonds (hydrogen bonds) are of greater importance for intramolecular cross-links in proteins than the salt-bridges (see Chapter VII, E). They are also of great significance for the binding of polar organic compounds, such as the carbohydrates, which have a large number of polar hydroxyl groups. The mutual attraction between proteins and sugars is so strong that the coagulation of denatured proteins is prevented by high concentrations of sugars.

The combination of proteins with nonproteins, particularly with dyes or with substances which have absorption bands in the ultraviolet region, is frequently accompanied by a shift in the wave length and intensity of the absorption maxima. This is attributed to changes in the distribution of the electrons and to electronic sharing between protein and nonprotein. Complexes of this type have been designated as *charge transfer complexes* (A-5). Thus, the green color of lipoyl dehydrogenase has been ascribed to charge transfer between the apoenzyme, diphosphopyridine nucleotide and FAD (A-6).

The third type of bond to be discussed is the bond between hydrophobic *nonpolar groups*, type (c) above. We find this bond in solid and in liquid hydrocarbons. Originally, it was attributed to short-period oscillations of the electrons and was designated as London dispersion effect (A-3). According to more recent views, the nonpolar groups of the proteins are surrounded by an aqueous medium in which the attraction between the nonpolar groups and $H_2O$ molecules is weaker than the mutual attraction of water molecules on the one hand, and the mutual attraction of nonpolar groups on the other. Accordingly, the state of lowest energy is one in which clusters of nonpolar groups and of polar residues or water (icebergs) are formed, but hardly any stable bonds between polar and nonpolar groups. In view of the essential role of water in their formation, the bonds between nonpolar groups have been designated *hydrophobic bonds* (A-4; Chapter VII, E). The attraction between nonpolar groups is weaker than that between ionic or polar groups. This is evident from the low melting and boiling points of the lower members of the paraffin series. The mutual attraction between nonpolar groups becomes more effective with increasing chain length, for instance, in the elongated paraffin chains of the higher fatty acids. The flexibility of these chains allows large portions of the chains to come close enough to one another for attractive forces to result.

In addition to the three main types of intermolecular bonds discussed, *intermediary types* exist. Electrostatic forces are effective not only between 2 ions or 2 dipoles but also between an ion and an adjacent dipole. The strength of the ion-dipole bond lies between that of the ion-ion and the dipole-dipole bond. Similarly, intermediary types of linkage may exist between polar groups on the one hand, and nonpolar, but polarizable, groups on the other hand.

## B. *Combination of Proteins with Inorganic Ions*

Previous discussion of the isoelectric point of proteins (Chapter VI, B) mentioned that proteins combine with multivalent anions and cations, and that even univalent ions are bound to a certain extent by some proteins. The existence of *ion-protein complexes* can be demonstrated for example by equilibrium dialysis. In this method the protein solution containing the ion to be examined is dialyzed against a protein-free solution of the same ion (B-1). If the dialyzable ion is not bound to the protein, its total concentration inside the membrane will be equal to its outside concentration. If binding occurs, the total concentration in the protein solution will be higher, since a certain amount of the ion must migrate from the outside to the inside until equilibrium is established between the free ions on both sides of the membrane. The linkage of ions to protein molecules can also be demonstrated by electrophoresis experiments. When there occurs a transfer of cations such as $Ca^{++}$ to the anode, or of anions such as phosphate to the cathode, it must be concluded that the ions are linked to protein molecules of an opposite net charge and that the resulting complexes migrate as such.

As mentioned earlier (Chapter VI, B), the fixation of ions to proteins increases with their valence. The univalent ions $Na^+$, $K^+$, and $Cl^-$ are either free or bound to a minor extent to certain proteins. Their binding can be demonstrated by electrometric titration and by determination of the conductance. We still do not know why chloride, which is bound neither by pepsin, fibrinogen, casein, gelatin, nor hemoglobin (B-4), is bound by bovine serum albumin. Iodide is 4 times more firmly bound by BSA than chloride; for thiocyanate and trichloroacetate the affinity constant is 20 times higher than for chloride (B-2). Calcium, magnesium, and phosphate ions are also bound by the proteins of the body fluids to a much greater extent than chloride ions.

The binding of *calcium* has been investigated with the most thoroughness (B-5). About 40% of the calcium of muscle juice and about one-third of the Ca of the blood serum is bound to protein molecules. One g of the serum proteins is capable of binding 0.062 mmole of calcium. The combination of protein with $Ca^{++}$ takes place in accordance with the law of mass action. In milk a considerable part of the calcium is bound to the phosphate groups of casein. Similarly, *magnesium ions* in muscle juice and serum are attached in considerable amounts to protein molecules (B-6).

The combination of *phosphate* with protein molecules is of small importance in the physicochemical equilibrium of the blood serum because the phosphate content of the serum is very low. Greater importance is attributed to the combination of the serum proteins with *carbonic acid.* Amino acids combine with carbonates in alkaline solutions to form *carbaminates* (B-7)

$$H_2N \cdot R \cdot COO^- + CO_3^{--} \rightarrow {}^-O \cdot CO \cdot NH \cdot R \cdot COO^- + OH^-$$

In contrast to the carbonates, the carbaminates do not give a precipitate with barium hydroxide. It is a prerequisite for the carbaminate formation that the amino group of proteins or amino acids is in the $NH_2$ form and not present as the cation $—NH_3^+$. For this reason carbaminate formation increases in alkaline solutions (B-8). However, carbamino derivatives of amino acids and proteins are also formed in the physiological pH range of pH 7.0–7.4. The carbamino derivatives of the amino acids can be precipitated by mercuric acetate (B-9). In the blood serum, $CO_2$ combines with the protein moiety of hemoglobin to form *carbhemoglobin* (B-10 to 13). The combination of carbon dioxide with hemoglobin is of great importance in the physicochemical equilibrium of blood, because the affinity of hemoglobin for oxygen is lowered at lower pH values, a phenomenon well known in physiology as the Bohr effect (B-14).

The binding of heavy metals to proteins is of great biological importance since many of the heavy metal complexes of proteins act as enzymes which catalyze hydrolytic cleavage or oxidoreduction. The heavy metal ion forms the active group in these enzymes. Often it mediates the binding of substrates to the enzyme (B-15). Among the heavy-metal complexes of proteins are many of the natural copper proteins discussed in Chapter XI. *In vitro*, copper ions combine with the carboxyl groups of proteins or other negatively charged groups. If the reaction proceeds in an alkaline solution the purplish complex which is responsible for the biuret test is formed. In this complex the copper ion seems to be bound to the enolized peptide bonds (B-16). The binding of copper ions to serum albumin is an exergonic reaction in which $\Delta F$ decreases from an initial value of $-5.18$ kcal for the first $Cu^{++}$ ion to lower values for subsequent copper ions; the $\Delta F$ value for the 16th copper ion is $-1.27$ kcal (B-17). The simultaneous increase in entropy indicates that the copper ions displace water molecules from their linkage with the protein (B-17).

## C. Combination of Proteins with Organic Dyes

Although proteins combine with a great variety of biologically important organic substances, most of the investigations in this field have been done with the colored ions of organic dyes, particularly with the anions of acidic dyes. The principal reason for their use is the ease with which the partition of a dye between the two sides of a membrane can be measured colorimetrically in equilibrium dialysis. The equilibrium between free and bound dye can also be determined by rocking the mixture of protein and dye with hexanol and hexane and measuring the dye concentration in the aqueous and the organic layer (C-1). The dye most frequently used in such analyses is methyl orange, which is an azo dye whose anion has the structure $(CH_3)_2N \cdot C_6H_4 \cdot N_2 \cdot C_6H_4 \cdot SO_3^-$. The colored anion is bound to cationic groups of proteins, in particular to the $\epsilon$-ammonium groups of lysine (C-2). Bovine serum albumin is able to bind 22 equivalents of methyl orange per molecule. If pH is raised above 12, the cationic groups of lysine are converted into uncharged $NH_2$ groups and the amount of bound dye decreases.

Spectrophotometric analysis reveals that the absorption spectrum of the bound dye anions is different from that of the free dye ions; a shift in the absorption maximum is frequently observed. This change in the absorption spectrum may have different causes, some of which are: (1) disturbance of the equilibrium between undissociated dye and dye anion, (2) dimerization or polymerization of the dye (C-3,4), and (3) interaction of the protein with the chromophoric groups of the free dye by means of charge transfer. Although binding of the dye to the protein molecule is initiated by the long-range electrostatic forces, the short-range van der Waals' forces are more efficient after binding has been established; the latter are thus responsible for the irreversible binding of numerous dyes to proteins (C-5). A thermodynamic analysis of the reversible binding of methyl orange to protein reveals that the free energy change in the binding process is $\Delta F = -6.4$ kcal/mole of the dye at 25°C and that $\Delta H$ is $-2.1$ kcal (C-6). If the number of dye ions bound per protein molecule, $r$, is plotted against the molar concentration of the free dye, $c$, it is found that $r$ is always higher for denatured than for native proteins (C-7 to 9). Evidently, more cationic groups are accessible to the dye anions in the denatured than in the native protein. Since binding of the dye involves the neutralization of cationic groups of the protein, the isoelectric point of the protein-dye complex is shifted towards lower pH values (C-10).

It was mentioned above that the color change of the bound dye ion indicates participation of short-range van der Waals' forces in the dye-protein interaction. The high temperature coefficient of the dissociation constant of some dye-protein complexes (C-9), and the fact that the pK of the dimethylamino group after binding to bovine serum albumin is shifted from 4.27 to 1.67 indicate the formation of hydrophobic bonds and of icebergs (see Section A) (C-11). The binding of methyl orange by human and by bovine serum albumin is the same at pH 6.8; however, at pH 9 it is higher with the human protein (C-12). These observations also indicate that the combination of methyl orange with serum albumin involves more than merely the mutual neutralization of a sulfonate and an ammonium ion. Indeed the binding is highly specific; although methyl orange combines readily with serum albumin and β-lactoglobulin, it does not combine with pepsin, trypsin, insulin, or γ-globulin (C-6).

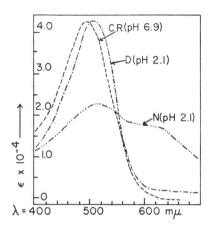

Fig. X-1. Absorption spectra of 0.0013% Congo red in the absence of protein (CR) and in the presence of 0.040% native (N) or denatured (D) ovalbumin (C-7).

A particularly striking color change is observed when Congo red is bound to native or denatured ovalbumin or serum albumin. The dye is attached so firmly to the denatured proteins that it remains in the red anionic form even at pH 2 although the free dye is converted into a blue form when pH is less than 4. If a blue solution of native serum albumin and Congo red is denatured by heating, the color turns red, indicating denaturation of the protein (C-7). The reaction has been used for determining the rate of denaturation. Figure X-1 shows the absorption spectrum of mixtures of Congo red with native or denatured ovalbumin at pH 2.1. The absorbancy in the presence of denatured ovalbumin is similar to that of the free Congo red anion at pH 6.85.

Intravenously injected Congo red combines with serum albumin and therefore circulates for a long time in the vascular system. It disappears rapidly in amyloidosis, a pathological state in which amyloid, a glycoprotein, is deposited in the parenchymatous organs. Congo red has a high affinity to amyloid and can be used to stain amyloid in tissue sections. This staining is particularly interesting in view of the fact that amyloid becomes birefringent by combination with Congo red (C-13).

Proteins combine not only with acidic dyes, but also with the very weakly acid nitrophenols (C-14) and with neutral dyes such as aminoazobenzene (C-15). The combination depends on steric factors as demonstrated by the fact that serum albumin combines with only 6 molecules of $o$-nitrophenol, but with 24 and 25 molecules of $m$- and $p$-nitrophenol, respectively (C-14). The phenolic OH group is not necessary for the formation of the protein-dye complex as shown by the binding of 2,4-dinitrophenylmethane to bovine serum albumin (C-16). Serum albumin is also able to solubilize reversibly fat-soluble dyes such as dimethylaminoazobenzene (C-17) and fluorescent dyes which are not bound by serum globulin and other proteins (C-18). This ability of serum albumin to bind many dyes which are not bound by other proteins is not yet satisfactorily explained. It indicates a high "adaptability" of the serum albumin molecule to the different dyes and its ability to assume a large number of equilibrium conformations of approximately equal energy content (C-19). It has been suggested that the basic binding groups of the serum albumin molecule are in a lipophilic adaptable environment (C-20).

## D. Combination of Proteins with Other Organic Substances

Serum albumin and many of the other proteins combine readily with short-chain *fatty acids* such as butyric, caproic, or caprylic acid (D-1). Binding of fatty acids can be demonstrated by electrophoresis (D-2) and also by the fact that the fatty acids prevent the coagulation of proteins by heat or by other methods of denaturation (D-3). The affinity of protein to fatty acid increases with increasing chain length of the latter (D-4,30).

Numerous investigations deal with the combination of proteins with *anionic detergents*. These consist of long-chain alkylsulfonates, $RSO_3^-Na^+$, or alkyl sulfuric acid esters, $ROSO_3^-Na^+$; frequently SDS (sodium dodecyl sulfate, $C_{12}H_{25}OSO_3Na$) has been used. At pH values lower than the isoelectric point of the protein, the anionic detergents combine with cationic groups of the protein and cause precipitation (D-5). If an excess of the detergent is added, the amount of detergent bound to the protein increases to a multiple of that required for the neutralization of cationic groups. This is attributed to the binding of multimolecular micelles of the detergent to the protein (D-5,6) and also to interaction of the paraffin chains of the detergent with nonpolar groups of the protein (D-7). Thus, the reaction taking place between $\beta$-lactoglobulin and octylbenzenesulfonate proceeds in three stages; in the first, a few molecules of the detergent are bound "tail first," i.e., by their paraffin chains, which combine with hydrophobic groups of the protein (D-31). If more detergent is added, 22 molecules of it are bound "head first," i.e., by the sulfonate anion; finally, micelles are bound (D-8). Solubilization in an excess of the detergent is attributed to the formation of large aggregates which are kept soluble by anionic groups on their surface (D-5).

*Cationic detergents* (invert soaps) have the general structure $R_4N^+X^-$, where $R$ represents various alkyl or aryl residues and $X^-$ is usually a chloride ion. Some of the cationic detergents are cetylpyridinium chloride, dodecyltrimethylammonium chloride (Zephiran), or tolyldodecyltrimethylammonium methosulfate (Desogen). The invert soaps are very effective disinfectants. They precipitate proteins in the alkaline pH range where the carboxyl groups of the proteins have their maximum charge (D-9,10). The precipitated complex of hemoglobin and Desogen contains about 300 molecules of the invert soap per hemoglobin molecule (D-11). Similarly, up to 100 molecules of dodecylamine, $C_{12}H_{25}NH_{25}$ are bound per molecule of ovalbumin (D-12). Evidently, micelles of the detergent are bound to the protein.

Like the anionic detergents, the cationic detergents combine not only with ionic groups of the opposite charge but also with other, nonpolar groups of the proteins. They precipitate not only mucin and other mucoproteins (D-13) but also neutral polysaccharides such as starch (D-14). On the other hand, the proteins combine not only with the anionic and cationic detergents, but also with neutral, nonionic detergents such as Triton X100; the formation of a protein-detergent complex is indicated by an increase in the viscosity of the solution and a decrease in the electrophoretic mobility of either component (D-15).

Other macromolecular anions which combine with proteins are the anions of *polysaccharidic acids* such as heparin, alginic acid, or the polysaccharides of plant gums (D-16). They prevent heat coagulation of proteins at pH 7–8, but form precipitates in the interisoelectric range (D-17). The precipitates of these polysaccharide-protein complexes frequently have a loose, gel-like structure and form microscopic droplets. This form of precipitation, which has been called *coacervation* (D-18), is due to the immobilization of large volumes of water between the chains of the macromolecules involved. The reactions taking place between proteins and the macromolecular anions serve as a model of the combination of proteins with *nucleic acids* which will be discussed in Chapter XI, I.

Although the combination of proteins with other substances is, in many or most instances, initiated by the mutual electrostatic attraction of ionic groups of opposite charge, it has been mentioned above that the binding also depends on the short-range van der Waals' forces which act between *polar, uncharged groups*. The binding of simple sugars and of numerous drugs by proteins is due to the interaction of polar groups of these substances with polar or with ionic groups of the proteins. Even nonpolar hydrocarbons are bound to proteins. Thus, a 2% solution of edestin in 10% sodium chloride is able to hold in solution 5000 molecules of pentane per protein molecule (D-19). Most probably the nonpolar side chains of the amino acids leucine, isoleucine, valine and phenylalanine are involved in combination with the nonpolar hydrocarbons, and also with the paraffin chains of fats, fatty acids, or detergents (D-20).

Mutual precipitation of acidic and basic macromolecules occurs not only between proteins and nonproteins, but also *between acidic and basic proteins* (D-28). The precipitation takes place in the *interisoelectric range*, i.e., at pH values between the isoelectric points of the acidic and the basic protein. Thus many of the plasma proteins form precipitates with acidic azoprotein at pH 4–5 where the plasma proteins have a net positive charge and the azoproteins containing azophenylsulfonate groups have a net negative charge (D-21). Similarly, many of the denatured proteins give precipitates with the strongly basic protamines at pH 10 where the proteins have a net negative charge and the guanidino groups (pK = 12) of the protamines still carry positive charges (D-22). Soluble protein-protein complexes are formed between fibrinogen and serum albumin or γ-globulin (D-23), and also by combination of the basic protein lysozyme with serum albumin (D-24), catalase, or α-casein (D-25). The soluble complexes reveal their presence in paper electrophoresis. The lysozyme-protein and the similar protamine-protein complexes are soluble in an excess of the basic protein component. The formation of the protein-protein complexes is inhibited by neutral salts since their ions are bound by ionic groups of the opposite charge on the surface of the proteins. Proteins also combine with urea, guanidine salts, and urethane as proved by equilibrium dialysis (D-29). However, no binding of neutral amino acids to the plasma proteins has been observed (D-26,27).

The interaction between proteins and other proteins or nonproteins discussed in this chapter is largely a nonspecific interaction although the formation of the bond between protein and nonprotein, in some instances, is different for proteins of different species. A much higher influence of specificity is observed in the otherwise similar interaction of enzyme proteins with their protein or nonprotein substrates, and in the mutual binding of antigens and antibodies. These two types of highly specific protein complexes will be discussed in the Chapters XII and XIV.

## REFERENCES

### Section A

**A-1.** G. E. Perlmann and H. Hermann, *BJ* 32: 926, 931(1938).   **A-2.** D. Pressman and L. Pauling, *JACS* 71: 2893(1949).   **A-3.** F. London, *Z. physik. Chem.* B11: 222 (1930); J. C. Slater and J. G. Kirkwood, *Phys. Rev.* 31: 682(1931).   **A-4.** I. M. Klotz, *Brookhaven Symposia in Biol.* 13: 25(1960).   **A-5.** C. A. Coulson, "Valence." Oxford Univ. Press, New York and London, 1952; H. A. Harbury and K. A. Foley, *Proc. Natl. Acad. Sci. U.S.* 44: 662(1958); R. S. Mulliken, *JACS* 74: 811(1952).   **A-6.** V. Massey and G. Palmer, *BJ* 81: 22P(1961).

### Section B

**B-1.** P. Rona *et al.*, *Biochem. Z.* 149: 393(1924).   **B-2.** G. Scatchard *et al.*, *JACS* 79: 12(1957).   **B-3.** W. C. Stadie and F. Sundermann, *JBC* 91: 227(1931).   **B-4.** C. W.

Carr, *ABB* 46: 417, 424(1953).   **B-5.** D. M. Greenberg, *Adv. in Protein Chem.* 1: 121 (1944).   **B-6.** H. Benjamin *et al.*, *JBC* 103: 383, 629(1933).   **B-7.** M. Siegfried, *Ber. deut. chem. Ges.* 39: 397(1906).   **B-8.** A. Jensen and C. Faurholt, *Acta Chem. Scand.* 6: 385, 395(1952).   **B-9.** C. Neuberg *et al.*, *ABB* 58: 159(1955).   **B-10.** O. Henriques, *Biochem. Z.* 260: 58(1933).   **B-11.** A. Ferguson and F. J. W. Roughton, *J. Physiol.* 88: 40(1936).   **B-12.** D. D. Van Slyke *et al.*, *Proc. Natl. Acad. Sci. U.S.* 19: 828(1933). **B-13.** W. C. Stadie and H. O'Brien, *JBC* 117: 439(1937).   **B-14.** F. J. W. Roughton, *Physiol. Revs.* 15: 241(1936).   **B-15.** I. M. Klotz and W. C. Loh-Ming, *JACS* 76: 805 (1954).   **B-16.** J. Nyilasi and Z. Kovats, *Acta Chim. Acad. Sci. Hung.* 4: 11(1954). **B-17.** I. M. Klotz and H. Curme, *JACS* 70: 939(1948).

### Section C

**C-1.** F. Karush, *JACS* 73: 1246(1951).   **C-2.** I. M. Klotz, *Chem. Revs.* 41: 373(1947). **C-3.** R. K. Burkhardt *et al.*, *JACS* 75: 2977(1953).   **C-4.** J. Sponar and Z. Vodrazka, *Collection Czechoslov. Chem. Communs.* 22: 1232(1957).   **C-5.** M. Singer, *Intern. Rev. Cytol.* 1: 135(1952).   **C-6.** I. M. Klotz *et al.*, *JACS* 69: 1609(1947); 71: 847(1949); *J. Phys. Chem.* 53: 100(1949).   **C-7.** F. Haurowitz *et al.*, *JACS* 74: 2265(1952).   **C-8.** L. L. Uzman, *Nature* 171: 653(1953).   **C-9.** J. R. Colvin, *Can. J. Chem.* 30: 320(1952). **C-10.** I. G. Longsworth and C. F. Jacobsen, *J. Phys. Chem.* 53: 126(1949).   **C-11.** I. M. Klotz and H. A. Fiess, *BBA* 38: 57(1960).   **C-12.** I. M. Klotz *et al.*, *J. Phys. Chem.* 56: 77(1952).   **C-13.** P. Ladewig, *Nature* 156: 81(1945).   **C-14.** J. D. Teresi and J. M. Luck, *JBC* 174: 653(1938).   **C-15.** I. M. Klotz and J. Ayers, *JACS* 74: 6178(1952). **C-16.** M. E. Carsten and H. E. Eisen, *JACS* 75: 4451(1953).   **C-17.** B. Caroll, *J. Colloid Sci.* 6: 103(1950).   **C-18.** G. Weber and D. J. R. Laurence, *BJ* 56: XXXI (1954).   **C-19.** F. Karush, *J. Phys. Chem.* 56: 70(1952).   **C-20.** D. J. R. Laurence, *BJ* 51: 168(1952).

### Section D

**D-1.** A. Chanutin and R. R. Curnish, *ABB* 89: 218(1960).   **D-2.** G. A. Ballou *et al.*, *JBC* 159: 111(1945).   **D-3.** R. G. Rice *et al.*, *JBC* 158: 609(1945).   **D-4.** J. D. Teresi and G. M. Luck, *JBC* 194: 823(1952).   **D-5.** F. W. Putnam, *Adv. in Protein Chem.* 4: 80(1948).   **D-6.** J. Yang and J. F. Foster, *JACS* 75: 5560(1953).   **D-7.** F. Karush and M. Sonenburg, *JACS* 71: 1369(1949).   **D-8.** R. M. Hill and D. R. Briggs, *JACS* 78: 1590(1956).   **D-9.** K. H. Schmidt, *Z. physiol. Chem.* 277: 117(1943).   **D-10.** W. G. Jaffe, *JBC* 148: 185(1943).   **D-11.** F. Haurowitz, *Bull. fac. méd. Istanbul* 12: 183 (1949); *Discussions Faraday Soc.* 6: 58(1949).   **D-12.** S. N. Timasheff and F. F. Nord, *ABB* 31: 309(1951).   **D-13.** S. Tsuiki *et al.*, *JBC* 236: 2172(1961).   **D-14.** M. M. Yenson, *Nature* 159: 813(1947).   **D-15.** R. M. Douben and W. R. Koehler, *ABB* 93: 496(1961).   **D-16.** D. Hamer, *BJ* 56: 610(1954).   **D-17.** D. Dervichian and C. Magnant, *Bull. soc. chim. biol.* 29: 655, 660(1947).   **D-18.** H. G. B. de Jong and P. Teunissen, *Kolloidchem. Beih.* 47: 254(1938).   **D-19.** D. L. Talmud, *Acta Physicochim. U.S.S.R.* 14: 562(1941).   **D-20.** S. J. Przylecki *et al.*, *Biochem. Z.* 282: 362(1935). **D-21.** F. Haurowitz *et al.*, *ABB* 11: 515(1946).   **D-22.** F. Haurowitz *et al.*, *Z. physiol. Chem.* 315: 127(1959).   **D-23.** J. H. Quastel and S. F. Van Straten, *Proc. Soc. Exptl. Biol. Med.* 81: 6(1952).   **D-24.** R. F. Steiner, *ABB* 47: 56(1953).   **D-25.** T. Nagumo, *J. Biochem. (Tokyo)* 49: 379(1961).   **D-26.** J. M. Hunter and S. L. Commerford, *JACS* 77: 4857(1955).   **D-27.** A. Lietze *et al.*, *ABB* 76: 255(1958).   **D-28.** D. F. Waugh, *Adv. in Protein Chem.* 9: 326(1954).   **D-29.** A. G. Pasynskii and R. C. Cherniak, *Kolloid. Zhur.* 12(6): 460(1950).   **D-30.** D. S. Goodman, *JACS* 80: 3892(1958). **D-31.** B. Jirgensons, *Makromol. Chem.* 51: 137(1962).

# Chapter XI

## Conjugated Proteins

### A. Survey on Composition and Properties (A-1,2)

Conjugated proteins are defined as natural complexes formed by the combination of proteins with nonproteins. In the older literature, gluco-proteins, lipoproteins, chromoproteins, nucleoproteins, and phosphopro-teins were considered the most important groups of conjugated proteins. Most of these complexes are formed by the combination of 1 molecule of a *prosthetic group* with 1 molecule of the protein. The *phosphoproteins*, however, contain numerous phosphoric acid residues bound by ester bonds to the hydroxyl groups of serine and threonine. They can be considered as phosphoric acid esters of simple proteins. The typical phosphoproteins of milk and eggs have been discussed with other milk and egg proteins in Chapter VIII, Sections D and E. Whereas the prosthetic groups of the glycoproteins (mucoproteins) are polysaccharides which are bound to the protein by covalent bonds, some of the lipoproteins contain a mixture of lipids instead of a well defined prosthetic group. In this case the term *prosthetic group* loses its original meaning. Difficulties of classification arise also with the so-called *chromoproteins*, whose colored components may be heme derivatives, bile pigments, heavy metals, or carotenoids. Many of the heme proteins are important enzymes and will be discussed in more detail in Chapter XII. The carotenoid proteins will be treated in this chapter in Section B, with other lipoproteins, the metal proteins in Section D, and hemoglobin and its derivatives in Section E. Although all of these conjugated proteins occur as such in the organism, we are not sure whether the *nucleoproteins* prepared by various methods exist as such in cells. They may be artifacts, formed *in vitro* by the combination of nucleic acids with proteins with which they did not combine *in vivo*. It is evident from this survey of conjugated proteins that we are dealing with a very heterogeneous class of protein derivatives, and that their treatment in a single chapter does not signify any essential relation between these substances.

Although the various conjugated proteins are very different, they share one common property, namely the stabilization of their protein component by combination with their prosthetic group. Most probably, the large prosthetic groups of the lipo-, glyco-, or chromoproteins cover a large portion of the surface of the globular protein component and thereby prevent unfolding of the peptide chains and loss of the native structure. In a similar manner the heavy metal of the metalloproteins may form the coordination center of a number of peptide bonds or side chains and may thus stabilize the protein moiety. Removal of the prosthetic group, even if accomplished by mild methods, usually results in increased lability of the isolated protein component of the conjugated proteins and frequently in its denaturation.

## B. Lipoproteins (B-1 to 3b)

Lipoproteins occur in the organism as either soluble or insoluble complexes. Some of the insoluble lipoproteins are found in the cytoplasmic granules as lipo-ribonucleoproteins, others in the nervous system as insoluble lipoproteins, very rich in lipids. The presence of lipoproteins in the blood serum was first recognized by Macheboeuf who designated them as "cenapses" (B-4b). In contrast to free lipids, the protein-bound lipids cannot be extracted quantitatively by ether. The loose linkage between lipid and protein in the lipoproteins can be cleaved by treatment with high concentrations of alcohol or acetone, or by deep-freezing at $-60°C$ (B-4a). Denaturation of the protein moiety by the organic solvents can be avoided by extracting at a temperature of $-20°C$ or less. In contrast to the lipoproteins of other organs, those of the nervous system, particularly the lipoproteins of the white matter of the brain, are soluble in methanol-chloroform; they have been designated as proteolipids (B-26). The solubility of these complexes in organic solvents has been attributed to the formation on their surface of a lipid mantle which covers the protein moiety in such a manner that the hydrophilic groups of the protein are buried inside the molecule.

The *normal human blood serum* contains about 0.5–0.7 % lipid. The serum owes its transparency to the bonding of the normal lipids to proteins. The *lipoproteins* are salted out by half-saturation with ammonium sulfate. In the alcohol fractionation method (Chapter VIII, B) they are found in the fractions III and IV which are precipitated by 17–25 % ethanol. Upon electrophoresis the former migrates with the β-globulins, the latter with the α-globulins; accordingly, these lipoproteins have been designated as β- and α-lipoproteins, respectively. In paper electrophoresis the lipoproteins can be made visible by staining with Sudan Red, Sudan Black, Oil Red O, or by other lipid stains (B-5). *β-Lipoprotein* forms approximately 5 % of the plasma proteins and contains about 70 % of the plasma lipids (B-6). It consists of 23 % protein, 29.3 % phospholipids, and 39.1 % cholesterol and its esters (B-6). In addition to these lipids, the β-lipoprotein contains all of the carotene of the serum (B-7) as well as small amounts of other carotenoids and estrogens (B-6). Evidently it is not a well-defined, uniform substance, but a complex formed by a protein or a protein mixture with a mixture of lipids. Although the β-lipoproteins are soluble in serum and in isotonic salt solutions, they become insoluble in distilled water (B-9), behaving in this respect like euglobulins. If a solution of the β-lipoprotein is repeatedly frozen and thawed, it is irreversibly denatured and becomes insoluble; denaturation is also brought about by lyophiliaztion of the β-lipoprotein (B-10). *α-Lipoprotein* has a molecular weight of $2 \times 10^5$; it contains about 35 % lipid and 65 % protein and forms 3 % of the total plasma proteins (B-11).

The physicochemical properties of the lipoproteins are different from those of artificial emulsions of lipids and proteins. It is not yet clear whether the lipid is occluded between the peptide chains of the protein portion or whether it forms the center or the surface of a lipid-protein micelle (B-12). Nothing definite is known about the bonds which link the lipids to protein. In addition to bonds between nonpolar groups of the lipids and nonpolar side chains of some of the amino acid residues, electrostatic forces may be involved in the mutual attraction of polar groups of the two components. The importance of the positively charged amino groups of the phospholipids is manifested by the fact that lipoproteins are precipitated by dextran sulfates (B-8).

Because of their high lipid content, the density of the lipoproteins, especially that of the β-type, is much lower than that of other proteins. For this reason they sediment very slowly on ultracentrifugation and may even rise to the surface of the centrifuged solution, particularly if its density is increased by the addition of salts. The sedimentation constant (see Chapter V, D and E) then has a negative value. It is customary to replace the negative sedimentation constant by the analogous flotation constant, $S_f = -S$, and to differentiate the lipoproteins according to their $S_f$ values as high-density lipoproteins (HDL), low density, and very low density lipoproteins (B-13). The densities of these three groups are 1.063–1.21, 1.019–1.063, and 1.0006–1.019 respectively (B-15). They can be separated from each other by preparative ultracentrifugation at increasing densities produced by adding increasing amounts of sodium chloride, other neutral salts, or $D_2O$. The sedimentation constant and also the flotation constant depend on the density of the solvent (B-14).

Most of the analyses have been performed at densities of 1.063 or 1.21 g/ml. At the higher value, 1.21, the high-density lipoproteins rise to the surface and have an $S_f$ value of 0 to 16, whereas at the lower density of

1.063 sedimentation takes place. At a density of 1.063 two classes of lipoproteins, in addition to the HDL, have been differentiated; their flotation constants, $S_f$, are 0–20 and 20–400. The composition of these two lipid fractions and that of the HDL fraction in per cent of the total lipid of each of the fractions is shown below (B-3a):

| $S_f$ | 20–400 | 0–20 | HDL |
|---|---|---|---|
| Triglycerides | 55 | 14 | 17 |
| Phospholipids | 20 | 25 | 44 |
| Cholesterol and esters | 13 | 60 | 34 |

The lightest lipid particles are the chylomicrons, small lipid droplets coated by a thin layer of protein. Their flotation constant is higher than 400; their particle size corresponds to a molecular weight of many millions. Their lipids consist predominantly of triglycerides. Their increase in concentration after the ingestion of large amounts of lipids causes turbidity of the blood plasma.

This turbidity is cleared up by the injection of heparin, a polysaccharide sulfate. Heparin seems to activate lipoprotein lipase (B-16), an enzyme which hydrolyzes the triglycerides of the chylomicrons and thus furthers their conversion into high-density lipoproteins. During the last few years much attention has been paid to the distribution of the lipoproteins between the various $S_f$ classes; a correlation between this distribution and the incidence of atherosclerosis has been claimed by some authors (B-17) but contested by others (B-18).

An important group of lipoproteins includes those present in structural elements of the cells such as the cytoplasmic granules. The lipid content by weight of the liver mitochondria amounts to 15–20%, that of the microsomes to 40–50%. In the microsomes the lipids are bound to ribonucleoproteins. They can be solubilized by cholate or deoxycholate. Some of the enzymes of the electron transport system of the mitochondrial fraction have been purified and dissociated in this manner (B-19,20). The lipoproteins of the central nervous system consist most probably of leaflets, each of them formed by a monomolecular protein film of 10 Å thickness coated by two phospholipid layers. In the lipid layers (25 Å each) the phospholipid molecules are bound by their polar groups to the protein film and by their nonpolar groups to each other (B-27,28). Similar leaflets may be arranged concentrically around the nerve fibers.

Proteins also combine with *carotenes* to form colored lipoproteins. The color of crustaceans is due to crustacyanin, a lipoprotein containing the carotenoid astaxanthin, a 5,5'-dihydroxy-4,4'-diketo-β-carotene. Crustacyanin is extracted from lobster shells by citric acid (B-21). A similar lipoprotein is the green pigment of lobster eggs, ovoverdin; the molecular weight of this lipoprotein, which contains 1 molecule of astaxanthin per protein molecule, is about 300,000; its isoelectric point is pH 6.7 (B-22). While it exists in combination with protein, the carotene is safeguarded against oxidation; upon heating, the protein is denatured and the carotenoid is liberated as a red pigment. Lipoproteins containing carotenoids have also been found in green grasshoppers (B-23) and are probably present in other insects.

The most interesting carotenoid-protein compounds are the visual pigments and in particular the *visual purple*, which is present in the retinal rods but not in the cones, and is necessary for night vision (Kühne, 1879). To extract visual purple from the retina, substances are used which contain hydrophilic as well as lipophilic groups in their molecules, such as bile acids, digitonin, or cationic detergents. Stable solutions are obtained by using 75% aqueous glycerol. Solutions of visual purple must be prepared in the dark since the solutions are bleached instantaneously by light; the rose-colored solution first becomes yellow, then colorless. Prior to 1940 only two visual pigments were known, rhodopsin and porphyropsin, both occurring in the rods of the retina which are the sensory cells for night vision. Similar pigments were later found in the cones, the sensory receptors for daylight vision (B-24). The protein moieties of the visual pigments of rods and cones have been designated as scotopsin and photopsin, respectively (B-24). The composition of the four visual pigments is shown below:

| Visual pigment | Protein component | Carotenoid component | Absorption maximum (mμ) |
|----------------|-------------------|---------------------|-------------------------|
| Rhodopsin | Scotopsin | Retinene 1 | 500 |
| Porphyropsin | Scotopsin | Retinene 2 | 522 |
| Iodopsin | Photopsin | Retinene 1 | 562 |
| Cyanopsin | Photopsin | Retinene 2 | 620 |

Retinene 1 has been found in the eyes of terrestrial and marine animals, retinene 2 in fresh water fish. Retinene 1 is an aldehyde formed from vitamin A by oxidation of the primary alcohol group to an aldehyde group. Retinene 2 has the same carbon skeleton as retinene 1, but has in addition to the 5,6-double bond another double bond in 3,4-position. Retinene occurs in the visual purple in the 11-*cis* form shown below. By the action of light it is isomerized to form the all-*trans* form which does not combine with scotopsin (B-25). Evidently, the combination of retinene with scotopsin is highly specific. Nothing is yet known concerning the groups

of the protein component, scotopsin, which mediate the combination
of the two components of rhodopsin. The intense color change which

Retinene (retinal, retinene 1, vitamin
A aldehyde, 11-*cis*-retinene, neoreti-
nene *b*)

accompanies the combination of the two compounds and the inability
of scotopsin to combine with vitamin A suggest that the aldehyde
group is involved in this reaction. Most probably, it combines with an
amino group and thus gives rise to the formation of a methylene imine:
$R \cdot CHO + H_2NX \rightarrow R \cdot CH:NX$. The double bond of the methylene
imine would elongate the chain of conjugated double bonds and would
thus cause deepening of the color. In addition to an amino group a cystine
or cysteine residue seems to be involved in binding retinene since the
bleaching reaction is accompanied by the liberation of a sulfhydryl group
(B-24). Regeneration of rhodopsin consists of several reactions, viz.,
reduction of *trans*-retinene to the *trans*-isomer of vitamin A, stereoisomer-
ization to the *neo-b* isomer, reoxidation to the aldehyde form, and combination with the respective opsin. The oxidation of vitamin A (retinol) to
retinene (retinal) is catalyzed by a DPN enzyme which can also act as an
ethanol dehydrogenase (B-24).

## C. Glycoproteins (Mucoproteins) (C-1 to 5)

The terms glycoproteins and mucoproteins are used here to designate
conjugated proteins containing carbohydrates. In some instances the
carbohydrate is present as a single polysaccharide (C-28a), in others as
several oligosaccharide units or as a large number of disaccharide residues
(C-28b,c). Most if not all proteins contain small amounts of carbohydrate (see Chapter VIII, Table VIII-3). It is difficult therefore to draw
a sharp borderline between these proteins and the typical mucoproteins.
The nomenclature in this field is extremely confusing.

The term *glycoprotein* is used by some authors for those proteins which contain
less than 4% carbohydrate (C-1), chiefly hexoses, hexosamines, and sialic acid,
other authors use the term glycoprotein for complexes in which the carbohydrate
is bound to the protein by covalent bonds (C-3,4). Since many of the typical

protein-carbohydrate complexes are secreted by the mucosa of the gastro-intestinal tract or the respiratory mucosa, the glycoproteins occurring in these secretions were originally called *mucins*. Later, similar viscous proteins were detected in the synovia of the joints, in the vitreous body, and in other body fluids or tissues. They were designated as *mucoids*. The term mucoid is used by some authors to denote covalent protein-carbohydrate complexes (C-1), by others for those glycoproteins which are not precipitated by acidification (C-3). The designation *mucoprotein* is applied sometimes to loose, easily dissociable complexes of proteins with mucopolysaccharides which contain hexoses and hexuronic acids (C-1,3).

Since some of the well-defined hydrolytic breakdown products of the protein-carbohydrate complexes are quite generally designated as *glycopeptides*, the term *glycoproteins* is used here for their mother substances, the typical protein-carbohydrate complexes of body fluids or tissue extracts. The polysaccharides present in glycoproteins contain various monosaccharides, particularly galactose and mannose, acetylhexosamines, and sialic acid. The latter is a designation used for $N$-acetyl or glycolyl derivatives of neuraminic acid, a $C_9$-acid of the formula

$$\overset{\displaystyle \ulcorner\!\!-\!\!-\!\!-\!\!-\!\!O\!\!-\!\!-\!\!-\!\!-\!\!\urcorner}{CH_2OH \cdot CHOH \cdot CHOH \cdot CH \cdot CHNH_2 \cdot CHOH \cdot CH_2 \cdot COH \cdot COOH}$$

$N$-acetyl-neuraminic acid has been prepared by the condensation of 2-mannosamine with pyruvic acid in the presence of suitable enzymes. The carbohydrate content of various human plasma proteins is shown in Table XI-1, that of some other glycoproteins in Table XI-2.

Table XI-1 must not give the impression that all of the plasma proteins are typical glycoproteins. The carbohydrate content of most of the other plasma proteins is less than 4%. Thus fibrinogen contains in 1 molecule (mol. wt. 350,000) only 45 monosaccharide residues or 8100 g carbohydrate in 350,000 g protein, i.e., 2.3% carbohydrate. About 30% of the protein-bound carbohydrate consists of neuraminic acid, another 30% of hexosamines, and 40% of hexoses (C-8).

Many of the glycoproteins are relatively stable compounds. Thus far it has not been possible to dissociate them and to separate the unchanged carbohydrate from the native protein. Although the glycosidic bond is stable to alkali, ester bonds and other labile bonds may be hydrolyzed by alkali. One of the best methods for the isolation of the carbohydrate is the degradation of the protein by proteolytic enzymes. The polysaccharide of ovalbumin has been obtained free of amino acids by hydrazinolysis of the protein (C-28e). The isolated polysaccharide can be fractionated by ethanol in the presence of calcium or barium acetate; the barium salts of polysaccharide sulfuric acid esters are insoluble in alcohol, whereas other polysaccharides are soluble (C-9,10). Fractionation is also possible by chromatography on cellulose columns or by the addition of cationic detergents which form insoluble complexes with the acidic mucopolysaccharides (C-4). The *sugars* which form the polysaccharide are frequently determined by colorimetric reactions. Since most of these are carried out in strong solutions of sulfuric acid, it is not necessary to hydrolyze the polysaccharide prior to colorimetry. The *hexosamines* (glucosamine, galactosamine) give the Elson-Morgan reaction in which the material is first heated with acetylacetone and sodium carbonate and then with a solution of Ehrlich's aldehyde (*p*-dimethylaminobenzaldehyde) in

TABLE XI-1
CARBOHYDRATE CONTENT OF SOME HUMAN PLASMA PROTEINS[a]

| Protein | Mol. wt. ($\times 10^{-3}$) | Galactose | Mannose | Fucose | $N$-acetylhexosamine | Siliac acid | Hexuronic acid |
|---|---|---|---|---|---|---|---|
| | | (Moles/mole protein) | | | | | |
| $\alpha_1$-Seromucoid[b] | 41 | 18[d] | 18[d] | 3[d] | 33[d] | 16 | 1 |
| $\alpha_2$-Macroglobulin | 846 | 85 | 85 | 7 | 111 | 49 | 6 |
| $\beta_1$-Metalloprotein[c] | 88 | 8 | 4 | tr | 8 | 4 | tr |
| Fibrinogen | 350 | 7 | 14 | 0 | 14 | 7 | 3 |
| $\gamma$-Pseudoglobulin | 161 | 3 | 7 | 2 | 8 | 1.5 | 2 |
| $\gamma$-Euglobulin | 161 | 3 | 6 | 2 | 8 | 1.5 | 1 |

[a] From (C-7).

[b] Also designated as orosomucoid and $\alpha_1$-acid glycoprotein.

[c] Also designated as transferrin or siderophilin.

[d] From (C-28d).

TABLE XI-2
CARBOHYDRATE CONTENT OF VARIOUS PROTEINS[a]

| Protein | Mol. wt. ($\times 10^{-3}$) | Carbohydrate content (%) | | |
|---|---|---|---|---|
| | | Hexose | Hexosamine | Sialic acid |
| Thyrotropic hormone (cattle) | 10 | 3.5 | 2.5 | — |
| Gonadotropic hormone (mare serum) | 23 | 18.6 | 13.5 | 10.4 |
| Chorionic gonadotropin (human) | 30 | 11.0 | 8.7 | 8.5 |
| Interstitial cell-stimulating hormone (sheep) | 40 | 4.5 | 5.8 | — |
| Follicle-stimulating hormone (sheep) | 70 | 1.2 | 1.9 | — |

[a] From (C-5).

HCl (C-11). Colorimetry of the red solution allows the determination of the hexosamines quantitatively. If the determination is made in the presence of borate, glucosamine can be differentiated from galactosamine (chondrosamine) (C-12). Sialic acid gives the color reaction with Ehrlich's aldehyde directly without the necessity of previous treatment with alkali and acetylacetone. Hexuronic acids are determined by colorimetry of the purplish red solution obtained after heating the polysaccharide with concentrated sulfuric acid in the presence of carbazole (C-13), or by means of their color reaction with orcinol and HCl (C-14). Fucose, like other methyl-pentoses, gives a characteristic color reaction with sulfuric acid and cysteine and can be determined colorimetrically (C-15). Some of the reactions for the detection of sugars in glycoproteins can be carried out on paper chromatograms or after paper electrophoresis (C-16,17).

If a tissue extract or a body fluid contains polysaccharides with acidic groups, acidification frequently produces a precipitate consisting of protein and the polysaccharide. Such precipitates contain hyaluronic acid, a polymer of glucuronyl-3-glucosamine, or sulfuric acid esters of chondroitin or mucoitin. Chondroitin sulfate A and C consist of the same polysaccharide with sulfate ester groups at the carbon atoms 4 and 6, respectively, of the $N$-acetylgalactosamine residues (C-56). In chondroitinsulfate B, the glucuronyl residue of A is replaced by an iduronyl residue (C-56). The combination of these polysaccharides with proteins is attributed in part to the neutralization of the acidic groups of the polysaccharides by basic groups of the proteins. Complexes of this type have been discussed in Chapter X, D. Glycoproteins can be detected by electrophoresis in the presence and absence of borate (C-18). The latter combines with carbohydrates and thus increases their negative charge and anodic mobility. If the mobility of a protein rises on addition of borate, it can be assumed that it contains protein-bound carbohydrate.

As can be seen in Table XI-1, carbohydrate is present in many of the *plasma proteins*. The highest carbohydrate content, amounting to 37%, has been found in a protein designated as $\alpha_1$-acid glycoprotein, $\alpha_1$-*seromucoid* or *orosomucoid* (C-19). Although the orosomucoid content of the serum is small, its carbohydrate forms 10% of the protein-bound carbohydrate of the blood serum. Orosomucoid is not precipitated by heat, nor by trichloroacetic or perchloric acid; it is found in the soluble fraction VI of the alcohol fractionation procedure of the plasma proteins (C-20) and has been isolated as a crystalline lead salt (C-21). Its carbohydrate component contains the $\beta$-galactosyl-1,4-$N$-acetylglucosaminyl-1,4-mannosyl-1,4-$N$-acetylglucosamine residue (C-28d).

Another glycoprotein is *haptoglobin* (see Chapter VIII, B) which contains 5.1% $N$-acetylneuraminic acid, 5.4% glucosamine, 1% fucose, 0.5% glucose, and 8.5% galactose + mannose (C-22). The terminal residue of the polysaccharide is formed by neuraminic acid linked to the C-atom No. 3 of an adjacent galactose residue (C-23). Finally, *fetuin*, a glycoprotein from calf serum (see Chapter VIII, B) may be mentioned since it has been obtained in a highly purified state (C-24,25). Its molecular weight is 48,400, it contains 72% protein, 13 molecules of sialic acid, 15 of hexosamine, 12 of galactose, and 8 of mannose (C-26,25). If blood serum is exposed to the amylolytic action of saliva, free hexoses are split off from the serum proteins (C-27).

A well-defined *glycopeptide* has been obtained from human γ-globulin by the action of proteolytic enzymes. Its amino acid sequence is —Glu.-Glu.AspN.Tyr.Glu.Asp with a *C*-terminal aspartyl residue; a polysaccharide consisting of 3 galactose, 5 mannose, 2 fucose, 8 glucosamine, and 1 sialic acid residue is bound by covalence to a carboxyl group of the aspartyl residue (C-28a). Other glycopeptides have been isolated from bovine γ-globulin (C-54) and from ovalbumin (C-29,28e,58). In all these glycopeptides the carbohydrate is bound to the aspartyl residue, probably to its β-carboxyl group.

The glycoproteins of saliva, gastric juice, intestinal juice, and other secretions are frequently called *mucins* (C-30). They are responsible for the high viscosity of these secretions and consist chiefly of glucuronic acid, mannose, acetylglucosamine, and acetylgalactosamine (C-31,32). Small amounts of gluconic acid have been discovered in the mucin of the submaxillary gland (C-1). In the bovine submaxillary glycoprotein, the prosthetic groups contain sialyl-*N*-acetylgalactosamine bound to the carboxyl groups of aspartyl or glutamyl residues (C-33,34).

A highly viscous glycoprotein, called *ovomucoid*, has been isolated from the white of hens' eggs. It can be separated from the other proteins since it is not coagulated by heat. After removal of the denatured ovalbumin and ovoglobulin, ovomucoid is precipitated from the filtrate by ethanol. Ovomucoid contains about 20% carbohydrate. The *N*-terminal amino acid is alanine (C-35), the *C*-terminal amino acid, phenylalanine (C-36). The carbohydrate moiety is formed by *N*-acetylglucosamine, mannose, and galactose in the ratio 7:3:1 (C-37). The polysaccharide is bound to the protein by an ester bond between a hydroxyl group of one of the sugars and a carboxyl group of the protein.

Mannose and glucosamine have also been found in *avidin*, a glycoprotein of egg white (C-38); galactose occurs in casein (C-39). The whey proteins (see Chapter VIII), contain both hexoses and hexosamines (C-39).

It has been mentioned earlier (Chapter IX, B) that collagen combines in the organism with a polysaccharide. The polysaccharide content is particularly high in cartilage which consists essentially of collagen and chondroitinsulfuric acid. It is difficult to decide whether the two components are linked to each other merely by saltlike bonds between the sulfate groups and basic groups of the protein, or whether their combination is more specific (C-1). *In vitro* chondroitinsulfuric acid combines with proteins at pH values below 4.7 (C-40). The water-soluble extract of cartilage contains a conjugated protein having a molecular weight of 1–5 millions; 75% of the molecule is formed by chondroitinsulfuric acid chains having an average molecular weight of 50,000; the reducing groups of the polysaccharide are free; the chondroitinsulfuric acid chains are linked to each other by the protein (C-41,42). Both collagen and chondroitinsulfuric acid are also present in some of the other connective tissues.

A particularly high content of carbohydrate, consisting of mannose, glucose, galactose, and fucose, has been found in the reticular fibers of the connective tissue, for instance in the skin or the lung; it is not certain yet whether this carbohydrate is bound to collagen or forms another glycoprotein, frequently designated as *reticulin* (C-43). In the skin, some of the proteins combine also with hyaluronic acid. This polysaccharide is found also in the umbilical cord, in the synovial fluid of the joints, and in the vitreous humor of the eye; it is responsible for the viscosity and the gelatinous consistency of these fluids. Large amounts of substances of this type can be isolated from pathological ovarian cysts, which sometimes attain a volume of more than a liter. None of these protein-polysaccharide complexes have been obtained as well defined substances of a constant composition. They may be merely saltlike compounds formed by the combination of hyaluronic acid with various proteins. The same may be true for the small amounts of mucoid excreted in the normal urine. Under pathological conditions, particularly in chronic inflammatory processes, a substance is deposited in the liver and in other parenchymatous organs which gives a purplish color with iodine and has been called *amyloid*. Although sulfuric acid esters have been identified in amyloid (C-44), it has not yet been determined whether amyloid is a glycoprotein containing chondroitinsulfuric acid.

Polysaccharides have also been found in the *blood group substances* of the types A, B, O(H), and Le$^a$. The blood group substances are glycopeptides consisting of 50% or more of carbohydrate (C-45,46). They are responsible for the agglutination of red blood cells by agglutinins present in the blood serum of persons of other blood group types. The four types of blood group substances have a very similar composition; each of them contains glucose, glucosamine, galactosamine, and fucose. The ratios of the two hexosamines vary considerably from one type to the other and may thus determine the type specificity (C-47,48). The peptide moiety of the blood group substances is particularly rich in threonine and proline, and also contains glycine, alanine, valine, leucine and/or isoleucine, serine, arginine, lysine, aspartic and glutamic acid, traces of histidine, but no cystine, methionine, nor any of the aromatic amino acids. The total amino acid content is 15–18 per cent (C-55). The polysaccharide is free of sialic acid. The blood group substances occur in red blood cells in very small amounts. Similar substances have been found in higher concentrations in saliva, in hog stomach, and in commercial peptone (C-49,50). They can be extracted from the mucosa by 90% phenol and are precipitated by ethanol. The determinant group of the blood group substance A is formed by a terminal $\alpha$-N-acetylgalactosamine-1,3-$\beta$-galactosyl-1,3-N-acetylglucosamine residue (C-57).

A new type of glycopeptides has been discovered in the cell walls of certain bacteria. They contain *muramic acid* which is an ether formed by the alcohol group of lactic acid and the 3-C atom of *N*-acetyl-glucosamine. In the cell wall of Staphylococcus aureus, muramic acid is bound to a peptide of the composition Ala.D-Glu.Lys.D-Ala.D-Ala (C-51) which contains the D-isomers of alanine and glutamic acid. Protein-polysaccharide complexes of a different type have been found in the cell wall of yeast; they consist of a water-soluble mannan-protein, a glucan-mannan-protein soluble in ethylene diamine, and an insoluble complex of unknown composition (C-52). The insoluble cuticula of insects consists likewise of protein and a carbohydrate, chitin. It is not yet clear whether the two materials form an interpenetrating lattice or whether they are linked to each other by co-valent bonds (C-53).

## D. Metalloproteins (D-1 to 4)

The metalloproteins are complexes formed by the combination of proteins with heavy metals. In most of these the metal is loosely bound and can be removed by treatment with dilute mineral acids. However, in the heme proteins, which contain iron-porphyrin complexes, the iron is very firmly bound and cannot be split off by dilute HCl. Hemoglobin and its derivatives will be discussed in Sections E, F, and G of this chapter; the enzymatically active heme and metallo-proteins will be treated in Chapter XII.

One of the typical metalloproteins is *ferritin* (D-5), a crystalline protein containing about 20 % iron. It was first isolated by Laufberger (D-6) from liver and spleen. Ferritin contains iron in the ferric state; the elemental composition corresponds to almost 1 iron atom per amino acid residue (D-7). If ferritin is reduced by sodium dithionite, cysteine, or ascorbic acid (D-2), iron-free apoferritin is obtained (D-8), a protein of the molecular weight 465,000 (D-9) which consists of 24 subunits (D-26b). Its amino acid composition is shown in Table XI-3. In ferritin the apoferritin molecule combines with the polymer of an inorganic ferric compound of the composition $(FeO \cdot OH)_8(FeO \cdot OPO_3H_2)$ (D-8) to form a clathrate (inclusion compound) in which chains of $O{=}Fe{-}OH \cdots O{=}Fe{-}OH \cdots$ , some of them containing phosphate, are embedded between the peptide chains, and in which the iron combines with the nitrogen atoms of the peptide bonds (D-2). Ferritin can be regenerated from apoferritin and ferrous salts. We do not yet know the structural peculiarity which endows apoferritin with the ability to combine specifically with ferric hydroxide or phosphate. Ferritin is a storage form of iron in the animal organism. Its amount increases after the oral administration of iron salts (D-10). The vasodepressive material produced by the liver is identical with ferritin (D-11).

Another storage form of iron, called *hemosiderin*, has been known for many years to histologists. In contrast to ferritin, hemosiderin is insoluble in water. It is an iron protein complex which contains about 25% of nucleotides and carbohydrate (D-2). In the tissues, where it occurs chiefly in reticuloendothelial cells, it can be detected, by means of the Prussian Blue reaction, after exposure of the tissue sections to HCl and potassium ferrocyanide. The claim that hemosiderin is an insoluble form of ferritin is disproved by the finding that the amino acid composition of hemosiderin is quite different from that of ferritin (Table XI-3).

The blood plasma of man and many mammals contains a soluble iron-protein complex called *transferrin* or *siderophilin*. It is a $\beta_1$-globulin which contains 0.13% iron. As mentioned in Chapter VIII, B, several transferrins, determined by genetic factors, have been discovered in human

TABLE XI-3
AMINO ACID COMPOSITION OF SOME METALLOPROTEINS[a]

| Amino acids | Apo-ferritin[b] | Apo-siderin[c] | Hem-erythrin[d] | Erythro-cuprein[e] |
|---|---|---|---|---|
| Glycine | 3.4 | 7.23 | 2.19 | 7.85 |
| Alanine | 1.9 | 6.70 | 4.57 | 4.75 |
| Valine | 4.3 | 1.70 | 5.85 | 6.87 |
| Leucine | 19.1 | 12.1 | 13.85 | 6.07 |
| Isoleucine | 1.4 | — | | 3.74 |
| Proline | 1.5 | 3.87 | 4.74 | 3.25 |
| Phenylalanine | 6.1 | 1.96 | 14.8 | 4.37 |
| Tyrosine | 5.0 | 1.47 | 5.38 | 0.95 |
| Tryptophan | 1.2 | 1.89 | — | 0.34 |
| Serine | [f] | 7.79 | 3.21 | 5.18 |
| Threonine | 4.3 | 3.77 | 3.80 | 5.54 |
| Cystine[g] | 1.7 | 3.10 | 0.30 | 3.51 |
| Methionine | 1.9 | 4.57 | 0.43 | 0 |
| Arginine | 9.1 | 7.24 | 3.28 | 4.27 |
| Histidine | 4.8 | 1.49 | 5.19 | 5.97 |
| Lysine | 7.8 | 8.88 | 9.07 | 7.86 |
| Aspartic acid | 6.8 | 4.42 | 13.8 | 14.32 |
| Glutamic acid | 17.2 | 21.7 | 9.08 | 10.97 |
| Ammonia | 0.9 | — | — | 1.18 |

[a] Gram amino acid 100 g protein.

[b] From (D-17).

[c] Aposiderin is the protein moiety of hemosiderin; it forms 17.5% of the dry weight of hemosiderin. The inorganic iron salts form 57.5% and the nucleotide component 25% (D-2).

[d] From (D-48).

[e] From (D-49); erythrocuprein contains 1.63% hexosamine, 0.88% hexose, and 1.78% sialic acid.

[f] Not determined.

[g] Cystine + cysteine.

serum. In the alcohol fractionation procedure, transferrin is found in fraction IV (see Figure VIII-2). The molecular weight of transferrin is 90,000 (D-12). The molecule contains 2 iron atoms and is the physiological iron carrier of the organism. If the iron is removed by lowering pH, the iron-free protein can also bind 2 atoms of copper (D-12). The color of transferrin is salmon-pink, its absorption maximum is in the visible spectrum at 460–470 m$\mu$. Transferrin is free of sulfhydryl groups. The iron may be bound to the hydroxyl groups of tyrosine residues (D-4). A very similar iron complex is formed by the combination of ferric ions with conalbumin from hens' eggs (D-13). Both proteins, transferrin and the iron complex of conalbumin, have the same color and a very similar absorption spectrum; both also combine with copper or zinc.

Human and other mammalian blood sera contain, in addition to transferrin (siderophilin), another metalloprotein called *ceruloplasmin* (or coeruloplasmin). It contains 0.34% copper and has been isolated in crystals (D-14). On electrophoresis ceruloplasmin migrates with the $\alpha_1$-globulins; it is a blue cupric complex and has an absorption maximum at 610 m$\mu$ (D-4). On reduction to the cuprous state, ceruloplasmin loses its color. The cupric ion is bound in ceruloplasmin so firmly that it does not react with diethyl-dithiocarbamate (D-4). Ceruloplasmin has weak catalytic activity as an oxidase (D-15). In Wilson's disease (hepato-lenticular degeneration) large amounts of copper are found in the liver and in other organs (D-16). This is attributed to congenital deficiency of ceruloplasmin (D-14).

*Hemerythrin*, an iron-containing respiratory protein, occurs only in a few species of marine invertebrates, principally in the sipunculoids which form a class of the annelids, marine worms. The pigment is found in brown cells of the perienteric fluid and can be obtained from these in crystalline form by cytolysis (D-18). The hemerythrins of different invertebrates contain 0.8–0.99% of iron (D-19). The molecular weight, determined by osmometry or ultracentrifugation, is 66,000 (D-20,21). Hence, the molecule contains approximately 10 iron atoms. Hemerythrin is a respiratory pigment which, like hemoglobin, can bind molecular $O_2$; the maximum amount of $O_2$ bound is 1 molecule of $O_2$ per 2 atoms of Fe. The oxygen molecule probably forms a bridge between 2 ferrous ions (D-25). The oxygenated pigment is brown-red and has an absorption maximum at 500 m$\mu$. On deoxygenation the pigment is bleached; the color of the reduced form is pale yellow (D-22). Oxidation of hemerythrin with ferricyanide converts the pigment into a yellow ferric form called methemerythrin (D-23). The iron in oxygenated hemerythrin is presumably in the ferrous state; however, partial oxidation to the ferric form cannot be excluded (D-24). Hemerythrin is free of porphyrins. The iron seems to be bound to sulfur atoms of the protein moiety (D-24) since it can be displaced by mercury compounds. Hemerythrin, like hemoglobin, shows a strong Bohr effect, i.e., a decrease in its affinity for $O_2$ at increasing $CO_2$ pressure (D-26a).

*Hemocyanin*, the copper-containing respiratory protein of numerous mollusks and arthropods, resembles hemerythrin in the lack of porphyrins or any other prosthetic group. In contrast to hemerythrin, hemocyanin does not occur in blood cells but is present in a soluble form in the hemolymph of the animals. Crystallization has been accomplished by dialyzing hemocyanin against solutions of low ionic strength (D-27 to 29). Hemocyanin behaves in this respect like a euglobulin. Crystals of hemocyanin have also been obtained by salting-out with ammonium sulfate (D-30). Hemocyanin combines with $O_2$ to form a blue oxygenated compound which contains 1 molecule of $O_2$ for every 2 copper atoms (D-27). Under reduced pressure oxygen is given off and the slightly yellowish, oxygen-free compound is formed. The copper atom is split off by the action of hydrogen cyanide (D-30) or by hydrochloric acid.

The hemocyanins are species-specific, as shown by the differences in the shape of the crystals obtained from different animals. While the hemocyanins of lobster, crayfish, and other arthropods contain 0.17–0.18% copper, those of the snail *Helix pomatia*, the squid *Loligo*, and of other mollusks contain 0.24–0.26% copper (D-31,32). In the deoxygenated compound, which can be obtained by removal of oxygen in the vacuum, the copper is in the cuprous state. On oxygenation the $O_2$ molecule is bound to 2 copper ions between which it probably forms a bridge. The valence state of Cu has been discussed by several authors (D-24,25), but it is not yet clear whether it is in the cupric or cuprous state. On combination with $O_2$ the paramagnetic susceptibility of hemocyanin disappears as shown by measurements of the electron spin resonance (D-33). Oxyhemocyanin is diamagnetic. The copper ions cannot be displaced by silver or mercury although it has been suggested that they are bound to sulfur atoms of cysteine residues (D-24); they may be bound to imidazole groups of histidine (D-34,35). On treatment with $H_2O_2$, hemocyanin loses its capacity for binding $O_2$, probably due to oxidation of Cu to the cupric state (D-36,51).

The molecular weight of the hemocyanins is extremely high. Sedimentation measurements made with the ultracentrifuge give values varying from 500,000 to 10,000,000 (D-37). These values, the highest ever observed for any animal protein, were confirmed by osmometry and by the scattering of light by hemocyanin. Electron micrographs of the hemocyanin isolated from the horseshoe crab, *Limulus polyphemus*, revealed it to consist of almost spherical particles with an average diameter of 200 Å (D-38). The hemocyanin of *Busycon caniculatum*, on the other hand, is formed by parallel bundles of 4 rodlike subunits (D-39,40) and the hemocyanin of *H.pomatia* by a cylinder which consists of 6 disklike subunits (D-52).

In concentrated solutions of urea the hemocyanins undergo disaggregation into smaller particles (D-41); reversible disaggregation of hemocyanin molecules into subunits is also brought about by acids or by alkaline solutions. This splitting occurs under conditions which ordinarily are considered too mild to cause denaturation. The split products recombine to form hemocyanin. Since the equivalent weight of the hemocyanins of mollusks is approximately 25,000 per copper atom and that of arthropods 37,000, it is evident that each molecule contains a large number of copper atoms and is capable of combining with many oxygen molecules.

Copper has also been found in the enzymes laccase, tyrosinase, and ascorbic acid oxidase (D-29,42,43) and in the metalloproteins *hemocuprein* (erythrocuprein) and *hepatocuprein* which were isolated from the red blood cells and the liver respectively (D-44). They can be separated from most of the other proteins by treatment with ethanol and chloroform which renders other proteins insoluble. The copper proteins are precipitated from the aqueous solution by lead acetate and extracted from the precipitate by a solution of potassium phosphate. Hemocuprein has been obtained in crystals. Both hemocuprein and hepatocuprein contain 0.34 % Cu which is also the copper content of ceruloplasmin. The molecular weight of hemocuprein is 35,000. Its copper ion is present in the cupric form as demonstrated by electron spin resonance (D-42). The molecule contains two Cu atoms (D-53).

*Hemovanadin* is a vanadium-protein complex which occurs in the vanadocytes, blood cells of the tunicates, marine animals which are covered by a translucent mantle of cellulose. It can be isolated from the yellowish-green vanadocytes by hemolysis as a brown red complex of molecular weight 24,000 which contains 24 atoms of vanadium (D-2,45). The brown-red $V^{3+}$ complex is oxidized on exposure to air to a blue green $V^{4+}$ complex. In these complexes the protein seems to be bound by saltlike bonds to a coordination complex of vanadium in which the ligands are sulfate ions and tertiary amino groups (D-2,45). Since the tunicates accumulate relatively large amounts of vanadium from sea water which contains only traces of this metal, they seem to contain a protein with a specific affinity for this rare metal.

*Metallothionein.* This protein occurs in the cortical zone of the horse kidney as cadmium or zinc complex (D-46). It contains 9.3 % of sulfur, 95 % of which is present in the form of cysteine. Approximately one-third of all amino acid residues of metallothionein are cysteine residues. The isolated metallothionein contains 5.9 % cadmium and 2.2 % zinc. Each of the heavy metal ions is bound to 3 cysteine residues. The absorption maxima of the Cd- and Zn-cysteine complexes are at 250 and 215 m$\mu$, respectively (D-46). The observation of these maxima is made easy by the absence of tyrosine and tryptophan whose high absorbance at 250 to 300 m$\mu$ would otherwise interfere with the measurement of the absorbancy of the cysteine complexes. Zinc-cysteine complexes have also been found in the tapetum lucidum of the eye of various carnivorous animals, particularly in the eyes of seals (D-47). They contain 1 zinc atom per cysteine.

Zinc is an integral constituent of carbonic anhydrase, several dehydrogenases, and other enzymes. These and other enzymatically active metalloproteins will be treated in Chapter XII.

## E. Hemoglobin: General Properties (E-1 to 6,43)

Hemoglobin is the principal oxygen carrier of the vertebrates, but it occurs also in the blood or the lymph of numerous invertebrates. It has been found in some of the protozoa (e.g., *Paramecium caudatum, Tetrahymena*) and molds (yeast, *Neurospora, Penicillium notatum*) (E-7). Mammalian red cells contain about 30–34% hemoglobin.

In order to prepare hemoglobin, whole blood is centrifuged, the plasma removed, and the red cells are washed with isotonic saline solution. The cells are hemolyzed by the addition of water, diethyl ether, or toluene (E-40). The stromata of the red cells are removed by centrifuging. If ether or toluene has been used, a considerable portion of the stromata remains suspended between the aqueous and the organic solvent layer from under which the hemoglobin solution can be removed by syphoning off. Hemoglobin is precipitated from this solution by cautiously adding ethanol at low temperatures. Some hemoglobins such as those of the rat, horse, or dog, are barely soluble in salt-free water and crystallize on dialysis or are precipitated by passing a current of oxygen and carbon dioxide through the solution, the latter serving to keep the reaction slightly acidic. The prosthetic group of hemoglobin is the same for all hemoglobins and myoglobins; it is protoheme, the ferrous compound of protoporphyrin:

$$CH_3C\!\!-\!\!-\!\!-\!\!-CCH\!\!=\!\!CH_2$$

**Protoheme**

In the presence of air, heme is instantaneously oxidized to a ferric porphyrin complex called hemin. The heme-free protein component of hemoglobin is called *globin*. In contrast to hemoglobin, globin is a very labile protein, easily denatured by dilute acid or alkali. Native globin was first obtained when hemoglobin was carefully acidified with HCl at 0°C and the free hemin extracted with ether (E-41). Better results are obtained by acidification of hemoglobin in the presence of high concentrations of acetone (E-8 to 10). In contrast to the denatured globin, the native globin combines with protohemin at pH 8–9 in the presence of a reducing agent to form hemoglobin (E-41). Suitable reducing agents are sodium dithionite ($Na_2S_2O_4$) or ammonium sulfide; they convert the added hemin to heme which then combines with globin. The combination of heme with globin is a reversible reaction; hence heme can be transferred from one hemoglobin molecule to another (E-11).

Whereas denatured globin is insoluble between pH 8 and 9, native globin is soluble over the pH range 5–9. Owing to the high histidine content of globin (see Table XI-4) its isoelectric point at pH 6.8–7.0 is at more alkaline pH values than that of the soluble plasma proteins. Whereas heme is rapidly oxidized to hemin by the oxygen of the air, hemoglobin combines with $O_2$ but remains in the ferrous state. Evidently globin prevents the oxidation of iron from the ferrous to the ferric state thus rendering possible the formation of oxyhemoglobin, $HbO_2$. The regenerated hemoglobin, prepared by coupling native globin with heme, has the same affinity for $O_2$ (E-13) and behaves in many respects in the same way as native hemoglobin (E-14,15).

Determinations of the molecular weight of mammalian hemoglobins by osmometry or ultracentrifugation give values of 66,000–68,000 (E-16). The complete resolution of the hemoglobin molecule into all of its amino acids (Fig. XI-6), revealed that these values were too high and that the molecular weight of human hemoglobin is 64,450 (E-17). The molecule contains 4 iron atoms bound to 4 protoporphyrin residues. If their equivalent weights are deducted, we obtain for the protein the value 61,990. However, urea (E-18), salts (E-19), or acids (E-20) lower the molecular weight to one-half of the original value. The cleavage of hemoglobin into smaller subunits is demonstrated by the increase in the interfacial pressure when hemoglobin films are acidified to pH 3 (E-21 to 23), and by an instantaneous increase in the number of titratable basic groups (E-24). About 22 masked imidazole groups per molecule of hemoglobin or its CO or cyanide derivatives are released on acidification to pH 2 (E-25). Similarly, 6 masked imidazole groups are released from myoglobin at pH 3–4 (E-42).

The structure around the iron atom of heme in hemoglobin is shown in Fig. XI-1. Hemoglobin, according to this structure, is an aquo compound with an Fe-bound water molecule (E-12). In oxyhemoglobin the water molecule is replaced by an $O_2$ molecule, in carbon monoxide hemoglobin by a molecule of CO. Oxidation of the ferrous to ferric iron yields methemoglobin in which the third positive charge of the ferric ion is neutralized by a hydroxyl ion. In all these hemoglobin derivatives the iron is linked by 4 coordinate bonds to the 4 pyrrole nitrogen atoms. The sixth ligand is a group of the globin molecule. It has been claimed for many years that the heme iron is bound to the imidazole nitrogen of a histidine residue. This view is based on the change in the affinity of hemoglobin for $O_2$ in the pH range 5–8 where imidazole groups are titrated (E-2,26). Moreover, the base binding capacity of hemoglobin increases on oxygenation (E-2). Coordination of Fe with imidazole is also strongly indicated by X-ray diffraction which reveals one of the histidine residues close to the

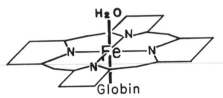

Fig. XI-1. Ligands of iron in hemoglobin.

heme group (E-34). On the other hand, imidazole complexes of heme are unstable at pH 9.5 and higher pH values (E-27) whereas the hemoglobins of many mammals are quite stable even at higher pH ranges (E-28). Therefore a weakly acidic group such as a carboxyl or a sulfhydryl group has been suggested as the sixth ligand of the heme iron (E-12; E-29 to 32). The role of a sulfhydryl group in the binding of $O_2$ to heme was demonstrated by the effect of $N$-ethylmaleimide which increases the affinity of hemoglobin for $O_2$ (E-29,31). It will be shown later (Section G, this chapter) that each globin molecule consists of two peptide chains and that each of these contains a histidyl-cysteinyl residue in proximity to the heme molecule (E-17,34; Fig. XI-6). Neither the vinyl groups of protohemin nor its propionic acid side chains are involved in the combination of heme with globin. This is demonstrated by the formation of typical hemoglobin when globin combines with mesohemin (E-14) in which the vinyl groups are converted to ethyl groups, or with hemin esters in which the carboxyl groups of heme are methylated (E-14,33).

The 4 heme groups of a hemoglobin molecule are located near the surface of the molecule as shown by X-ray diffraction (E-34). This explains the ease with which globin combines with heme in an aqueous solution. Investigation by electron spin resonance reveals that the planes of the 4 heme molecules are inclined at angles of 32–36° to the direction of the peptide chains (E-35).

Since the small ethylisocyanide combines more rapidly with the hemoglobin iron than the larger tertiary butylisocyanide, it had been concluded that the heme residues are buried in a crevice of the globin molecules; other interpretations which are compatible with the X-ray analysis have been given for this apparent contradiction (E-36). The nature of the linkage of the iron atoms in hemoglobin and its derivatives has been elucidated by measuring the *magnetic susceptibility*. The ferrous ion has 4 unpaired electrons, the ferric ion 5. Since each of these has a "spin," a magnetic moment results which can be measured by means of the electron spin resonance method or with a magnetic balance. It is customary to express the paramagnetic susceptibility in Bohr magnetons. In metal complexes, where the orbital magnetic moment is small in comparison to the spin moment, the magnetic susceptibility is approximately $\sqrt{n(n+2)}$ magnetons where $n$ is the number of unpaired electrons in a molecule or ion. Hence a magnetic moment of 4.9 magnetons is characteristic for ionic ferrous compounds, and 5.9 magnetons for ferric ions if orbital contributions are neglected. The values found for heme and hemin are 4.7 and 6.0 magnetons, respectively (E-37,38). Accordingly, both contain ionic iron, whereas ferrocyanide is diamagnetic, i.e., devoid of a magnetic moment. This is usually attributed to penetration of the cyanide ligands into the electron orbits of the ferrous ion and filling up of the gaps in the $d$-orbital of the iron atom, depriving it in this manner of the unpaired electrons. Hemoglobin has a paramagnetic susceptibility of 5.43 magnetons, indicative of ferrous ions and an additional orbital contribution, whereas oxyhemoglobin and carbon monoxide hemoglobin are diamagnetic (E-38). This suggests that all the unpaired electrons of hemoglobin are involved in the transition to $HbO_2$ or HbCO and also that the bond between iron and globin loses its ionic character.

All changes in the state of the iron atom and in the bond between iron and globin are reflected in changes of the intense absorption spectrum of the hemoglobin derivatives. Two absorption bands in the green region of the spectrum are characteristic for the diamagnetic compounds $HbO_2$ and HbCO and are also found in the hemochromogenes, complexes of ferroporphyrins with pyridine, imidazole, or other bases (E-39).

## F. Hemoglobin: Combination with $O_2$ and Other Ligands; Conformation

The combination of hemoglobin with $O_2$ or CO affects the physical and chemical behavior of the hemoglobin molecule as a whole. It has been mentioned before that the acidity or base-binding capacity of hemoglobin increases on oxygenation. As a consequence of this change, the pH is almost the same in the arterial as in the venous blood. The higher amount of $CO_2$ in the venous blood is compensated for by the higher acidity of oxyhemoglobin in the arterial blood. The combination of hemoglobin with $O_2$ is frequently represented by the equation $Hb + O_2 \rightleftharpoons HbO_2$, where Hb stands for 1 heme group and the equivalent portion of the globin molecule. However, since hemoglobin is an aquo compound, it is preferable to represent the equilibrium by the equation:

$$Hb(H_2O) + O_2 \rightleftharpoons HbO_2 + H_2O \qquad (XI\text{-}1)$$

If Hb is deprived of its water by drying at room temperature *in vacuo*, the broad absorption band of hemoglobin in the green spectral region is replaced by a typical two-banded hemochromogen spectrum (F-1,2). Evidently the hemoglobin iron, after loss of the water molecule, is bound to some basic group of an adjacent globin molecule and may thus be bound to 2 globin molecules. The dehydrated compound has been designated as *anhydro-hemoglobin* (F-3). In contrast to hemoglobin it is diamagnetic (F-4). The necessity for a sixth ligand of the iron in hemoglobin is demonstrated by the fact that dry oxyhemoglobin does not release its oxygen *in vacuo* (F-3) although solutions of oxyhemoglobin are easily deoxygenated by evacuation. The oxygen is released only if it can be replaced by water or some other substituent. The affinity of hemoglobin for $O_2$ is reduced by acidification and is a reversal of the phenomenon described above, the increase in acidity on oxygenation (Bohr effect, see Fig. XI-2). Both phenomena are attributed to the presence in globin of an ionic group in the vicinity of the iron binding group (E-26).

According to Eq. (XI-1) the equilibrium constant of the oxygenation of hemoglobin is equal to

$$\frac{[HbO_2][H_2O]}{[Hb(H_2O)][O_2]}$$

where the brackets indicate concentrations of the respective compounds. Since the concentration of water is very high and practically constant, we can write

$$K = \frac{[HbO_2]}{[Hb(H_2O)][O_2]}$$

or, if we use the symbol Hb for the aquo compound hemoglobin

$$K = \frac{[HbO_2]}{[Hb][O_2]}$$

This is a second order reaction which is represented by a hyperbola. However, sigmoid curves such as those shown in Fig. XI-2 are obtained when the per cent saturation of oxyhemoglobin is plotted against the oxygen pressure or its logarithm (F-5). The expected hyperbola has been found when the oxygen affinity of hemoglobin of the lamprey (F-6) or of myoglobin was investigated; both these pigments have the molecular weight 17,000 and contain only a single heme residue. Evidently, the sigmoid curve obtained with the mammalian hemoglobins is due to a mutual interaction of the 4 heme groups within the hemoglobin molecule. If oxygen combines with one of the 4 hemes, the affinity of the 3 others for $O_2$ is changed, and the molecule becomes saturated with oxygen in preference to other hemoglobin molecules. The intermediates $Hb_4O_2$, $Hb_4O_4$, and $Hb_4O_6$ cannot be detected by spectroscopy, but reveal their presence by striking changes in the dielectric increment (see Chapter VI, H) (F-29). A kinetic analysis of the oxygenation of $Hb_4$ and deoxygenation of $Hb_4O_8$

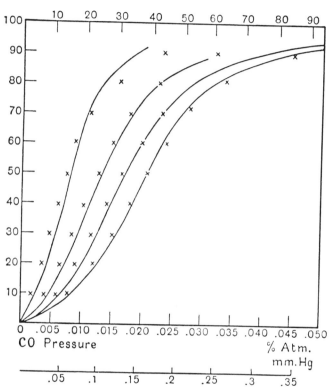

Fɪɢ. XI-2. Equilibrium curves of CO-hemoglobin (full lines) and $O_2$-hemoglobin (crosses) at 0, 19, 41, and 79 mm $CO_2$ pressure (from left to right). Abscissa at bottom: CO pressure; abscissa at top: $O_2$ pressure in mm Hg; ordinate: per cent saturation. From J. Barcroft, "The Respiratory Function of the Blood," Part II. Cambridge Univ. Press, London and New York, 1928.

by means of the rapid flow method has made it possible to determine the velocity constants for the first and last step of oxygenation and deoxygenation and to estimate the constants for the intermediate reactions in sheep blood hemoglobin (F-7). At pH 7.1 the ratio of the 4 equilibrium constants $K_1$, $K_2$, $K_3$, and $K_4$ is 1.0:0.4:0.24:9.3 (F-7). It can be seen from these figures that the equilibrium in the last step $Hb_4O_6 \rightarrow Hb_4O_8$ is shifted far to the right and that the affinity of $Hb_4O_6$ for $O_2$ is higher than the affinity of $Hb_4$ or of the intermediates.

The observations mentioned in the preceding paragraph indicate clearly some *interaction between the 4 heme groups* of the $Hb_4$ molecule. That the whole hemoglobin molecule is affected by the combination with $O_2$ is evident from the change in the shape of hemoglobin crystals on oxygenation shown in Fig. XI-3 where oxygen diffuses from the air bubble at the

Fig. XI-3. Hexagonal plates of horse hemoglobin (bottom) and formation of oxyhemoglobin needles near the air bubble (top left). From F. Haurowitz (F-30).

top of the microphotograph to the six-sided platelets of reduced hemoglobin, transforming them into long oxyhemoglobin needles. A difference in the solubility of hemoglobin and oxyhemoglobin is also responsible for the "sickling" in erythrocytes of patients suffering from sickle cell anemia. At low oxygen pressures the reduced hemoglobin of these patients, which is less soluble than their oxyhemoglobin, crystallizes inside the red cells and thus causes the sickling phenomenon (F-8).

It is well known that $O_2$ is displaced from its combination with hemoglobin by carbon monoxide, CO. The absorption spectrum, the crystalline shape, and many other properties of *carbon monoxide hemoglobin* are similar to those of oxyhemoglobin. The main differences are: (1) CO-hemoglobin is much more stable in that the dissociation into CO and hemoglobin proceeds at a slower rate than the dissociation of oxyhemoglobin (F-9); (2) in contrast to Hb and $HbO_2$, CO-hemoglobin has no observable absorption band in the near infrared region at 0.9–1.0 $\mu$ (F-10); (3) CO-hemoglobin is split into its components by irradiation with visible light as was discovered by Haldane in 1896 (F-11); one quantum is required for each CO molecule (F-12); similar photolysis of oxyhemoglobin and other ferrous hemoglobin complexes has been detected by means of short but intense light flashes (F-13). The quantum yield for the photolysis of oxyhemoglobin is only 0.8% of that for CO-hemoglobin.

Carbon monoxide hemoglobin is differentiated from oxyhemoglobin by the bright red color of its solutions even after treatment with copper sulfate, sodium hydroxide, or tannic acid; oxyhemoglobin under the same conditions is converted to a brown compound. All these tests are based on the higher stability of COHb and on the formation of bright red CO-heme complexes, whereas the decomposition of oxyhemoglobin gives brown hemin derivatives. Hemoglobin combines in a similar manner with nitric oxide, NO, to give NO-hemoglobin. It also forms less stable compounds with nitrosobenzene and other ligands. The physiological importance of *carbhemoglobin*, in which $CO_2$ is bound to amino groups of the globin moiety, has been discussed in Chapter X, B.

By the action of oxidizing agents the ferrous iron of hemoglobin is oxidized to ferric iron. The oxidation product is *methemoglobin* (F-14), sometimes called hemiglobin. It behaves like an indicator, is brown in acid and red in alkaline solution (F-15); its pK value is approximately 8.5 (F-16). The equilibrium between the two forms can be represented by the symbols

$$\diagdown\!\!\underset{\diagup}{Fe^+(H_2O)} \rightleftharpoons \diagdown\!\!\underset{\diagup}{FeOH} + H^+$$

where $Fe^+$ is the ferric ion bound by two of its valences to pyrrole nitrogen atoms (F-17). If the acid form of methemoglobin is dried, it loses the iron-bound water molecule and is converted into a "parahematin" (F-17). The water molecule of the acid form of methemoglobin can be replaced by the fluoride, cyanide, sulfide, azide, and other ions or by $H_2O_2$ (F-18, 19). Each of these derivatives displays a specific absorption spectrum and can be prepared in crystalline form.

When hemoglobin solutions are treated with hydrogen sulfide and oxygen gas or hydrogen peroxide, sulfhemoglobin, a green compound, is formed (F-20). Its formation is an irreversible process which seems to cause substitution of the methine groups of heme with sulfur containing groups. Whereas all these derivatives contain iron in the ferrous or ferric state, it has been claimed that the peroxide derivative of methemoglobin is a ferryl compound containing $Fe^{(IV)}$ (F-21). This view is based on titrimetric analyses with molybdocyanide or other univalent oxidants. However, in such titrations it is difficult to avoid reactions of the oxidizing agent with the methine groups of heme or with sulfhydryl groups of the globin moiety (F-22,31).

*Myoglobin*, the red pigment of muscles, like hemoglobin contains protoheme as its prosthetic group (F-23). However, the molecular weight of myoglobin is only 17,000 and it contains only one heme per molecule (F-24). Myoglobin has a much higher affinity to $O_2$ than hemoglobin (F-25). Its absorption maxima are at the same wave length as those of hemoglobin and it is much more resistant to denaturation by alkali (E-28). Myoglobin is obtained in crystalline form when muscle extract is dialyzed against a concentrated solution of ammonium sulfate (F-24). The amino acid composition of myoglobin is different from that of hemoglobin of the same species (Table XI-4) (F-26). It contains no cystine nor cysteine. Cysteine has been found in fish but not in mammalian myoglobin (F-34).

The *conformation* of the peptide chain of myoglobin is shown in Fig. VII-3 (Chapter VII), in which the resolution was brought down to 6 Å. Further improvement of the method has increased the resolving power to distances of 2Å which are close to the distances between 2 carbon atoms in organic molecules (F-27). It has been possible in this manner to build models of myoglobin which allow us not only to see the conformation of the peptide chain but also to recognize most of the amino acids, particularly those with bulky side chains. It is thus possible to determine the sequence of many of the amino acids without the necessity of partial hydrolysis and chromatography. The great importance of this method can hardly be overemphasized. At the same time it must be kept in mind that analyses of this kind, even though they are carried out with electronic computers, are extremely time-consuming and difficult. It seems also that the X-ray diffraction analysis is not yet able to detect small differences between myoglobins of different animal species, although such differences are demonstrable by two-dimensional chromatography of enzymatic digests of these myoglobins (F-32). X-ray diffraction (E-43) as well as measurements of the optical rotatory dispersion (F-33) indicate clearly that the myoglobin and hemoglobin molecules have the same high α-helix content in their aqueous solutions as in their crystals. The molecules are extremely compact, held together by hydrophobic bonds between nonpolar side chains of their amino acids. The polar side chains are in the hydrated surface of the molecules (E-43). The interior of the myoglobin and hemoglobin molecules contains only very few water molecules.

The X-ray diffraction analysis of *horse hemoglobin* reveals that the molecule consists indeed of 4 subunits and that the conformation of each

TABLE XI-4

AMINO ACID COMPOSITION OF HEMOGLOBIN, MYOGLOBIN, AND PHYCOCYANIN
(Gram amino acid/100 g protein)

| Amino acid | Myoglobin (Human)[a] | Human hemoglobins[b] | | Phycocyanin[c] |
|---|---|---|---|---|
| | | A | F | |
| Glycine | 6.08 | 4.52 | 4.48 | 5.77 |
| Alanine | 5.82 | 10.20 | 9.64 | 12.15 |
| Valine | 4.64 | 11.09 | 9.60 | 7.87 |
| Leucine | 13.67 | 15.22 | 15.20 | 12.18 |
| Isoleucine | 5.27 | (0.32) | 1.83 | 6.28 |
| Proline | 5.40 | 5.00 | 4.29 | 4.10 |
| Phenylalanine | 8.22 | 7.93 | 7.99 | 3.21 |
| Tyrosine | 2.19 | 4.40 | 3.60 | 8.27 |
| Tryptophan | 3.40 | — | — | |
| Serine | 4.43 | 5.50 | 6.90 | 8.99 |
| Threonine | 2.85 | 6.10 | 7.30 | 5.67 |
| Cystine[d] | 0 | 1.03 | 0.98 | 1.73 |
| Methionine | 2.69 | 1.60 | 2.05 | 4.16 |
| Arginine | 2.47 | 3.43 | 3.31 | 8.90 |
| Histidine | 7.79 | 8.47 | 7.45 | 0.41 |
| Lysine | 19.09 | 10.60 | 10.60 | 4.01 |
| Aspartic acid | 8.27 | 10.60 | 10.60 | 12.52 |
| Glutamic acid | 16.17 | 7.20 | 7.56 | 13.34 |
| Ammonia | 1.22 | 1.10 | 1.10 | 1.70 |

[a] A. Rossi Fanelli *et al.*, *BBA* 17: 377(1955).

[b] P. C. van der Schaaf and T. H. J. Huisman, *BBA* 17: 81(1955).

[c] From (H-8); phycocyanin was prepared from *Bangacea Porphyra tenera* Kjelm.

[d] Including cysteine.

of these is very similar to that of myoglobin (F-28). The analysis also shows that the 4 subunits consist of two pairs of identical peptide chains and that the 4 subunits are arranged at the corners of an irregular tetrahedron. Each of the subunits seems to be adjusted spatially to the adjacent one so that the four subunits form a rather compact spherelike molecule. The *C*-terminal amino acid of each of the subunits is close to the *N*-terminal amino acid of the adjacent subunit. The hemoglobin model also shows that a histidine residue is in proximity to each of the heme iron atoms and that sulfhydryl groups are in key positions close to the heme-linked histidyl residues (F-28). The chain conformation of the hemoglobin subunits is very similar to that of myoglobin as shown by Fig. VII-3 (Chapter VII) although the sequence of amino acids is not quite the same in the two proteins.

## G. *The Specificity of Hemoglobins*

It has been known for many years that most proteins are species-specific and can be differentiated from each other by serological tests (see Chapter XV). It was no surprise, therefore, that hemoglobins of different species proved to be serologically different (G-1,2). The species-specificity of the hemoglobins has also been demonstrated very impressively by the differences in their crystalline shapes (G-3,4). Since all vertebrate hemoglobins yield the same protohemin chloride on heating with glacial acetic acid and NaCl, their differences must be attributed to differences in their globin moieties.

A different heme has been found in *chlorocruorin*, a green pigment which replaces hemoglobin in the marine worm spirographis and in related species (G-5). It contains a heme in which the vinyl group in the 2-position of protoporphyrin is replaced by a formyl group —CHO(G-6). The molecular weight of chlorocruorin is 2.8 million; in urea or alkali it dissociates into units of the molecular weight 35,000 (G-69). The hemoglobins of the other invertebrates contain the same protohemin as those of the vertebrates. The molecular weight of those hemoglobins which occur in red blood cells of invertebrates are low, varying from 16,000 to 56,500 (G-7,8). In other invertebrates, hemoglobin is dissolved in the lymph; the molecular weights of these hemoglobins are very high and vary from 350,000 to several million (G-8,9). These macromolecular hemoglobins have been designated by some authors as *erythrocruorins*. All these hemoglobins, including chlorocruorin, combine with one $O_2$ per iron atom. In the root nodules of legumes a hemoglobin-like pigment has been discovered and named *leghemoglobin* (G-10). The amino acid composition of crystalline leghemoglobin differs considerably from that of the animal hemoglobins (G-11).

Differences between the hemoglobins of the vertebrates can be easily detected by their very different resistance to *denaturation by alkali*. If dilute solutions of oxyhemoglobin, containing about 0.1% of the pigment, are made alkaline by adding 0.05 mmoles of NaOH per milliliter, the bright red oxyhemoglobin is denatured to a brown product (G-12). The rate of this change, which can be measured easily by means of a spectrophotometer, is quite typical for each animal species, as is shown in Fig. XI-4 (G-13). The time required for 90% decomposition varies from about 1 min in adult human hemoglobin to more than 24 hr in bovine hemoglobin. The striking change in color and absorption spectrum reveals differences in the stability of the peptide chains which in a colorless protein would not be so easily detectable. It must also be remembered that hemoglobin is much more stable than globin and that heme in the alkaline denaturation is more than just an indicator of the reaction which takes place in the protein. It slows down the denaturation velocity in some cases only very little, in others very considerably. One can reasonably assume that a close fit between the rigid heme molecule and the surface of globin will inhibit the denaturation and slow it down, whereas a loose fit will hardly reduce the rate of denaturation.

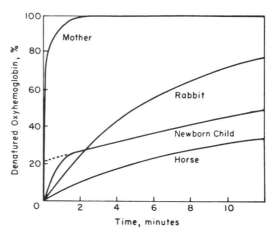

Fig. XI-4. Denaturation of various hemoglobins by adding one-fifth volume of 0.25 N sodium hydroxide to their dilute solution (G-15).

The alkali denaturation method revealed also that human blood may contain more than a single hemoglobin. This was first observed when the blood of a newborn child was compared with that of his mother. Whereas the latter was denatured within 1 min., that of the child remained bright red for about 1 hr (G-14). As Fig. XI-4 shows, the fetal blood contains about 20% of the unstable maternal hemoglobin A and about 80% of the stable fetal hemoglobin F. Similar ratios were found in numerous blood samples of newborn children (G-15). Hemoglobin F differs from the adult hemoglobin in many other properties. It has a lower affinity for oxygen (G-19), a lower mobility in electrophoresis at pH 6.94–7.0 (G-16) and has only 2 N-terminal valine residues whereas the hemoglobin A has 4 N-terminal valine residues (G-17). The hemoglobins A and F differ also from each other by their X-ray diffraction pattern (G-18) and the shape of their crystals as shown by Fig. XI-5 (G-19). The amino acid analysis reveals great differences between the composition of the two hemoglobins. Hemoglobin A is free of isoleucine whereas hemoglobin F contains 1.47% of this amino acid (G-23). In normal children hemoglobin F disappears slowly and cannot be found after the seventh month of life. In persons suffering from Cooley's disease (Mediterranean anemia) hemoglobin F is found later in life (G-20), particularly in the severe form which is called Thalassemia major. Smaller amounts of HbF are found in the milder form, Thalassemia minor (G-21). The formation of the two hemoglobins A and F is determined by two different genes. In homozygous individuals all of

Fig. XI-5. Hemoglobin of adult human (G-19). The fetal hemoglobin crystallizes in six-sided (hexagonal?) platelets.

the hemoglobin may be hemoglobin F and the symptoms those of Thalassemia major. Heterozygous individuals have a mixture of hemoglobins A and F and suffer from Thalassemia minor (G-22).

The fact that hemoglobin F contains only 2 $N$-terminal valine residues whereas hemoglobin A has 4 such residues suggests that 2 of the 4 peptide chains which form the molecule may be identical in both hemoglobins. As mentioned earlier, it is possible to dissociate hemoglobin into 2 half-molecules by acidification to pH 3–4, by urea or by salts or alkalinization. Dissociation can also be accomplished by sodium dodecyl sulfate or other detergents (G-24). The 2 half-molecules are different; they can be separated by electrophoresis (G-25), counter-current distribution (G-33b), or by precipitation with trichloroacetic acid or 50% acetone (G-68); only the electrophoretically fast-moving chain is precipitated (G-26). Using electrophoresis, Schroeder *et al.* (G-27) found that the more slowly moving half-molecule is similar or identical in hemoglobin A and F; it has the $N$-terminal sequence Val.Leu- and has been designated as $\alpha_2$, the subscript indicating that it consists of 2 $\alpha$-chains. The fast-moving component of hemoglobin A is called $\beta_2$; its $N$-terminal sequence is Val.His.-Leu-. Hemoglobin F has, in addition to the 2 $\alpha$-chains, 2 chains which are similar to, but not identical with, the $\beta$-chains of hemoglobin A. They have $N$-terminal Gly.His.Phe- residues and have been designated as $\gamma$-chains (G-28). In contrast to the $\alpha$-chains, the $\gamma$-chains do not pass through dialysis membranes (G-28). According to these results we can represent hemoglobin A by the symbol $\alpha_2^A \beta_2^A$, and hemoglobin F by the symbol $\alpha_2^A \gamma_2^F$. The dissociation of hemoglobins into their half-molecules occurs predominantly in an asymmetric manner so that $\alpha_2$ and $\beta_2$ halves are formed, but very little if any $\alpha\beta$ units (G-29,30). The reaction is reversible. Indeed, hybrid hemoglobins have been produced by mixing the dissociation products of human and canine hemoglobins (G-31).

One of the most important results in this area of research is the complete elucidation of the amino acid sequence in the $\alpha$ and $\beta$ peptide chains of the adult human hemoglobin A by Braunitzer (G-32) and independently in the $\alpha$-chain by Hill and Konigsberg (G-33a). A comparison of the 2 chains (Fig. XI-6) shows that they contain many identical peptide sequences, and indicates strongly that they were evolved from a common ancestor molecule, probably a simpler hemoglobin with a single peptide chain (G-32).

Very valuable information has been gained by comparing the amino acid sequence with the results of X-ray diffraction (Fig. VII-3) which shows the presence of 7 or 8 straight helical portions of the peptide chains (G-33c,33d). The helical stretches are marked in Fig. XI-6 by the capital letters A—A, B—B, etc., the broken lines indicating the presence of helical structure. The figure shows that most of the prolyl residues are near the turn of the helices or in the nonhelical portions. It will be recalled (see Chapter VII, B) that prolyl residues cannot form the typical $\alpha$-helix because of the lack of an N-bound hydrogen atom which is indispensable for hydrogen bond formation with the adjacent turn of the helix. Figure XI-6 also shows that the histidyl residues of both chains, which are close to the heme molecule, have adjacent cysteinyl groups, and that the D-helix is missing in the $\alpha$-chain.

The assumption that the normal adult and fetal blood contains only the hemoglobins A and F, respectively, is an oversimplification. Adult blood always contains one or two *minor hemoglobins*, designated A$_2$ and A$_3$, as well (G-34,36). They form less than 10–15 % of the total hemoglobin. Hemoglobin A$_2$, in contrast to hemoglobin A, contains isoleucine (G-37); the $\beta^A$ chains are replaced in hemo-

```
         A------
α  Val-    Leu-Ser-Pro-Ala-Asp-Lys-Thr-AsN-Val-Lys-Ala-Ala-Try-Gly-Lys-Val-Gly-Ala-His-
            1    2   3   4   5   6   7   8   9  10  11  12  13  14  15  16  17  18  19  20
β  Val-His-Leu-Thr-Pro-Glu-Glu-Lys-Ser-Ala-Val-Thr-Ala-Leu-Try-Gly-Lys-Val-AsN-
     1   2   3   4   5   6   7   8   9  10  11  12  13  14  15  16  17  18  19
   B------                                                               ---BC---
α  -Ala-Gly-Glu-Tyr-Gly-Ala-Glu-Ala-Leu-Glu-Arg-Met-Phe-Leu-Ser-Phe-Pro-Thr-Thr-Lys-Thr
     21  22  23  24  25  26  27  28  29  30  31  32  33  34  35  36  37  38  39  40  41
β  -Val-Asp-Glu-Val-Gly-Gly-Glu-Ala-Leu-Gly-Arg-Leu-Val-Val-Tyr-Pro-Try-Thr-GlN-Arg-
     20  21  22  23  24  25  26  27  28  29  30  31  32  33  34  35  36  37  38  39  40
   ---C----                                                ----D------
                                                                    -DE--
α  -Tyr-Pro-His-Phe-    -Asp-Leu-Ser-His-            -Gly-Ser-Ala-GlN-Val-Lys-
     42  43  44  45  46   47  48  49  50              51  52  53  54  55  56
β  -Phe-Phe-Glu-Ser-Phe-Gly-Asp-Leu-Ser-Thr-Pro-Asp-Ala-Val-Met-Gly-AsN-Pro-Lys-Val-Lys-
     41  42  43  44  45  46  47  48  49  50  51  52  53  54  55  56  57  58  59  60  61
                                                            --------E---
α  -Gly-His-Gly-Lys-Lys-Val-Ala-Asp-Ala-Leu-Thr-AsN-Ala-Val-Ala-His-Val-Asp-Asp-Met-Pro-
     57  58  59  60  61  62  63  64  65  66  67  68  69  70  71  72  73  74  75  76  77
β  -Ala-His-Gly-Lys-Lys-Val-Leu-Gly-Ala-Phe-Ser-Asp-Gly-Leu-Ala-His-Leu-Asp-AsN-Leu-Lys
     62  63  64  65  66  67  68  69  70  71  72  73  74  75  76  77  78  79  80  81  82
   -----F----                                                     ---G----
α  -AsN-Ala-Leu-Ser-Ala-Leu-Ser-Asp-Leu-His-Ala-His-Lys-Leu-Arg-Val-Asp-Pro-Val-AsN-Phe-
     78  79  80  81  82  83  84  85  86  87  88  89  90  91  92  93  94  95  96  97  98
β  -Gly-Thr-Phe-Ala-Thr-Leu-Ser-Glu-Leu-His-Cys-AsN-Lys-Leu-His-Val-AsN-Pro-Glu-Asp-Phe
     83  84  85  86  87  88  89  90  91  92  93  94  95  96  97  98  99 100 101 102 103
                                                                        ---H------
α  -Lys-Leu-Ser-His-Cys-Leu-Leu-Val-Thr-Leu-Ala-Ala-His-Leu-Pro-Ala-Glu-Phe-Thr-Pro-
     99 100 101 102 103 104 105 106 107 108 109 110 111 112 113 114 115 116 117 118 119
β  -Arg-Leu-Leu-Gly-AsN-Val-Leu-Val-Cys-Val-Leu-Ala-His-His-Phe-Gly-Lys-Glu-Phe-Thr-Pro-
    104 105 106 107 108 109 110 111 112 113 114 115 116 117 118 119 120 121 122 123 124
                                                                        ---HC----
α  -Ala-Val-His-Ala-Ser-Leu-Asp-Lys-Phe-Leu-Ala-Ser-Val-Ser-Thr-Val-Leu-Thr-Ser-Lys-Tyr-Arg
    120 121 122 123 124 125 126 127 128 129 130 131 132 133 134 135 136 137 138 139 140 141
β  -Pro-Val-GlN-Ala-Ala-Tyr-GlN-Lys-Val-Val-Ala-Gly-Val-Ala-AsN-Ala-Leu-Ala-His-Lys-Tyr-His
    125 126 127 128 129 130 131 132 133 134 135 136 137 138 139 140 141 142 143 144 145 146
```

Fig. XI-6. Amino acid composition and helical structure of human hemoglobin A(G-32,33c,33d,66,67). The α- and β-chains are shown in such a manner that similar sequences appear in identical positions. The gaps do not indicate absence of amino acids. Thus the N-terminal peptide residue of the α-chain is valyl-leucyl-seryl-. The helical portions are shown by broken lines and are marked by capital letters. Absence of the broken line means a nonhelical conformation. AsN: asparagine; GlN: glutamine.

globin $A_2$ by 2 other chains which are genetically independent of $\beta$ and have been designated as $\delta^A{}_2$(G-38). Minor hemoglobins have also been found in fetal blood. During the first three months of the intrauterine life the human fetus contains a *primitive hemoglobin* (G-39) which disappears during later phases of the fetal life (G-40,41). Different minor hemoglobins F have been found by means of electrophoresis or fractionated precipitation with salts (G-42,43). Nothing is known concerning their nature. The high resistance of hemoglobin F to alkali denaturation cannot be attributed to the $\gamma$-chains since these are labile to alkali in Bart's hemoglobin (see below). It is not yet quite clear whether the $\alpha$-chains of hemoglobin F are identical with, or merely similar to, those of hemoglobin A (G-44,45). Although the great difference between fetal and maternal hemoglobins in humans is unique, differences in the affinity for $O_2$ have also been found in the blood of other animals such as the ox or the goat (G-46). Such differences need not always be caused by differences in the hemoglobins. Thus solutions of the human hemoglobin A have a higher affinity for $O_2$ than those of human hemoglobin F, whereas the reverse phenomenon is observed in the red blood cells containing the same hemoglobins (G-19,47).

In pernicious anemia, normal hemoglobin A is formed, not hemoglobin F (F-15). However, in *sickle cell anemia*, a form of anemia which occurs only in Negroes, another hemoglobin, named *hemoglobin S*, has been discovered by its lower anodic mobility at pH 8.5 (F-8). If blood with hemoglobin S is deoxygenated under the microscope, either by a reducing agent or by evacuation, hemoglobin S crystallizes within the red blood cells and appears inside the cells as a purplish red birefringent sickle (G-48). Partial digestion with trypsin and subsequent electrophoresis and chromatography ("finger-printing") reveals that 1 out of 30 peptide fragments differs in its $R_f$ value from the analogous peptide fragment of hemoglobin A. It is the $N$-terminal peptide of the $\beta$-chain which has in hemoglobin A the amino acid sequence Val.His.Leu.Thr.Pro.*Glu*.Glu.Lys. In hemoglobin S the *Glu* residue is replaced by *Val* (G-49). Hence, hemoglobin S can be represented by the symbol $\alpha_2{}^A\beta_2{}^S$ (G-38). The replacement of a glutamyl residue by a valyl residue satisfactorily explains the lower anodic mobility of hemoglobin S. The genotypes of a homozygous person having sickle cell anemia are $\alpha^A/\alpha^A$ and $\beta^S/\beta^S$; those of a heterozygous person suffering from sickle cell trait are $\alpha^A/\alpha^A$ and $\beta^A/\beta^S$ (G-38).

Another modification of the $\beta$-chains has been found in *hemoglobin C*, which has a more positive charge than hemoglobin S (G-50). In HbC the $N$-terminal peptide of the $\beta$-chain, sometimes called peptide No. 4, has a lysyl residue instead of the glutamyl residue of hemoglobin A (G-51). The composition of hemoglobin C is $\alpha_2{}^A\beta_2{}^C$. Its appearance does not involve anemia. If the changes in the amino acid composition involve the iron-bound histidyl residues, the ferrous ion in hemoglobin loses its protection against oxidation by $O_2$(G-71). Consequently, conversion into methemoglobin takes place *in vivo*. Hemoglobins of this type have been designated as hemoglobin M (G-65).

Subsequent to the discovery of the hemoglobins A, F, S, and C, a large number of other modifications of the hemoglobin molecule has been discovered. At present (1962) about 30 human hemoglobins are known. Some of them occur only among African or Asiatic populations. Thus hemoglobin E, in which a glutamyl residue in the tryptic peptide No. 26 in the $\beta$-chain of hemoglobin A is replaced by a lysyl residue, has been found only in Thailand and Cambodia (G-52,53). Among the other human hemoglobins, hemoglobin H, occurring in Algiers, should be mentioned. It contains 4 $\beta$-chains. Its symbol is $\beta_4$ (G-54). In contrast to other hemoglobins, it shows no Bohr effect (see Chapter X, D) (G-55). Similarly, Bart's hemoglobin, whose composition is $\gamma_4$, differs from hemoglobin F($\alpha_2\gamma_2$) by its lability toward alkali (G-56,57). Evidently, the Bohr effect of Hb A and the alkali resistance of Hb F cannot be properties of the $\beta$- or $\gamma$-chains, respectively, but must be caused by the interaction between $\alpha$- and $\beta$-chains in Hb A and $\gamma$- and $\alpha$-chains in Hb F (G-70). One of the hemoglobins with a modified $\alpha$-chain is hemoglobin I, whose composition is $\alpha_2^I\beta_2^A$ (G-38). If a mixture of hemoglobin I, whose $\alpha$-chains are abnormal, and hemoglobin S, whose $\beta$-chains are modified, is dissociated by acidification and then neutralized, hemoglobin A and another artificial hemoglobin are formed (G-58). The reaction proceeds in the following manner:

$$\alpha_2^I\beta_2^A + \alpha_2^A\beta_2^S = \underline{\alpha_2^A\beta_2^A} + \alpha_2^I\beta_2^S$$

The underlined product is identical with hemoglobin A. The electrophoretic mobilities of the various human hemoglobins at pH 8.6 decrease in the following order: H > I > J > A > P > G > S = D > E > C (G-59). Two or more of these hemoglobins can occur in a single blood sample (G-60). A survey on the possible genotypes and hemoglobin phenotypes and a discussion of the genetical aspects of the occurrence of multiple hemoglobins will be found in special articles (E-43,G-38,59,61 to 64).

## H. *Other Chromoproteins*

As mentioned earlier, in Chapters VII and X, serum albumin has a great affinity for a number of nonprotein molecules, particularly for anions. It is a matter of definition whether these complexes are considered as conjugated proteins or merely as transport form of the respective molecules carried by serum albumin. Similar complexes are also formed between various pigments and either serum albumin or other proteins. Under physiological conditions many of them are quite stable and can be precipitated and redissolved without dissociation. In view of this stability and because of their biological importance, some of these complexes will be discussed here.

It is well known that the protein precipitate obtained from the blood serum of patients suffering from jaundice has an intense yellow color caused by *bilirubin*. This bile pigment is firmly bound to serum albumin, but can be extracted from the protein precipitate by ethanol. Each molecule of serum albumin is able to bind 3 molecules of bilirubin (H-1). On electrophoresis, bilirubin migrates with serum albumin (H-2). The clinicians used to differentiate bilirubin, which gives directly a red color with diazotized sulfanilic acid from ."indirect bilirubin," which gives this color reaction only after the addition of ethanol. The "direct" bilirubin is bilirubin glucuronide (H-3,4). Another conjugation product of the bile pigments with protein is *choleglobin*, the first product of the physiological degradation of hemoglobin. Its protein component is globin, its green prosthetic group the ferric salt of biliverdin (H-5). The latter is a linear tetrapyrryl compound formed by oxidative cleavage of the protoporphyrin ring and loss of the methine groups, $=CH—$, which connects the two pyrrole rings carrying the vinyl groups. In contrast to hemoglobin, choleglobin loses its iron atom by treatment with dilute HCl. Similar green *"verdoglobins"* are produced *in vivo* in persons treated with large amounts of sulfa drugs. They are formed *in vitro* by exposing hemoglobin to reducing substances, such as ascorbic acid, in the presence of $O_2$. Conjugated proteins containing bilirubin derivatives have also been found in the red and blue-green algae (H-6,7). *Phycoerythrin* from the red algae and *phycocyanin* from blue-green algae can be separated from each other by chromatography (H-17). Their amino acid composition (H-8) does not differ strikingly from that of the proteins of higher plants. The molecular weights of phycoerythrin and phycocyanin are close to 300,000. Only 89% of the phycoerythrin molecule are formed by amino acid. Phycoerythrin is a glycoprotein whereas phycocyanin consists only of amino acids. Both proteins contain small amounts of the bile pigments mesobilierythrin and mesobiliviolin (H-9). The amino acid composition of phycocyanin is shown in Table XI-4.

A dark brown conjugated protein containing mesobilifuscin, a dipyrryl derivative of bilirubin, occurs in the brown blood serum of persons suffering from myodystrophia, a rare muscle disease (H-10). Porphyrins occur in the body fluids of healthy individuals only in traces. When their amount increases under pathological conditions or due to the action of certain drugs, the porphyrins are bound to the albumin fraction of the blood serum (H-11). The green pigment of chloromas, lymphatic tumors of the bone marrow, is protoporphyrin (H12). It is bound to insoluble protein and is not extracted directly by ethyl ether.

Although chlorophyll, the green pigment of plants, is soluble in benzene and insoluble in water, it cannot be extracted from green leaves by benzene. Green aqueous extracts can be obtained from the grana of chloroplasts. The green protein-chlorophyll complex has been called *chloroplastin*

(H-18). It seems that chlorophyll in these complexes is loosely adsorbed to the protein. It does not migrate with the protein in paper electrophoresis (H-19). Since chloroplastin always contains considerable amounts of lipid, it has been suggested that the chlorophyll molecules form a monomolecular layer at the interphase, their porphyrin rings adsorbed to the protein phase, their phytyl residues oriented toward the lipid phase (H-20).

Very little is known about the *melanins*, brown and black pigments of the hair, the skin, and certain pathological tumors. We know that tyrosinase, DOPA oxidase, and other similar enzymes catalyze the oxidation of tyrosine, tryptophan, dihydroxyphenylalanine, and their derivatives, to red, brown, or dark quinoid substances (H-13,23). However, it is not yet clear whether this oxidation takes place while these amino acids are still bound in the peptide chain, or whether the melanins are true conjugated proteins in which a colorless protein is bound to the polymerized quinoid oxidation products of aromatic amino acids, perhaps by means of SH groups (H-22). If melanotic tumors are treated with pancreatin, the melanin dissolves and can be precipitated by acetic acid (H-14). It is free of sulfur and contains 8.8% nitrogen (H-14). In contrast to this soluble melanin, the dark melanokeratin of brown or black hair is insoluble in sodium hydroxide (H-15). The melanoid portion of the melanins forms 30–56% of the molecule, the protein portion 44–70% (H-16). The yellow and red pigments of the hair and probably also the analogous pigments of the skin are oxidation products of tryptophan or kynurenine (H-21).

## I. Nucleoproteins (*I-1 to 3*)

Nucleoproteins are the most important of the conjugated proteins since the nucleic acids, which form their prosthetic group, occur ubiquitously in living matter and have a paramount role in the transmission of genetic information. In spite of this great importance of the nucleoproteins, our knowledge of their nature is still very fragmentary. This is largely due to our inability to decide whether the "nucleoproteins," which we isolate from cells and tissues, exist as such in the organisms or whether they are artifacts formed during the isolation procedures. Since the nucleic acids are polyanions, they combine like the anionic detergents or other polyanions (see Chapter X, D) with numerous proteins to form soluble or insoluble complexes which are very similar to the "nucleoproteins" isolated from living matter.

It cannot be our purpose to discuss here the chemistry of the nucleic acids, which has been treated in a number of excellent books (I-2,4,5). It may be useful, however, to describe briefly the structure of the nucleic acids in relation to their combination with proteins. The two principal classes of nucleic acids are ribonucleic acid (RNA) and deoxyribonucleic acid (DNA). Ribonucleic acid consists of long chains of ribose-5-phosphate units in which each phosphate residue is linked by an ester bond to the carbon atom No. 3 of the adjacent ribose residue. The purine bases adenine and guanine, and the pyrimidine bases uracil and cytosine are linked by glycosidic bonds to the carbon atom No. 1 of the ribose-5-phosphate units. Hence, the structure of RNA is represented by the following diagram in which P represents phosphate residue; R, ribose; and B, the various purine and pyrimidine bases.

In DNA, the ribose residue is replaced by 2-deoxyribose. Otherwise, the linkage of phosphate and bases is the same as in RNA. In contrast to RNA, DNA is free of uracil. Instead, it contains thymine (5-methyl-uracil).

Although no other sugar but ribose has been found in RNA, it is quite possible that small amounts of other $C_5$-sugars occur. For this reason, some authors prefer to use the term pentosenucleic acid for RNA. During the last few years small amounts of methylated or otherwise modified purine or pyrimidine bases have been found in various nucleic acids. One of these, 5-hydroxymethylcytosine, is found in bacteriophages in large amounts. Other bases, which occur in much smaller amounts, are 5-methylcytosine, 2-methyladenine, and 6-methyladenine. The ribose and deoxyribose residues of the nucleic acids are linked by a glycoside bond of their carbon atom No. 1 to the nitrogen atom No. 3 of the pyrimidines or No. 9 of the purines. However, in a newly discovered compound, called pseudouridine, the carbon atom No. 1 of ribose is linked directly to the carbon atom No. 5 of uracil (I-6). Since this compound has no glycoside bond, it is not a typical nucleoside. The discovery of new purine or pyrimidine bases and of new types of chemical bonds increases the number of possible links in the polynucleotide chains and thereby also the number of specific sequences of nucleotides.

Both ribo- and deoxyribonucleoproteins can be detected by the intense absorbancy of their nucleic acids at 260 mμ, which is caused by their purine and pyrimidine bases. Although proteins also absorb in that spectral region, their absorbancy is much lower; the protein concentration can be calculated from the absorbancy at 280 mμ where that of the nucleic acids is low. It is customary to express the absorption of nucleic acids and their derivatives in terms of the absorbancy per gram-atom of phosphorus per liter; this unit is represented by the symbol $\epsilon(P)$ (I-7). It depends on pH, ionic strength, and the type of ions present, and is always much lower than that calculated for a mixture of the two purine and two pyrimidine bases which occur in each of the nucleic acids. Thus the $\epsilon(P)$ of DNA in a molar solution of NaCl is approximately 6400 whereas the calculated value for an equimolar mixture of adenosine, guanosine, cytidine, and thymidine is approximately 11700 (I-8). The hypochromism (low absorbancy) of the nucleic acids is caused by charge transfer between ring electrons of those purine and pyrimidine bases which are linked to each other by hydrogen bonds (I-79). Hypochromism is observed when hydrogen bonds link adenine to thymine or guanine to cytosine (I-9). Similar lowering of the ultraviolet absorption is also observed in mixtures of synthetic polyadenylic acid and polyuridylic acid (I-10).

The high intensity of the ultraviolet absorption of nucleic acids has been used for their detection in tissue sections (I-11). Since in the cells all of the nucleic acids are bound to proteins, intense ultraviolet absorption indicates the presence of nucleoproteins. Both *deoxyribonucleoprotein* (*DNPr*) and *ribonucleoprotein* (*RNPr*) have a high UV absorbancy. Both are also stained by basic dyes. However, since the concentration of DNA in the nucleus is usually much higher than that of RNA in the cytoplasm, the nucleus is stained predominantly by the basic dyes. Most of the histological methods are based on this principle. Indeed, the designation "nucleic acid" is due to the original belief that only the nuclei contain the acid which combines with basic dyes. DNA can be differentiated from RNA by means of the Feulgen reaction (I-12) in which the tissue section or cell suspension is treated with $N$-hydrochloric acid and then with Schiff's aldehyde reagent, i.e., a solution of fuchsin which has been reduced and bleached by sulfurous acid. DNA gives a red color, typical for aldehydes, whereas RNA does not give this reaction. DNPr and RNPr in tissue sections are frequently differentiated also by staining with basic dyes before and after treatment with RNase (I-13).

The physical and chemical properties of ribonucleoproteins (RNPr) are in many respects similar to those of the deoxyribonucleoproteins (DNPr). It is advantageous, therefore, to separate nuclei from cytoplasm before the isolation of nucleoproteins. This can be accomplished by homogenization and centrifugation at low speed at which only the nuclei pass into the sediment whereas mitochondria, microsomes, and the cell sap remain in the supernatant solution. *Ribonucleoproteins* (I-14) are soluble in isotonic saline solution (0.14 $M$) and can be extracted by this solution from tissue homogenates whereas the deoxyribonucleoproteins are insoluble and are not extracted. When the RNPr extract is acidified to pH 5, most of the RNPr is precipitated (I-15,16). Alternatively RNPr can be salted out by ammonium or sodium sulfate. The bulk of the ribonucleoprotein is derived from the ribosomes, i.e., RNPr particles of the microsome fraction (I-17). A smaller part of the RNPr is extracted from the mitochondria; a third portion occurs as such in the cellular fluid and is merely diluted by the saline solution. The particulate RNPr can be separated from the soluble complexes by high-speed centrifugation at 105,000 $g$. The supernatant solution still contains ribonucleic acid which can be precipitated by acidification to pH 5. This RNA has been designated as sRNA (soluble RNA), acceptor RNA, or transfer RNA; it forms complexes with protein at pH less than 7 and at ionic strength of less than 0.1 (I-18).

For a long time it has been assumed that RNA preparations were always single-stranded. However, X-ray analysis of the soluble RNA revealed that it had a hairpin structure (I-81). Its double-helical chains coil back "like Aesculapius' snake" (E-43). The RNA double helix is less stable than the DNA double helix. Whereas aqueous solutions of the latter require temperatures of 80–100°C for the cleavage of the hydrogen bonds between its two strands, the double-stranded RNA undergoes an analogous change at 50–60°C. This transition temperature has been designated by the symbol $T_m$ and is frequently called the "melting point" of the nucleic acids. The low $T_m$ of the soluble RNA indicates that only parts of it consists of a double helix. Its role in amino acid transfer will be treated in Chapter XVI, where another type of RNA, denoted as messenger RNA or informational RNA, will also be discussed.

In the microsomes the RNPr occurs as a lipo-RNPr complex from which the lipids can be removed by deoxycholate (I-19). Since, in addition to RNPr, many cells also contain ribonuclease which degrades the ribonucleic acid, all preparations must be carried out at low temperature. In this manner ribonucleoproteins have been prepared from rat liver (I-20), yeast (I-21), and other sources. The RNPr isolated from the ribosomes probably contains the same protein to which the RNA is bound in the ribosomes (I-22). However, we do not know whether the bonds between RNA and protein in the isolated RNPr are the same as *in vivo*. When RNPr is prepared from whole tissues or organs, we are not even sure whether the protein of the isolated RNPr is the same as the protein with which the RNA combined in the cell. None of these RNPr preparations is crystalline. In mammalian cells RNA forms dissociable complexes with proteins (I-78). The X-ray diffraction pattern of these complexes is not very different from that of the free proteins (I-80).

The soluble RNPr fraction from rat liver, which contains the enzymes for the incorporation of terminal nucleotides into sRNA, has been fractionated into three ribonucleoproteins designated $\alpha$, $\beta$, and $\gamma$ (I-23). Crystalline ribonucleo-tropomyosin has been isolated from mucle. In spite of the anionic character of RNA, RN-tropomyosin migrates electrophoretically at the same rate as tropomyosin (I-24). This seems to indicate that the charged groups of RNA are not exposed and that the RNA moiety of the molecule is covered by a layer of the protein tropomyosin. Nucleoproteins, like other proteins are precipitated by trichloroacetic acid (TCA). Ribonucleoproteins, in contrast to deoxyribonucleoproteins, can be extracted from the TCA-precipitate by ethanol (I-25).

The plant viruses and many of the animal viruses (e.g., the polio and the influenza virus) consist chiefly or exclusively of protein and RNA. They can be considered as ribonucleoproteins. Tobacco ring spot virus contains 40% RNA. One of the best investigated viruses is the *tobacco mosaic virus* (TMV) (I-26a) which was crystallized for the first time by Stanley (I-26b) by fractionation with ammonium sulfate. In electron micrographs the TMV appears in the form of cylindrical rods; X-ray diffraction shows that these rods are hollow (I-27,28). The X-ray diffraction pattern of the virus is almost the same as that of the RNA-free virus since the RNA forms only 5.1% of the total mass of the virus; the latter corresponds to a molecular weight of $39 \times 10^6$. Accordingly, the molecular weight of the RNA is 2 million or, if the RNA consists of subunits, less than 2 million. The large size of the virus molecule made it seem almost hopeless to get more insight into its structure. However, determination of the C-terminal amino acids of this giant protein revealed the presence of more than 2000 C-terminal threonine residues (I-29,30). Evidently it contains more than 2000 peptide chains. Further analysis demonstrated that these chains are identical and that their N-terminus is formed by N-acetyl-seryl-tyrosyl residues (I-31). Within a short period of time the complete amino acid sequence of the TMV protein subunits was elucidated (I-32,33). It is shown in Fig. XI-7.

From X-ray diffraction (I-28), electron microscopy, and infrared dichroism (I-34), the following picture of the tobacco mosaic virus emerges. About 2130 protein subunits are arranged in a large spiral, wound around the hole in the cylindrical rod which forms the virus. The ribonucleic acid is present as a long chain which is embedded between the turns of the spiral and thus likewise forms a spiral around the cylinder (I-35). The length of the cylinder is 3000 Å, its diameter 170–180 Å, the diameter of the hole about 35–40 Å, and the pitch of the spiral 20–25 Å (I-36).

If TMV is dissociated into RNA and protein, neither of these components is active as a virus. Viral activity is restored to a considerable extent on reconstitution of TMV from its separated components (I-37). The infectivity is a property of the RNA (I-38). The protein is responsible for the antigenicity of the virus. Proteins with similar serological specificity, but devoid of infectivity, are found in the cell sap of infected tobacco leaves; they are precipitated by antibodies to TMV (I-39). A comparison of the amino acid composition of different mutants or strains of TMV reveals small but distinct and typical differences (I-40). Reviewing these results, it may be asked whether the giant virus particle should be called a nucleoprotein molecule or rather an organism consisting of a protein mantle and an RNA spiral. We will be confronted with the same problem when we describe the bacteriophages.

*Deoxyribonucleoproteins* (DNPr) in contrast to RNPr, are insoluble in isotonic saline solutions, but soluble at very low ionic strength. They are prepared from tissues from which RNPr has been removed by 0.14 $M$ saline solution; subsequent extraction with distilled water yields DNPr (I-41). To prevent enzymatic degradation, it is necessary to work at low temperatures and to use enzyme inhibitors such as citrate, fluoride, or arsenate. The DNPr is precipitated from the aqueous extract by raising the salt concentration to 0.14 $M$. The isolated DNPr is a highly viscous, gel-like mass. It is not very suitable for examination by optical methods, but seems to correspond to the state of the native DNPr in living cells (I-42). Solutions of DNPr can also be obtained by extraction of the RNPr-

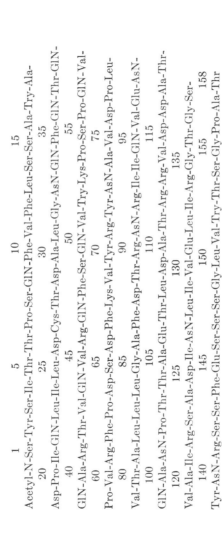

1                           5                           10                          15
Acetyl-N-Ser-Tyr-Ser-Ile-Thr-Thr-Pro-Ser-GlN-Phe-Val-Phe-Leu-Ser-Ser-Ala-Try-Ala-
                20                          25                          30                          35
Asp-Pro-Ile-GlN-Leu-Ile-Leu-Asp-Cys-Thr-Asp-Ala-Leu-Gly-AsN-GlN-Phe-GlN-Thr-GlN-
                40                          45                          50                          55
GlN-Ala-Arg-Thr-Val-GlN-Val-Arg-GlN-Phe-Ser-GlN-Val-Try-Lys-Pro-Ser-Pro-GlN-Val-
                60                          65                          70                          75
Pro-Val-Arg-Phe-Pro-Asp-Ser-Asp-Phe-Lys-Val-Tyr-Arg-Tyr-AsN-Ala-Val-Asp-Pro-Leu-
                80                          85                          90                          95
Val-Thr-Ala-Leu-Leu-Gly-Ala-Phe-Asp-Thr-Arg-AsN-Arg-Ile-Ile-GlN-Val-Glu-AsN-
                100                         105                         110                         115
GlN-Ala-AsN-Pro-Thr-Thr-Ala-Glu-Thr-Leu-Asp-Ala-Thr-Arg-Arg-Val-Asp-Asp-Ala-Thr-
                120                         125                         130                         135
Val-Ala-Ile-Arg-Ser-Ala-Asp-Ile-AsN-Leu-Ile-Val-Glu-Leu-Ile-Arg-Gly-Thr-Gly-Ser-
                140                         145                         150                         155         158
Tyr-AsN-Arg-Ser-Ser-Phe-Glu-Ser-Ser-Ser-Gly-Leu-Val-Try-Thr-Ser-Gly-Pro-Ala-Thr

FIG. XI-7. Amino acid sequence of the protein of tobacco mosaic virus (I-32). The molecular weight of the peptide chain is 17,500. The virus contains 2,130 of these chains. It is remarkable that the virus protein contains a single cysteine residue and that the amino acid sequence Nos. 1–40 is free of basic amino acids (I-76).

free nuclei with 5–10% NaCl (I-45). The DNPr is dissolved and can be precipitated from the concentrated salt solution by diluting it with water to a concentration of 0.14 $M$ NaCl. Since the high salt concentration causes not only dissolution but also dissociation of the DNPr into protein and nucleic acid, the reprecipitated material may consist of nonspecific precipitates of DNA and protein. Therefore, the extraction with solutions of low ionic strength is preferable for the preparation of native DNPr.

Separation of protein from DNA is more easily accomplished than the separation of protein from RNA. One of the simplest methods for this purpose consists of shaking the DNPr solution with chloroform and amyl alcohol (I-43). In another technique, the protein is precipitated by sodium dodecylsulfate or other detergents (I-44). Both methods yield good DNA preparations, but denatured protein. Native protein can be obtained by saturation with NaCl or by careful precipitation with ethanol. Most of the DNPr preparations from animal tissues consist of fibrous material insoluble in physiological saline solution, which indicates that none of these preparations is unchanged native deoxyribonucleoprotein. All of them may be artifacts produced by the unphysiological conditions of too high or too low ionic strength. A DNPr soluble in 0.14 $M$ NaCl has been isolated, however, from avian tubercle bacilli (I-45).

It is customary to classify the deoxyribonucleoproteins according to their protein component as nucleoprotamines, nucleohistones, and nucleoproteins. The histones can be differentiated from proteins by the deficiency of tryptophan, the protamines by the lack of tryptophan and of sulfur-containing amino acids (see Chapter VIII, G). *Nucleoprotamines* have been found in the sperm cells of fishes and birds. About 91% of the dry nuclei of trout sperm is formed by nucleoprotamine. Hence, the nuclei dissolve almost completely in 10% NaCl solution (I-46). The nucleoprotamine is precipitated from the filtered solution by dilution with water. The protamine preparations obtained from different species of fish have a different amino acid composition; they are species-specific (I-47). The ratio arginine:phosphorus in the nucleoprotamines is close to unity; this indicates the formation of saltlike bonds between the guanidino groups of arginine residues and the acidic phosphoryl groups of the DNA moiety. The nonbasic amino acids of the protamine may form loops in the protamine strand (I-48). Approximately 100 clupeine molecules are bound to each DNA molecule of herring sperm (I-47). The nucleus of the sperm cell contains 4–5 million nucleoprotamine molecules, or about 200,000 molecules per chromosome. X-ray diffraction indicates that in the nucleoprotamines the protamines are wrapped around a DNA helix (I-48).

The *nucleohistones* form a large portion of the deoxyribonucleoprotein in the nuclei of the somatic cells and particularly in the chromosomes. In the cells, the nucleohistones form gels which swell at low ionic strength and are solubilized by water extraction. Solubilization at high ionic strength is accompanied by dissociation into DNA and histone (I-49,50). Enzymatic degradation of nucleohistones during their preparation has been prevented by high concentrations of sucrose and glycerol (I-49). The molecular weight of the nucleohistone from thymus is approximately $19 \times 10^6$, the length of the molecule is 4200 Å, its radius of gyration 1700 Å; since the analogous values for the liberated DNA molecule are $8 \times 10^6$, 7100 Å, and 2900 Å, respectively, it has been concluded that the DNA chain is coiled up in the DNPr complex and thereby shortened (I-54). Histone bridges seem to be spaced 35–38 Å apart along the DNA molecules (I-51). The DNPr prepared from thymus glands is not homogeneous (I-42). Fractionation of the isolated histone reveals the presence of a lysine-rich fraction and another fraction rich in arginine (I-52). It is not yet clear whether the histone in the nucleohistone complex is bound to the DNA moiety by salt linkages between the phosphoric acid residues of DNA and basic groups of the histone, or by other, more specific bonds. The latter assumption is supported by the different chromatographic behavior of the extracted nucleohistone and the same material examined after dissociation and reassociation (I-53).

Very little is known about the *deoxyribonucleoproteins* from the nuclei of somatic cells. Their protein has been designated as "residual protein" of the nuclei (I-55,56). In contrast to the histones, this protein contains tryptophan. We do not know whether it is bound to DNA in the nuclei. DNA has been found in two crystalline proteins; one of these is avidin, the biotin-binding protein from egg white (I-57; see Chapter VIII, E), the other, cytochrome $b_2$ (yeast lactic acid dehydrogenase) (I-58). They contain 10% and 6% DNA, respectively. The nucleic acid can be removed without affecting the enzymatic action of cytochrome $b_2$. Its physicochemical analysis indicates that it is present as a single-strand DNA chain (I-59). Deoxyribonucleoproteins are also present in the $\kappa$-particles of paramecia of the killer type (I-62).

DNA is found in the *bacteriophages* (bacterial viruses) and in some of the animal viruses. The best investigated of the bacteriophages are the phages $T_2$, $T_4$, and $T_6$ of *Escherichia coli*. Under the electron microscope they show a hexagonal "head" to which a cylindrical "tail" is attached. Their mantle is formed by protein, their interior by DNA. The phage attaches itself by means of its tail to the host bacterium and "injects" its DNA through the tail into the *E. coli* cell. After a lag period of several minutes new phage particles are formed in the bacterial cell at a rapid rate; subsequently the cell is destroyed by disintegration. The DNA of the $T_2$, $T_4$, and $T_6$ phages is rich in 5-hydroxymethylcytosine; some of the

DNA molecules have a free hydroxyl group, others are substituted by single glucose residues, and a third type by gentiobiose which consists of 2 glucose residues (I-77). The percentages of monoglucoside in the 3 types of phages are 70, 100, and 3%, those of diglucoside 5, 0, and 72% (I-60, 61). Evidently, the DNA of the 3 types of bacteriophage is type-specific. It seems to consist of a single piece of DNA. Therefore we may call the bacteriophage a nucleoprotein molecule with a molecular weight of several hundred millions, or we may consider it the simplest type of organism, unable to reproduce itself without the help of a living host which contributes necessary enzymes. We encountered a similar situation in discussing TMV, which can multiply only in living tobacco plants.

Hybrid DNA-RNA and also DNA-RNA-protein complexes have been discovered in $T_2$-infected *E. coli* (I-63) and in the chromatin of plant nuclei (I-64) respectively. The presence of such hybrids had been overlooked because of the inadequacy of the methods used for the differentiation of RNA and DNA. The colorimetric methods, frequently used for this purpose, become unreliable when RNA is accompanied by a large excess of DNA or vice versa. Similarly, we cannot rely on the application of RNase, DNase, or of proteolytic enzymes, since each of the three components, RNA, DNA, and protein, is resistant in some of the complexes to the action of nucleases or proteolytic enzymes (I-65). Thus DNA inhibits the proteolytic action of trypsin (I-66) and chymotrypsin (I-67). DNA in electron micrographs is not digested by DNase unless the protein has been hydrolyzed by treatment with pepsin (I-68). Similarly the RNA-DNA-protein complex in the chromatin of plant cells resists the action of RNase, but is attacked by the enzyme after exposure to DNase or to heat denaturation (I-64). It has also been reported that DNA in the nuclear chromatin is degraded by chymotrypsin into fragments which have only one-fourth or one-eighth of the original molecular weight (I-69). These and other observations (I-70) show that nucleic acids combine with enzyme proteins in the same nonspecific manner in which they combine with other proteins, that they may inactivate enzymes, and that the results of exposure to RNase, DNase, or proteolytic enzymes cannot be used to prove the presence or absence of any of the nucleic acids or proteins (I-68). When tissues containing RNPr and DNPr are treated with warm solutions of dilute alkali hydroxide, ribonucleic acid is hydrolyzed to soluble nucleotides whereas DNA is resistant to alkali (I-71). Although this method does not allow us to isolate DNPr and RNPr, it makes it possible to estimate the amount of RNA and DNA.

Both of the nucleic acids are polyanions with numerous negatively charged phosphate residues. Only two of the three acidic groups of each phosphate residue are esterified; the third is free and negatively charged.

As polyanions, the nucleic acids combine nonspecifically with the basic side chains of the proteins, forming in this manner numerous saltlike bonds. The nonspecific nucleic acid-protein complexes are stable in the acid pH range. Their solubility depends on the ratio protein:nucleic acid. In addition to the saltlike bonds, hydrogen bonds may also be involved in the mutual combination of the two components. Thus the NH groups of the peptide bonds may form hydrogen bonds with carbonyl groups of guanine, uracil, cytosine, or thymine. Alternatively, the amino groups of adenine, guanine, or cytosine may form hydrogen bonds with the CO groups of the peptide bonds (I-72). Many other polar or nonpolar groups of the amino acid side chains may be involved in the mutual combination of proteins and nucleic acids.

The solutions of nucleoproteins are highly viscous, those of the deoxyribonucleoproteins being much more viscous than those of ribonucleoproteins; they also show strong birefringence. Both properties indicate a threadlike shape of the nucleoprotein molecules; their axial ratios have been found to have very high values of 40:1 or more (I-73). The gel-like property of the nucleoproteins had suggested that their nucleic acids might consist of branched chains. However, no indication for any ramification of the nucleic acid chains has been found, and it is believed at present that their chains are not branched. In most DNA preparations the ratios adenine:thymine, guanine:cytosine, and, for the same reason, the ratio purine:pyrimidine, are always equal to 1.0 (I-72). This and the results of X-ray diffraction have led to the ingenious idea that DNA occurs in antiparallel helical strands of two chains, and that one of these chains acts as a template for the formation of the other chain, adenine acting as a template for thymine formation, guanine for cytosine, and vice versa (I-74). The implications of this assumption for the mechanism of protein biosynthesis are obvious. Moreover, the assumption of a DNA twin helix is also of great importance for the problem of the structure of nucleoproteins. Single-strand DNA has been isolated from certain bacteriophages. In these DNA preparations the base ratios A/T and G/C are different from 1.0. We do not yet know whether the native deoxyribonucleoproteins contain single-strand DNA or the double helix of antiparallel DNA chains. Denaturation by exposure of native DNA to heat, acid, alkali, or other factors results in the cleavage of the specific hydrogen bonds between the adenine-thymine and guanine-cytosine pairs of the twin helix, and conversion into a nonspecifically bonded, disordered, double-strand conformation (I-75). This conformation may exist in the artificial nucleoproteins which are formed *in vitro* when solutions of proteins are mixed with those of nucleic acids.

In evaluating base ratios it must be kept in mind that most of the nucleoproteins isolated from tissues are heterogeneous and that the base ratio in such a mixture may be quite different from that of the pure components. Our limited knowledge about the structure of nucleoproteins is chiefly due to the lack of methods for the determination of nucleotide sequences in nucleic acids. As long as sequence analyses in nucleic acids are impossible, our insight into the structure of nucleoproteins and into the relations between nucleic acids and proteins will remain limited. In view of the recent discoveries of the role of nucleic acids as carriers of the genetic code (see Chapter XVI) this gap in our knowledge is particularly unfortunate.

## REFERENCES

### Section A

**A-1.** "Some Conjugated Proteins," Rutgers Univ. Press, New Brunswick, New Jersey, 1953. **A-2.** F. Haurowitz, *in* "Handbuch der Pflanzenphysiologie," Vol. 8, p. 333. Springer, Berlin, 1958.

### Section B

**B-1.** E. Chargaff, *Adv. in Protein Chem.* 1: 1(1944). **B-2.** F. R. N. Gurd, *in* "Lipide Chemistry" (D. J. Hanahan, ed.), pp. 208, 260(1960). **B-3a.** *Discussions Faraday Soc.* 6(1950). **B-3b.** F. T. Lindgren and A. V. Nichols, *in* "The Plasma Proteins" (F. W. Putnam, ed.), Vol. 2, p. 1(1960). **B-4a.** A. S. McFarlane, *Nature* 149: 439(1942). **B-4b.** M. Macheboeuf, *Exposés ann. biochim. méd.* 5: 71(1945). **B-5.** W. P. Jencks *et al.*, *J. Clin. Invest.* 34: 1437(1955). **B-6.** J. L. Oncley *et al.*, *JACS* 72: 458(1950). **B-7.** N. I. Krinsky *et al.*, *ABB* 73: 233(1958). **B-8.** P. Bernfeld *et al.*, *JBC* 226: 51(1957); J. L. Oncley *et al.*, *JACS* 79: 4666(1957). **B-9.** J. L. Oncley *et al.*, *J. Phys. Chem.* 56: 85(1952). **B-10.** J. L. Oncley *et al.*, *JACS* 71: 541, 1227(1949). **B-11.** J. L. Oncley *et al.*, *J. Phys. Chem.* 51: 184(1947). **B-12.** A. S. McFarlane, *in* "Lipide Chemistry," (D. J. Hanahan, ed.), p. 74. **B-13.** J. W. Gofmann *et al.*, *Physiol. Revs.* 34: 589(1954). **B-14.** O. de Lalla and J. W. Gofmann, *Methods of Biochem. Anal.* 1: 459(1954). **B-15.** S. Hayashi *et al.*, *JACS* 81: 3793(1959). **B-16.** P. F. Hahn, *Science* 98: 19(1943). **B-17.** J. W. Gofmann, *Ann. N.Y. Acad. Sci.* 64: 590(1956). **B-18.** K. Jahnke and W. Scholtan, "Die Bluteiweisskörper in der Ultrazentrifuge." Stuttgart, 1960. **B-19.** D. E. Green, *Discussions Faraday Soc.* 27: 206(1959). **B-20.** W. W. Wainio, *JBC* 234: 658(1959). **B-21.** G. Wald *et al.*, *Biol. Bull.* 96: 249(1948). **B-22.** K. G. Stern and K. Salomon, *JBC* 122: 461(1938). **B-23.** H. Junge, *Z. physiol. Chem.* 268: 179(1941). **B-24.** G. Wald *et al.*, *J. Gen. Physiol.* 38: 623(1955); 39: 923(1956); *Science* 118: 505(1953). **B-25.** R. Hubbard *et al.*, *J. Gen. Physiol.* 37: 381(1954); 39: 935(1956). **B-26.** J. Folch and M. Lees, *JBC* 191: 807(1951). **B-27.** K. J. Palmer and F. O. Schmitt, *J. Comp. Cellular Physiol.* 17: 385(1941); R. S. Baer *et al.*, *Ibid.*, 355. **B-28.** J. D. Robertson, *Biochem. Soc. Symposia* 16: 3(1959).

### Section C

**C-1.** K. Meyer, *Adv. in Protein Chem.* 2: 249(1945); "Some Conjugated Proteins," Rutgers Univ. Press (1953) p. 64. **C-2.** F. R. Bettelheim-Jevons, *Adv. in Protein Chem.* 13: 36(1958). **C-3.** W. Pigman *et al.*, *Ann. Rev. Biochem.* 28: 15(1959). **C-4.** G. Blix and S. Gardell, *in* "Hoppe-Seyler/Thierfelder's Handbuch der physiologisch-

und biologisch-chemischen Analyse," Vol. 2, p. 662. Springer, Berlin, 1960.   **C-5.** R. J. Winzler, *in* "The Plasma Proteins" (F. W. Putnam, ed.) Vol. 1, 309(1960).   **C-6.** A. Gottschalk, *Nature* 170: 662(1952).   **C-7.** H. E. Schultze *et al.*, *Biochem. Z.* 329: 490(1958).   **C-8.** Z. Stary, *Clin. Chem.* 4: 557(1957).   **C-9.** K. Meyer *et al.*, *JBC* 205: 611(1953).   **C-10.** H. Smith *et al.*, *BJ* 52: 15(1952).   **C-11.** L. A. Elson and W. T. J. Morgan, *BJ* 27: 1824(1933).   **C-12.** M. Y. Tracey, *BBA* 17: 159(1955).   **C-13.** Z. Dische, *JBC* 167: 189(1947).   **C-14.** J. X. Khym and D. G. Doherty, *JACS* 74: 3199(1952).   **C-15.** Z. Dische and L. Shettles, *JBC* 192: 579(1957).   **C-16.** H. Masamune and Z. Yosizawa, *Tokyo J. Exptl. Med.* 53: 155, 237(1950).   **C-17.** B. Drevon and R. Donikian, *Bull. soc. chim. biol.* 37: 1321(1956).   **C-18.** E. Goldwasser and M. B. Mathews, *JACS* 77: 3135(1955).   **C-19.** H. E. Weimer *et al.*, *JBC* 185: 561 (1950).   **C-20.** C. de Vaux St. Cyr *et al.*, *Bull. soc. chim. biol.* 40: 579(1958).   **C-21.** K. Schmid, *JACS* 75: 60(1953).   **C-22.** R. I. Cheftel *et al.*, *Bull. soc. chim. biol.* 42: 993 (1960).   **C-23.** E. A. Popenoe, *BBA* 32: 584(1959).   **C-24.** K. O. Pedersen, *Nature* 154: 575, 642(1944).   **C-25.** R. G. Spiro, *JBC* 235: 2860(1960).   **C-26.** E. Klenk and G. Uhlenbruck, *Z. physiol. Chem.* 311: 227(1958).   **C-27.** Z. Stary and F. Arat, *Naturwiss.* 44: 11(1957).   **C-28a.** J. W. Rosevear and E. L. Smith, *JBC* 236: 425 (1961).   **C-28b.** K. Schmid *et al.*, *BBA* 58: 80(1960).   **C-28c.** A. Gottschalk and H. A. McKenzie, *BBA* 54: 226(1961).   **C-28d.** E. H. Eylar, R. W. Jeanloz, *JBC* 237: 622 (1962).   **C-28e.** E. D. Kaverzneva and T. De-Fan, *Biokhimiya* 26: 782(1961).   **C-28f.** M. J. Spiro and G. R. Spiro, *JBC* 237: 453(1962).   **C-29.** P. G. Johannsen *et al.*, *BJ* 78: 518(1961).   **C-30.** H. Florey, *Proc. Roy. Soc.* B143: 147(1955).   **C-31.** G. Blix and O. Karlberg, *Z. physiol. Chem.* 240: 43, 55(1936).   **C-32.** F. Hisamura *et al.*, *J. Biochem. (Tokyo)* 28: 217, 473(1938).   **C-33.** A. Gottschalk, *Nature* 186: 949 (1960). **C-34.** A. Gottschalk, *BBA* 24: 649(1957).   **C-35.** H. Fraenkel-Conrat and R. R. Porter, *BBA* 9: 557(1952).   **C-36.** L. Penasse *et al.*, *BBA* 9: 551(1952).   **C-37.** M. Stacey and J. M. Woolley, *JCS* 1942: 550.   **C-38.** H. Fraenkel-Conrat *et al.*, *ABB*, 39: 80(1952).   **C-39.** Z. Stary and R. Cindi, *Naturwiss.* 43: 179(1956).   **C-40.** S. M. Partridge, *BJ* 43: 387(1948).   **C-41.** S. M. Partridge *et al.*, *BJ* 79: 15(1961).   **C-42.** M. B. Methews and I. Lozaityte, *ABB* 74: 158(1958).   **C-43.** R. E. Glegg *et al.*, *Science* 111: 614(1953).   **C-44.** G. A. Johansson and F. Wahlgren, *Acta Pathol. Microbiol. Scand.* 15: 358(1939).   **C-45.** E. A. Kabat, *Bacteriol. Revs.* 13: 189(1950). **C-46.** W. T. J. Morgan and W. M. Watkins, *Brit. Med. Bull.* 15: 109(1959).   **C-47.** W. T. J. Morgan *et al.*, *BJ* 52: 247(1951).   **C-48.** S. Leskowitz and E. A. Kabat, *JACS* 76: 4887(1954).   **C-49.** F. Schiff, *Z. Bakteriol.* (I) 98: 94(1930).   **C-50.** C. Landsteiner and R. A. Harte, *J. Exptl. Med.* 71: 551(1940); *JBC* 140: 673(1945).   **C-51.** J. L. Strominger and R. H. Threnn, *BBA* 33: 280(1959).   **C-52.** E. D. Korn and D. H. Northcote *BJ* 75: 12(1960).   **C-53.** G. Fraenkel and K. M. Rudall, *Proc. Roy. Soc.* B134: 111(1947).   **C-54.** C. Nolan and E. L. Smith, *JBC* 237: 453(1962).   **C-55.** M. E. Carsten and E. A. Kabat, *JACS* 78: 3083(1956).   **C-56.** P. Hoffmann *et al.*, *Science* 124: 1252(1956); J. A. Cifonelli *et al.*, *JBC* 233: 541(1958).   **C-57.** G. Schiffman *et al.*, *JACS* 84: 73(1962).   **C-58.** R. H. Nuenke and L. W. Cunningham, *JBC* 236: 2452(1961).

### Section D

**D-1.** F. R. N. Gurd and P. E. Wilcox, *Adv. in Protein Chem.* 11: 312(1956).   **D-2.** H. J. Bielig and E. Bayer, Hoppe-Seyler, Thierfelder Hdb. d. Physiol. Chem. Analyse 4: 714(1960).   **D-3.** "Chemical Specificity in Biological Interactions" (F. R. N. Gurd, ed.). New York, 1954.   **D-4.** C. B. Laurell, *in* "The Plasma Proteins" (F. W. Putnam, ed.), Vol. 1, p. 349. 1960.   **D-5.** L. Michaelis, *Adv. in Protein Chem.* 3: 53(1947).   **D-6.** V. Laufberger, *Bull. soc. chim. biol.* 19: 1575(1938).   **D-7.** R. Kuhn *et al.*, *Chem. Ber.*

73: 823(1940). **D-8.** S. Granick *et al.*, *Science* 95: 439(1942); *JBC* 155: 661(1944). **D-9.** A. Rothen, *JBC* 152: 679(1944). **D-10.** P. F. Hahn *et al.*, *JBC* 150: 407(1943). **D-11.** A. Mazur *et al.*, *JBC* 182: 607(1950); 187, 497(1955). **D-12.** B. A. Koechlin, *JACS* 74: 2649(1952). **D-13.** R. C. Warner and I. Weber, *JACS* 75: 5094(1953). **D-14.** C. G. Holmberg and C. B. Laurell, *Acta Chem. Scand.* 2: 550(1948); 5: 476, 921(1951). **D-15.** F. L. Humoller *et al.*, *J. Lab. Clin. Med.* 56: 222(1960). **D-16.** F. Haurowitz, *Z. physiol. Chem.* 190: 72(1930). **D-17.** A. Mazur *et al.*, *JBC* 187: 473 (1950). **D-18.** M. Florkin, "Les hémerythrines." Thesis University of Liège, 1933. **D-19.** J. Roche, "Pigments Respiratoires." Paris (1935). **D-20.** A. Roche and J. Roche, *Bull. soc. chim. biol.* 17: 1494(1935). **D-21.** I. M. Klotz *et al.*, *Nature* 195: 900(1962). **D-22.** E. Boeri and A. Ghiretti-Magaldi, *BBA* 23: 489(1957). **D-23.** W. E. Love, *BBA* 23: 465(1957). **D-24.** I. M. Klotz, *in* "Sulfur in Proteins" (R. Benesch, ed.), p. 127. Academic Press, New York, 1959; *Science* 121: 477(1955). **D-25.** C. Manwell, *Ann. Rev. Physiol.* 22: 191(1960). **D-26a.** C. Manwell, *Science* 132: 550(1960). **D-26b.** P. M. Harrison, *J. Mol. Biol.* 1: 69(1959). **D-27.** A. C. Redfield, *Biol. Revs.* 9: 175(1934); *JBC* 76: 185, 191(1928). **D-28.** Ch. Dhéré, *Rev. Suisse zool.* 35: 277(1928). **D-29.** C. R. Dawson and M. F. Malette, *Adv. in Protein Chem.* 2: 179(1945). **D-30.** F. Kubowitz, *Biochem. Z.* 299: 32(1938). **D-31.** F. Hernler and E. Philippi, *Z. physiol. Chem.* 216: 110(1933). **D-32.** J. Roche, *Skand. Arch. Physiol.* 69: 87(1934). **D-33.** T. Nakamura and H. S. Mason, *Biochem. Biophys. Research Communs.* 3: 297(1960). **D-34.** R. Leontie, *Clin. Chim. Acta* 3: 68 (1958). **D-35.** L. D. G. Thomson *et al.*, *ABB* 83: 88(1959). **D-36.** G. Felsenfeld and M. P. Printz, *JACS* 81: 6259(1959). **D-37.** B. Erickson-Quensel and T. Svedberg, *Biol. Bull.* 71: 498(1936). **D-38.** W. M. Stanley and T. F. Anderson, *JBC* 146: 25 (1942). **D-39.** A. Polson and W. R. G. Wyckoff, *Nature* 160: 153(1947). **D-40.** G. Schramm and G. Berger, *Z. Naturforsch.* 7b: 284(1952). **D-41.** N. F. Burk, *JBC* 133: 511(1940). **D-42.** B. G. Malmström, *FP* 20: 60(1961). **D-43.** I. H. Scheinberg and I. Sternlieb, *Pharmacol. Revs.* 12: 355(1950). **D-44.** T. Mann and D. Keilin, *Proc. Roy. Soc.* B126: 303(1938/39). **D-45.** H. J. Bielig *et al.*, *Pubbl. staz. zool. Napoli* 25: 26(1953); 29: 109(1957); *Ann. Chem.* 580: 134(1953). **D-46.** J. H. R. Kägi and B. L. Vallee, *JBC* 236: 2435(1961); B. L. Vallee, *Adv. in Protein Chem.* 10: 318(1955). **D-47.** G. Weitzel *et al.*, *Z. physiol. Chem.* 304: 1(1956). **D-48.** J. W. Holleman and G. Biserte, *Bull. soc. chim. biol.* 40: 1417(1958). **D-49.** J. R. Kimmel *et al.*, *JBC* 234: 46(1959). **D-50.** P. M. Harrison, *J. Mol. Biol.* 1: 69(1959). **D-51.** E. Bayer and H. Fiedler, *Liebigs Ann. Chem.* 653: 149(1962). **D-52.** E. F. J. Van Bruggen *et al.*, *J. Mol. Biol.* 4: 1, 8(1962). **D-53.** P. O. Nyman, *BBA* 45: 387(1960).

### Section E

**E-1.** F. Haurowitz and R. L. Hardin, *in* "Proteins" (H. Neurath and K. Bailey, eds.), Vol. 2, p. 279. Academic Press, New York, 1954. **E-2.** J. Wyman, Jr., *Adv. in Protein Chem.* 4: 410(1948). **E-3.** F. Haurowitz, *Tabulae Biologicae* 10: 18(1934). **E-4.** H. Itano, *Adv. in Protein Chem.* 12: 216(1957). **E-5.** D. L. Drabkin, *Physiol. Revs.* 31: 345(1952). **E-6.** C. Manwell, *Ann. Rev. Physiol.* 22: 191(1960). **E-7.** D. Keilin and J. F. Ryley, *Nature* 172: 451(1953). **E-8.** A. Hamsik, *Z. physiol. Chem.* 187: 229(1930). **E-9.** M. L. Anson and A. E. Mirsky, *J. Gen. Physiol.* 13: 469(1930). **E-10.** M. Laporta, *Arch. sci. biol.* 16: 575(1931). **E-11.** A. Rossi-Fanelli and E. Antonini, *JBC* 235: PC4(1960). **E-12.** F. Haurowitz, *Z. physiol. Chem.* 232: 146 (1936). **E-13.** A. Rossi-Fanelli and E. Antonini, *ABB* 80: 299, 308(1959). **E-14.** F. Haurowitz and H. Waelsch, *Z. physiol. Chem.* 182: 82(1929). **E-15.** J. Roche and M. Chouaiech, *Bull. soc. chim. biol.* 22: 263(1925). **E-16.** G. S. Adair, *Proc. Roy. Soc.* A109: 292(1925). **E-17.** G. Braunitzer *et al.*, *Z. physiol. Chem.* 325: 283(1961). **E-18.**

J. Steinhardt, *Nature* 138: 800(1936).   **E-19.** T. Svedberg and K. O. Pedersen, "The Ultracentrifuge." Oxford Univ. Press, London and New York, 1940.   **E-20.** G. S. Adair *et al.*, *Compt. rend.* 195: 1433(1933).   **E-21.** F. Haurowitz *et al.*, *ABB* 59: 52 (1955).   **E-22.** F. Haurowitz *et al.*, *Nature* 180: 437(1957).   **E-23.** N. Benhamou and G. Weill, *Compt. rend.* 243: 2054(1956).   **E-24.** E. M. Zaiser and J. Steinhardt, *JACS* 73: 5568(1951).   **E-25.** J. Steinhardt *et al.*, *Biochemistry* 1: 29(1962); Z. Vodracka and J. Cejka, *BBA* 49: 502(1961).   **E-26.** J. Wyman, Jr., *JBC* 127: 1(1939).   **E-27.** R. W. Cowgill and W. M. Clark, *JBC* 198: 33(1952).   **E-28.** F. Haurowitz, *Z. physiol. Chem.* 232: 125(1935).   **E-29.** A. Riggs, *JBC* 236: 1948(1961).   **E-30.** J. E. O'Hagan, *Nature* 184: 1652(1959).   **E-31.** R. Benesch and R. E. Benesch, *JBC* 236: 405(1961). **E-32.** H. Theorell and A. Ehrenberg, *Acta Chem. Scand.* 5: 823(1951).   **E-33.** J. E. O'Hagan, *BJ* 74: 417(1960).   **E-34.** M. F. Perutz *et al.*, *Nature* 185: 416(1960). **E-35.** D. J. E. Ingram *et al.*, *Nature* 178: 906(1956).   **E-36.** S. Ainsworth *et al.*, *Proc. Roy. Soc.* B152: 331(1960).   **E-37.** F. Haurowitz and H. Kittel, *Ber. deut. chem. Ges.* 66: 1047(1933).   **E-38.** L. Pauling and C. D. Coryell, *Proc. Natl. Acad. Sci. U.S.* 22: 159(1936).   **E-39.** H. Theorell, *Arkiv Kemi Mineral. Geol.* 16A: No. 3(1941).   **E-40.** M. Heidelberger, *JBC* 53: 34(1922).   **E-41,** R. Hill and H. F. Holden, *BJ* 20: 1326 (1928).   **E-42.** E. Breslow and F. R. N. Gurd, *JBC* 237: 371(1962).   **E-43.** M. F. Perutz, "Proteins and Nucleic Acids," p. 52. Elsevier, Amsterdam, 1962.

### Section F

**F-1.** R. von Zeynek, *Nowiny Lekarskie (Poland)* 38: 10(1926).   **F-2.** F. Haurowitz, *in* "Hemoglobin" (Barcroft Memorial Conference), p. 53. 1949.   **F-3.** F. Haurowitz, *JBC* 193: 443(1951).   **F-4.** R. Havemann and W. Haberditzl, *Z. physik. Chem. (Leipzig)* 210: 267(1959).   **F-5.** W. H. Forbes and F. J. W. Roughton, *J. Physiol.* 71: 229(1931).   **F-6.** A. Riggs and G. Wald, *FP* 10: 109(1951).   **F-7.** F. J. W. Roughton *et al.*, *Proc. Roy. Soc.* B144: 29(1955).   **F-8.** L. Pauling *et al.*, *Science* 110: 543(1949); M. F. Perutz *et al.*, *Nature* 167: 929(1951).   **F-9.** H. Hartridge and F. J. W. Roughton, *J. Physiol.* 64: 405(1928).   **F-10.** B. L. Horecker, *JBC* 148: 173(1943).   **F-11.** J. Haldane and J. B. Smith, *J. Physiol.* 20: 405(1896).   **F-12.** T. Bücher and E. Negelein, *Naturwiss.* 29: 672(1941).   **F-13.** Q. H. Gibson and S. Ainsworth, *Nature* 180: 1416 (1957).   **F-14.** O. Bodansky, *Pharmacol. Revs.* 3: 144(1951).   **F-15.** F. Haurowitz, *Z. physiol. Chem.* 138: 68(1924); 194: 98(1931).   **F-16.** C. D. Coryell *et al.*, *JACS* 59: 633(1937).   **F-17.** D. Keilin and E. F. Hartree, *Nature* 170: 161(1952).   **F-18.** F. Haurowitz, *Z. physiol. Chem.* 232: 159(1935).   **F-19.** D. Keilin and E. F. Hartree, *Proc. Roy. Soc.* B117: 1(1935).   **F-20.** H. O. Michel, *JBC* 126: 323(1938); F. Haurowitz, *JBC* 137: 771(1941).   **F-21.** P. George *et al.*, *Science* 117: 220(1953); *J. Colloid Sci.* 11: 327(1956).   **F-22.** H. Schiller, *Z. physik. Chem. (Leipzig)* 216: 84(1961). **F-23.** R. Schoenheimer, *Z. physiol. Chem.* 180: 144(1929).   **F-24.** H. Theorell, *Biochem. Z.* 252: 1(1932); H. Theorell and C. de Duve, *ABB* 12: 113(1947); A. Rossi-Fanelli, *Science* 108: 15(1948).   **F-25.** R. Hill, *Nature* 132: 897(1933).   **F-26.** A. Rossi-Fanelli *et al.*, *BBA* 17: 377(1955).   **F-27.** J. C. Kendrew *et al.*, *Nature* 185: 422(1960).   **F-28.** M. F. Perutz *et al.*, *Nature* 185: 416(1960).   **F-29.** S. Takashima and R. Lumry, *JACS* 80: 4238(1958).   **F-30.** F. Haurowitz, *Z. physiol. Chem.* 254: 268(1938).   **F-31.** A. S. Brill and R. J. P. Williams, *BJ* 78: 253(1961).   **F-32.** A. Stockell, *J. Mol. Biol.* 3: 362(1961).   **F-33.** S. Beychok and E. R. Blout, *J. Mol. Biol.* 3, 769 (1961).   **F-34.** S. Konosu *et al.*, *Bull. Jap. Soc. Sci. Fish.* 24: 563 (1958).

### Section G

**G-1.** M. Heidelberger and K. Landsteiner, *J. Exptl. Med.* 38: 561(1923).   **G-2.** F. Haurowitz *et al.*, *Rev. fac. sci. univ. Istanboul* 9: 120(1944).   **G-3.** A. Reichert and A. Brown, *Carnegie Inst. Wash. Publ. No.* 116: (1909). **G-4.** D. L. Drabkin, *JBC* 164: 703(1946). **G-5.** M. Fox, *Proc. Roy. Soc.* B115: 451(1934). **G-6.** H. Fischer and C. von Seemann, *Z. physiol. Chem.* 242: 133(1936). **G-7.** J. Roche and R. Combette, *Bull. soc. chim. biol.* 19: 613(1937). **G-8.** T. Svedberg and B. Ericksson-Quensel, *JACS* 55: 2834(1933); 56: 700(1934). **G-9.** J. Roche and M. Chouaiech, *Compt. rend. soc. biol.* 130: 562(1939). **G-10.** A. Virtanen *et al.*, *Acta Chem. Scand.* 1: 90(1949). **G-11.** N. Ellfolk, *Acta Chem. Scand.* 13: 595(1959); 15: 545(1961). **G-12.** F. v. Krüger, *Z. vergleich. Physiol.* 2: 254(1925). **G-13.** F. Haurowitz *et al.*, *Z physiol. Chem.* 183: 78(1929); *J. Phys. Chem.* 58: 103(1954). **G-14.** Wakulenko, *Sci. Papers Univ. Tomsk* 2(1910, in Russian); quoted by F. v. Krüger and W. Gerlach, *Z. ges. exptl. Med.* 53: 233(1926). **G-15.** F. Haurowitz, *Z. physiol. Chem.* 186: 141(1930). **G-16.** H. H. Zinsser, *ABB* 38: 195(1952). **G-17.** R. R. Porter and F. Sanger, *BJ* 42: 287(1948); M. S. Masari and K. Singer, *ABB* 58: 407(1955). **G-18.** J. C. Kendrew and M. F. Perutz, *Proc. Roy. Soc.* A194: 375(1948). **G-19.** F. Haurowitz, *Z. physiol. Chem.* 232: 125(1935). **G-20.** A. M. L. Liquori and F. Bertinotti, *Ricerca sci.* 21: 1200(1951). **G-21.** J. Roche *et al.*, *Compt. rend. soc. biol.* 141: 771(1953). **G-22.** K. Betke, "Der menschliche Rote Blutfarbstoff bei Fetus und reifem Organismus." Springer, Berlin, 1954. **G-23.** W. H. Stein *et al.*, *BBA* 24: 640(1957). **G-24.** V. M. Ingram, *Nature* 183: 1795(1959). **G-25.** R. T. Jones *et al.*, *JACS* 81: 4749(1959). **G-26.** H. Hayashi, *J. Biochem. (Tokyo)* 50: 70(1961). **G-27.** H. S. Rhinesmith *et al.*, *JACS* 80: 3358(1958). **G-28.** W. A. Schroeder *et al.*, *Proc. Natl. Acad. Sci. U.S.* 47: 811(1961); 48: 284(1962); *BBA* 54: 181(1961). **G-29.** S. J. Singer and H. A. Itano, *Proc. Natl. Acad. Sci. U.S.* 45: 174(1959). **G-30.** J. Vinograd and W. D. Hutchinson, *Nature* 187: 216(1959). **G-31.** E. A. Robinson and H. A. Itano, *Nature* 188: 798 (1961). **G-32.** G. Braunitzer *et al.*, *Z. physiol. Chem.* 325: 283(1961); *Nature* 190: 480 (1961). **G-33a.** W. Konigsberg, *JBC* 236: PC 55(1961). **G-33b.** R. J. Hill and L. C. Craig, *JACS* 81: 2272(1959). **G-33c.** M. F. Perutz *et al.*, *Nature* 185: 416 (1960). **G-33d.** J. C. Kendrew *et al.*, *Nature* 185: 422(1960). **G-34.** H. G. Kunkel *et al.*, *J. Clin. Invest.* 36: 1615(1957). **G-35.** G. Modiano, *Italian J. Biochem.* 6: 334 (1957). **G-36.** T. H. J. Huisman, *in* "Sulfur in Proteins" (R. Benesch, ed.), p. 153. Academic Press, New York, 1960. **G-37.** A. Rossi-Fanelli *et al.*, *JBC* 236: 391, 397 (1961). **G-38.** V. M. Ingram and A. O. W. Stretton, *BBA* 63: 20(1962). **G-39.** A. C. Allison, *Science* 122: 640(1955). **G-40.** K. Betke, *Schweiz. med. Wochschr.* 88: 1005(1958). **G-41.** H. Zilliacus, *Nature* 188: 1202(1961). **G-42.** Y. Derrien *et al.*, *Bull. soc. chim. biol.* 42: 519(1960). **G-43.** W. Küntzer, *Klin. Wochschr.* 38: 404 (1960). **G-44.** P. A. Charlwood *et al.*, *BBA* 40: 191(1960). **G-45.** T. H. J. Huisman, *BBA* 46: 384(1961). **G-46.** J. Barcroft, *Lancet* 225: 1021(1933). **G-47.** E. McCarthy, *J. Physiol.* 102: 55(1943). **G-48.** H. A. Itano and L. Pauling, *Blood* 4: 66 (1949). **G-49.** V. M. Ingram, *Nature* 180: 326(1957). **G-50.** H. A. Itano and J. V. Neel, *Proc. Natl. Acad. Sci. U.S.* 36: 613(1950). **G-51.** J. A. Hunt and V. M. Ingram, *BBA* 42: 409(1960). **G-52.** A. I. Chernoff *et al.*, *Science* 120: 605(1954). **G-53.** J. A. Hunt and V. M. Ingram, *Nature* 184: 870(1959). **G-54.** R. Cabannes *et al.*, *Compt. rend. soc. biol.* 149: 914(1955); R. T. Jones *et al.*, *JACS* 81: 3161(1959). **G-55.** R. E. Benesch *et al.*, *JBC* 236: 2926(1961). **G-56.** J. A. Hunt and V. M. Ingram, *Nature* 184: 870(1959). **G-57.** R. A. Kekwick and H. Lehmann, *Nature* 187: 158(1960). **G-58.** H. A. Itano and E. A. Robinson, *Nature* 183: 1799(1959). **G-59.** H. A. Itano, *Ann. Rev. Biochem.* 25: 331(1956). **G-60.** C. Baglioni and V. M. Ingram, *BBA* 48: 243 (1961). **G-61.** K. Plotner and K. Betke, *in* "Handbuch der allgemeinen Pathologie," Vol. IV, Pt. 2, p. 245. Springer, Berlin, 1957. **G-62.** C. B. Anfinsen, "The Molecular Basis of Evolution." Wiley, New York, 1959. **G-63.** H. A. Itano and E. A. Robinson, *Proc. Natl. Acad. Sci. U.S.* 46: 1492(1960). **G-64.** V. M. Ingram, *Nature* 189: 704

1961. **G-65.** M. F. Perutz, *Nature* 194: 914(1962). **G-66.** K. Hilse and G. Braunitzer, *Z. physiol. Chem.* 329: 113(1962). **G-67.** W. Konigsberg and R. J. Hill, *JBC* 237: 2547(1962). **G-68.** K. Satake and T. Take, *J. Biochem. (Tokyo)* 52: 304(1962). **G-69.** E. Antonini *et al.*, *ABB* 97: 336(1962). **G-70.** R. E. Benesch and R. Benesch, *Biochemistry* 1: 735(1962). **G-71.** H. C. Watson and J. C. Kendrew, *Nature* 190: 670(1961).

## Section H

**H-1.** E. J. Cohn *et al.*, *JACS* 69: 750(1947). **H-2.** C. Gray and R. Kekwick, *Nature* 161: 274(1948). **H-3.** B. H. Billing *et al.*, *BJ* 65: 774(1957). **H-4.** R. Schmid *et al.*, *ABB* 70: 285(1957). **H-5.** R. Lemberg *et al.*, *BJ* 35: 328ff(1941). **H-6.** F. Haxo *et al.*, *ABB* 54: 162(1955). **H-7.** T. Fujiwara, *J. Biochem. (Tokyo)*, 44: 723(1957); 48: 317 (1960); 49: 361(1961). **H-8.** J. R. Kimmel and E. L. Smith, *Bull. soc. chim. biol.* 40: 2049(1958). **H-9.** R. Lemberg and J. W. Legge, "Hematin Compounds and Bile Pigments." Interscience, New York, 1949. **H-10.** G. Meldolesi *et al.*, *Z. physiol. Chem.* 259: 137(1939). **H-11.** H. Bennhold, *Kolloid-Z.* 85: 171(1938). **H-12.** P. Thomas, *Bull. soc. chim. biol.* 20: 1059(1938). **H-13.** H. S. Mason, *JBC* 172: 83 (1948); *Nature* 175: 771(1955); 177: 79(1956). **H-14.** J. P. Greenstein, *Ann. N.Y. Acad. Sci.* 4: 433(1948). **H-15.** Z. Stary and R. Richter, *Z. physiol. Chem.* 253: 159(1938). **H-16.** J. A. Serra, *Nature* 157: 771(1946). **H-17.** A. A. Krasnovskii *et al.*, *Doklady Akad. Nauk S.S.S.R.* 82: 947(1952). **H-18.** A. Stoll, *Fortschr. chem. Forsch.* 2: 538(1952). **H-19.** D. R. Anderson *et al.*, *BBA* 15: 298(1954). **H-20.** J. J. Walkin and F. A. Schwertz, *J. Gen. Physiol.* 37: 111(1953). **H-21.** W. Montagna and R. A. Ellis, "Hair Growth." Academic Press, New York, 1958. **H-22.** H. Burton and J. L. Stoves, *Nature*, 165: 569(1950). **H-23.** I. W. Sizer, *Adv. in Enzymol.* 14:129 (1953).

## Section I

**I-1.** J. P. Greenstein, *Adv. in Protein Chem.* 1: 209(1944). **I-2.** E. Chargaff and J. Davidson, "The Nucleic Acids," 3 vols. Academic Press, New York, 1955–60. **I-3.** "Solvay Conference on Chemistry. Nucleoproteins" (11th Conference, 1959). Interscience, New York, 1960. **I-4.** J. N. Davidson, "The Biochemistry of the Nucleic Acids." Methuen, London, 1960. **I-5.** D. O. Jordan, "The Chemistry of Nucleic Acids," Butterworth, London, 1960. **I-6.** W. E. Cohn, *JBC* 235: 1488(1960). **I-7.** E. Chargaff and S. Zamenhof, *JBC* 173: 327(1948). **I-8.** D. O. Jordan, *in* "The Chemistry of Nucleic Acids," p. 223. Butterworth, London, 1960. **I-9.** J. D. Watson and F. H. C. Crick, *Cold Spring Harbor Symposia Quant. Biol.* 18: 123(1953). **I-10.** G. Felsenfeld and A. Rich, *BBA* 26: 461(1957). **I-11.** T. Caspersson, *Symposia Soc. Exptl. Biol.* 1: 127(1947). **I-12.** R. Feulgen and K. Imhaeuser, *Z. physiol. Chem.* 148: 1(1925). **I-13.** J. Brachet, *Compt. rend. soc. biol.* 133: 88(1940). **I-14.** B. Magasanik, *in* "Nucleic Acids" (E. Chargaff and J. Davidson, eds.), Vol. 1, p. 373. Academic Press, New York, 1955. **I-15.** A. E. Mirsky and A. W. Pollister, *J. Gen. Physiol.* 30: 101(1946). **I-16.** S. E. Kerr and K. Seraidarian, *JBC* 180: 1203(1949). **I-17.** G. Zubay *et al.*, *J. Mol. Biol.* 2: 10, 105(1960). **I-18.** E. L. Hess *et al.*, *JBC* 236: 3020 (1961). **I-19.** J. W. Littlefield *et al.*, *JBC* 217: 111(1956). **I-20.** M. L. Petermann and M. G. Hamilton, *JBC* 224: 725(1957). **I-21.** F. Chao and H. K. Schachman, *ABB* 61: 220(1956). **I-22.** G. E. Pallade and P. Siekevitz, *J. Biophys. Biochem. Cytol.* 2: 171(1956). **I-23.** E. Herbert and E. Cannelakis, *BBA* 42: 363(1960). **I-24.** G. Hamoir, *BJ* 48: 146(1951). **I-25.** P. R. Venkataraman and C. U. Lowe, *BJ* 72: 430(1959). **I-26a.** A. Gierer, *Biophys. J.* 2: Pt. 2, 5(1962). **I-26b.** W. M. Stanley, *Science* 81: 644(1935). **I-27.** R. E. Franklin, *Nature* 177: 928(1956); *Discussions*

*Faraday Soc.* 25: 197(1958).   **I-28.** D. L. D. Caspar, *Nature* 177: 928(1956).   **I-29.** G. Braunitzer, *Z. Naturforsch.* 9b: 675(1954).   **I-30.** C. A. Knight, *JBC* 214: 231(1955). **I-31.** A. Tsugita, *BBA* 38: 145(1960).   **I-32.** A. Tsugita *et al., Proc. Natl. Acad. Sci. U.S.* 46: 1463(1960).   **I-33.** F. A. Anderer *et al., Z. Naturforsch.* 15b: 79(1960). **I-34.** M. Beer, *BBA* 29: 423(1958).   **I-35.** A. Klug and D. L. D. Caspar, *Adv. in Virus Research* 7: 223(1960).   **I-36.** H. Fraenkel-Conrat and L. K. Ramachandran, *Adv. in Protein Chem.* 14: 175(1959); A. Rich and D. W. Green, *Ann. Rev. Biochem.* 30: 93(1961).   **I-37.** H. Fraenkel-Conrat and R. C. Williams, *Proc. Natl. Acad. Sci. U.S.* 41: 690(1955).   **I-38.** A. Gierer and G. Schramm, *Nature* 177: 702(1956).   **I-39.** B. Commoner *et al., J. Gen. Physiol.* 38: 459ff(1955).   **I-40.** C. A. Knight, *JBC* 197: 241 (1952).   **I-41.** C. F. Crampton *et al., JBC* 206: 499(1954).   **I-42.** A. L. Dounce and M. O'Conell, *JACS* 80: 2013(1958).   **I-43.** M. Sevag *et al., JBC* 124: 425(1938). **I-44.** A. L. Dounce *et al., JACS* 74: 1724(1952).   **I-45.** E. Chargaff and H. F. Saidel, *JBC* 177: 417(1949).   **I-46.** K. Felix, *Z. physiol. Chem.* 287: 224(1951).   **I-47.** K. Felix, *Experientia* 8: 312(1952).   **I-48.** M. H. F. Wilkins, in **I-3** p. 45ff.   **I-49.** J. St. L. Philpot and J. E. Stanier, *BJ* 63: 214(1956).   **I-50.** K. V. Shooter *et al., BBA* 13: 192(1954).   **I-51.** M. H. F. Wilkins *et al., J. Mol. Biol.* 1: 179(1960).   **I-52.** J. A. Lucy and J. A. V. Butler, *BBA* 16: 431(1955).   **I-53.** F. C. Crampton and J. F. Scheer, *JBC* 227: 495(1957).   **I-54.** P. Doty and G. Zubay, *JACS* 78: 6207 (1956). **I-55.** A. E. Mirsky and H. Ris, *J. Gen. Physiol.* 34: 475(1951).   **I-56.** I. B. Zbarskii and S. S. Debov, *Doklady Akad. Nauk. S.S.S.R.* 72[N.S.]: 795(1948); *Biokhimiya* 16: 390(1951); cited in Zbarskii, *Proc. Intern. Congr. Biochem., 5th Congr., Moscow* (1961), Symposium 2.   **I-57.** H. Fraenkel-Conrat *et al., ABB* 39: 80(1952).   **I-58.** C. A. Appleby and R. K. Morton, *BJ* 75: 258(1960).   **I-59.** H. R. Mahler, *J. Mol. Biol.* 4: 211(1962).   **I-60.** I. R. Lehman and E. A. Pratt, *JBC* 235: 3254(1960).   **I-61.** J. Lichtenstein and S. S. Cohen, *JBC* 235: 1134(1960).   **I-62.** T. Sonneborn, *Nature* 175: 1100(1955).   **I-63.** M. Hayashi and S. Spiegelman, *Proc. Natl. Acad. Sci. U.S.* 47: 1564(1961).   **I-64.** J. Bonner *et al., Proc. Natl. Acad. Sci. U.S.* 47: 1548(1961).   **I-65.** Y. Tashiro, *J. Biochem. (Tokyo)* 45: 802(1958); A. Kleczkowski and A. van Kammen, *BBA* 53: 181(1961).   **I-66.** A. L. Oparin *et al., Doklady Akad. Nauk S.S.S.R.* 116: 270 (1957).   **I-67.** B. H. J. Hofstee, *Biochem. Biophys. Research Communs.* 4: 9(1961). **I-68.** E. Leduc *et al., Compt. rend.* 250: 2948(1960).   **I-69.** L. F. Cavalieri *et al., Biochem. Biophys. Research Communs.* 1: 124(1959).   **I-70.** J. P. Greenstein *et al., JBC* 182: 607(1950); *J. Natl. Cancer Inst.* 6: 219(1946).   **I-71.** G. Schmidt and S. J. Thannhauser, *JBC* 161: 83(1945).   **I-72.** E. Chargaff, in "Symposium on the Chemical Basis of Heredity" (B. Glass, ed.), p. 521. Johns Hopkins Press, Baltimore, Maryland, 1957.   **I-73.** Q. Van Winkle and W. G. France, *J. Phys. Colloid Chem.* 52: 207 (1948).   **I-74.** J. D. Watson and F. H. C. Crick, *Nature* 171: 737(1953).   **I-75.** D. O. Jordan, in "The Chemistry of Nucleic Acids," p. 265. Methuen, London, 1960. **I-76.** C. A. Knight, *Brookhaven Symposia in Biol.* 13: 232(1960).   **I-77.** S. Kuno and I. R. Lehman, *JBC* 237: 1266(1962).   **I-78.** E. L. Hess *et al., JBC* 236: 3020(1961). **I-79.** A. M. Michelson, *BBA* 55: 841(1962).   **I-80.** G. Zubay and M. H. F. Wilkins, *J. Mol. Biol.* 4: 444(1962).   **I-81.** M. H. F. Wilkins *et al., Nature* 194: 1014(1962).

# Chapter XII

## Enzymes: Proteins with
## Enzymatic Properties

### A. The Function of Proteins as Enzymes and Apoenzymes (A-1)

Enzymes are macromolecular catalysts of biological origin. This definition does not include any statement concerning the protein nature of enzymes. Originally the protein content of enzymes was attributed to their contamination by proteins. Willstätter and his school tried in vain to remove the protein portion of enzyme preparations. It was finally proved by Sumner (A-2) and Northrop (A-3) that urease and pepsin, two crystallized enzymes, were proteins, and that their enzymatic activity was inseparably connected with their protein substance. Nevertheless, we cannot exclude the possibility that protein-free macromolecules might be catalytically active.

Thus it has been claimed that the alkaline phosphatase from hog kidney is free of proteins and amino acids and consists essentially of carbohydrates and pyrimidines (A-4). Although this has not yet been confirmed (A-5), it would not be surprising to find a protein-free phosphatase, since it has been reported that the trivalent ferric and aluminum ions, and particularly the tetravalent cerium and lanthanum ions, can also act as phosphatases (A-6).

The number of enzymes discovered and purified during the last few years has increased at a rapid rate. This is concomitant with the discovery of new metabolic pathways. Each of the new metabolic intermediates is also a new substrate and allows us to search for new enzymes which are able to catalyze its oxidation, reduction, hydrolysis, or any other metabolic reaction. It is important to keep in mind that we cannot detect any enzyme unless we know at least one of its substrates. Indeed, many of the proteins described in the older literature as nonenzymatic proteins were later found to be enzymes. Thus the soluble protein fraction of muscle extracts, which previously was designated as myogen, contains phosphorylase, aldolase, glyceraldehydephosphate dehydrogenase, and many other enzymes, which form at least 40% of the "myogen" (A-23). Since it has been found that the imidazole residues of histidine catalyze to some extent the hydrolysis of nitrophenyl esters, most if not all of the proteins may have some esterase activity. Indeed such activity has been demonstrated in human and bovine serum albumin (A-28).

It has been realized for many years that some of the enzymes possess catalytically active prosthetic groups which can be removed by dialysis or by passing the enzyme solution through a Sephadex column (A-7). If the dialyzable portion of the enzyme is an organic molecule, it is designated as a *coenzyme*. The term coenzyme (A-12) is usually not applied to metal ions although these have frequently the same function as the typical coenzymes. The nondialyzable protein portion of the conjugated enzymes is denoted as *apoenzyme*. Most of the enzymes which catalyze oxidation-reduction have well defined coenzymes and are classified according to the chemical structure of these coenzymes. Many of the enzymes which catalyze hydrolyses and the transfer of acyl, glycosyl, or other residues consist either only of amino acids or contain one of the bivalent metal ions in their active site. In both types of enzymes the protein moiety is essential for the catalytic action and specificity of the enzyme. This is clearly demonstrated by the loss of the enzymatic activity of numerous enzymes after their denaturation by heat or by chemical agents. However, some enzymes of low molecular weight, e.g., ribonuclease, are highly resistant to heat (A-8). These thermostable enzymes undergo no drastic changes of their conformation when their solutions are heated and seem to return to the native conformation on cooling.

If the protein moiety of urease is carefully denatured by the cautious addition of ethanol, the enzyme remains catalytically active (A-9). Likewise, spreading of enzymes in monomolecular films does not always inactivate them (A-10,11).

It is a familiar fact that the apoenzymes are usually much less stable than the holoenzyme (= apoenzyme + coenzyme). Evidently the coenzymes exert a protective action on the apoenzymes as long as they are bound to these. They may act like a clamp which prevents the unfolding of the peptide chains of the apoenzyme thereby preventing its denaturation as well (A-12,26).

Some of the enzymes are dissolved in the body fluids, such as the blood plasma, the gastric juice, and the secretion of the pancreas. Most of these soluble enzymes are typical globular proteins which can be purified by fractionation with ammonium sulfate. Their isolation from other proteins is facilitated by their enzyme action which usually is quite specific and permits the investigator to determine very small amounts of the enzyme (A-24,25). Similar soluble enzymes can be extracted from the cells after destruction of the cell structure by the methods described in Chapter II. The same methods are applied to the extraction of enzymes from the intracellular granules (mitochondria, microsomes) or from nuclei. Most of the hydrolyzing enzymes are of the soluble globular type, whereas many of the enzymes involved in oxidation-reduction are bound to the cellular structure and cannot be easily extracted. Some of these enzymes can be solubilized by treatment of the homogenates with cholate, deoxycholate, or with certain detergents which form water-soluble complexes with lipids (A-13 to 15). This indicates that lipids are involved in the binding of enzyme proteins to the intracellular structures. The cleavage of ATP by myosin demonstrates that enzymatic activity may also be displayed by fibrous proteins (A-16).

The colloidal or insoluble state of the enzymes seems to be essential for their action in the cells. If the molecular weights of the enzymes were very low, some of them might diffuse through the cellular membranes and would thus leak out of the cells. A further consequence of the insoluble or colloidal state of the enzymes is the presence of a boundary between the aqueous solutions or body fluids on the one hand, and the insoluble or colloidal protein particle or molecule on the other. The heterogeneous character of such a two-phase system frequently enhances catalytic actions, since substrate molecules in the phase boundary undergo adsorption and orientation of their molecules, and, also quite often, induction of electron shifts, resonance effects, and bond distortion (A-17). Thus, linoleic acid is rapidly oxidized by $O_2$ in the water-oil interphase in the presence of hemin, but is not oxidized when the same system is made homogeneous by the addition of ethanol or of bile salts (A-18).

Although the prosthetic group (coenzyme) of the conjugated enzymes frequently has a slight catalytic activity, the protein moiety (apoenzyme) is essential for the full activity of the enzymes. This is due partly to the high molecular weight of the proteins and their ability to create a system of heterogeneous phases, and partly to a specific action of the apoenzymes which cannot be replaced by other proteins. Each apoenzyme is specific in that it combines only with a certain type of coenzyme and thus endows the coenzyme with the ability to catalyze only specific reactions. On the other hand, many of the coenzymes can combine with various apoenzymes. Thus hemoglobin, catalase, and peroxidase contain the same protoheme as active group, but different apoenzymes. Likewise, we know a great many apoenzymes with which NAD (=DPN) can combine. The mode of linkage between coenzyme and apoenzyme differs in various enzymes. In some of them, strong covalent bonds are involved; in others, we find very labile linkage by electrostatic attraction or by van der Waals forces. Cytochrome c is an example of covalent bonding between coenzyme and apoenzyme; its protein moiety is bound to the protoheme residue by thioether bonds between the sulfhydryl of cysteine residues and the unsaturated vinyl groups of protoheme

$$\text{—CH:CH}_2 + \text{HS·R} \rightarrow \text{—CH}_2\text{·CH}_2\text{·S·R}$$

The thioether bond is very resistant to heat, acids, and other reagents, but can be split by treatment with silver salts (A-19). The dehydrogenases which contain DPN or TPN are examples of a very labile bond between apo- and coenzymes. They dissociate spontaneously into their two components and can be obtained free of coenzyme by dialysis.

The apoenzyme may combine not only with the coenzyme but also with the substrate. This may be a short, temporary combination, but may be essential for the specificity of the catalytic action. The bivalent metals which have been found in many enzyme proteins act as coordination centers of a complex in which both the apoenzyme and the substrate are bound to the metal (A-20,27). In the resulting substrate-metal-protein complex, the substrate is brought so close to the active group of the protein that the catalytic action can easily take place.

The catalytic activity of enzymes has given rise to speculations on the formation of excited electronic states of amino acid residues in their peptide chains or in $\alpha$-helices. There is very little evidence for such excited states (A-17), or for semiconductivity (A-21) or long-range forces (A-22).

Although the amino acid analysis of numerous enzymes and apoenzymes has demonstrated that they consist only of proteins, we have to be cautious in excluding the presence of nonprotein material. Investigations during the last few years have revealed in many of the hydrolyzing enzymes the presence of small amounts of metal ions, particularly zinc and manganese. If a protein molecule of the molecular weight 80,000 contains 1 or 2 zinc ions, their percentage in the protein amounts to only one- or two-tenths % and thus does not significantly affect the results of the amino acid analysis. The inactivation of amylase by dialysis and its reactivation by the addition of various chlorides demonstrates that even chloride ions may function as activators, although their mode of action may be different from that of the typical coenzymes. Evidently our differentiation between activators and coenzymes is to a certain extent arbitrary.


## B. *The Active Site of Enzymes (B-1)*

The idea of a limited active site in enzyme proteins is very old. It is based on an analogy to the relatively small size of the coenzymes in the large conjugated enzymes. The proof that the active site of at least some of the enzymes is indeed very small has been conclusively established only recently by the demonstration that chymotrypsin is inactivated by the combination with a single diisopropyl-phosphoryl group (B-2). The latter is supplied by diisopropyl-fluorophosphate (DFP or diisopropyl-phospho-fluoridate). Other organic phosphates can replace DFP. If an organic phosphate labeled by $P^{32}$ is used, the enzyme combines with *one* equivalent of $P^{32}$. Partial hydrolysis of the acylated chymotrypsin yields a radioactive peptide in which the diisopropylphosphoryl residue is bound to the side chain —$CH_2OH$ of serine in the sequence —Gly.Asp.Ser.Gly— (B-3).

It is highly significant that the same sequence has been found in trypsin, thrombin, and elastase, and a similar sequence, —Gly.Glu.Ser.Ala—, in the active site of liver aliesterase and pseudocholine esterase (B-4 to 6). In phosphoglucomutase the active site is —Thr.Ala.Ser.His.Asp— (B-7). In all these enzymes the alcohol group of serine is the acyl binding site. The reaction of chymotrypsin with DFP not only demonstrates the small size of the reactive site of the enzyme but also reveals its mode of action, namely the formation of acyl-enzyme as essential intermediate in the enzyme-substrate interaction. Thus, if chymotrypsin acts on $p$-nitro-phenyl acetate, acetyl-chymotrypsin is formed and $p$-nitrophenol liber-ated (B-8). The reaction can be investigated photometrically since $p$-nitrophenol, in contrast to its acetate, is intensely colored. In the presence of ethanol, part of the acetyl residues of acetyl-chymotrypsin are trans-ferred to ethanol and thus give rise to the formation of ethylacetate (B-9).

In model experiments it has been found that $p$-nitrophenyl acetate is also hydrolyzed by imidazole derivatives, for instance by histidyl peptides or histidine polymers (B-10,11). The *participation of imidazole* derivatives is also indicated by the pH dependence of the enzyme action which paral-lels the pH dependence of the dissociation of the imidazole group. It has been concluded, therefore, that an imidazole group in the vicinity of the serine residue takes part in the enzyme action. As shown above, such an imidazole residue is linked to serine in phosphoglucomutase (B-7). Various reaction mechanisms have been suggested (B-12 to 14,38). One of these is shown in Fig. XII-1. Most probably, the acyl ester of serine, which is formed as an intermediate, is hydrolyzed by the catalytic action of the adjacent imidazole residue. Such a *cooperative action* of adjacent groups requires a rigid conformation of the peptide chains in the active site. The fluctuation of charges in the protein molecule (see Chapter VI, B) may cause limited distortions in the otherwise rigid peptide chains and may in this manner enhance the catalytic action of the imidazole residues. Transfer of phosphate residues in phosphoglucomutase may occur in an analogous manner, leading first to the formation of a phosphorylated en-zyme, then to release of the phosphate by hydrolysis (see Section G of this chapter).

All these observations indicate the formation of a short-lived but demonstrable acyl-enzyme and the subsequent transfer of the acyl residue to the molecule of water or to another acyl acceptor. Not all of the hydrolyzing enzymes contain serine in their active site. No acyl-binding seryl residue has been detected in ribonuclease in which two histidyl residues may act as proton donor and acceptor, respectively (B-15). The proteolytic enzymes which contain serine in their active site have been designated serine proteinases (B-16). They include chymotrypsin, trypsin, subtilisin,

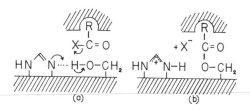

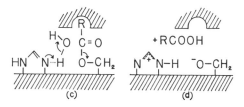

FIG. XII-1. Chymotrypsin-catalyzed reaction involving general basic catalysis by imidazolyl in both the acylation and deacylation steps. From Sturtevant (B-32).

and thrombin. Serine phosphate has also been found in the active site of human acid phosphatase(B-34). Very little is known about the role of tryptophan in the active sites of trypsin, chymotrypsin (B-35), lysozyme (B-36), and $\alpha$-amylase (B-37).

In numerous enzymes the active site contains either a *sulfhydryl group* of cysteine or a bivalent *metal ion*. The latter is frequently bound to the sulfur atom of a cysteine residue. The enzymes which contain thiol groups in their active site are inhibited by *p*-chloromercuribenzoate (PCMB), heavy metal ions, particularly zinc and cadmium, by the alkylating agents iodoacetamide and *N*-ethylmaleimide, and by iodosobenzoate (B-17). The last-mentioned reagent oxidizes the SH groups. Papain, ficin, some of the cathepsins, and proteolytic enzymes of plants and molds belong to this group of thiol enzymes. The thiol enzymes also include some of the dehydrogenases, such as 3-phosphoglyceraldehyde dehydrogenase (see Section H, this chapter). In a large number of enzymes, zinc has been found to be present and essential for the enzymatic activity. They include carboxypeptidase (B-18), carbonic anhydrase, yeast alcohol dehydrogenase (B-19), liver alcohol dehydrogenase (B-20), lactic acid dehydrogenase (B-21), and glutamic acid dehydrogenase (B-22). All of these enzymes or apoenzymes are inhibited by 1,10-phenanthroline, which forms chelates with zinc and other heavy metal ions. Consequently, it is assumed that the metal is an essential part of the active site of the enzyme (B-23). Many of the peptidases are activated by bivalent metal ions such as $Co^{++}$, $Mn^{++}$, and $Zn^{++}$. Manganese is required for the activation of prolinase and prolidase, enzymes which catalyze the hydrolysis of the peptide bonds formed by the imino or carboxyl group of proline (B-24). Animal and vegetable amylases have been found to contain 1 calcium atom per molecule (B-25). Calcium stabilizes these enzymes and protects them against proteolytic enzymes. Ethylenediamine tetraacetate (EDTA) inhibits the amylases by its chelating action on the calcium ions. Magnesium ions are essential for the action of alkaline phosphatase (B-26).

It has been known for a long time that some of the oxidases contain copper ions. Copper has been detected in laccase, uricase, and also in cytochrome c oxidase. It is not yet quite clear whether the metal acts as an electron carrier, being alternately reduced to the cuprous and reoxidized to the cupric state, or whether it acts merely as a coordination center in analogy to the action of zinc ions. A change in valence is indicated by changes in the electron spin resonance after addition of an excess of the substrate *in vitro* to laccase (B-27,28) or cytochrome c oxidase (B-29). In uricase the cupric ion forms the coordination center of a ternary complex of enzyme, uric acid, and oxygen; it seems to remain in the cupric state during the catalytic action (B-30). In xanthine oxidase and some of the flavoproteins molybdenum ions act as coordination centers. Investigation of the electron spin resonance during the action of xanthine oxidase yields a signal which corresponds to a valence change of molybdenum (B-31). Manganese has been found in enolase and creatine phosphokinase; the metal seems to form a coordination center with the enzyme, the substrates, and water as ligands (B-33).

The preceding paragraphs are based on the assumption that each enzyme has a localized active site which is responsible for the catalytic action. Although this is true for some of the enzymes, attempts to find a localized active site in other enzymes have failed. Thus, different parts of the folded peptide chain which forms the ribonuclease molecule are responsible for the catalytic action of ribonuclease. If any of these parts of the molecule is altered by substitution or scission, the catalytic activity of ribonuclease is impaired or destroyed. Evidently the three-dimensional structure of the ribonuclease molecule is essential for its catalytic activity. We must conclude that different parts of the molecule cooperate in bringing about the catalytic action by "proximity effects" or "orientation effects" (B-1). Since the ribonuclease molecule consists only of a single peptide chain and since no coenzyme is known, the active site must be formed by amino acid residues which are arranged in such a way that their surface fits closely the surface of the substrate. We still do not know which of the amino acids of ribonuclease are the "contact amino acids" and which are merely "auxiliary amino acids" (B-1). The picture which emerges from these considerations is not limited to those enzymes which consist only of amino acids. The conformation of the peptide chain is also responsible for the specific combination of apoenzymes such as the dehydrogenases with the pyridine nucleotides or with flavins.

## C. Specificity of Enzymes

Enzymes possess two types of specificity: (a) substrate specificity and (b) specificity of origin. *Substrate specificity* is very narrow in some enzymes, very wide in others. Urease has an extremely high specificity for urea and cannot hydrolyze any other urea derivative, whereas leucine-aminopeptidase can catalyze the hydrolysis of very different peptide bonds. In Chapter IV, the great importance of the specificity of proteolytic enzymes for the elucidation of amino acid sequences has been emphasized. We find in this respect great differences between various proteolytic enzymes (C-1). The high substrate specificity of trypsin, which hydrolyzes only peptide bonds formed by the carboxyl groups of the basic amino acids lysine and arginine, has been of great importance in the clarification of amino acid sequences. Chymotrypsin, on the other hand, hydrolyzes preferentially peptide bonds formed by the aromatic amino acyl residues, but is not as specific as trypsin, since it also catalyzes the hydrolysis of peptide bonds formed by the carboxyl groups of methionine, asparagine, glutamine, leucine, and other amino acids (C-2). Both enzymes, trypsin as well as chymotrypsin, are inhibited by DFP and contain a seryl residue in their active group. Evidently, their specificity is determined by amino acyl residues in the immediate environment of the active group. We do not yet know the structures which are responsible for the specificity of these two enzymes. One might speculate that chymotrypsin contains some aromatic residues in the vicinity of its seryl residue and that these combine by van der Waals forces with aromatic residues of the substrate and thus labilize the peptide bond formed by the aromatic amino acids of the substrate. The affinity of trypsin to the basic amino acid side chains of the substrate might be due to acidic groups in the vicinity of the catalytically active site of trypsin.

The older classification of enzymes was based on their substrate specificity since nothing was known about the protein nature, let alone the composition, of the active site of the enzymes. Thus the hydrolyzing enzymes were classified as proteinases, carbohydrases, and esterases. In the meantime it has been found, however, that both trypsin and chymotrypsin catalyze the hydrolysis not only of peptide bonds but also of ester bonds (C-3). The hydrolysis of esters actually proceeds more rapidly than that of peptide bonds. One of the simplest substrates of trypsin, which is rapidly hydrolyzed, is toluenesulfonyl-arginine methyl ester (TAME). Chymotrypsin can hydrolyze not only peptide and ester bonds, but even the 2,3-carbon-carbon bond in the ethyl ester of 5-hydroxyphenyl-3-ketovaleric acid (C-4):

$$HO \cdot C_6H_4 \cdot CH_2 \cdot CH_2 \cdot CO \cdot CH_2 \cdot CO \cdot OC_2H_5 \rightarrow$$
$$HO \cdot C_6H_4 \cdot CH_2 \cdot CH_2 \cdot COOH + CH_3 \cdot CO \cdot OC_2H_5$$

These experiments have severely shaken the basis of the older enzyme classification. From the structure of the substrates of trypsin and chymotrypsin it is evident that for the action of these enzymes the steric arrangement and the shape of the substrate molecules are more important than the chemical nature of the bond which is hydrolyzed by the enzyme. If the spatial arrangement, the shape, and the configuration of enzyme and substrate allow the two molecules to come very close to each other, the activation exerted by the active site of the enzyme may be so powerful that the hydrolysis not only of C-N or C-O bonds, but also that of C-C bonds is rendered possible. The great importance of steric arrangement and configuration of the substrate for the formation of the enzyme-substrate complex is demonstrated by the high stereospecificity of most enzymes. Most of the proteolytic enzymes hydrolyze only peptide bonds formed by L-amino acids, not those formed by D-amino acids. Likewise, L-amino acid oxidase catalyzes the oxidation of L- but not of D-amino acids. The latter are oxidized by another enzyme, D-amino acid oxidase. The differentiation by enzymes of the D- from the L-configuration is not surprising since the enzyme proteins, owing to their content of L-amino acids and their specific conformation, are stereospecific.

The great importance of stereospecificity for the combination of enzyme with substrate can be demonstrated in model experiments. Let us assume that the combining site of the enzyme is an area (Fig. XII-2) indicated by ABC where the three letters stand for different chemical groups of a molecule or for parts of different molecules. Let the combining surface of the substrate be formed by the three groups X, Y, and Z, all three linked to an asymmetric carbon atom. Let us further assume that combination and catalysis can only occur if A combines with X, B with Y, and C with Z. This is obviously impossible if the L compound, shown in Fig. XII-2 at the left, is replaced by its D stereoisomer shown at the right side. It is not necessary to postulate the three points of attachment between enzyme and substrate as shown in Fig. XII-2. Even if only two-point attachment

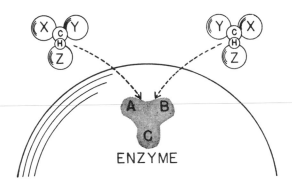

Fig. XII-2. Stereospecificity of enzymes. The large sphere represents the enzyme, the shaded area its combining site. The substrate is shown at the left side, its stereo-isomer at the right side. (C-6).

occurs, the enzyme-substrate complex formed by an L-substrate, because of the stereospecific structure of the enzyme, differs from the complex formed by the D-isomer (C-5).

The great influence of optically active substances present during a catalytic reaction on its stereospecific course has been demonstrated in a classical experiment by Bredig (C-7) who found that benzaldehyde, $C_6H_5$·CHO, and HCN, in the presence of an optically active base, combine asymmetrically to form an optically active cyanhydrin, $C_6H_5$·C*HOH·CN, which on hydrolysis gives optically active mandelic acid, $C_6H_5$·C*HOH·COOH. A complex formed by the asymmetric base and the cyanhydrin has been postulated as an intermediate (C-8). Similarly, optically active products are obtained from racemic mandelic acid by hydrogenation with platinum when the solvent contains gum arabic, an asymmetric polysaccharide (C-9).

The phenomena discussed in the preceding paragraphs pertain to the substrate specificity of enzyme proteins. In addition to substrate specificity, enzymes also exhibit *species specificity* and *organ specificity*. In other words, enzymes with an identical type of activity may differ from each other by the structure of their proteins. Thus, the alcohol dehydrogenase of yeast is different from that prepared from the liver of rats, and the latter again different from that of other animals.

Five different forms of lactic acid dehydrogenases have been isolated from the rabbit and five from the human organism (C-10). The pancreatic α-amylase of hogs is serologically identical with that found in saliva and the serum, but different from hepatic α-amylase (C-12). Genetically determined different forms of glucose-6-phosphate dehydrogenase have been found in human mutants; these enzymes differ from each other in their Michaelis constants and their pH optima (C-13). Enzymes of this type which have identical active groups and identical functional activities but varied protein moieties have been designated as *isozymes* (C-11,18). Since the amino acid sequence of the isozymes mentioned in the preceding paragraph are not yet known, it cannot be decided whether the isozymes differ from each other in their amino acid composition, in the amino acid sequence, in conformation, or in more than one of these factors. The amino acid sequence of bovine ribonuclease, as shown later in this chapter, has been completely unraveled. Ovine ribonuclease differs from the bovine enzyme merely by small changes in the amino acid sequence: the serine residue in position No. 3 has been replaced by threonine, a glutamyl residue (No. 37) by lysine, and the sequence of amino acids Nos. 99–104 by another sequence (C-14).

It has been known for a long time that formation of apparently new types of enzymes can be brought about in bacteria or molds by offering them an unusual substrate as the only nutrient. Thus microorganisms which are not able to ferment galactosides, produce a galactosidase when they are grown on a medium containing galactosides. Originally this had been attributed to the formation of a new type of enzyme, called "adaptive enzyme." However, formation of galactosidase can also be brought about by thiomethyl-galactoside, which cannot by hydrolyzed or fermented by microorganisms. Evidently, we are not dealing here with an adaptation of the microorganisms to a new nutrient but rather with the *induction* of enzyme formation by an inducer molecule which may, but need not, be a substrate (C-15). In all instances investigated it was found that very small amounts of the "induced" enzyme are present even before induction and that the inducer merely increases the rate of enzyme formation. Thus the induced penicillinase, induced by penicillin in *Bacillus cereus*, crystallizes in the same way as the small amounts of constitutive penicillinase and does not differ from the constitutive enzyme in any property (C-16). In induced *E. coli*, aspartic transcarbamylase constitutes almost 7 % of the total bacterial protein, but it is found in wild *E. coli* only in trace amounts of approximately 50 molecules per cell (C-17). The rate of synthesis of the enzyme in the induced cells is about 2,000 times faster than in the wild type cells. This is attributed to the presence of a repressor in the wild type cells, and to release of the repression by added inducer (C-17). The experiments on enzyme induction are extremely valuable for an understanding of protein biosynthesis, since enzyme proteins, by reason of their catalytic activity, can be detected in much smaller amounts than proteins which are devoid of catalytic actions.

## D. The Enzyme-Substrate Complex (D-1)

Our concepts concerning chemical reactions and catalysis are based on the supposition that *collisions* between reacting moelcules are a prerequisite of chemical reaction. According to this concept the substances A and B cannot react to give C and/or D unless the complex AB is formed at least temporarily. Similarly, we must postulate that the enzyme-substrate complex, usually denoted by ES, is unstable, so that it is very short-lived. If the enzyme combines with a substance to give a stable complex, this substance does not undergo any further reaction. Some of the enzyme inhibitors combine in this manner irreversibly with the enzymes. If the inhibitor is reversibly bound to the same site as the substrate of the respective enzyme, or if it prevents the formation of ES, we call it a *competitive inhibitor*, since it competes with the substrate for the active site.

It has been mentioned earlier that DFP is bound to the same seryl residue of chymotrypsin to which acyl residues of various substrates can be bound. Hence DFP inhibits specifically acyl transfer. DFP and similar organophosphorus compounds are considered as enzyme poisons. Since the acyl residues inhibit enzyme action, they protect proteolytic enzymes against autodigestion. Thus trypsin can be stabilized by acylation with succinic acid (D-2). The action of enzymes can be inhibited not only by specific inhibitors, such as DFP, but also by the reaction products of the catalytic reaction. If we represent the hydrolytic cleavage of a substrate AB by the equation

$$AB + H_2O \rightleftharpoons A \cdot H + B \cdot OH$$

the enzyme-substrate complex is represented by the symbol E(AB) where E stands for the enzyme part of the complex. If the hydrolysis is inhibited by an excess of A·H, but not by B·OH, we may safely assume that E combines with the A moiety of AB and not with the B moiety. Thus the hydrolysis of sucrose by yeast invertase is inhibited by fructose and not by glucose. This indicates that the enzyme combines with the fructose moiety of sucrose and that the first phase of the enzymatic reaction is represented by the formation of E-AB where A is the fructosyl and B the glucosyl residue of sucrose (D-3,4).

The assumption of *enzyme-substrate complexes* was based originally on the results of kinetic experiments (see below, Section E). If, however, the enzyme is colored and has a measurable absorption spectrum, the formation of the ES complex usually involves a change in the absorption spectrum which can be measured photometrically. In this manner the enzyme-substrate complexes formed by the combination of catalase or peroxidase with $H_2O_2$ have been directly "seen." Both enzymes undergo typical changes in their color and their absorbancy when they combine with hydrogen peroxide. At first a green complex is formed which later is converted into a red complex (D-5). Similarly, the formation of colorless complexes from apodehydrogenase and DPN or TPN can be determined in this manner since the absorption as well as the fluorescence of the pyridine nucleotides undergo significant changes when they combine with the apoenzyme. Formation of the apoenzyme-DPN complex can also be proved by measuring the depolarization of polarized ultraviolet light (see Chapter V, J). Whereas the free pyridine nucleotide molecules, because of their small size and rapid rotations, depolarize polarized ultraviolet light and emit depolarized fluorescence, the protein-bound pyridine nucleotides, owing to the large size of their protein carrier and to its slow rotational motion, do not depolarize the incident polarized ultraviolet light and emit polarized fluorescent light. The extent of polarization of the emitted light is a measure of protein-bound coenzyme (D-6). If the absorbancy of an enzyme-substrate complex or an apoenzyme-coenzyme complex is the same as that of the free enzyme, the concentration of the complexes cannot be determined by photometric methods. The concentration of the enzyme-substrate complex can then be determined by the indirect method of kinetic analysis, discussed in Section E, below.

In the preceding paragraphs it has been pointed out that the enzyme or apoenzyme protein combines not only with its substrate but, if the coenzyme-apoenzyme bond is dissociable, also with the coenzyme. In such a case a *ternary complex* of apoenzyme, substrate, and coenzyme is formed. If the catalyzed reaction is reversible, more complicated complexes are formed. Thus, glutamic acid dehydrogenase, which converts glutamic acid into $\alpha$-ketoglutaric acid and ammonia, combines not only with TPN and glutamic acid but also with $\alpha$-ketoglutarate and ammonium ions in a definite compulsory order (D-7). Frequently the ES complex also combines with hydrogen ions. In this case the rate of the catalytic reaction depends on the hydrogen ion concentration; the reactive intermediate is here $ESH^+$ and not ES (D-8).

In the activated enzyme-substrate complex the substrate sometimes may be only slightly distorted, or its electrons may have undergone a slight shift due to an exchange of resonance energy. In some instances however, the substrate becomes bound to the enzyme by covalent bonds, as shown by the binding of acyl residues to the active seryl group of the serine proteinases (Section B, this chapter). It has been assumed for a long time that enzyme and substrate fit to each other like lock and key. In this comparison only the steric arrangement of the two components has been taken into consideration, not the change which the substrate undergoes as a result of the combination with the enzyme. However, we must assume mutual interaction of the two partners, not only "activation" of the substrate, but also a slight deformation of the enzyme and/or an alteration in the distribution of its electrons. The enzyme must not be considered as a rigid protein rack but rather as a "dynamic rack" (D-9). The changes in the enzyme structure which are induced by the action of the substrate have been designated as "induced fit" and considered in some instances as essential for the action and specificity of the enzyme (D-10).

The forces involved in the formation of the enzyme-substrate complexes are the *short-range forces* which are also responsible for the formation of other complexes of proteins with nonproteins. They have been discussed in Chapter X. Mutual attraction of charged groups of opposite sign, hydrogen bond formation, van der Waals forces, and hydrophobic bonds (see Chapter VII, E) may contribute to the formation of the enzyme-substrate complex. The specificity of this combination is based principally on steric prerequisites, a certain complementariness in shape which allows the substrate to approach the active site of the enzyme so closely that water is displaced and mutual attraction by means of short-range forces becomes possible.

Although the catalytic action of platinum, palladium, nickel, and other metals in oxidation-reduction has been known for a long time, *models for the enzymatic hydrolysis* were not available. More recently it has been found, however, that organic sulfonic acids and also sulfonate resins are quite active as catalysts in the hydrolysis not only of amides but also of peptides and proteins. Thus proteins undergo partial hydrolysis with very dilute dodecylsulfonate at 65°C (D-11,12). Similarly glycyl-glycine is hydrolyzed by Dowex 50, a sulfonate resin (D-13). The molar concentrations of these sulfonates are much lower than those of sulfuric acid required for the same hydrolytic action. Evidently, the sulfonate groups are "activated" by the organic skeleton of the resin, which can be compared with the apoenzyme. The high efficiency of the polysulfonate resins may be due to the accumulation of numerous sulfonate groups within a restricted space on the surface of the resin and to their immobility in the resin. Similarly an accumulation of acidic or basic amino acids within a limited part of a protein molecule might enhance the ability of such a protein to catalyze the hydrolysis of certain substrates.

It is evident from these considerations that the combination of the enzyme with the substrate depends mainly on the conformation and the structural arrangement of the reacting groups. Denaturation of the enzyme alters its conformation and may thus prevent approach and binding of the substrate. Alternately, denaturation may destroy the active site of the enzyme. However, some of the enzymes of low molecular weight are remarkably resistant to denaturation. Thus ribonuclease is rather resistant to heat and exerts its enzymatic activity even after exposure to 8 *M* urea (D-14).

## E. Kinetics of Enzyme Reactions (E-1)

The conversion of a substrate $S$ into a product $P$ by means of an enzyme $E$ can be represented by the following series of reactions

$$E + S \underset{k_2}{\overset{k_1}{\rightleftharpoons}} ES \overset{k_3}{\rightarrow} E + P \tag{XII-1}$$

where $E$ is the free enzyme, $ES$ is the enzyme-substrate complex, and $k_1$, $k_2$, and $k_3$ are velocity constants. Let us first consider the combination of enzyme with substrate: $E + S = ES$. The velocity, $v_1$, depends on the molar concentrations of $E$ and $S$ and is defined by the equation $v_1 = k_1[E][S]$, where $k_1$ is the velocity constant and $[E]$ and $[S]$ are molar concentrations of $E$ and $S$, respectively. Similarly we can write $v_2 = k_2[ES]$. At equilibrium $v_1$ is equal to $v_2$. Hence we write $k_1[E][S] = k_2[ES]$, and:

$$\frac{k_2}{k_1} = \frac{[E][S]}{[ES]} = K_S \tag{XII-2}$$

where $K_S$ is the dissociation constant of the enzyme-substrate complex. However, since part of $ES$ is converted into $E + P$, we are not dealing with a true equilibrium. The velocity of the conversion of $ES$ into $E + P$ is $v_3 = k_3[ES]$. The rate of decomposition of $ES$ is, therefore, not $k_2$ but is $k_2 + k_3$. The term $(k_2 + k_3)/k_1$ is denoted as the Michaelis constant and is represented by the symbol $K_M$. The relation between $K_S$ and $K_M$ is expressed by the equation $K_M = K_S + k_3/k_1$. Usually the velocity of the reaction $ES \rightarrow E + P$ is much smaller than $v_1$ and $v_2$. Consequently $V$, the over-all velocity of the catalyzed reaction, is equal to $v_3$. Under these conditions $k_3$ is very small and $K_M$ almost identical with $K_S$.

The concentrations of $E$ and $ES$ can rarely be measured directly. However, $K_M$ can be determined without knowing $[E]$ or $[ES]$, as shown by the following considerations. Let the molar concentration of the total enzyme be $[E_t]$, that of the free enzyme $[E]$, and that of the enzyme-substrate complex $[ES]$. Then $[E] = [E_t] - [ES]$. Substituting this term in Eq. (XII-2) we obtain

$$K_S = \left(\frac{[E_t] - [ES]}{[ES]}\right)[S] = \frac{[E_t][S]}{[ES]} - [S] = \left(\frac{[E_t]}{[ES]} - 1\right)[S] \tag{XII-3}$$

The velocity of the over-all reaction increases with increasing concentrations of the substrate because this causes conversion of free enzyme into $ES$ and an increase in $v_3$ which is proportional to $[ES]$. The maximum velocity, $V_{max}$, is attained when all of the enzyme is converted into $ES$, and $[ES] = [E_t]$. The velocity $V$ of (XII-1), as mentioned above, is determined by the slow reaction $ES \to E + P$ and is therefore equal or close to $v_3$. Accordingly we can write $V = k_3[ES]$. Remembering that the introduction of $k_3$ requires replacement of $K_S$ by $K_M$, we obtain from (XII-3) the analogous Eq. (XII-4)

$$K_M = \left(\frac{[E_t]}{[ES]} - 1\right)[S] \qquad \text{(XII-4)}$$

Since the velocity $V = k_3[ES]$, the maximum velocity, when $[ES] = [E_t]$,

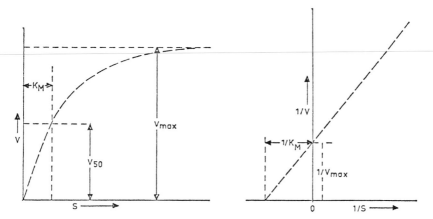

FIG. XII-3. Determination of $K_M$. Left: by plotting $V$ against $[S]$. Right: by plotting $1/V$ against $1/[S]$. The value $1/K_M$ has a negative sign. From F. Haurowitz, "Progress in Biochemistry since 1949," p. 72. Interscience, New York, 1959.

is $V_{\max} = k_3[E_t]$. Accordingly $V_{\max}/V = k_3[E_t]/k_3[ES] = [E_t]/[ES]$. Substituting $V_{\max}/V$ for $[E_t]/[ES]$ in Eq. (XII-4), we finally obtain

$$K_M = \left(\frac{V_{\max}}{V} - 1\right)[S] = \frac{V_{\max}[S]}{V} - [S] \qquad \text{(XII-5)}$$

The importance of these correlations was first stressed by Michaelis (E-2). $K_M$ can be determined from Eq. (XII-5) by measuring $V$ at different concentrations of the substrate $S$. When $V$ is one-half of $V_{\max}$, the ratio $V_{\max}/V$ becomes 2, and $K_M = (2 - 1)[S] = [S]$. In other words, $K_M$ is equal to the molar concentration of the substrate when $V$ is one-half of the maximum velocity. This is shown in Fig. XII-3. It is difficult to determine $K_M$ in this manner since the velocity $V$ increases asymptotically and extrapolation to the maximum value is inaccurate. Better results are obtained when Eq. (XII-5) is written in the form

$$K_M + [S] = \frac{V_{\max}[S]}{V}$$

and transformed into Eq. (XII-6):

$$V = \frac{V_{\max}[S]}{K_M + [S]} \tag{XII-6}$$

Taking the reciprocal values we obtain Eq. (XII-7):

$$\frac{1}{V} = \frac{K_M + [S]}{V_{\max}[S]} = \frac{K_M}{V_{\max}[S]} + \frac{1}{V_{\max}} \tag{XII-7}$$

If $1/V$ is plotted against $1/[S]$, a straight line is obtained whose slope is equal to $K_M/V_{\max}$. The intercept of this line with the ordinate gives $1/V_{\max}$ (E-3). Hence $K_M$ can be computed from the slope.

Equation (XII-4) shows that $K_M$ is high when the concentration of $ES$ is low and vice versa. When $[ES]$ is small and $K_M$ high, $V$ increases with increasing concentrations of $S$ and the reaction proceeds as a first-order reaction at low substrate concentrations. If, on the other hand, $[S]$ is sufficiently high, and practically all of the enzyme is converted to $ES$, the concentrations of $ES$ and $V$ remain nearly constant, and we are then dealing with a zero-order reaction (E-4). Most catalyses are neither strictly first-order reactions nor strictly zero-order, but proceed according to some intermediary kinetics (E-5).

In the treatment given above certain simplifying assumptions have been made. One of these is the disregard of $k_4$, the velocity constant of the reaction $E + P \rightarrow ES$. If the enzyme catalyzes a reaction in which the conversion of $S$ into $P$ is readily reversible and where the equilibrium constant for $S \rightleftharpoons P$ is not very far from unity, $v_4$ and $k_4$ must be taken into consideration and a more complicated treatment is necessary. If the solution contains substances which are able to combine with the enzyme, a competition for the enzyme ensues between these substances and the substrate. Accordingly these substances will act as *inhibitors*. The kinetics of the enzyme reaction in the presence of inhibitors has been calculated in the same way as the kinetics of the enzyme-substrate reaction (E-3). Reactions of a higher order result when the enzyme reacts with 2 or more substrates or products (E-13).

When we compare the enzymatic hydrolysis with the analogous reactions brought about *in vitro* by the action of strong acids or bases, we find that the enzymatic reactions proceed at much lower temperatures. The reaction path in enzymatic hydrolysis is different from that in hydrolysis by acids or bases. The activation energy can be calculated for both types of hydrolysis from the reaction rates found at different temperatures as shown in Chapter VII, Section I. Calculations of this type lead to values of approximately 10–13 kcal/mole for the acid- or base-catalyzed ester hydrolysis, and about 20 kcal/mole for the hydrolysis of peptide bonds. The analogous values for enzyme-catalyzed hydrolyses are about 4 and 12–14 kcal/mole, respectively (E-6).

The activation process can be treated from the point of view of thermodynamics like any other chemical reaction. If the symbols $\Delta H^{\ddagger}$, $\Delta F^{\ddagger}$ and $\Delta S^{\ddagger}$ are used for the heat, the free energy, and the entropy of activation, respectively, we can write in analogy to Eq. (VII-12): $\Delta F^{\ddagger} = \Delta H^{\ddagger} - T\,\Delta S^{\ddagger}$. Hence, the free energy of activation decreases when either $\Delta H^{\ddagger}$ decreases or when the entropy of activation, $\Delta S^{\ddagger}$, increases. We have every reason to assume that enzyme action involves an increase in the entropy of activation. Although the thermodynamic analysis provides us with quantitative data, it does not give us any insight into the mechanism of the enzymatic catalysis. Such insight can be gained chiefly by the kinetic method.

The rate of the enzymatic catalysis depends not only on the concentrations of enzyme, substrate, and enzyme-substrate complex, but also on that of the hydrogen ions and of various activators or inhibitors. Dependence on pH indicates that ionizable groups of the enzyme or substrate are involved in the activation process. The conditions for kinetic analysis are simplified to a certain extent when the investigated reaction is of the type $AB + A^* \rightleftharpoons A^*B + A$, where $A^*$ is an isotopically labeled form of $A$ which contains, e.g., $N^{15}$ instead of $N^{14}$, or radioactive $C^{14}$ instead of $C^{12}$. Reactions of this type have been designated as *virtual reactions* (E-7). Thus, if Cbo-tyrosyl-glycinamide (Cbo = carbobenzoxy— = $C_6H_5 \cdot CH_2O \cdot CO$—) is incubated with chymotrypsin in the presence of isotopically labeled glycinamide, exchange of the glycinamide residues can be measured and the rate of this reaction determined (E-8). Similarly it can be proved that Cbo-phenylalanine in a solution of $H_2O^{18}$ incorporates isotopically labeled OH groups into its carboxyl group (E-9):

$$C_6H_5 \cdot CH_2O \cdot CO \cdot NH \cdot \underset{\underset{\underset{C_6H_5}{|}}{\overset{|}{CH_2}}}{CH} \cdot C \overset{O}{\underset{OH}{\diagup}} + H_2O^{18} \rightleftharpoons$$

$$C_6H_5 \cdot CH_2O \cdot CO \cdot NH \cdot \underset{\underset{\underset{C_6H_5}{|}}{\overset{|}{CH_2}}}{CH} \cdot C \overset{O}{\underset{O^{18}H}{\diagup}} + H_2O \quad \text{(XII-8)}$$

Equation (XII-8) reveals clearly the essential mechanism of the chymotrypsin action, namely the activation of the carboxyl group which makes
the OH group labile and exchangeable. Since chymotrypsin combines
readily with acyl groups to form stable acylchymotrypsin derivatives (see
Section B, this chapter), there can be no doubt but that the activation of
Cbo-phenylalanine by chymotrypsin consists in a labilization of the bond
between C and O. The simplest formulation would be to represent the
Cbo-phenylalanyl residue as a positively charged carbonium ion and a
negatively charged hydroxyl ion. Evidently, a hydroxyl ion, $HO^-$, is
liberated and then exchanged for $HO^{18}$. However, we do not know whether
activation results in the formation of a free carbonium ion of the activated
amino acid. More probably, the carbon atom of the labilized C—O bond
combines with the seryl—OH group of the enzyme before the hydroxyl
ion is released. Reaction (XII-8) demonstrates also that water is not
merely an indifferent solvent. The water molecule has a very significant
role in enzymatic reactions. It acts as an acyl acceptor and can thus compete with the enzyme for the acyl residues of amino acids, peptides,
amides or esters.

Reaction (XII-8) deviates from that shown in (XII-1) by its complete reversibility. The velocity constant $k_4$ cannot be neglected since its magnitude is similar
to that of $k_1$; similarly $k_3$ is of the same order as $k_2$. Although $O^{18}$ is significantly
heavier than $O^{16}$, the velocities of the replacement of $HO^{18}$ and $HO^{16}$ are of the
same order of magnitude. It is evident from this discussion that the "virtual
reactions" allow us to measure the kinetics of enzyme action under conditions
which are simpler than those in reactions where substrate and product have
different chemical structure. In the virtual reactions the structure of substrate
and product is the same. The only different parameter is their molecular weight
which is higher in the material labeled by the heavy oxygen isotope.

Neither the kinetics nor thermodynamical data permit us to describe
unequivocally the *activated state*. It may be a short-lived, loose complex
formed by the exchange of resonance energy between the active site of
the enzyme and the activated group of the substrate (E-10;A-17). As
mentioned earlier, a stable enzyme-substrate complex may be formed in
some enzymatic reactions, such as reaction (XII-8). In such cases, the
*ES* complex can be represented by a depression in the peak of the curve
which demonstrates the activation process (see Fig. VII-7). The activated complexes *ES\** and *EP\** which are formed in the forward and the
reverse reaction would be represented by additional maxima. According
to this view the activated states (transition states) are different from the
stable enzyme-substrate complexes (E-5, 11). If more than one stable
intermediate is formed, the number of possible activated states also
increases.

## F. *Proteinases and Peptidases*

Proteinases and peptidases belong to the large group of enzymes which are frequently called hydrolases. Most if not all of these enzymes act as transferases, catalyzing the transfer of a part of a donor molecule to an acceptor molecule. If the acceptor is a water molecule or a hydroxyl ion, the enzyme acts as a hydrolase (F-1). Hence hydrolysis is merely a special case of a transfer reaction. The term transferase is not used when the transferred material consists merely of electrons, in some instances accompanied by protons (F-1).

It is not the purpose of the following sections of this chapter to discuss systematically all classes of enzymes, nor even to treat all of the important enzymes. Since this book is devoted to the chemistry and function of proteins, only those enzymes will be treated in which some relation between their structure and their catalytic function has been found.

It is surprising that the correlation between structure and function is best known in those enzymes which have the most complicated substrates, the *proteinases*. In Section B of this chapter the mechanism of action of *chymotrypsin* (F-2) has been discussed. It will be recalled that the active site of chymotrypsin contains a serine group whose side chain acts as acceptor for acyl or amino-acyl groups; an imidazole group in the vicinity of the serine residue seems to be essential for the enzyme action. Chymotrypsin shares with trypsin, pepsin, and other proteolytic enzymes the property of being present in the tissue as an inactive proenzyme (zymogen). Both *chymotrypsinogen* and chymotrypsin have been prepared from pancreas in crystalline form (F-3). *Activation of chymotrypsinogen* can be accomplished by treatment with trypsin. One mg of trypsin is sufficient for the activation of about 3 g of chymotrypsinogen. The molecular weight of chymotrypsin is approximately 23,000. In aqueous solutions dimerization has been observed (F-4). Chymotrypsinogen seems to consist of a single peptide chain with an *N*-terminal half-cystine residue (F-5) and a *C*-terminal asparagine. The amino acid composition is shown in Table XII-1. Chymotrypsinogen B, another proteolytic enzyme, differs from chymotrypsinogen by a much higher content of serine and amide-*N* and lower values for methionine and proline (F-55). During the conversion of chymotrypsinogen to chymotrypsin, *seryl-arginine* and *threonyl-asparagine* are removed from the peptide chain shown below:

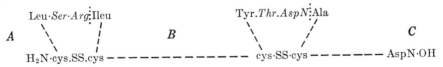

In addition to the two cystine residues, shown above as cys·SS·cys, the molecule contains 2 more cystine residues. The 4 cystine residues hold the molecule together when seryl-arginine and threonyl-asparagine have been removed (F-6). The portion of the molecule designated by *A* contains 13 amino acid residues, *B* about 140, and *C*, 50 amino acid residues (F-2). The limits of *B* are shown in the diagram by dotted lines. Activation is brought about not only by trypsin but also by other proteolytic enzymes. Since these reactions can occur in different sequences, several intermediate chymotrypsins are formed which have been designated by the Greek letters $\pi$ and $\delta$. The end-product is $\alpha$-chymotrypsin. The inactivity of chymotrypsinogen is attributed to the inaccessibility of the active site in the folded structure of the chymotrypsinogen molecule or, more probably, to an inadequate mutual orientation of parts of the active site (e.g., the seryl and the imidazole group) which is unfavorable for catalytic action. Cleavage of the two peptide loops between *A* and *B* and between *B* and *C* may cause rearrangement accompanied by formation of a catalytically active site (F-2). X-ray diffraction at 5 Å resolution indicates that the dimensions of the chymotrypsinogen molecule are approximately 50 × 40 × 40 Å and that its content in $\alpha$-helices is very low (F-63).

It has been mentioned before that chymotrypsin catalyzes the hydrolysis not only of peptide bonds but also of ester bonds. One of the simplest substrates is ethyl lactate. The enzyme combines with the lactyl residue and transfers it then from its binding groups to water (F-7). Chymotrypsin catalyzes the transfer of various aminoacyl or peptide residues from one to another acceptor. It can thus bring about the formation of "plastein," an insoluble polypeptide, in peptic digests (F-8,9). If radioactive amino acids containing $C^{14}$ are added to the peptic digest, they are not incorporated into plastein, whereas the esters of phenylalanine, tyrosine, methionine, and some of the other amino acids are incorporated (F-9). The ease with which these transpeptidations take place *in vitro* at 37°C and pH 4–5 suggests that they may also occur *in vivo* in some cases and may be involved in some steps of protein biosynthesis.

*Trypsin* (F-10)

Crystalline trypsinogen is extracted from pancreas by dilute sulfuric acid at low temperatures and is precipitated by the fractional addition of ammonium sulfate (F-11). The first fractions contain chymotrypsinogen; trypsinogen is precipitated by further addition of ammonium sulfate. The molecular weight of trypsin is about 34,000 (F-12). The activation of *trypsinogen* to *trypsin* involves hydrolysis of a single peptide bond and release of a hexapeptide from the *N*-terminal end of the molecule. The composition of the hexapeptide is: Val.Asp$_4$.Lys (F-13). The hydrolysis of the lysyl bond is brought about by autolysis or by the enzyme enterokinase. In trypsinogen the lysyl group is bound to the $N$-terminus of trypsin which is formed by the sequence Ileu.Val.Gly—. The amino acid composition of trypsin is shown in Table XII-1. Partial hydrolysis reveals

TABLE XII-1

AMINO ACID COMPOSITION OF SOME PROTEOLYTIC ENZYMES

(A = gram amino acid in 100 g enzyme; B = amino acid residues per enzyme molecule)

| Amino acid | Chymo-trypsino-gen[a] Mol. Wt. 25,000 B | Tryp-sino-gen[b] 23,700 B | Papain[c] 20,900 B | Leucine-amino-pepti-dase[d] — A | Pepsin[e] 36,422 B | Car-boxy-pepti-dase[f] 34,400 B |
|---|---|---|---|---|---|---|
| Glycine | 23 | 21 | 23 | 4.3 | 38 | 23.2 |
| Alanine | 22 | 13 | 13 | 6.4 | 18 | 20.0 |
| Valine | 22 | 15 | 15 | 6.6 | 21 | 16.4 |
| Leucine | 19 | 12 | 10 | 8.8 | 28 | 24.7 |
| Isoleucine | 10 | 12 | 10 | 5.7 | 27 | 20.1 |
| Proline | 9 | 7 | 9 | 4.3 | 15 | 11.0 |
| Phenylalanine | 6–7 | 4 | 4 | 6.1 | 14 | 14.9 |
| Tyrosine | 4 | 9 | 17 | 2.7 | 18 | 19.7 |
| Tryptophan | 7 | ? | 5 | — | 6 | 6.1 |
| Serine | 30 | 38 | 11 | 6.4 | 44 | 33.1 |
| Threonine | 23 | 9–11 | 7 | 5.5 | 28 | 26.6 |
| Cystine/2 | 10 | 12 | 6 | 2.4 | 6 | 4.02 |
| Methionine | 2 | 1 | — | 1.2 | 5 | 1.02 |
| Arginine | 4 | 2 | 10 | 5.5 | 2 | 10.0 |
| Histidine | 2 | 3 | 2 | 2.1 | 1 | 7.7 |
| Lysine | 13 | 14 | 9 | 8.3 | 1 | 18.4 |
| Aspartic acid | 22(14) | 24 | 17 | 10.2 | 44 | 30.3 |
| Glutamic acid | 14(11) | 10 | 17 | 11.0 | 27 | 25.0 |
| Ammonia | 25 | 23 ± 3 | 19 | 1.9 | 36 | 19.4 |

[a] From (F-10).　　[d] From (F-51).
[b] From (F-50).　　[e] From (F-52).
[c] From (F-29).　　[f] From (F-53).

the presence of similar amino acid sequences in trypsin and chymotrypsin (F-14). It will be recalled that both enzymes are formed in the pancreas.

The active center of trypsin is inhibited by DFP and contains seryl as well as histidyl residues (F-15). If the SS bonds in DFP-trypsin are oxidized by performic acid and the resulting product is further hydrolyzed by trypsin, a peptide of the composition —Asp.Ser.Cys(SO$_3$H).Glu.Gly.-Asp.Ser.Gly.Pro.Val.Cys(SO$_3$H).Gly.Lys— is obtained (F-15). Trypsin, like chymotrypsin, catalyzes transpeptidations of aminoacyl or peptide residues; the enzyme combines quite specifically with the carbonyl group of lysyl or arginyl residues; thus it transfers the dilysyl group from dilysyl ethyl ester to tyrosine-amide (F-16). One of the simplest substrates of trypsin is benzoyl-arginine-amide which is hydrolyzed to benzoyl arginine and ammonia. Benzoylarginine amide is bound not only to trypsin, but also to trypsinogen (F-17). Evidently, the binding site is accessible also in trypsinogen. Its activation by other parts of the molecule (e.g., an imidazole residue) does not seem to be possible in the proenzyme. Autodigestion of trypsin yields a dialyzable fragment which still has enzymatic activity but is rather unstable (F-18). The action of trypsin is inhibited by a natural inhibitor which occurs in pancreas, egg white, lima beans, and other natural sources (F-10).

*Thrombin*, which catalyzes the clotting of fibrinogen, and *plasmin*, which hydrolyzes fibrin, resemble trypsin in their ability to hydrolyze arginine and lysine esters (F-19). The same substrates are also hydrolyzed by cathepsin B (F-20). None of these enzymes has yet been obtained in a pure state, nor has any mono-*acyltrypsin*, comparable to monoacylated chymotrypsin, been isolated. When thrombin acts on fibrinogen, only 4 out of approximately 1000 peptide bonds are hydrolyzed (F-66).

## *Pepsin* (F-21)

Pepsin differs from trypsin and chymotrypsin by its very low isoelectric pH value. In 0.1 $N$ hydrochloric acid, pepsin still migrates anodically. Its high acidity is due either to the adsorption of chloride ions or to one or more abnormally strong carboxyl groups. The molecular weight of pepsin is approximately 36,000 (F-22). The molecule consists of a single peptide chain (F-23). In the gastric mucosa the enzyme is present as inactive *pepsinogen* which has been isolated in crystalline form (F-27). Its molecular weight is 42,000. On activation of pepsinogen a basic peptide consisting of 34 amino acid residues is split off; it owes its basicity to 8 lysyl and 4 arginyl residues (F-23) and acts as an inhibitor of pepsin. Its $N$-terminal octadecapeptide has the amino acid sequence Leu.Val.Leu.Glu.Pro.Ala.-Glu.Phe.Ser.Leu.Lys.Asp.Gly.Val.Lys.(Asp,Pro).Leu (F-65). The amino acid composition of pepsin is shown in Table XII-1. Like chymotrypsin, pepsin is able to catalyze transpeptidation at pH 4; it is not yet clear whether this action is due to the presence of another enzyme called gastricsin (F-24). In contrast to trypsin and chymotrypsin, pepsin activates not only the carbonyl part of peptide bonds (F-61) but also the NH groups, particularly those contributed by aromatic amino acid residues (F-25). Poly-$\alpha$-L-glutamic acid, but not its D-isomer, is hydrolyzed by pepsin (F-62).

The dialyzate of autolyzed pepsin contains an active fragment of the enzyme which is still able to hydrolyze hemoglobin or casein (F-26). Like the other proteolytic enzymes, pepsin contains dithio bonds which maintain the single peptide chain of its molecule in a definite folded conformation; in addition to three SS bonds, the molecule also contains 1 phosphoric acid residue. It is not yet known whether this residue is involved in the catalytic action of the enzyme. Reduction of the disulfide bonds by mercaptoethanol causes inactivation of pepsin (F-33). The active site of pepsin contains probably an aspartyl or glutamyl residue in close vicinity to a tyrosine residue; this would result in the interaction of the free carboxyl group with the phenolic hydroxyl group (F-64).

## Papain

Papain has been prepared from the latex of the papaya tree in crystalline form (F-28). The molecular weight of papain is approximately 21,000 (F-29). Table XII-1 shows the amino acid composition of the enzyme. Papain is a thiol enzyme, inhibited by *p*-chloromercuribenzoate. If the Hg-enzyme is digested by leucineaminopeptidase, 109 of the 185 amino acids can be split off from the *N*-terminus without loss of enzymatic activity (F-30). However, the Hg ion must be removed for restoration of the enzyme activity. The enzyme is active only in the reduced state (F-31). Evidently, the thio group in the decapeptide —Glu.Leu.Leu.-Asp.Cys(SH).Asp.Arg.Arg.Ser.Try— is essential for the catalytic action (F-54). This thiol group may act as an acyl acceptor and form short-lived thiol esters as intermediates. One of the simplest reactions catalyzed by papain is the virtual reaction taking place between hippuryl-amide and ammonia (F-32):

$$C_6H_5 \cdot CO \cdot NH \cdot CH_2 \cdot CONH_2 + N^{15}H_3 \rightleftharpoons C_6H_5 \cdot CO \cdot NH \cdot CH_2 \cdot CON^{15}H_2 + NH_3$$

*Bromelin* from pineapple and *ficin* from the fig tree are similar to papain. They are activated by cyanides or reducing agents.

In the preceding paragraphs it has been shown repeatedly that a deeper insight into the structure of proteolytic enzymes can be obtained by their partial hydrolysis. The latter is produced by other proteolytic enzymes or by peptidases. A special type of limited hydrolysis is accomplished by using *elastase* (F-34) or *collagenase* as hydrolyzing agent. Collagenase hydrolyzes only peptide bonds formed by glycine and either proline or hydroxyproline (F-35). Peptide bonds, which are not hydrolyzed by the typical proteolytic enzymes from animal or vegetal sources, may be cleaved by bacterial proteolytic enzymes such as subtilisin from *Bacillus subtilis* (F-36).

The proteinases discussed in the preceding paragraphs act as *endo-*

*peptidases.* In other words, they hydrolyze peptide bonds remote from the terminal amino acids. However, the activation of Cbo-phenylalanine by chymotrypsin (reaction XII-8) proves that chymotrypsin can activate the carbonyl groups of terminal amino acid residues.

Among the exopeptidases *carboxypeptidase* has been used quite frequently for the stepwise breakdown of peptide chains from the carboxyl end of the chain. The enzyme is prepared from the fluid exuded from frozen pancreas by precipitation with ammonium sulfate (F-37) and has been obtained in crystals. The inactive precursor of this enzyme, procarboxypeptidase, has a molecular weight of 96,000; on activation by trypsin a fragment of molecular weight 34,000 was found to be the active enzyme (F-38). The remaining $\frac{2}{3}$ of procarboxypeptidase have a different enzyme activity. Carboxypeptidase contains 1 zinc atom per molecule. Like the proteinases, carboxypeptidase hydrolyzes not only peptide but also ester bonds. It attacks peptide bonds formed by the aromatic amino acids faster than other peptide bonds. *Carboxypeptidase B* seems to be identical with *protaminase,* an enzyme which catalyzes the hydrolysis of peptide bonds formed by $N$-terminal arginine or lysine residues (F-39).

*Leucine aminopeptidase* is found in the intestinal juice and forms part of the enzyme complex which previously had been called erepsin (F-40). The enzyme degrades proteins and peptides stepwise from the $N$-terminal amino acid residue. Its amino acid composition is shown in Table XII-1. It is a metal enzyme, activated by magnesium or manganese. Its action can be measured by means of L-leucinamide which is hydrolyzed by the enzyme.

Leucine aminopeptidase is inhibited by cystine and therefore does not attack proteins which are rich in SS bonds such as serum albumin, ribonuclease, or lysozyme; if the SS bonds are cleaved by performic acid, the proteins are attacked by the enzyme (F-41). The *metal activated peptidases* include also dipeptidase (which hydrolyzes glycyl-glycine and is activated by cobalt or manganese), glycyl-leucine peptidase, prolinase, and prolidase. The three last-mentioned enzymes are also activated by manganese (F-41). The dipeptidase from rat liver catalyzes not only the hydrolysis of dipeptides but also their synthesis from the free amino acids (F-43). It may finally be mentioned that transpeptidation involving the $\gamma$-carboxyl group of glutamic acid is catalyzed by specific enzymes (F-44).

*Rennin* (rennet, lab enzyme) is the milk-coagulating enzyme of the gastric juice. It has been isolated in crystalline form (F-45). It is not yet known whether rennin acts as an exo- or endopeptidase. The presence of *protective enzymes* (defensive enzymes) in the blood and urine of animals injected with foreign proteins was claimed by Abderhalden (F-46). The author and his coworkers were unable to produce protective enzyme by injecting rabbits with foreign plasma globulins (F-47).

*Urease* is mentioned here because, like the peptidases, it catalyzes the hydrolysis of —CO·NH— bonds. It was the first crystalline enzyme, and is obtained from Jack bean or soy bean meal by extraction with 32% acetone (F-48). The molecular weight of urease is 483,000 (F-49); its action is highly specific and is limited to the hydrolysis of urea. The active site of urease is still unknown.

During the last few years *insoluble forms of the proteolytic enzymes* have been used as very valuable agents for the partial hydrolysis of proteins. Thus, insoluble papain or trypsin are obtained by coupling a diazotized copolymer of leucine and *p*-aminophenylalanine with papain (F-56) or with a complex formed by trypsin and polytyrosine (F-57). Pepsin and carboxypeptidase have been made insoluble by coupling with diazotized poly-*p*-aminostyrene (F-58). Trypsin and chymotrypsin were also coupled to diazobenzylcellulose (F-59). The water-insoluble preparations are more stable than the water-soluble forms since they cannot so easily undergo autodigestion. The partial hydrolyzate obtained is free of soluble enzyme and contains only the breakdown products of the hydrolyzed protein. The insoluble enzyme preparations were found to be particularly valuable in the limited degradation of antibodies into their fragments I, II, and III (see Chapter XV) (F-60).

## G. Other Nonoxidative Enzymes

Enzymes which catalyze the transfer of phosphate residues from a donor to an acceptor molecule have been particularly thoroughly investigated (G-1). If the donor is a "high energy phosphate" compound, i.e., a compound with highly reactive phosphate, such as ATP, the reaction is practically irreversible. *Hexokinase*, which catalyzes the transfer of phosphate from ATP to the 6-position of glucose, has a molecular weight of 96,000 and has been obtained in crystals (G-2). Another enzyme important in the glycolytic breakdown is *phosphoglucomutase* (see Table XII-2); its molecular weight is 74,000 (G-3). Phosphoglucomutase catalyzes the conversion of glucose-1-phosphate into glucose-6-phosphate. The active site of phosphoglucomutase has the amino acid sequence —Asp.Ser.Gly.Glu.Ala.Val— which is similar to that found in chymotrypsin and other transacylating enzymes (G-4). The enzyme is a phosphate acceptor and the phosphorylated enzyme a phosphate donor. Glucose-1,6-diphosphate acts in this transfer reaction as a coenzyme (G-5 to 7). The reaction can be represented by the following symbols:

$$\begin{matrix} C_1{-}P \\ | \\ C_6{-}P \end{matrix} + E \rightarrow \begin{matrix} C_1 \\ | \\ C_6{-}P \end{matrix} + E{-}P \qquad (XII\text{-}9)$$

$$E{-}P + \begin{matrix} C_1{-}P \\ | \\ C_6 \end{matrix} \rightarrow E + \begin{matrix} C_1{-}P \\ | \\ C_6{-}P \end{matrix} \qquad (XII\text{-}10)$$

The overall reaction, obtained as the sum of Eqs. (XII-9) and (XII-10), is: Glucose-1-P → glucose-6-P. *Phosphoglyceric mutase*, which catalyzes the conversion of 3-phosphoglycerate into 2-phosphoglycerate, seems to act in a similar manner.

The enzymes that hydrolyze esters in which the phosphoric acid residue is bound to two alcohol groups are called phosphodiesterases. The most thoroughly investigated of these is *ribonuclease* (see Table III-1). The enzyme is remarkably stable to heat; it remains active when exposed to 100°C for 5 min (G-8). It has been obtained from pancreas in crystals (G-9) and has a low molecular weight of approximately 13,000. In contrast to other enzymes, ribonuclease remains active in concentrated solutions of urea (G-10). The amino acid sequence of ribonuclease has been unravelled completely (G-11,12). Figure XII-4 shows that the molecule has 4 dithio bonds. If these are cleaved by reducing agents, the enzymatic action is lost. However, part of it is restored when the reduced enzyme is carefully reoxidized (G-13). Evidently, the molecule refolds under favorable conditions in such a manner that the original, apparently most stable, conformation is restored. The picture shown in Fig. XII-4 has nothing to do with the three-dimensional conformation of native ribonuclease; it is merely an attempt to show the entire peptide chain and the 4 SS bonds in one plane. The three-dimensional conformation is not yet known, but some attempts have been made to resolve it by indirect methods (G-14).

Native RNase is not attacked by trypsin (G-42). When ribonuclease is exposed to the action of the bacterial proteolytic enzyme subtilisin, a single peptide bond is hydrolyzed and the molecule split into the *N*-terminal peptide of 20 amino acids, called S-peptide, and S-protein containing the other 104 amino acids (G-15). Although each of the two components is devoid of enzymatic activity, the S-protein is able to bind the cyclic cytidine-2′,3′-phosphate, a substrate of RNase (G-42). Surprisingly, a mixture of S-protein and S-peptide has RNase activity. It appears that interaction between two charged groups, one in the S-peptide and one in the S-protein, is necessary for the action of ribonuclease and that this interaction can take place in a mixture of the two components in the same manner as in the intact ribonuclease molecule (G-15). Pancreatic ribonuclease consists of two components, A and B, which differ from each other by one COOH group (G-16). Ovine ribonuclease differs from the bovine enzyme by minor changes in the amino acid sequence shown in Fig. XII-4; serine in the 3-position is replaced by threonine, lysine in No. 37 by glutamic acid; another alteration has been found in the residues 99–104 (G-17).

The substrate of ribonuclease is ribonucleic acid which is hydrolyzed by the enzyme. A simpler substrate is the cyclic cytidine-2′,3′-phosphate in which the phosphoryl residue is bound to the 2′- and 3′-position of the ribose residue. The enzyme catalyzes the reaction: cytidine-3′-phosphate alkyl ester → alcohol + cytidine-2′,3′-phosphate; + $H_2O$ ⇌ cytidine-2′-phosphate (G-18). In spite of our extensive knowledge of the structure of

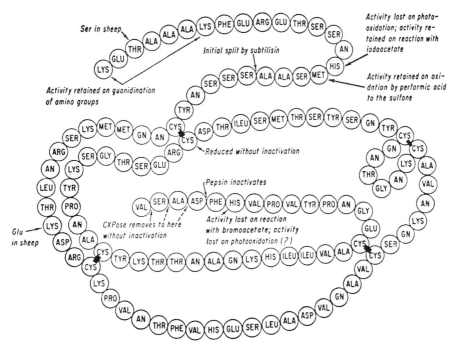

FIG. XII-4. Amino acid sequence of bovine ribonuclease; also illustrated are certain relations between structure and enzymatic activity. AN = asparagine, GN = glutamine. More recent analyses have revealed that the sequence of the amino acid residues 11–18 (Glu.Thr.Ser.Ser.AspN.His.Met.Ser.) has to be replaced by the following sequence: GluN.His.Met.Asp.Ser.Ser.Thr.Ser. (G-46). From C. B. Anfinsen, *in* "The Enzymes" (P. D. Boyer, H. Lardy, and K. Myrbäck, eds.), 2nd ed., Vol. 5, p. 111. Academic Press, New York, 1961.

ribonuclease, we still do not know the conformation of the molecule. The active center seems to contain two histidyl residues (G-41).

*Phosphorylases* are enzymes which catalyze the transfer of glucosyl residues from glucose-1-phosphate to glycogen or starch according to the following reversible reaction:

$$nC_6H_{11}O_5 \cdot OPO_3H_2 \rightarrow (C_6H_{10}O_5)_n + nH_3PO_4 \tag{XII-11}$$

A small amount of glycogen or starch is necessary as a *primer* in this reaction. The glucosyl residues are bound to the 4-C atom of the terminal glucose residue of the polysaccharide which serves as a primer so that its chain length increases. Crystalline phosphorylase has been prepared from potatoes (G-19) and from muscle (G-20). The muscle phosphorylase occurs in two forms called *a* and *b*. The *a* form is a dimer of the *b* form. The latter contains two residues of pyridoxal phosphate. When the enzyme combines with radioactive phosphate partial hydrolysis yields the peptide Lys.GluN.Ileu.P-Ser.Val.Arg (G-43). The role of pyridoxal phosphate in the enzymatic action is not yet known.

The *indispensability of the primer* (glycogen or starch) is attributed to its role as an acceptor for the transferred glucosyl residues. In addition to this function, the primer in other enzymatic reactions frequently acts as a template which determines the sequence of the polymer formed by transfer reactions. A primer is usually required for the action of *polynucleotide phosphorylase* which catalyzes the reaction: $n$BRPP $\rightarrow$ (BRP)$_n$ + $nP_i$, where BRPP is a nucleoside pyrophosphate (e.g., adenosine pyrophosphate), (BRP)$_n$ a polynucleotide and $P_i$ inorganic phosphate. If a mixture of the four nucleoside pyrophosphates ADP, GDP, UDG, and CDP is exposed to the action of the enzyme, the composition of the produced polynucleotide depends to a certain extent on the primer used. The role of the primer is not yet quite clear since the enzyme does not require any primer for the formation of polyadenylic acid. It contains a small amount of oligonucleotide which may act as a primer in this case. Synthesis of the other polynucleotides takes place much more slowly than that of polyadenylic acid. The primer does not seem to act here as a template (see Chapter XVI, K); its role is still obscure (G-21).

Although many of the *carbohydrases* have been known for many years, very little is known about the composition and mode of action of these enzymes. Most of them act as transglycosylases, transferring monosaccharide residues from oligo- or polysaccharides to other carbohydrates (G-22). In many of these enzymes uridine-diphosphosugars act as coenzymes or as intermediates (G-23). It is not yet understood why certain uridine derivatives should activate sugars whereas certain cytidine derivatives activate the enzymatic synthesis of phospholipids, and guanosine derivatives the incorporation of amino acids into proteins. This *specific role of each of the nucleotides* is very surprising in view of their structural similarity; each of them contains ribose and phosphate; they differ from one another only in their purine or pyrimidine residues.

*Amylases*, the enzymes which catalyze the hydrolytic degradation of starch or glycogen, have been obtained in the crystalline state not only from animals and plants, but also from bacteria (G-24). Kinetic analysis of the degradation of starch to maltose by $\beta$-amylase leads to the view that the enzyme has at least two active groups, one with a pK value of 4.3, the other at 7.1 (G-25). They are, probably, a carboxyl and an imidazole group; a sulfhydryl group appears also to be involved in the action of $\beta$-amylase. Native $\alpha$-amylase is resistant to proteolytic enzymes; it is digested after denaturation. Inhibition experiments indicate that the lock and key picture does not explain the specificity of amylase, and that the assumption of an "induced fit" is more satisfactory; see (D-10).

Lysozyme (see Chapter VIII, E) hydrolyzes certain *N*-acetyl-oligosaccharides (G-47). The amino acid sequence of lysozyme has been completely resolved (G-48). Only the position of the disulfide bonds is not yet clear.

A thermostable amylase has been isolated from *Bacillus stearothermophilus* (G-26). The enzyme has a molecular weight of 15,700 and does not undergo any changes in its properties in urea, guanidine, or at 75°C. It is very rich in proline and glutamic acid.

Very little is known on the chemical properties of *lipases* (G-27), enzymes which hydrolyze triglycerides or form these from glycerol and fatty acids. In contrast to other hydrolyzing enzymes, the lipases act in a heterogeneous system, an emulsion of fat in an aqueous medium. Their activity decreases if the medium is homogenized by suitable solvents. It is well known that proteins form monomolecular films in the interface between aqueous and nonpolar phases (G-28). Lipase activity may be limited to this interface. It has been suggested, therefore, that the substrate concentration in the interface be used for calculating the Michaelis constant (see Section E) (G-29).

One of the simplest chemical reactions, the interconversion of carbon dioxide and bicarbonate, is catalyzed by the enzyme *carbonic anhydrase*:

$$CO_2 + H_2O \rightleftharpoons H^+ + HCO_3^-$$

Carbonic anhydrase is a zinc enzyme. Its amino acid composition has been determined (Table XII-2). Although the mechanism of its action is not yet known $CO_2$ and/or bicarbonate are presumably bound as ligands to the protein-bound zinc atom, thus forming a coordination complex. The zinc atom is bound to the sulfur atom of an SH group in the enzyme (G-45).

As mentioned before, the myogen fraction of muscle consists of a large number of enzymes which can be separated from each other. Many of these have been isolated in the crystalline state. One of them is *enolase* which catalyzes the dehydration of 2-phosphoglycerate: 2-phosphoglycerate $\rightleftharpoons$ phosphopyruvate $+ H_2O$. Enolase has been prepared from muscle extracts by fractionation with ammonium sulfate (G-31), from yeast extracts by precipitation with nucleic acid and mercury salts (G-32). It is a metal enzyme containing magnesium and/or manganese (B-33). The molecular weight of enolase is 66,000 (G-33), corresponding to about

550–600 amino acid residues. However, 150 of these residues can be split off from the amino end by leucine aminopeptidase or from the carboxyl end by carboxy-peptidase without loss of the enzymatic activity (G-34). Another important component of the myogen fraction is *aldolase* which

TABLE XII-2

AMINO ACID COMPOSITION OF SOME HYDROLYZING ENZYMES

(A = gram amino acid in 100 g protein; B = amino acid residues per molecule)

| Amino acid | Enzyme: Human salivary amylase[a]<br>Mol. Wt. —<br>A | α-Amylase from A. oryzae[b]<br>51,860<br>B | β-Galactosidase from E. coli[c]<br>—<br>A | Deoxyribonuclease[d]<br>61,566<br>B | Phosphorylase[e]<br>—<br>A | Phosphoglucomutase[e]<br>—<br>A | Carbanhydrase[f]<br>31,000<br>B |
|---|---|---|---|---|---|---|---|
| Glycine | 6.82 | 39 | 3.33 | 26 | 2.3 | 3.25 | 20 |
| Alanine | 4.43 | 34 | 4.30 | 35 | 5.4 | 7.8 | 13 |
| Valine | 6.89 | 27 | 5.41 | 42 | 5.8 | 6.1 | 13–14 |
| Leucine | 5.77 | 30 | 13.70 | 41 | 6.7 | 7.1 | 23–26 |
| Isoleucine | 5.80 | 25 | | 21 | | | 4 |
| Proline | 3.6 | 19 | 4.64 | 19 | 5.0 | 5.8 | 15 |
| Phenylalanine | 7.20 | 13 | 4.73 | 20 | 8.9 | 14.6 | 10 |
| Tyrosine | 5.51 | 31 | 2.81 | 29 | 7.6 | 7.8 | 6 |
| Tryptophan | 7.2 | 10 | 6.00 | 5 | 2.05 | 0.5 | — |
| Serine | 7.8 | 30 | 3.20 | 62 | 4.8 | 5.5 | 17 |
| Threonine | 4.5 | 37 | 4.67 | 36 | — | 2.8 | 14 |
| Cystine/2 | 4.4 | 9 | 1.68 | 6 | 2.85 | 3.8 | 0 |
| Methionine | 2.4 | 7 | 2.61 | 12 | — | — | 3 |
| Arginine | 8.75 | 9 | 7.19 | 26 | 13.5 | 10.5 | 10–12 |
| Histidine | 3.24 | 6 | 2.92 | 22 | 1.4 | 0.6 | 10 |
| Lysine | 6.33 | 17 | 2.52 | 33 | 9.0 | 8.2 | 16–17 |
| Aspartic acid | 19.3 | 62 | 12.77 | 65 | 14.6 | 15.6 | 30–31 |
| Glutamic acid | 9.6 | 28 | 10.59 | 43 | 14.35 | 11.8 | 22 |
| Ammonia | — | 51 | 1.66 | 83 | — | — | 32–35 |

[a] From (G-36).          [d] From (G-39).
[b] From (G-37).          [e] From rabbit muscle (G-40).
[c] From (G-38).          [f] From (G-30,44).

in a reversible reaction catalyzes the formation of fructose-1,6-diphosphate from 3-glyceraldehyde phosphate and dihydroxy-acetone phosphate. From the muscles of 20–30 rats about 1 g of the crystalline enzyme has been obtained (G-35). The molecular weight of aldolase is 149,000 (G-3). In contrast to muscle aldolase, yeast aldolase is a metal enzyme.

Yeast *carboxylase,* an enzyme important in alcoholic fermentation, contains thiamine pyrophosphate as coenzyme and requires magnesium ions for its action.

Bacterial decarboxylases, which split $CO_2$ from tyrosine and other amino acids, contain pyridoxal phosphate as their prosthetic group. Very little is known about the protein components of these enzymes.

## H. Enzymes Involved in Oxidoreduction

The enzymes to be discussed here differ in several respects from those treated in the two preceding sections. Most if not all of the enzymes which catalyze oxidoreductions have an easily detectable colored or fluorescent active group or coenzyme. This is understandable since color and fluorescence indicate the presence of labile, easily excitable electrons and thus parallel the ease of electron shifts and electron transfer. The question might be raised whether the apoenzymes have any significant role in the electron transport. The apoenzymes are indeed *indispensable* for the enzymatic action. Moreover, the apoenzyme determines the specificity of the reaction. Whereas one and the same pyridine nucleotide can act as a transferring agent in the dehydrogenation of various substrates such as lactic acid, ethanol, or glycerophosphate, the apoenzyme determines which of these is dehydrogenated.

The enzymes involved in oxidation-reduction differ from the hydrolyzing enzymes and transferases not only in their content of well-defined coenzymes, but also in their requirement for a series of electron carriers as catalysts. The principal types of enzyme proteins involved in oxidoreduction belong to three classes: (1) the dehydrogenases which have pyridine nucleotides as coenzymes, (2) flavoproteins, and (3) heme proteins. Since the electrons are usually carried from the organic substrates to oxygen through all three types of enzymes in the indicated sequence: pyridine-nucleotide enzyme $\rightarrow$ flavoprotein $\rightarrow$ heme proteins $\rightarrow O_2$, we will discuss these three types in this same sequence.

### Pyridine Nucleotide Enzymes (DPN or NAD, and TPN or NADP)

The coenzymes of these catalysts are loosely bound to the apoenzymes. One of these coenzymes was detected in the dialyzate of the enzyme mixture which causes fermentation in yeast. It was originally called cozymase. Later, when a similar coenzyme was found in red blood cells, the coenzyme of alcoholic fermentation was designated as coenzyme I. It is a dinucleotide, formed by the combination of adenosine monophosphate with nicotinamide-ribosephosphate. Its composition is shown by the linear arrangement N-R-P-P-R-Ad, where N is nicotinamide; R, ribose; P, phosphate; and Ad, adenine. Coenzyme II, which was discovered as a coenzyme of glucose-6-phosphate dehydrogenase, has another phosphate

residue attached to the adenine-bound ribose. Hence its formula is represented by

$$N—R—P—P—R\overset{\displaystyle P}{\underset{\displaystyle Ad}{\diagdown}}$$

The designations DPN (diphosphopyridine nucleotide) and TPN (triphosphopyridine nucleotide) have been replaced by the more logical
designation NAD (nicotinamide adenine dinucleotide) and NADP
(nicotinamide adenine dinucleotide phosphate) according to a decision of
the International Union of Biochemistry (H-1). The catalytically active
portion of NAD and NADP is their nicotinamide ring (ADP = adenosine
diphosphate):

$$
\begin{array}{c}
\text{H} \\
\text{|} \\
\text{C} \\
\text{HC}_5{}^4{}_3\text{C—CONH}_2 \\
\text{|}_6 {}_2\text{||} \\
\text{HC}_1\quad\text{CH} \\
{}_{+}\text{N} \\
\text{|} \\
\text{ribose–ADP}
\end{array}
\;+\; \text{XH}_2 \;\rightleftharpoons\;
\begin{array}{c}
\alpha\text{H}\quad\text{H}\beta \\
\diagdown\text{C}\diagup \\
\text{HC}\quad\text{C—CONH}_2 \\
\text{||}\quad\text{||} \\
\text{HC}\quad\text{CH} \\
\diagdown\text{N}\diagup \\
\text{|} \\
\text{ribose–ADP}
\end{array}
\;+\; \text{X} \;+\; \text{H}^+ \quad \text{(XII-12)}
$$

Equation (XII-12) shows that the substrate XH$_2$, which may be
ethanol, lactic acid, glucose, glutamic acid, etc., is dehydrogenated to X.
One proton and two electrons are bound to the nicotinamide residue, and
one proton is liberated in the reaction. The proton which combines with
the nicotinamide ring is bound to carbon atom No. 4, in p-position to the
ring-N atom. The two electrons saturate one of the double bonds. Before
discussing the role of the apoenzyme in this dehydrogenation it is important to point out that the hydrogen atom which is transferred from the
substrate XH$_2$ to the pyridine ring is not exchangeable with water molecules. If deuteriated ethanol is used as a substrate, the carbon-bound
deuterium is transferred to the pyridine ring and is found as carbon-bound
deuterium in the isolated reduced NAD (H-2). Most probably, the transferred unit is a hydride ion, H$^-$, which consists of a proton and two electrons. In view of the direct transfer there can be no doubt that the
hydrogen supplying group of the substrate is in closest proximity to the
carbon atom No. 4 of the pyridine ring during the catalytic oxidoreduction. The apoenzyme determines the stereospecificity of this reaction.
Some of the dehydrogenases catalyze transfer of the hydrogen from the
substrate to the α-position (see the formula XII-12), others to the β-posi-
tion at the 4-C atom (H-3).

Although the bond between apoenzyme and coenzyme is loose, the
apoenzyme, frequently designated as dehydrogenase, is indispensable for

the enzymatic reaction and for its specificity. Since the apoenzyme determines which substrate can and which cannot be dehydrogenated, it must be concluded that it has a specific site which combines with the substrate. On the other hand, we know that the substrate is also bound by the coenzyme. Apparently a ternary complex is formed, in which the sub-

TABLE XII-3

AMINO ACID COMPOSITION OF SOME ENZYMES INVOLVED IN
OXIDATION-REDUCTION

(A = gram amino acid in 100 g protein; B = amino acid residues per molecule)

| Enzyme: Amino acid | Yeast alcohol dehydro- genase[a] Mol. Wt. — A | Lactic acid dehydro- genase[b] 126,000 B | Cyto- chrome $b_2$ — —[c] | Cyto- chrome c from beef heart[d] 12,400 B | Cata- lase (Beef liver)[e] 225,000 B | Peroxi- dase (japan. radish)[f] 54,500 B | Eno- lase[g] 67,225 B |
|---|---|---|---|---|---|---|---|
| Glycine | 7.52 | 95 | 41.4 | 16.5 | 98 | 6 | 55.1 |
| Alanine | 7.34 | ? | 39.5 | 6.6 | 108m | 64 | 88.4 |
| Valine | 11.67 | 136 | 42.1 | 2.7 | 118 | 20 | 47.1 |
| Leucine | 16.54 | 129 | 55.2 | 6.0 | 148 | 36 | 59.7 |
| Isoleucine | | 82 | 27.8 | 5.6 | 66 | 15 | 29.1 |
| Proline | 3.88 | 66 | 32.9 | 3.5 | 89 | 17 | 18.7 |
| Phenylalanine | 7.51 | 23 | 14.0 | 3.8 | 104 | 18 | 23.5 |
| Tyrosine | 6.17 | 19 | 13.8 | 3.1 | 85 | 3 | 15.1 |
| Tryptophan | 3.69 | 22 | 3.4 | 1 | 40 | 2 | 5.3 |
| Serine | 5.29 | 90 | 26.5 | 1.0 | 67 | 51 | 53.4 |
| Threonine | 4.65 | 34 | 22.1 | 8.4 | 63 | 27 | 31.1 |
| Cystine/2 | 3.05 | 16 | 18.5 | 0.1[h] | 20[i] | 10 | 0 |
| Methionine | 1.80 | — | 0 | 1.6 | 43 | 4 | 8.0 |
| Arginine | 3.69 | 39 | 21.0 | 2.9 | 103 | 11 | 18.0 |
| Histidine | 4.10 | 16 | 6.2 | 2.8 | 112 | 4 | 13.7 |
| Lysine | 9.07 | 87 | 45.7 | 18.7 | 152 | 7 | 53.4 |
| Aspartic acid | 11.13 | 113 | 59.5 | 9.2 | 193 | 51 | 77.4 |
| Glutamic acid | 10.98 | 78 | 59.1 | 13.5 | 149 | 26 | 50.2 |
| Ammonia | 0.84 | — | 81.0 | 8.5 | 286 | 50 | 55.8 |

[a] From (H-66).

[b] From rat liver (H-67).

[c] Amino acid residues per heme residue (H-68).

[d] From (H-69).

[e] From (H-70).

[f] Contains also 12 hydroxyproline, 1 heme, 35 mannose, 11 xylose, 4 arabinose, and 15 hexosamine, residues (H-71).

[g] From (H-71).

[h] Two SH.

[i] According to (H-72) 8 SH.

strate is bound to both apoenzyme and coenzyme. Many of the dehydrogenases contain Zn (H-4). It has been suggested therefore that the three
components of the ternary complex are held together by coordination
with the zinc ion. In the coordination complex of alcohol dehydrogenase
(see Table XII-3) the zinc atom is probably bound to the apoenzyme by
the sulfur of a cysteinyl residue, and to the substrate (ethanol) by combination with the hydroxyl group of the alcohol (H-5). The linkage between apo- and coenzyme seems to be formed by combination of the zinc
ion with the N-atom No. 7 of the adenine residue (B-20). The direct

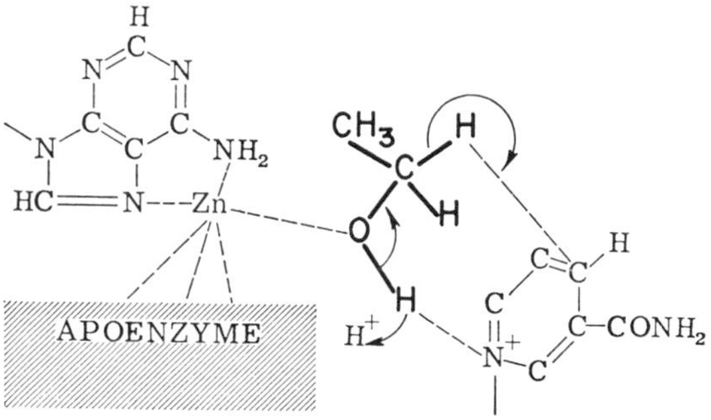

FIG. XII-5. Alcohol dehydrogenase (B-20). The ethanol molecule is shown by bold
letters.

hydrogen transfer from ethanol to the nicotinamide ring necessitates
juxtaposition of the C-1 atom of ethanol and the C-4 atom of the nicotinamide ring. Accordingly, the ternary apoenzyme-coenzyme-substrate
complex may be represented by Fig. XII-5 (B-20). The arrows in Fig.
XII-5 indicate which of the hydrogen atoms is transferred to the nicotinamide ring and which is released as a hydrogen ion. Similar complexes
may be formed in the dehydrogenation of lactate, malate, isocitrate,
glucose, glycerophosphate, and other compounds in which an alcohol
group is dehydrogenated to an aldehyde or keto group. The molecular
weight of liver alcohol dehydrogenase is 73,000, that of yeast alcohol
dehydrogenase, 150,000; the former contains 2, the latter 4, zinc atoms;
both enzymes also contain for each zinc atom one equivalent of pyridine
nucleotide. Evidently, the enzymes are polymers of subunits; these seem
to be held together by the zinc ions (H-6).

Glyceraldehyde phosphate dehydrogenase forms almost 10% of the
soluble muscle proteins (H-7). A similar enzyme has also been isolated
from yeast. The molecular weight of both enzymes is close to 120,000
(G-3). In the conversion of 3-glyceraldehyde phosphate into 1,3-diphos-
phoglycerate the aldehyde first forms a thioacetal by condensation with a

sulfhydryl group of the protein. Oxidation of the thioacetal results in the formation of a thiolester which on phosphorolysis yields the acyl phosphate (H-8).

$$R \cdot CHO + R'SH \rightarrow R \cdot C \underset{SR'}{\overset{H}{\underset{|}{\overset{|}{\diagdown}}} OH} \xrightarrow{-2H} R \cdot C \underset{SR'}{\diagdown} = O \xrightarrow{+H_3PO_4} R \cdot C \underset{OPO_3H_2}{\diagdown} = O \qquad (XII\text{-}13)$$

Glyceraldehyde phosphate dehydrogenase catalyzes not only the oxidation of glyceraldehyde-3-phosphate, but also the hydrolysis of $p$-nitrophenyl acetate (H-9). It is five times more efficient than chymotrypsin in this respect; the active sites for dehydrogenation and hydrolysis seem to be different but close to each other.

In contrast to the oxidized pyridine nucleotides, the reduced forms have an intense ultraviolet absorption maximum at 340–360 m$\mu$ and show blue fluorescence if excited by ultraviolet radiation. As mentioned earlier (Section D), measurement of the depolarization of the fluorescent light has been used to determine the ratio of bound to free pyridine nucleotide (H-10). This ratio can also be determined from a shift of the absorption maximum from 340 to 326 m$\mu$. However, the fluorescence method is about a hundred times more sensitive than measurement of the ultraviolet absorption (H-11).

Some of the apoenzymes of the dehydrogenases combine with NAD(DPN), others with NADP (TPN), quite specifically. Enzymes called transhydrogenases have been discovered in various materials; they catalyze the exchange of electrons between the two pyridine nucleotides (H-12a).

Dehydrogenation of pyruvic and $\alpha$-ketoglutaric acid requires the presence not only of NAD but also of thiamine pyrophosphate and lipoic acid. In both reactions the first step is the loss of $CO_2$ and the formation of an "active" aldehyde, i.e., a loose complex of thiamine pyrophosphate with acetaldehyde or succinic semialdehyde. In a second step the active aldehyde is oxidized by the disulfide form of lipoic acid to yield acetyl or succinyl-dihydrolipoic acid and free thiamine pyrophosphate. In a last phase of the reaction the acyl-dihydrolipoic acid reacts with coenzyme A to yield acyl-coenzyme A and dihydrolipoic acid. The latter is then oxidized by $NAD^+$. In this sequence of reactions the apoenzymes of pyruvic and $\alpha$-ketoglutaric dehydrogenase combine with lipoic acid. Partial hydrolysis of lipoyl-pyruvic acid dehydrogenase yields the peptide Gly.Asp.Lys.Ala with a lipoyl residue attached to the $\epsilon$-amino group of lysine; similarly hydrolysis of lipoyl-$\alpha$-ketoglutaric acid dehydrogenase gives the peptide Thr.Asp.Lys.Val.(Val,Leu).Glu with the lipoyl residue attached to lysine (H-12b). The two $\alpha$-keto acid dehydrogenases apparently combine not only with the pyridine nucleotide (NAD) and the substrate (pyruvic or α-ketoglutaric acid), but also with lipoic acid and possibly also with thiamine pyrophosphate.

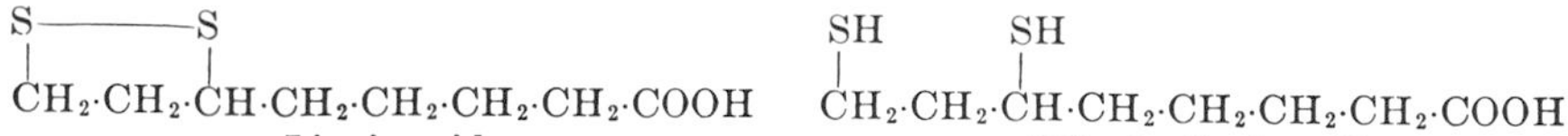

S————S
| |
$\dot{C}H_2 \cdot CH_2 \cdot \dot{C}H \cdot CH_2 \cdot CH_2 \cdot CH_2 \cdot CH_2 \cdot COOH$
Lipoic acid

SH    SH
|    |
$\dot{C}H_2 \cdot CH_2 \cdot \dot{C}H \cdot CH_2 \cdot CH_2 \cdot CH_2 \cdot CH_2 \cdot COOH$
Dihydrolipoic acid

*Flavoproteins*

The first yellow enzyme was discovered in yeast (H-13) and precipitated from the aqueous solution by acetone or methanol. On dialysis against dilute HCl the yellow coenzyme is split off; the apoenzyme is obtained in native form after neutralization of the acid solution (H-14). The coenzyme of this "old yellow enzyme" is riboflavin phosphate. Its molecule consists of dimethyl-isoalloxazine, ribitol, and phosphate.

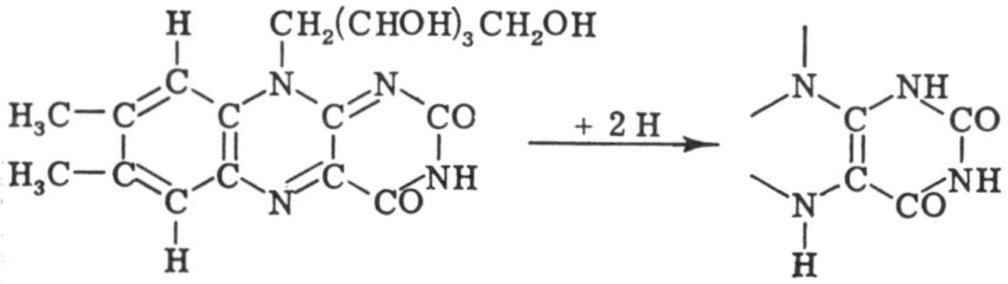

Riboflavin

Reduced form of riboflavin
(partial formula)

(XII-14)

The molecular weight of the "old yellow enzyme" is 52,000 (H-15). In aqueous solutions the enzyme dimerizes easily; hence higher molecular weights have been reported. The unit of the molecular weight 52,000 contains one molecule of riboflavin phosphate (FMN or flavin mononucleotide). At present 10 or more different flavoproteins are known. They include amino acid oxidases, xanthine oxidase (H-16), and enzymes which transfer electrons from the pyridine nucleotides to the cytochromes. Some of these enzymes contain flavin adenine dinucleotide (FAD) as their coenzyme. FAD is formed by the combination of FMN with AMP (adenosine monophosphate). Its structure is represented by F-Rl-P-P-R-Ad where F represents dimethyl-isoalloxazine; Rl, ribitol; P, phosphate; R, ribose, and Ad, adenine. Dihydroorotic acid dehydrogenase, one of the flavoproteins, contains one FMN and one FAD residue in its molecule of molecular weight 62,000 (H-17).

The reduced form of the flavoproteins, shown on the right side of (XII-14), is colorless, but is converted again into the oxidized form by the loss of 2 electrons and 2 protons. The great modifying influence of the apoenzyme on this process is shown by a shift of the oxidation-reduction potential from −0.060 volt (at pH 7) for the flavin to −0.185 volt for some flavoproteins (H-18). The great importance of the protein component of the flavoproteins is manifested in the pronounced specificity of these enzymes. Some of these have been mentioned in the preceding paragraph. However, the principal function of the flavoproteins seems to be the transfer of electrons from the pyridine nucleotides to the cytochromes. Only a limited number of substrates can be dehydrogenated by the flavoproteins directly. One of these is succinate which is converted into fumarate by a flavoprotein (H-19). The reduced flavoproteins are rapidly reoxidized by the cytochromes, but only slowly by molecular oxygen. The physiological pathway of the electrons seems to lead from the flavoproteins to the cytochromes.

In this connection it is important that metal ions (Fe, Cu, Mo) have been discovered in many of the flavoproteins. These metal ions seem to be essential for the channeling of electrons from the two-electron transfer system (substrate-pyridine nucleotide-flavoprotein) to the one-electron transfer system (flavoprotein-cytochrome c- cytochrome a- cytochrome $a_3$) (H-20). A further complication arises by the close association of the flavoproteins with lipids. DPN-cytochrome c reductase, the flavoprotein which carries electrons from the reduced pyridine nucleotides to the cytochromes, seems to be an artifact obtained by delipidation. Diaphorase, which for some time had been identified with DPN-cytochrome c reductase, is a dihydrolipoyl dehydrogenase which catalyzes the dehydrogenation of the sulfhydryl form of lipoic acid (H-21). Very little is known about the apoenzymes of the flavoproteins or their combination with FMN or FAD. It is quite possible that the metal ions present in many of these enzymes act as coordination centers and as links between protein and flavin, similar to the zinc ions of the pyridine nucleotide dehydrogenases.

*Heme Proteins* (H-55a)

The heme proteins which are involved in electron transport from the flavoproteins to the $O_2$ molecule have been designated as *cytochromes*. They have two narrow absorption bands in the green region of the visible spectrum when they are reduced to the ferrous state, and can thus be easily detected by spectroscopy. All cytochromes have also the intense Soret band in the near ultraviolet, between 400 and 450 m$\mu$. Spectroscopically three different cytochromes were detected by Keilin and Hartree (H-22). They were called cytochrome a, b, and c; their absorption maxima are: cytochrome a, 605, ?, 452 m$\mu$; cytochrome b, 564, 530, and 432 m$\mu$; cytochrome c, 550, 521, and 415 m$\mu$. For many years only cytochrome c was obtainable in crystalline form. It was extracted from heart muscle by trichloroacetic acid, precipitated by ammonium sulfate, and purified by dialysis (H-22–24). It can also be extracted from heart muscle by deoxycholate (H-25).

The molecular weight of *cytochrome c* is 13,000. It contains 1 iron atom per molecule, bound to protoporphyrin; hence the prosthetic group of cytochrome c is identical with protoheme, the prosthetic group of hemoglobin. It is linked to the protein portion of the molecule by two thioether bonds which are formed by the addition of a sulfhydryl group of cysteine across each of the 2 vinyl side chains of protoheme (H-26):

$$-CH=CH_2 + RSH \rightarrow -CH \begin{smallmatrix} CH_3 \\ \\ SR \end{smallmatrix}$$

Mammalian cytochrome c, owing to its high lysine content (see Table XII-3) is a basic protein with an isoelectric point close to pH 10. Digestion with pepsin and subsequent cleavage of the thioether bond with silver sulfate yields the peptide: —Val.Glu.Lys.Cys.Ala.GluN.Cys.His.Thr. Val.Glu.Lys— (H-27). The 2 cysteinyl residues of this heme-peptide chain form the links to the heme residue. The iron atom is linked to the histidine residue of this chain and to a lysine residue (H-28). The complete amino acid sequence is shown in Fig. XII-6.

```
                                                ┌──Heme──┐
                                                │
Acetyl.Gly.Asp.Val.Glu.Lys.Gly.Lys.Lys.Ileu.Phe.Val.GluN.Lys.Cys.Ala.GluN.Cys-
                                    10
His.Thr.Val.Glu.Lys.Gly.Gly.Lys.His.Lys.Thr.Gly.Pro.AspN.Leu.His.Gly.Leu.Phe-
    20                                       30
Gly.Arg.Lys.Thr.Gly.GluN.Ala.Pro.Gly.Phe.Thr.Tyr.Thr.Asp.Ala.AspN.Lys.AspN-
    40                                       50
Lys.Gly.Ileu.Thr.Tyr.Lys.Glu.Glu.Thr.Leu.Met.Glu.Tyr.Leu.Glu.AspN.Pro.Lys.Lys-
                60                                       70
Tyr.Ileu.Pro.Gly.Thr.Lys.Met.Ileu.Phe.Ala.Gly.Ileu.Lys.Lys.Lys.Thr.Glu.Arg.Glu-
                80                                       90
Asp.Leu.Ileu.Ala.Tyr.Leu.Lys.Lys.Ala.Thr.AspN.Glu
                100                      104
```

Fig. XII-6. Amino acid sequence of horse heart cytochrome c (H-54). Note that 12 of the 19 lysine residues occur in groups of 2 or 3 basic amino acids. The molecule is free of serine. In human heart cytochrome c only 12 of the 104 amino acids are different from those shown above (H-74).

Identical heme peptides have been isolated from horse, hog, and beef cytochrome c. In the cytochrome of the salmon, 1 of the lysine residues is missing; in chicken cytochrome, alanine is replaced by serine, and in the silkworm, lysine by arginine (H-29). Somewhat different peptides are obtained from the cytochromes of bacteria and molds; however, each of them contains 2 cysteinyl residues separated from each other by 2 amino acyl residues. The cytochrome c of *Pseudo-*

*monas aeruginosa* has much less lysine than the mammalian pigment; its isoelectric point is at pH 7 (H-30).

In contrast to hemoglobin, cytochrome c cannot combine with molecular oxygen, nor can it be oxidized to the ferric state by $O_2$. Evidently, the iron is bound so firmly to the other ligands that the $O_2$ molecule cannot displace them. It is not yet quite clear whether cytochrome c under physiological conditions accepts electrons directly from the reduced form of the flavoproteins. A number of other electron transferring molecules may be involved in this process, particularly cytochrome $c_1$ and b, and coenzyme Q (ubiquinone). Cytochrome $c_1$, in contrast to cytochrome c, is thermolabile and is identical with a substance earlier called cytochrome e (H-31). The absorption maxima of *cytochrome $c_1$* are shifted towards longer wave lengths by 3–4 m$\mu$ in comparison to those of cytochrome c. *Cytochrome b* can be extracted from beef heart by cholate (H-32). Particles isolated in this manner have a molecular weight of 350,000 and contain both cytochromes $c_1$ and b (H-33). After exposure of these particles to proteolytic enzymes a soluble preparation of cytochrome b (mol. wt., 175,000) is obtained.

Several cytochromes of the b-type have been isolated from various sources and have been differentiated by subscripts. One of these, cytochrome $b_2$ from yeast, acts as a lactic acid dehydrogenase; the crystalline enzyme contains deoxyribonucleic acid, two equivalents of a flavin, and two cytochromes whose absorption maxima indicate the b-type; their prosthetic group is protoheme (H-34). Cytochrome b and $c_1$ and coenzyme Q (ubiquinone) may form a complex which acts as an electron carrier from the flavoproteins to cytochrome c (H-35).

The designation coenzyme Q (H-36) for the lipid-soluble polyisoprenoid ubiquinone (H-37), a quinone of ubiquitous occurrence, suggests its participation in enzymatic electron transfer. However, we do not know at present anything about an analogous apoenzyme. The solubility of this quinone in lipids may be sufficient to make it suitable for electron transfer in the lipid-rich medium of the electron transfer particles which will be discussed later in this section.

Neither cytochrome c nor cytochrome b is reoxidized to the ferric state by $O_2$. This reaction requires another heme protein, designated by Warburg as the respiratory enzyme (H-38). It was later called cytochrome c oxidase (H-39) and is identical with *cytochrome a* (H-40). It is, however, not yet clear whether it is a homogeneous cytochrome of the a-type or a mixture of two components, designated a and $a_3$ (H-55c). Our lack of knowledge is caused by the difficulties encountered in the preparation of the cytochromes of a-type. They cannot be extracted from heart muscle by the customary solvents and seem to be bound to insoluble lipoproteins. A dispersion of the enzyme can be obtained by extraction with bile salts (cholate, deoxycholate) (H-41). Subsequent treatment with proteolytic enzymes yields a water-soluble preparation of the cytochromes a (H-42). The structure of porphyrin a differs from that of protoporphyrin by the substituents in 2-, 4-, and 8-position. The vinyl group in 2-position is replaced by —CHOH·CO·R, that in 4-position by —CH:CHR', where R and R' are alkyl residues with a total number of 13 carbon atoms. The methyl group in position 8 is replaced by a formyl group —CHO (H-43). In addition to heme a, cytochrome a also contains 1 Cu atom for each atom of iron (H-44), and also a certain amount of lipid (H-45). The copper is formyl-bound, and the activity of the enzyme is claimed to be proportional to its Cu content, although this view is not shared by all workers in this field (H-46). Nothing is known concerning the protein moiety of cytochrome a or $a_3$. The formyl group of their porphyrin ring seems to be bound to the $\epsilon$-amino group of a lysine residue of cytochrome c, forming in this manner a complex of the cytochromes a and c (H-55b).

Cytochrome a is not the only enzyme which can reduce $O_2$. The same capacity is encountered in *copper containing oxidases* (H-47). The occurrence of these in plant material has been known for many years. However, only the availability of isotopic $O^{18}_2$ made it possible to elucidate their mode of action. Experiments with the heavy oxygen isotope revealed that tyrosinase and other phenolases, which are typical copper enzymes, catalyze the direct transfer of isotopic oxygen atoms from $O_2$ to the aromatic or heterocyclic rings of the substrates (H-48). In agreement with an old claim of Warburg, molecular oxygen is here activated by a metal enzyme. Therefore, these enzymes have been designated as oxygen transferases (H-49). The copper ion of the copper containing oxidases is split off by dialysis against cyanide, but combines again with the protein in the absence of cyanide. It is not yet clear to which group of the apoenzyme the copper is bound in the active oxidase. As mentioned earlier (Chapter XII, B) the valence state of the copper ion in these enzymes is still debated.

Peroxidase and catalase are heme proteins containing protoheme as the prosthetic group. In contrast to hemoglobin, they cannot combine with $O_2$ to form a stable oxygenated complex. The decided influence of the protein component on the specificity of catalytic action is well illustrated by this comparison of catalase and peroxidase with hemoglobin. *Catalase* has been isolated in crystals from liver (H-50) by extraction with aqueous dioxane and precipitation by higher concentrations of dioxane. Its molecular weight is approximately 250,000, with 4 protoheme groups per molecule. The iron-porphyrin can be removed by dialysis against dilute HCl. The colorless apoenzyme is inactive, but is reactivated by combining with protohemin (H-51). Catalase can be considered a specific *peroxidase*.

The general reaction catalyzed by the latter type of enzyme is (XII-15):

$$H_2O_2 + \text{dihydrosubstrate} \rightarrow 2H_2O + \text{substrate} \qquad \text{(XII-15)}$$

where the substrate can be benzidine, pyrogallol or some other electron donor. In the reaction (XII-16), catalyzed by catalase, the electron (and hydrogen) donor is a second molecule of $H_2O_2$:

$$H_2O_2 + H_2O_2 \rightarrow 2H_2O + O_2 \qquad \text{(XII-16)}$$

A sharp distinction between the two enzymes cannot be made. In the presence of suitable substrates such as methanol or other alcohols catalase acts as a peroxidase and catalyzes the dehydrogenation of the alcohols (H-52). Crystalline peroxidase has been prepared from horse radish (H-53). Peroxidatic activity is exerted not only by the enzyme peroxidase, but by all iron porphyrins and also by hemoglobin. However, the peroxidatic activity of these substances is much lower than that of peroxidase. Combination of protohemin with the apoenzyme of peroxidase increases the peroxidatic activity by several orders of magnitude. The structural prerequisites for this increase in activity are still unknown.

## The Electron Transport System

In the preceding paragraphs various distinct enzymes involved in oxidation-reduction have been discussed: the pyridine nucleotide enzymes, the flavoproteins, and the different types of heme proteins. It was also indicated that the organic metabolites are usually oxidized by the oxidized form of the pyridine nucleotides, the latter by the oxidized flavoproteins, and the reduced form of the last by the ferric form of the cytochromes. The picture which emerges from this description is that of a complicated conveyer belt system in which the electrons are shifted stepwise from the electron donor through a series of electron acceptors to the last acceptor, usually the oxygen molecule. Since this book deals principally with proteins, we should ask what role the proteins play in this mechanism. In answering this question one cannot generalize. The proteins have various functions. Some of them, the apoenzymes of the pyridine nucleotide dehydrogenases, determine the substrate specificity of the dehydrogenase; they also bind substrate and coenzyme and orient them thereby in such a manner that the reaction between these two components proceeds smoothly. In the flavoproteins the coenzyme is bound more firmly to the apoenzyme; in some of these enzymes the protein still determines the substrate specificity. In the cytochromes the prosthetic group is very firmly bound. Their principal action is the reoxidation of the reduced flavoproteins. Hence, their specificity is not as limited as that of the pyridine nucleotide dehydrogenases. Cytochromes of the b, c, and a types may form a catalytic unit. The species specificity of cytochrome c has been discussed earlier (H-29, 30).

The complicated gearing work of these multiple components, each of them acting as a catalyst and at the same time as electron carrier, can be *investigated kinetically* much more easily than other protein systems because each of the components has either a typical absorption or fluorescence spectrum. Changes from the oxidized to the reduced form of each component can be measured simultaneously when suitable regions of the absorption or emission spectrum are selected and scanned during the process of oxidation-reduction. Experiments of this type have been performed with an ingenious equipment which records automatically difference spectra or translates them by means of a computing system into data which express directly the percentage of oxidized and reduced participant (H-56a). It is a great advantage of this method that it records changes not only in the soluble enzyme systems but also in those which like cytochrome a are bound to the structural elements. Obviously, corrections have to be made in such instances for scattering of light and similar phenomena.

The idea of electron transport along a chain of five or more electron acceptors suggests for obvious reasons a comparison with the transport of electrons in conductors or semiconductors. Since the biological systems do not contain metals in their elemental form, true electron conduction cannot be involved in the biological electron transport. However, some of the electron acceptors may indeed act as semiconductors. It may be recalled that the porphyrins and their metal complexes, like many of the other deeply colored organic pigments, have a metallic luster. This as well as their deep color indicate the presence of loose electrons and make understandable the participation in oxidation-reduction reactions. The finding of non-heme iron, of copper and molybdenum in flavoproteins (H-20) supports the idea of oxidation-reductions within units of closely packed oxidation-reduction catalysts (H-56b). The presence of such *"electron transferring particles"* is further supported by the observation that disintegration of these complexes into their components can be brought about by bile salts such as cholate or deoxycholate (H-56c). In some instances it has been possible to reconstitute the original electron transport system from the fractions of the disintegrated particles (H-57). The higher efficiency of the ETP (electron transferring particle) in comparison to the disintegrated system is attributed to a favorable orientation of the enzymes in the ETP which allows the electrons to be channelled smoothly through the chain of enzymes whereas in a solution of the same components only few of the frequent collisions of reacting components would bring the reacting groups into the required juxtaposition. Obviously, the closely packed unit, in which the electron losing and electron accepting groups are lined up as a series of semiconductors, is much more efficient than the aqueous solution of the same semiconductors in random orientation.

The protein components (apoenzymes) of the enzymes involved may have an important role in lining up the enzymes in the right sequence. They also prevent the polymerization of the iron porphyrins which, in aqueous solutions, takes place in the absence of proteins. In addition to the proteins, lipids seem to be involved in the formation of the ETP. This is evident not only from the presence of large amounts of lipids, particularly phospholipids (H-65), but also from the disintegrating action of the bile salts which form complexes with lipids. It is imaginable that the electron transport system is enclosed in an insulating layer of lipoproteins which protects it against the action of proteolytic enzymes or other disintegrating agents, and also prevents "short circuits" between neighboring systems. The high lipid content may also be favorable to the penetration of $O_2$ molecules into the system since the solubility of oxygen in organic solvents is much higher than that in aqueous systems.

Within the chain of cytochromes *single electrons* can be transferred by migration from the ferrous to the ferric form. However, the reduction of the pyridine nucleotides, flavins, and quinones (ubiquinone, vitamin E or K) requires an *electron pair*. We still do not know whether this transfer of two electrons takes place in two steps, each involving a single electron, or in one step. If this reaction took place in two steps, a *radical* would be formed, i.e., a molecule containing an odd number of electrons. Since single electrons, owing to their spin, have a paramagnetic moment, whereas electron pairs are diamagnetic (devoid of a magnetic moment), radicals have a paramagnetic susceptibility. Michaelis (H-58), who claimed that biological oxidations take place in single-electron transfers, was able to demonstrate the formation of radicals as intermediates in model experiments. The classical method which was based on the use of a magnetic balance is not able to detect traces of radicals. The development of instruments which measure electron spin resonance renders possible the detection of very small amounts of radicals (H-59). Radical formation has been observed (H-60) when sulfite was oxidized in illuminated solutions containing a dye as sensitizer and also during the action of xanthine oxidase in the presence of luminol (H-61). The radical forms of the flavins are intensely colored and can be produced *in vitro* at unphysiological pH values when a large excess of the reducing agent is added. It is not clear at present whether significant amounts of radicals are formed under physiological conditions when only small amounts of the normal metabolites are available as reducing agents (H-62). As pointed out earlier, the transfer of hydrogen from substrate to the pyridine nucleotides probably takes place as hydride transfer, i.e., as transfer of a unit consisting of a proton and an electron pair. Hence we need not invoke single-electron transfer and formation of a radical. Indeed, if the nicotinamide ring, even temporarily, were converted into a radical, with an odd electron in 4-position (see XII-12), the stereospecificity of the $\alpha$- and $\beta$-specific dehydrogenases would most probably get lost and random transfer to the $\alpha$- and $\beta$-position would occur. The high stereospecificity of the hydrogen transfer does not support radical formation.

Radical formation would also be expected if the reduction of molecular oxygen by cytochrome a or a$_3$ were to take place as a single step reduction. Model experiments indicate that the oxidation of heme derivatives by molecular oxygen does not involve the transfer of single electrons but two-electron steps (H-73). Free radicals are formed, however, by irradiation with X-rays or with ultraviolet light. The peroxide radicals produced by radiation react with water to yield hydroxyl radicals

$$\text{HO}\dot{} + \text{H}_2\text{O} \rightarrow \text{H}_2\text{O}_2 + \text{H}\dot{\text{O}}$$

Since such effects are not seen in the absence of radiation, it is doubtful whether significant amounts of radicals are formed under physiological conditions.

The author has earlier suggested that the heme iron acts primarily as a coordination center which binds the substrate (for instance O$_2$) and activates it by polarization so that it is reduced by an electron pair in a single step (H-63,64). In this mechanism the metal atom would not undergo any change in its valence. Indeed, no such change is observed when H$_2$O$_2$ is decomposed by hemin. Nor are radicals formed when the oxidation of H$_2$S by O$_2$ is catalyzed by heme (H-63). The importance of the coordinating power of the heme iron is demonstrated by the fact that in hemoglobin it combines with O$_2$ without being oxidized. Activation of the O$_2$ molecule by copper enzymes has been mentioned earlier in this section (H-48,49). Hence, the coordinating and chelating capacity of the metal ions may suffice to explain their catalytic activity, and it would not be necessary to invoke alternating oxidation and reduction of the metal ion. It is evident from these remarks that we are still far from understanding completely the role of proteins in electron transport systems and in the multimolecular electron transferring particles.

## *REFERENCES*

**Section A**

**A-1.** "The Enzymes" (P. D. Boyer, H. Lardy, and K. Myrbäck, eds.), 2nd ed., Vols. 1–5. Academic Press, New York, 1959–61. **A-2.** J. B. Sumner, *JBC* 69: 435

(1926). **A-3.** J. H. Northrop, *J. Gen. Physiol.* 13: 739(1929/1930). **A-4.** F. Binkley, *JBC* 236: 735(1961). **A-5.** E. K. Patterson, *JBC* 234: 2327(1959). **A-6.** E. Bamann and H. Trappman, *Adv. in Enzymol.* 21: 169(1959). **A-7.** R. L. Kisliuk, *BBA* 40: 531 (1960). **A-8.** R. J. Dubos and R. H. S. Thompson, *JBC* 124: 501(1938). **A-9.** J. B. Sumner, *Science* 108: 410(1948). **A-10.** I. Langmuir and V. J. Schaefer, *JACS* 60: 1351(1938). **A-11.** H. Sobotka and E. Bloch, *J. Phys. Chem.* 45: 9(1941). **A-12.** S. Shifrin and N. O. Kaplan, *Adv. in Enzymol.* 22: 337(1960). **A-13.** G. Hübscher *et al., Biochem. Z.* 325: 223(1954). **A-14.** B. Mackler *et al., BBA* 15: 437(1954). **A-15.** G. V. Marinetti *et al., JBC* 224: 819(1957). **A-16.** V. A. Engelhardt and M. N. Ljubimova, *Nature* 144: 668(1939). **A-17.** R. Lumry *in* "The Enzymes (P. D. Boyer, H. Lardy, and K. Myrbäck, eds.), 2nd ed., Vol. 1, p. 158. Academic Press, New York, 1959. **A-18.** F. Haurowitz, and P. Schwerin, *Enzymologia* 9: 193(1940). **A-19.** K. G. Paul, *Acta Chem. Scand.* 5: 389(1951). **A-20.** E. L. Smith and L. T. Hanson, *JBC* 176: 997(1948); 179: 803(1949). **A-21.** D. E. Koshland, *in* "The Enzymes" (P. D. Boyer, H. Lardy, and K. Myrbäck, eds.), 2nd ed., Vol. 1, p. 305. Academic Press, New York, 1959. **A-22.** A. Rothen, *J. Phys. Chem.* 63: 1929(1959). **A-23.** R. Czok and T. Bücher, *Adv. in Protein Chem.* 15: 315(1960). **A-24.** S. Schwimmer and A. B. Pardee, *Adv. in. Enzymol.* 14: 375(1953). **A-25.** F. Turba, *Adv. in Enzymol.* 22: 417 (1960). **A-26.** K. Okunuki, *Adv. in Enzymol.* 23: 29(1961). **A-27.** M. Cohn and J. S. Leigh, *Nature* 193: 1037(1962). **A-28.** S. B. Tove, *BBA* 57: 230(1962).

*Section B*

**B-1.** D. E. Koshland, *Adv. in Enzymol.* 22: 45(1960); H. Lindley, *Adv. in Enzymol.* 15: 271 (1954). **B-2.** A. K. Balls and E. F. Jansen, *Adv. in Enzymol.* 13: 321(1952). **B-3.** N. K. Shaffer *et al., JBC* 225: 197(1957). **B-4.** D. E. Koshland and M. J. Erwin, *JACS* 79: 2657(1957). **B-5.** H. N. Rydon, *Nature* 182: 928(1958). **B-6.** M. A. Naughton *et al., BJ* 77: 149(1960). **B-7.** C. Milstein and F. Sanger, *BJ* 79: 456(1961). **B-8.** T. Spencer and J. M. Sturtevant, *JACS* 81: 1874(1959). **B-9.** A. K. Balls and H. N. Wood, *JBC* 219: 245(1956). **B-10.** E. Katchalsky *et al., ABB* 88: 361(1960). **B-11.** R. B. Merrifield and D. W. Woolley, *FP* 17: 275(1958). **B-12.** I. W. C. Cunningham, *Science* 125: 1145(1957). **B-13.** H. A. Saroff, *Enzymologia* 21: 101(1957). **B-14.** T. C. Bruice and J. M. Sturtevant, *BBA* 30: 208(1958). **B-15.** D. Findlay *et al., Nature* 190: 781(1961). **B-16.** B. S. Hartley, *Ann. Rev. Biochem.* 29: 45(1960). **B-17.** P. D. Boyer, *in* "The Enzymes" (P. D. Boyer, H. Lardy, and K. Myrbäck, eds.), 2nd ed., Vol. 1, p. 517. Academic Press, New York, 1959. **B-18.** B. L. Vallee *et al., JBC* 236: 2244(1961). **B-19.** J. H. R. Kägi and B. L. Vallee, *JBC* 235: 3188(1960). **B-20.** H. Theorell *et al., FP* 20: 967(1961). **B-21.** B. L. Vallee and W. E. C. Wacker, *JACS* 78: 1771(1956). **B-22.** C. Frieden, *BBA* 27: 431(1958). **B-23.** B. C Malmström and A. Rosenberg, *Adv. in Enzymol.* 21: 131(1959). **B-24.** E. L. Smith, *in* "The Enzymes" (P. D. Boyer, H. Lardy, and K. Myrbäck, eds.), 2nd ed., Vol. 4, p. 1. Academic Press, New York, 1960. **B-25.** B. L. Vallee *et al., JBC* 234: 2901 (1959). **B-26.** H. A. Fadl and E. J. King, *BJ* 44: 435(1949). **B-27.** K. Nakamura, *BBA* 30: 640(1958). **B-28.** B. M. Malström *et al., Nature* 183: 321(1959). **B-29.** R. H. Sands and H. Beinert, *Biochem. Biophys. Research Communs.* 1: 175(1959). **B-30.** H. R. Mahler, *Science* 124: 705(1956). **B-31.** R. C. Bray, *BJ* 81: 196(1961). **B-32.** J. M. Sturtevant, *Brookhaven Symposia Quant. Biol.* 13: 161(1960). **B-33.** M. Cohn, J. S. Leigh, *Nature* 193: 1037(1962). **B-34.** H. Greenberg and D. Nachmansohn, *Biochem. Biophys. Research Communs.* 7: 189(1962). **B-35.** K. Simon, *Z. Naturforsch.* 17b: 371(1962). **B-36.** G. J. S. Rao and L. K. Tamachandran, *BBA* 59: 507(1962). **B-37.** K. Onoue *et al. J. Biochem. (Tokyo)* 51: 443(1962). **B-38.** M. L. Bender, *JACS* 84: 2582(1962); T. C. Bruice *et al. JACS* 84: 3012(1962).

## Section C

**C-1.** J. S. Fruton and M. Bergmann, *JBC* 145: 253(1942). **C-2.** P. Desnuelle, *in* "The Enzymes" (P. D. Boyer, H. Lardy and K. Myrbäck, eds.), 2nd ed., Vol. 4, 93. Academic Press, New York, 1960. **C-3.** G. W. Schwerdt *et al.*, *JBC* 172: 221(1948); H. Neurath *et al.*, *JBC* 175: 893(1948). **C-4.** D. G. Doherty, *JACS* 77: 4887(1955). **C-5.** P. Schwartz and H. E. Carter, *Proc. Natl. Acad. Sci. U.S.* 40: 499(1954). **C-6.** F. Haurowitz, "Progress in Biochemistry since 1949," p. 31. Interscience, New York, 1959. **C-7.** G. Bredig and F. Gerstner, *Biochem. Z.* 250: 414(1932). **C-8.** A. and E. Albers, *Z. Naturforsch.* 9b: 122(1954). **C-9.** A. A. Balandin *et al.*, *Doklady Akad. Nauk S.S.S.R.* 127: 557(1959). **C-10.** P. G. W. Plageman *et al.*, *JBC* 235: 2282(1960). **C-11.** J. Paul and P. Fottrell, *BJ* 78: 418(1961); *Ann. N.Y. Acad. Sci.* 94: 655(1961); C. L. Markert and F. Meller, *Proc. Natl. Acad. Sci. U.S.* 45: 753(1959). **C-12.** R. L. McGeachin and J. M. Reynolds, *BBA* 39: 531(1960). **C-13.** H. N. Kirkman *et al.*, *Proc. Natl. Acad. Sci. U.S.* 46: 938(1960). **C-14.** C. Anfinsen *et al.*, *JBC* 234: 1118 (1959). **C-15.** J. Monod *et al.*, *BBA* 7: 585(1951). **C-16.** M. R. Pollock *et al.*, *BJ* 62: 387, 391(1956). **C-17.** M. Shepherdson and A. B. Pardee, *JBC* 235: 3233(1960). **C-18.** T. Wieland and G. Pfleiderer, *Angew. Chemie* 74: 261(1962).

## Section D

**D-1.** B. Chance, *Adv. in Enzymol.* 12: 153(1951). **D-2.** L. Terminiello *et al.*, *ABB* 73: 171(1958). **D-3.** R. Kuhn, *Z. physiol. Chem.* 135: 1(1924). **D-4.** L. Michaelis and P. Rona, *Ber. deut. chem. Ges.* 60: 62(1914). **D-5.** B. Chance and R. R. Ferguson, *in* "The Mechanism of Enzyme Action," (W. McElroy and B. Glass, eds.), p. 389. Johns Hopkins, Baltimore, Maryland, 1954. **D-6.** S. F. Velick, *JBC* 233: 1455 (1958). **D-7.** C. Frieden, *JBC* 234: 2891(1959). **D-8.** H. B. Bull and B. T. Currie, *JACS* 71: 2758(1949). **D-9.** R. Lumry, *in* "The Enzymes" (P. D. Boyer, H. Lardy, and K. Myrbäck, eds.), 2nd ed., Vol. 1, p. 158. **D-10.** J. A. Thoma and D. E. Koshland, Jr., *JACS* 82: 3329(1960); D. E. Koshland, Jr., ref. A - 1, Vol. 1, p. 305. **D-11.** G. Schramm and J. Primosigh, *Z. physiol. Chem.* 283: 34(1948). **D-12.** H. Hartmann and L. Hübner, *Z. Elektrochem.* 155: 225(1951). **D-13.** L. Lawrence and W. J. Moore, *JACS* 73: 3973(1952). **D-14.** C. B. Anfinsen *et al.*, *BBA* 17: 141(1955).

## Section E

**E-1.** R. A. Alberty, *Adv. in Enzymol.* 17: 1(1956). J. Z. Hearon *et al.*, Vol. 1, p. 49. 1959. **E-2.** L. Michaelis and M. Menten, *Biochem. Z.* 49: 333(1913). **E-3.** H. Lineweaver and D. Burk, *JACS* 56: 658(1934). **E-4.** E. Elkins-Kaufman and H. Neurath, *JBC* 175: 893(1948). **E-5.** H. Neurath and G. W. Schwerdt, *Chem. Revs.* 46: 88(1950). **E-6.** H. Lineweaver, *JACS* 61: 403(1939). **E-7.** M. Dixon, *Discussions Faraday Soc.* 20: 9(1956). **E-8.** R. B. Johnston *et al.*, *JBC* 187: 205(1950). **E-9.** D. B. Sprinson and D. Rittenberg, *Nature* 167: 484(1951). **E-10.** G. Weber, *Discussions Faraday Soc.* 20: 156(1956). **E-11.** H. B. Bull and B. T. Currie, *JACS* 71: 2758(1949). **E-12.** A. H. Mehler, "Introduction to Enzymology," p. 9. Academic Press, New York, 1957. **E-13.** R. A. Alberty *et al.*, *JACS* 84: 4367ff(1962); W. W. Cleland, *BBA* 67: 104 (1962).

## Section F

**F-1.** O. Hofmann-Ostenhof, *Ann. Rev. Biochem.* 29: 73(1960). **F-2.** P. Desnuelle, *in* "The Enzymes" (P. D. Boyer, H. Lardy, and K. Myrbäck, eds.), 2nd ed., Vol. 4, 93. **F-3.** J. H. Northrop and M. Kunitz, *J. Gen. Physiol.* 18: 433(1935). **F-4.** G. W. Schwert, *JBC* 159: 655(1949). **F-5.** F. R. Bettelheim and H. Neurath,

*JBC* 212: 241(1955). **F-6.** M. Rovery *et al.*, *Bull. soc. chim. biol.* 38: 1101(1956); *BBA* 17: 565(1955); W. Dreyer and H. Neurath, *JBC* 217: 527(1955). **F-7.** I. Tinocco, Jr., *ABB* 76: 148(1958). **F-8.** H. Tauber, *JACS* 73: 4965(1951). **F-9.** J. Horowitz and F. Haurowitz, *BBA* 33: 231(1959). **F-10.** P. Desnuelle, *in* "The Enzymes" (P. D. Boyer, H. Lardy, and K. Myrbäck, eds.), 2nd ed., Vol. 4, p. 119. Academic Press, New York, 1960. **F-11.** J. H. Northrop and M. Kunitz, *Science* 73: 262(1931); 80: 505(1934). **F-12.** J. H. Northrup, "Crystalline Enzymes." Columbia Univ. Press, New York, 1939. **F-13.** E. W. Davie and H. Neurath, *JBC* 212: 515 (1955). **F-14.** F. Šorm *et al.*, *Collection Czechoslov. Communs.* 23: 985(1958). **F-15.** G. H. Dixon and H. Neurath, *FP* 16: 791(1957); *JBC* 233: 1373(1958). **F-16.** S. G. Waley and J. Watson, *BJ* 57: 529(1954). **F-17.** S. Nakamura and S. Takeo, *Z. physiol. Chem.* 320: 280(1960). **F-18.** E. Wainfan and G. P. Hess, *JACS* 82: 2069 (1960). **F-19.** W. Troll and S. Sherry, *JBC* 213: 881(1955). **F-20.** J. S. Fruton *et al.*, *JBC* 226: 173(1957). **F-21.** F. A. Bovey and S. S. Yanari, *in* "The Enzymes" (P. D. Boyer, H. Lardy, and K. Myrbäck, eds.), 2nd ed., Vol. 4, p. 63. Academic Press, New York, 1960. **F-22.** K. Heirwegh and P. Edman, *BBA* 24: 219(1957). **F-23.** H. Van Vunakis and R. M. Herriott, *BBA* 23: 600(1957). **F-24.** V. Richmond *et al.*, *BBA* 29: 453(1958). **F-25.** H. Neumann *et al.*, *BJ* 73: 22(1959). **F-26.** K. Tokuyasu and M. Funatsu, *J. Biochem.* (*Tokyo*) 49: 297(1961). **F-27.** R. M. Herriott, *J. Gen. Physiol.* 21: 501(1938). **F-28.** A. K. Balls and H. Lineweaver, *JBC* 130: 669(1939). **F-29.** E. L. Smith and J. R. Kimmel, *in* "The Enzymes" (P. D. Boyer, H. Lardy, and K. Myrbäck, eds.), 2nd ed., Vol. 4, p. 134. Academic Press, New York, 1960; *Adv. in Enzymol.* 19: 267(1957). **F-30.** R. L. Hill and E. L. Smith, *BBA* 19: 376(1956). **F-31.** E. Waldschmidt-Leitz *et al.*, *Naturwiss.* 18: 645(1930). **F-32.** R. B. Johnston and J. S. Fruton, *JBC* 185: 629(1950). **F-33.** O. O. Blumenfeld and G. E. Perlmann, *JBC* 236: 2472(1961). **F-34.** M. A. Naughton and F. Sanger, *BJ* 78: 156(1961). **F-35.** O. V. Kazakova *et al.*, *Doklady Akad. Nauk. S.S.S.R.* 122: 657(1958); W. Grassmann and A. Nordwig, *Z. physiol. Chem.* 322: 267(1960); S. Michaels *et al.*, *BBA* 29: 451(1958). **F-36.** B. Hagihara, *in* "The Enzymes" (P. D. Boyer, H. Lardy, and K. Myrbäck, eds.), 2nd ed., Vol. 4, p. 193. Academic Press, New York, 1960; K. Okunuki, *in* "Analytical Methods of Protein Chemistry" (P. Alexander and R. J. Block, eds.), Vol. 3, p. 31. Pergamon Press, New York, 1960. **F-37.** M. L. Anson, *J. Gen. Physiol.* 20: 663(1936). **F-38.** H. Neurath *et al.*, *JBC* 223: 457(1956); *in* "The Enzymes" (P. D. Boyer, H. Lardy, and K. Myrbäck, eds.), 2nd ed., Vol. 4, p. 11. Academic Press, New York, 1960. **F-39.** L. Weil *et al.*, *ABB* 79: 44(1959). **F-40.** E. L. Smith and R. L. Hill, *in* "The Enzymes" (P. D. Boyer, H. Lardy, and K. Myrbäck, eds.), 2nd ed., Vol. 4, p. 37. Academic Press, New York, 1960. **F-41.** R. L. Hill and E. L. Smith, *JBC* 228: 577(1957). **F-42.** E. L. Smith, Vol. 4, p. 1. **F-43.** M. Brenner and A. Vetterli, *Helv. Chim. Acta* 40: 937(1957). **F-44.** C. S. Hanes *et al.*, *Nature* 166: 288(1950); H. Waelsch *et al.*, *ABB* 27: 237(1950). **F-45.** N. J. Berridge, *Adv. in Enzymol.* 15: 423(1954). **F-46.** E. Abderhalden and S. Buadze, *Fermentforsch.* 14: 215(1934). **F-47.** F. Haurowitz *et al.*, *Bull. Fac. Med. Istanbul* 11: 30(1948). **F-48.** J. B. Sumner, *JBC* 69: 435(1926). **F-49.** J. B. Sumner *et al.*, *JBC* 125: 33(1938). **F-50.** P. Desnuelle, *in* "The Enzymes" (P. D. Boyer, H. Lardy, and K. Myrbäck, eds.), 2nd ed., Vol. 4, p. 121. Academic Press, New York, 1960. **F-51.** D. H. Spackman *et al.*, *JBC* 212: 255(1955). **F-52.** O. O. Blumenfeld and G. E. Perlmann, *J. Gen. Physiol.* 42: 553(1959). **F-53.** E. L. Smith and A. Stockell, *JBC* 207: 501(1954). **F-54.** E. L. Smith *et al.*, *Biochem. Soc. Symposia* 21: 88(1961). **F-55.** B. Kassell and M. Laskowski, *JBC* 236: 1996(1961). **F-56.** J. J. Cebra *et al.*, *JBC* 236: 1720(1961). **F-57.** A. Bar-Eli and E. Katchalski, *Nature* 188: 856(1960). **F-58.** N. Grubhofer and L. Schleith, *Z. physiol. Chem.* 297: 108(1954). **F-59.** M. A. Mitz and L. J. Summaria, *Nature* 189: 576(1961). **F-60.** K. Katchalski *et al.*, *JBC* 237: 1832(1962). **F-61.** N. Sharon *et al.*, *ABB* 97: 219(1962). **F-62.** E. R. Simmons *et al.*, *JBC* 236: 64(1961). **F-63.** J. Kraut *et al.*, *Proc. Natl. Acad. Sci. U.S.* 48: 1417(1962). **F-64.** R. M. Herriott, *J. Gen. Physiol.* 45, *Suppl.*, p. 57 (1962). **F-65.** L. S. Lokhshina, *Aktualnye Vopr. Sovrem. Biokhim.* 2: 71(1962) see *Chem. Abstr.* 57: 15503(1962). **F-66.** J. B. Clegg and K. Bailey, *BBA* 63: 525(1962).

### Section G

**G-1.** B. Axelrod, *Adv. in Enzymol.* 17: 159(1956). **G-2.** L. Berger *et al.*, *J. Gen. Physiol.* 29: 141(1946); M. McDonald and M. Kunitz, *J. Gen. Physiol.* 29: 142(1946). **G-3.** J. F. Taylor *et al.*, *BBA* 20: 109, 115(1956). **G-4.** D. E. Koshland, Jr. and M. J. Erwin, *JACS* 79: 2657(1957). **G-5.** V. Jagannathan and J. M. Luck, *JBC* 179: 569 (1949). **G-6.** J. B. Sidbury and V. A. Najjar, *JBC* 227: 517(1957). **G-7.** L. F. Leloir *et al.*, *ABB* 18: 201(1948); 19: 339(1948); 23: 55(1949). **G-8.** R. Dubos, *Science* 85: 549(1937). **G-9.** M. Kunitz, *J. Gen. Physiol.* 24:15(1940). **G-10.** C. B. Anfinsen *et al.*, *BBA* 17: 141(1955). **G-11.** C. H. W. Hirs, W. H. Stein, and S. Moore, *JBC* 235: 633(1960). **G-12.** C. B. Anfinsen, *FP* 16: 783(1957). **G-13.** C. B. Anfinsen *J. Comp. Biochem.* 4: 229(1962); *Proc. Natl. Acad. Sci. U.S.* 47: 1309(1961). **G-14.** H. A. Scheraga, *JACS* 82: 3947(1960). **G-15.** F. M. Richards and P. J. Vithayathil, *JBC* 234: 1459(1959); 236: 1380, 1386(1961). **G-16.** C. Tanford and J. D. Hauenstein, *BBA* 19: 535(1956). **G-17.** C. B. Anfinsen *et al.*, *JBC* 234: 1118(1959). **G-18.** R. Markham and J. D. Smith, *BJ* 52: 552(1952). **G-19.** G. A. Gilbert and A. D. Patrick, *BJ* 51: 181, 186(1952). **G-20.** C. F. Cori and B. Illingsworth, *Proc. Natl. Acad. Sci. U.S.* 43: 547(1957). **G-21.** S. Ochoa *et al.*, *JBC* 236: 3303(1961). **G-22.** R. A. Dedonder, *Ann. Rev. Biochem.* 30: 347(1961). **G-23.** L. F. Leloir *et al.*, *JBC* 179: 497(1949). **G-24.** K. Okunuki, *in* "Analytical Methods of Protein Chemistry" (P. Alexander and R. J. Block, eds.), Vol. 3, p. 31. Pergamon Press, New York, 1960. **G-25.** J. A. Thoma and D. E. Koshland, Jr., *FP* 19: 46(1960). **G-26.** L. L. Campbell *et al.*, *JBC* 236: 2952(1961). **G-27.** P. Desnuelle, *Adv. in Enzymol.* 23: 129(1961). **G-28.** F. Haurowitz *et al.*, *ABB* 59: 52(1955). **G-29.** L. Sarda and P. Desnuelle, *BBA* 30: 513(1958). **G-30.** M. Liefländer and H. Stegemann, *Z. physiol. Chem.* 325: 204(1961). **G-31.** C. F. Cori *et al.*, *JBC* 173: 591(1948). **G-32.** O. Warburg and W. Christian, *Biochem. Z.* 310: 384(1942). **G-33.** T. Bücher, *BBA* 1: 467(1947). **G-34.** O. Nylander and B. G. Malmström, *BBA* 34: 196(1959). **G-35.** O. Warburg and W. Christian, *Biochem. Z.* 314: 149(1943). **G-36.** J. Muus, *JACS* 76: 5163(1954). **G-37.** E. A. Stein *et al.*, *JBC* 235: 204(1961). **G-38.** K. Wallenfels and O. P. Malhotra, *in* "The Enzymes" (P. D. Boyer, H. Lardy, and K. Myrbäck, eds.), 2nd ed., Vol. 4, p. 413. Academic Press, New York, 1960. **G-39.** G. Gehrmann, and S. Okada, *BBA* 23: 621(1957). **G-40.** H. Boser, *Z. physiol. Chem.* 300: 1(1955). **G-41.** R. R. Rabin *et al.*, *BJ* 85: 127 ff.(1962). **G-42.** F. M. Richards *et al.*, *Brookhaven Symposia Quant. Biol.* 13: 135(1960); *Biochemistry* 1: 295(1962). **G-43.** R. C. Hughes *et al.*, *JBC* 237: 40(1962). **G-44.** M. Liefländer, *Z. physiol. Chem.* 329: 289 (1962). **G-45.** E. E. Rickli and J. T. Edsall, *JBC* 236: PC77(1961). **G-46.** E. G. Smyth *et al.*, *JBC* 237: 1845(1962); C. B. Anfinsen, *Ibid.* p. 1851; E. Gross and B. Witkop, *Ibid.* p. 1856. **G-47.** M. Wenzel, *Z. physiol. Chem.* 327: 13(1961). **G-48.** J. Jolles and P. Jolles, *Compt. rend.* 253: 2773(1962).

### Section H

**H-1.** M. Dixon, *Science* 132:1548(1960). **H-2.** B. Vennesland and F. H. Westheimer, *in* "Mechanism of Enzyme Action," (W. D. McElroy and B. Glass, eds.), pp. 321, 357.

Johns Hopkins, Baltimore, Maryland, 1954. **H-3.** B. Vennesland, *Discussions Faraday Soc.*, 20: 240(1956); *JBC* 222: 685(1956). **H-4.** B. L. Vallee *et al.*, *JACS* 78: 5879(1956). **H-5.** H. R. Mahler and J. Douglas, *JACS* 79: 1159(1957); K. Wallenfels and H. Schüly, *Biochem. Z.* 329: 59(1957). **H-6.** C. Frieden, *BBA* 27: 431(1958). **H-7.** R. Caputto and M. Dixon, *Nature* 156: 630(1945). **H-8.** E. Racker and I. Krimsky, *JBC* 198: 731(1952); H. Holzer and E. Holzer, *Z physiol. Chem.* 291: 67(1952). **H-9.** J. H. Park *et al.*, *JBC* 236: 136(1961). **H-10.** S. F. Velick, *JBC* 233: 1455(1957). **H-11.** P. Greengard, *Nature* 178: 632(1956). **H-12a.** N. O. Kaplan *et al.*, *JBC* 195: 107(1952); *Adv. in Enzymol.* 22: 337(1960). **H-12b.** K. Deigo and L. J. Reed, *JACS* 84: 666(1962). **H-13.** O. Warburg and W. Christian, *Biochem. Z.* 266: 377(1933); 298: 150(1938). **H-14.** H. Theorell, *Biochem. Z.* 272: 155(1934). **H-15.** H. Theorell and A. Akeson, *ABB* 65: 439(1956). **H-16.** E. C. de Renzo, *Adv. in Enzymol.* 17: 293(1956). **H-17.** H. C. Friedman and B. Vennesland, *JBC* 235: 1526(1960). **H-18.** K. Laki, *Z. physiol. Chem.* 249: 61, 63(1937). **H-19.** T. P. Singer, *Adv. in Enzymol.* 18: 65(1957). **H-20.** H. R. Mahler, *Adv. in Enzymol.* 17: 233(1956). **H-21.** V. Massey, *BBA* 37: 314(1960); *JBC* 235: PC 47(1960). **H-22.** D. Keilin and E. F. Hartree, *Proc. Roy. Soc.* B122: 298(1937); 127: 167(1939). **H-23.** B. Hagihara *et al.*, *Nature* 178: 629(1956). **H-24.** K. Okunuki, *in* **F-36. H-25.** E. G. Ball and O. Cooper, *JBC* 226: 755(1957). **H-26.** H. Theorell and A. Akeson, *Science* 90: 67(1939); *JACS* 63: 1809(1941). **H-27.** H. Tuppy *et al.*, Monatshefte 85: 807(1954); *Acta Chem. Scand.* 9: 353(1955). **H-28.** R. Margoliasch *et al.*, *BJ* 71: 559(1959). **H-29.** H. Tuppy *et al.*, *Naturwiss.* 46: 35(1959); *Science* 13: 641(1959). **H-30.** M. D. Kamen and Y. Takeda, *BBA* 21: 518(1956). **H-31.** D. Keilin and E. F. Hartree, *Nature* 176: 200(1955); R. A. Bomstein *et al.*, *BBA* 50: 527(1961). **H-32.** J. Sekuzu and K. Okunuki, *J. Biochem. (Tokyo)* 43: 107(1956). **H-33.** D. Feldman and W. W. Wainio, *JBC* 235: 3635(1960). **H-34.** R. K. Morton, *Nature* 192: 727(1961). **H-35.** Y. Hatefi *et al.*, *Biochem. Biophys. Research Communs.* 4: 447(1961). **H-36.** R. L. Lester *et al.*, *JACS* 80: 4751(1958). **H-37.** R. A. Morton *et al.*, *Nature* 182: 1764(1958). **H-38.** O. Warburg and E. Negelein, *Biochem. Z.* 202: 202(1929). **H-39.** D. Keilin and E. F. Hartree, *Proc. Roy. Soc.* B125: 171(1938). **H-40.** R. Lemberg, *Adv. in Enzymol.* 23: 265(1961). **H-41.** W. W. Wainio, *Science* 106: 471(1947); D. E. Griffith and D. C. Wharton, *JBC* 236: 1857(1961). **H-42.** L. Smith and E. Stotz, *JBC* 209: 819(1954). **H-43.** M. Morrison *et al.*, *JBC* 235: 1202 (1960). **H-44.** S. Takemori, *J. Biochem. (Tokyo)* 47: 382(1960). **H-45.** W. W. Wainio *et al.*, *JBC* 234: 2433(1959); 235 PC 12(1960). **H-46.** T. Yonetani, *Biochem. Biophys. Research Communs.* 3: 549(1960); R. H. Sands and H. Beinert, *Biochem. Biophys. Research Communs.* 1: 175(1959). **H-47.** J. M. Nelson and C. R. Dawson, *Adv. in Enzymol.* 4: 99(1944). **H-48.** O. Hayaishi *et al.*, *JBC* 229: 889(1957). **H-49.** H. S. Mason, *Adv. in Enzymol.* 19: 79(1957). **H-50.** J. B. Sumner and A. L. Dounce, *JBC* 121: 417(1937). **H-51.** K. Agner, *Z. physiol. Chem.* 235: II(1935). **H-52.** D. Keilin and E. F. Hartree, *BJ* 39: 283, 289(1945). **H-53.** H. Theorell, *Arkiv Kemi* 15B, No. 24; 16A, No. 2 (1942). **H-54.** E. Margoliash *et al.*, *Nature* 192: 1125(1961). **H-55a.** W. W. Wainio and S. J. Cooperstein, *Adv. in Enzymol.* 17: 329(1956); E. F. Hartree, *Adv. in Enzymol.* 18: 1(1957). **H-55b.** Y. Orii, *J. Biochem. (Tokyo)* 51: 204 (1962); K. Okunuki *et al.*, *Koso Kagaku Shinpojiumu* 15: 251(1961); see *Chem. Abstr.* 56: 3804(1962). **H-55c.** E. R. Redfearn, *Ann. Repts. on Progr. Chem.* 57: 395(1961). **H-56a.** B. Chance *et al.*, *JBC* 235: 2426(1960). **H-56b.** D. E. Green *et al.*, *BBA* 15: 435(1954). **H-56c.** F. L. Crane and J. L. Glenn, *BBA* 24: 100(1957). **H-57.** D. Keilin and T. E. King, *Proc. Roy. Soc.* B152: 163(1960); Y. Hatefi, *et al.*, *JBC* 237: 2661(1962). **H-58.** L. Michaelis, *FP* 7: 513(1948). **H-59.** B. Commoner *et al.*, *Proc. Natl. Acad. Sci. U.S.* 44: 1100, 1110(1958); 47: 1355(1961); R. C. Bray *et al.*, *BJ* 81:

196(1961); E. Beinert, *JBC* 225: 465(1957). **H-60.** I. Fridovich and Ph. Handler, *JBC* 235: 1835(1960). **H-61.** J. R. Totter *et al., JBC* 235: 1839(1960). **H-62.** H. R. Mahler and L. Brand, *in* "Free Radicals in Biological Systems" (M. S. Blois et al., eds.), p. 75. Academic Press, New York, 1961. **H-63.** F. Haurowitz, *Enzymologia* 10: 141(1941). **H-64.** A. L. Dounce and N. E. Fischer, *Enzymologia* 17: 182(1956). **H-65.** S. Fleischer *et al., JBC* 236: 2936(1961). **H-66.** K. Wallenfels and A. Arens, *Biochem. Z.* 332: 217(1961). **H-67.** D. M. Gibson *et al., JBC* 203: 397(1953). **H-68.** C. A. Appleby *et al., BJ* 75: 72(1960). **H-69.** K. Takahashi *et al., J. Biochem (Tokyo)* 46: 1323 (1959). **H-70.** Y. Morita and K. Kameda, *Bull. Agr. Soc. Japan* 23: 28(1959). **H-71.** B. G. Malmström *et al., JBC* 234: 1108(1959). **H-72.** A. Pihl *et al., Acta Chem. Scand.* 15: 1271(1961). **H-73.** J. H. Wang and W. S. Brinigar, *Proc. Natl. Acad. Sci. U.S.* 46: 958(1960). **H-74.** H. Matsubara and E. L. Smith, *JBC* 237: 3575(1962).

# Chapter XIII

## Proteins with Hormone Action

### A. General Remarks (A-1,4)

Hormones are products of the endocrine (ductless) glands and are secreted by these directly into the blood stream by which they are transported to the site of their action. The hormones of the thyroid gland, the parathyroids, the pancreas, and the hypophysis are proteins or protein derivatives. They consist of the well-known amino acids. It has been known for many years that the biological action of the thyroid hormone is caused by its content of iodinated amino acids. Nothing was known, however, about the active site of the other protein hormones. In the last decade, the amino acid sequence of some of the hormones has been unravelled, and some information on the structures responsible for the hormonal activity has been gained. Nevertheless, we are still far from understanding the mechanism of hormone action, particularly the action of the hormone in its target organ. For a long time it was believed that the hormones, quite generally, affect the action of enzymes in the cells of their target organ. There is little experimental evidence for this view. It is more probable that some of the hormones affect the permeability of the cells of the target organ and thus interfere with the metabolic processes (A-2).

Hormones prepared from animal endocrine glands have been administered parenterally to human patients without any serious consequences. This may seem surprising since many of these hormones are proteins which are antigenic and give rise to the formation of antibodies if injected to a foreign species. Antibody formation is indeed observed after the injection of thyroglobulin, the hormone of the thyroid gland. Antihormones are occasionally formed in persons injected with insulin or with the gonadotropic hormones (A-3). However, severe anaphylactic reactions are rare. This may be attributed to the small size of many of the protein hormones, and to their structural similarity in different species. Thus the insulins of the mammalian species studied differ from each other in only 4 of their 51 amino acids.

### B. Thyroglobulin and the Thyroid Hormones (B-1)

The hormonal activity of the thyroid gland and the importance of its iodine content for this activity have been known for many years. An

*347*

iodinated protein called thyroglobulin can be extracted from the gland by dilute salt solutions; it is precipitated from the extracts by acidification with acetic acid or by half-saturation with ammonium sulfate (B-2).

When thyroglobulin is digested by trypsin, the hormonal activity increases (B-3). The iodinated derivatives of tyrosine and thyronine which have been found in the hydrolyzate of thyroglobulin are shown in the following tabulation (B-1,4,5).

| Iodinated amino acid | Activity (Thyroxine = 100) | Symbol (B-6) | Total iodine (%) |
|---|---|---|---|
| Thyroxine (3,5,3',5'-tetraiodothyronine) | (100) | $T_4$ | 20 |
| 3,5,3'-Triiodothyronine | 500 (B-5) | $T_3$ | Little |
| 3,3',5'-Triiodothyronine | 10 | $T'_3$ | Little |
| 3,3'-Diiodothyronine | 85 | $T_2$ | Little |
| Diiodotyrosine (iodogorgoic acid) | — | DIT | 50–60 |
| Monoiodotyrosine | — | MIT | 10–15 |

The molecular weight of thyroglobulin is approximately 700,000 and its iodine content 0.6%; hence each molecule contains about 33 iodine atoms, most of them present as DIT, some as $T_4$ and MIT. The amounts of $T_3$, $T'_3$, $T_2$, and iodinated histidine are very small. Since the mixture of the iodinated amino acids has a higher hormonal activity than thyroglobulin, the iodoamino acids are frequently considered as the true hormones.

Thyroxine, the principal hormone of the gland, has been isolated after hydrolyzing thyroglobulin with barium hydroxide (B-7). The detection of the other thyronine derivatives has been facilitated by the use of the radioactive isotope $I^{131}$. The various iodinated amino acids can be separated from each other by paper chromatography and detected on the paper by autoradiography. Since thyroglobulin does not combine with antibodies to the diiodotyrosine group (B-8), it seems that the iodinated amino acids of thyroglobulin are not present in the surface of the large molecule; they may be buried inside the molecule and may thus be inaccessible to the combining groups of antibodies.

Thyronine: 4-Hydroxyphenyl-tyrosine

The reason for the accumulation of iodine in the thyroid gland is not yet known. The first step in the utilization of ingested iodide is most probably the oxidation of iodide to iodine. The cell-free homogenate of the thyroid gland has the ability to form iodinated amino acids from iodide in the presence of cupric ions (B-9,10) and/or a peroxidase (B-20). Flavoproteins also seem to be involved in this process which is inhibited by goitrogens such as thiocyanate or thiourea derivatives (B-11).

Thyroxine is also formed *in vitro* when an alkaline solution of casein is incubated with iodine at 37°C (B-12,13) or when DIT is exposed to the action of $H_2O_2$ in alkaline solution (B-14). The first product formed from DIT is 3,5-diiodo-4-hydroxybenzaldehyde which, on condensation with another molecule of DIT, forms thyroxine (B-4). Thyroxine and the other thyronine derivatives are only slightly soluble in water; they can be extracted from the hydrolyzates of thyroglobulin by butanol. On electrophoresis of blood serum, thyroxine migrates with the pre-albumin and albumin fraction, evidently bound to the proteins of these fractions (B-15,16).

Injection of thyroxine or the other hormonally active iodoamino acids causes an increase in the basal metabolic rate, i.e., in the rate of oxidations. The mechanism of this action is not yet clear. It has been attributed to swelling of the mitochondria, changes in their permeability and uncoupling of oxidation from phosphorylation (B-17,18). The thyroid hormones also increase the incorporation of amino acids into protein in the cell-free rat liver homogenate (B-19). Deficiency of the hormone results in a slowing down of the metabolic rate.

## C. Insulin (C-1)

Insulin is produced in the endocrine tissue of the pancreas, in the islets of Langerhans. In certain classes of fish these islets form a separate organ. In these fish, insulin is found in the islet organ and not in the hepato-pancreas. In the higher vertebrates the pancreas is predominantly an exocrine gland which produces numerous enzymes such as trypsin, lipase, amylase, ribonuclease, and chymotrypsin. Hence, insulin is rapidly destroyed by proteolysis if the pancreas or its extracts are stored for a period of time. This loss of insulin by proteolytic cleavage is the reason why previous workers failed to obtain active solutions of the hormone. Highly active solutions of insulin were obtained when Banting and Best (C-2) discovered that insulin is resistant to dilute acids, and that its destruction by the proteolytic enzymes of the pancreas can be prevented by acidified ethanol. Similar protection can be accomplished by picric acid (C-3).

Crystals of insulin were first prepared by dissolving crude preparations in acetic acid and neutralizing them carefully with ammonia or pyridine

(C-4,5). Crystallization has also been accomplished by extraction with acid alcohol, precipitation with ether, dissolution in dilute sulfuric acid, conversion into the insoluble fibrous form (see below) and recrystallization (C-6). One mg of the purest preparations of insulin contains approximately 26.8 international units. In aqueous solutions, the insulin molecules aggregate and form polymers whose sizes increase with the insulin concentration. Extrapolation to zero concentration gives a molecular weight of 6,000 (C-8,9). The unit cell of crystalline insulin, according to X-ray measurements, contains 8 subunits of the molecular weight 6,000 (C-10).

The amino acid sequence of the bovine insulin molecule, as shown on page 351, has been completely clarified by a series of brilliant investigations of Sanger (C-11) who first cleaved the dithio bonds by oxidation with performic acid and thus obtained two chains. One of these, called the A-chain, had an N-terminal glycine residue and contained 4 residues of cysteic acid; the B-chain had an N-terminal phenylalanine residue, contained only 2 residues of cysteic acid, but contained all of the basic amino acids of the molecule (arginine, histidine, lysine). The two chains can be separated from each other by counter-current distribution with acetic acid-butanol-pyridine (C-7).

Whereas the oxidative cleavage of the disulfide bonds is irreversible, insulin can be partially regenerated from the A- and B-chains if the dithio bonds are split by sulfitolysis (see Chapter VII, E) (C-12). Sulfite in the absence of urea cleaves only the interchain dithio bonds, not the intrachain bond in chain A (C-13). Claims that a peptide chain of 8 amino acids can be removed without loss of the hormonal activity have not been confirmed; the deoctapeptide insulin is inactive (C-14). The antigenic activity of insulin is determined by its A-chain (C-30).

Inactivation of insulin is also brought about by reductive cleavage of the dithio bonds (C-15). The hormonal activity is, however, not lost by esterification with cold sulfuric acid (C-16). If insulin is heated in a weakly acidic solution, it is converted into a *fibrous insoluble material* which is devoid of hormonal activity, but regains this activity by cautious dissolution at neutral or weakly alkaline pH (C-17). The formation of the fibrils is attributed to the formation of bonds between the side chains of the numerous nonpolar amino acids. Insulin combines readily with bivalent metals; the *zinc complex* crystallizes more easily than the metal-free hormone. Like many other proteins, insulin forms insoluble precipitates with protamines (see Chapter VIII, G, and Chapter X, D). The hormonal activity of injected *protamine-zinc-insulin* complexes does not decline as rapidly as that of pure insulin (C-18). The liver contains *insulinase*, an enzyme which specifically destroys insulin but does not attack many of the other proteins (C-19).

The insulins of various species of mammals differ from each other only by the sequence of the amino acids Nos. 8, 9, and 10 in the A-chain. It is

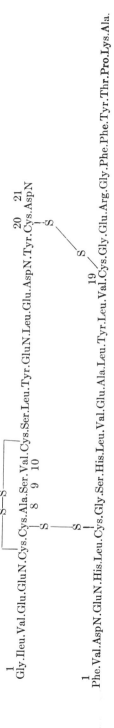

evident from the structure of insulin shown above that these 3 amino acids are members of a loop of 6 amino acid residues, 3 of these consisting of cystine half-residues, and that the loop is closed by a disulfide bond between the cystine residues Nos. 6 and 11 of the A-chain. The 8, 9, 10 sequence in cattle insulin is Ala.Ser.Val; in the sheep Ala.Gly.Val; in the horse Thr.Gly.Ileu; and in human, pig, and sperm whale insulin, Thr.Ser.Ileu (C-20). Differences have also been found in the C-terminal amino acid residue of the B-chain which is threonine in human and serine in rabbit insulin (C-29). In the insulin from the islet organ of Bonito, the N-terminal phenylalanine residue is replaced by alanine; the B-chain in this fish differs considerably from that of the mammalian insulin (C-21).

Insulin deficiency causes hyperglycemia and glycosuria. Injection of insulin lowers the blood sugar, prevents the loss of glucose through the urine, and prevents also the dangerous acidosis in diabetic patients. Very little is known about the mechanism of these very dramatic changes in the metabolism of carbohydrates and lipids. The best test object for insulin action *in vitro* seems to be the surviving rat diaphragm (C-22,23). In the presence of insulin the diaphragm converts glucose into glycogen (C-23,24), probably by activation of the UDPG-glycogen transglucosylase (C-25). Insulin increases also the incorporation of radioactive amino acids into the proteins of the surviving diaphragm, and lowers the release of amino acids from the diaphragm (C-26). The primary action of insulin may be a change in the permeability of the diaphragmatic cells (C-24). It may stimulate the respiration-dependent uptake of water by mitochondria (C-27), and thus interfere with the intermediary metabolism. However, there is no evidence for SS—SH exchange (see Chapter VII, E) between cystine residues of insulin and the proteins of its target organ (C-28).

## D. Oxytocin and Vasopressin (D-1)

Three different hormonal actions are attributed to the posterior lobe of the hypophysis: (1) excitation of uterine muscle contractions; (2) increase of the blood pressure; and (3) diuresis, i.e., increase in the volume of urine produced in the kidney. The substance causing uterine contractions is called oxytocin, while that which raises the blood pressure is termed vasopressin. Both are extracted from the gland by dilute acetic acid. They have been purified by fractionation with acetone and ether (D-2a) and by chromatography or electrophoresis. In the gland the two hormones may be loosely bound to a specific protein called neurophysin. The oxytocin-vasopressin-neurophysin complex can be salted out by sodium chloride (D-2b). The two hormones are extracted from the precipitate by trichloroacetic acid whereas neurophysin remains insoluble (D-2b).

The structure of oxytocin and vasopressin has been completely elucidated by du Vigneaud (D-3) and others (D-4,-5), and has been confirmed by their synthesis (D-6). The structure of oxytocin is shown below (D-1).

$$\overset{1}{\text{Cy}}\text{S.}\overset{2}{\text{Tyr.}}\overset{3}{\text{Ileu.}}\overset{4}{\text{Glu(NH}_2)}.\overset{5}{\text{Asp(NH}_2)}.\overset{6}{\text{Cy}}\text{S.}\overset{7}{\text{Pro.}}\overset{8}{\text{Leu.}}\overset{9}{\text{Gly(NH}_2)}$$

Oxytocin

It can be seen that oxytocin is an octapeptide, that it contains a ring formed by 4 amino acids (Nos. 2, 3, 4, and 5) and the 2 halves of a cystine molecule (Nos. 1 and 6), and that a tripeptide is attached to the carboxyl group of one of the half-cystine residues. The carboxyl group of the terminal glycine residue is substituted by ammonia. Vasopressin from human and many other mammalian glands is $Phe^3$-$Arg^8$-oxytocin, that is, an oxytocin in whose molecule the isoleucine residue in 3-position has been replaced by phenylalanine, and the 8-leucine by arginine (D-7). The vasopressin from hog glands is $Phe^3$-$Lys^8$-oxytocin. It has been designated as lysine-vasopressin (D-8).

The synthetic hormones have the same activity as the natural hormones. Oxytocin has an oxytocic activity of about 500 U.S.P. units/mg, and a weak pressor activity of approximately 7 units. The pressor activity of arginine-vasopression is approximately 600 U.S.P. units/mg, its oxytocic activity only 30 units (D-1). Attempts have been made to change the activities of the two pituitary hormones by modifying their structure. It is clear from these investigations that a basic amino acid (arginine or lysine) in 8-position is necessary for strong pressor activity. The oxytocic activity decreases if the tyrosine in 2-position is replaced by phenylalanine (D-9,10) or if asparagine is replaced by isoasparagine (D-11), but increases if the free amino group in the 1-position is replaced by a hydrogen atom (D-12).

The oxytocic activity decreases also when the side chain of the 3 amino acids 7, 8, and 9, is removed (D-13), when another tyrosine residue is inserted between the residues No. 1 and 2 (D-14), or when the isoleucine residue in 3-position is replaced by phenylalanine or valine (D-15). In addition to their oxytocic and pressor activities, oxytocin and vasopressin have also diuretic activity, milk ejecting action, and avian depressor activity. Modifications in the structure frequently result in the increase of some of these activities, but in the decrease of others. Moreover, the sensitivity of various animals to the modified hormones is different. Thus $Phe^3$-oxytocin is much more potent in the cat than in the rat uterus (D-10). A synthetic hybrid hormone (vasotocin) with the oxytocin ring and lysine in the side chain has only one-half of the pressor activity of vasopressin but 4 times its avian depressor activity (D-17). An analogous natural arginine-vasotocin has been found in the hypophysis of frogs (D-18). Synthetic $Arg^8$ oxytocin has a high oxytocic activity, whereas $His^8$-oxytocin has only $\frac{1}{200}$ of the activity of oxytocin (D-19).

If the dithio bond of the hormones is reduced, their activity decreases. Most of it is restored by reoxidation of the sulfhydryl groups to the dithio bond. The regeneration is not complete since reoxidation results partly in the formation of intermolecular dithio bonds and dimerization (D-20). After injection of the hormones, they are bound to the proteins of their target organs by covalent bonds (D-21). This combination with the tissue proteins is inhibited by sulfhydryl compounds which also cause dissociation of the hormones from the tissue proteins (D-16). Evidently, the combination of the hormones with tissue proteins is due to SH-SS interchange. A specific enzyme, oxytocinase, which destroys oxytocin has been found in the blood of pregnant women (D-22).

## E. The Melanocyte-Stimulating Hormone (MSH) or Intermedin (E-1)

The pars intermedia of the hypophysis, situated between the anterior and posterior lobe, contains a hormone which causes the expansion of the pigmented melanophores in the skin of batrachians such as the frog. In the acetic acid extracts of whole glands, the MSH is separated from corticotropin by adsorption of the latter to a very small amount of oxy-cellulose, subsequent adsorption of MSH to a large amount of the same adsorbent, and elution with dilute HCl (E-2). Purification is accomplished by chromatography on carboxymethyl-cellulose. The purified preparation contains two types of intermedins, designated as $\alpha$- and $\beta$-intermedin or as $\alpha$- and $\beta$-melanophore stimulating hormone. Both are peptides. The $\beta$-type has a free terminal $\alpha$-amino and also a free terminal $\alpha$-carboxyl group whereas in the $\alpha$-type the former is substituted by acetylation, the latter by amidation. The two MSH can be separated from each other by electrophoresis since the isoelectric point of the $\beta$-hormone is at pH 10–11 whereas that of the $\alpha$-hormone is close to pH 5.5. The structure of the two melanophore hormones from hog pituitary glands is shown below:

$\alpha$-MSH (E-3): Acetyl—Ser.Tyr.Ser.Met.Glu.His.Phe.Arg.Try.Gly.Lys.Pro.Val.$NH_2$
1   2   3   4   5   6   7   8   9   10  11

Asp.*Ser*.Gly.Pro.Tyr.Lys.Met.Glu.His.Phe.Arg.Try.Gly.Ser.Pro.Pro.Lys.Asp
$\beta$-MSH (E-4,5)

Bovine $\beta$-MSH differs from the porcine hormone by a glutamyl residue instead of the serine residue shown in italic (E-6). In the human $\beta$-MSH the $N$-terminal segment Ala.Glu.Lys.Lys.Asp.Glu.Gly.Pro— is bound to the tyrosine residue in the 1-position (E-7). In the β-MSH of the horse the C-terminal Pro.Lys.Asp chain is replaced by —Arg.Lys.Asp (E-8). Comparison of the amino acid sequences of porcine α- and β-MSH reveals that the positions of serine and lysine (Nos. 2 and 10, respectively) are exchanged in these two hormones. The sequence 3–9, —Met. Glu.His.Phe.Arg.Try.Gly—, occurs not only in all investigated MSH preparations but also in the corticotropins (adrenocortical hormones) of the pituitary gland (see Section F). This suggests that at least some steps in the biosynthesis of the two hormones are correlated. They may have common partial templates, or the sequence which occurs in both hormones may be transferred from a common precursor to the other portions of their molecules.

Peptides very similar to the intermedins have been prepared synthetically by several groups of workers (E-9 to 11) and have recently led to the synthesis of an α-MSH whose amino acid sequence was the same as that of porcine α-MSH, but whose lysyl residue was protected by a formyl group (E-12). Its activity was the same as that of the natural hormone.

## F. Adrenocorticotropic Hormone (ACTH) or Corticotropin (F-1)

The adrenocorticotropic hormone of the anterior lobe of the hypophysis has been studied by many investigators because of its alleviating action in rheumatoid arthritis. It was first observed by Hench et al. (F-2) that this disease is rare among pregnant women; on the other hand it has been known for a long time that the hormonal production of the hypophysis increases in pregnancy. The increased production of ACTH causes release of corticosteroids from the adrenal gland, thereby increasing their concentration in the blood plasma and thus preventing arthritic symptoms. In Sayers' test for ACTH the decrease of ascorbic acid in the adrenal gland is measured.

The hormones investigated in different laboratories were prepared from different sources by somewhat different methods. The preparation from porcine pituitary glands has been designated as corticotropin A (F-3), that from ovine glands as α-corticotropin (F-4). β-Corticotropin from hog glands was purified by counter-current distribution; about 10,000 transfers were required for purification of the hormone which consists of a single peptide chain of the following structure (F-5):

```
 1   2   3   4   5   6   7   8   9  10  11  12  13  14  15  16  17  18  19  20
Ser.Tyr.Ser.Met.Glu.His.Phe.Arg.Try.Gly.Lys.Pro.Val.Gly.Lys.Lys.Arg.Arg.Pro.Val.
21  22  23  24  25  26  27  28  29    30    31  32  33  34  35  36  37  38  39
Lys.Val.Tyr.Pro.Asp.Gly.Ala.Glu.Asp.GluN.Leu.Ala.Glu.Ala.Phe.Pro.Leu.Glu.Phe
```

The sequence No. 1–24 occurs in all corticotropins and has been confirmed by synthesis of the analogous tricosa- and tetracosa-peptides (F-6,13). In this structure the following groups are essential for the adrenocorticotropic action: the $N$-terminal serine residue, the phenol ring of tyrosine (Nos. 2 and 23), and the free amino group of 11-lysine. The sequence 4–10 is identical with the sequence 3–9 of the melanophore-stimulating hormone (see Section E).

The sequence 25–39 is different in different animal species and does not seem to be essential for the hormonal activity. In ovine α-corticotropin the amino acids 27–32 of the porcine hormone (see above) are replaced by Glu.Asp.Asp.Glu.Ala.Ser (F-7); in bovine corticotropin the sequence 25–33 is Asp.Gly.Glu.Ala.Glu.Asp.Ser.Ala.GluN (F-7). This chain contains all of the aspartyl and glutamyl residues. If the sequence of the residues Nos. 24–35 in α-corticotropin is compared with the common sequence 12–23, it is found that the basic residues of arginine and lysine in the sequence 12–23 are lined up in exactly the same order as the acidic residues in 24–35 (F-8):

Sequence 12–23　　　　　—Pro.Val.Gly.*Lys.Lys.Arg.Arg*.Pro.Val.*Lys*.Val.Tyr—
Sequence 24–35 (sheep α)　—Pro.Ala.Gly.*Glu.Asp.Asp.Glu*.Ala.Ser.*Glu*.Ala.Phe—

The conclusion has been drawn that there might exist two levels of information during the biosynthesis of these peptide chains; one which designs the general pattern of charge distributions and another which determines which of the individual amino acids is incorporated (F-8).

Due to their possession of the amino acid sequence of the melanocyte-stimulating hormones, the corticotropins have a weak melanocyte-stimulating activity. It increases by several orders of magnitude when the corticotropic activity is destroyed by exposure of the hormones to dilute NaOH at 100°C (F-9). Whereas the corticotropins are inactivated rapidly by hot alkali, they are relatively resistant to acids and can be briefly heated with dilute HCl without significant destruction (F-10).

The secretion of corticotropin from the pituitary gland seems to be dependent on the action of a corticotropin-releasing factor (F-11) which is a peptide and is degraded by pepsin. The designation "precorticotropin" has been used for a precursor of corticotropin which does not give Sayers' test but is activated by concentrated solutions of urea or by acidification to pH 3 (F-12).

## G. Other Hormones of the Anterior Pituitary Lobe

Like corticotropin, the other hormones of the anterior pituitary lobe have an indirect action in so far as they act by means of other endocrine glands. Thus the *thyrotropic hormone* causes increased passage of thyroid hormones from the thyroid gland into the blood (G-1). The bovine hormone is found in alkaline extracts of the pituitary gland. Its molecular weight is close to 28,000. The hormone has been purified by fractionation with ammonium sulfate and acetone, and by chromatography on columns of DEAE- and CM-cellulose (G-2). It is heterogeneous and consists of 6 or more components, each with thyrotropic activity (G-3). Thyrotropin is rich in cystine and tyrosine (see Table XIII-1) and contains about 5% carbohydrate consisting of glucosamine and galactosamine (G-3). The mode of action of the thyrotropic hormone in the thyroid gland is not yet understood.

*Prolactin* (lactogenic hormone) acts on the mammary gland. Ovine prolactin extracted from the pituitary gland by aqueous acetone is heat-stable. The hormone has been purified by counter-current distribution. Its molecular weight is approximately 24,000 (G-4). Its amino acid composition is shown in Table XIII-1. The hormone is inactivated if the dithio groups of its cystine residues are reduced to cysteine residues (G-5). Since the molecular weight is not changed by cleavage of the SS bonds, the hormone consists of a single peptide chain of approximately 200 amino acids.

The *growth hormone* (somatotropin) (F-1), if produced in abnormally large amounts, causes acromegaly, i.e., disproportionate rapid growth of the extremities. Deficiency of the hormone results in dwarfism. The bovine and human hormones have been obtained in the crystalline state (G-6,7). The molecular weight and the amino acid composition of bovine growth hormone is shown in Table XIII-1. The bovine hormone has one *C*-terminal phenylalanine residue, but two *N*-terminals (alanine and phenylalanine). The molecule is not degraded when the 4 dithio groups are cleaved by oxidation to cysteic acid. It has been suggested, therefore, that the molecule consists of a single peptide chain and that the *C*-terminal residue of one of these forms a peptide bond with an ε-amino group of lysine (F-1). The activity of the hormone is hardly affected by partial acetylation, guanidination, or dinitrophenylation, and is even retained if about 30% of the peptide bonds are hydrolyzed by enzymes. Evidently, not all parts of the molecule are required for the hormonal activity. The mode of action of the hormone is not yet well understood. Its administration results in an increased uptake of sulfate into the cartilage (G-9).

The *interstitial cell-stimulating hormone* (ICSH) of the pituitary gland of the sheep has been prepared from the acetone powder of the gland by extraction with 0.5% NaCl, subsequent fractionation with sodium salicylate and ammonium sulfate, chromatography on Amberlite IRC-50, and zone electrophoresis (G-10). The hormone is a glycoprotein which contains about 12% carbohydrate (G-11). Its molecular weight is close to

30,000 (G-10). The hormone causes luteinization in the ovary and stimulates the interstitial cells of the male sex gland. The *follicle-stimulating hormone* (FSH) of the pituitary gland can be separated from thyrotropin, somatotropin and ICSH by salting out these three hormones with 1.9 *M* ammonium sulfate. In the supernatant solution FSH is salted out by raising the ammonium sulfate concentration to 2.2 *M* (G-12). Like ICSH, the follicle-stimulating hormone is also a glycoprotein containing 7–8% carbohydrate (G-13). The molecular weight is approximately 29,000.

A gonadotropin whose actions are similar to those of the pituitary gonadotropins is produced in the placenta during pregnancy. This *chorionic gonadotropin* is of great clinical importance because it passes into the blood and the urine and permits an early diagnosis of the pregnancy. If urine containing the chorionic gonadotropin is injected into infantile female mice or rats, it causes maturation of the ovary with increase in the size of the uterus (Achheim-Zondek test). The urinary gonadotropin can be easily adsorbed from the urine to tungstic acid, benzoic acid (G-14), uranyl phosphate, or alumina (G-15). Since the hormone is not denatured by ethanol or acetone, the benzoic acid used as adsorbent can be removed by these organic solvents. The hormone can be eluted from alumina or uranyl phosphate by dilute alkali. It has also been adsorbed by charcoal and eluted by phenol (G-14). Its carbohydrate content is approximately 18% (G-16). The purified gonadotropin from the serum of pregnant mares contains 14.1% hexosamine and 10.4% sialic acid; it also contains galactose, mannose, fucose, glucose, and rhamnose (G-17). The carbohydrates seem to be essential for the hormonal activity since this activity is lost on incubation of the hormone with salivary amylase (G-18).

## H. Other Protein Hormones and Peptides with Hormone-like Action

The hypoglycemia caused by the injection of insulin is sometimes preceded by a short increase in the blood sugar level. This is attributed to contamination of insulin preparations by *glucagon*, another hormone of the pancreas. Glucagon has been obtained in crystalline form (H-1). It is a peptide of the formula: His.Ser.GluN.Gly.Thr.Phe.Thr.Ser.Asp.Tyr. Ser.Lys.Tyr.Leu.Asp.Ser.Arg.Arg.Ala.Glu.AspN.Phe.Val.GluN.Try.Leu. Met.AspN (H-2). Glucagon mobilizes the liver glycogen and thereby increases the blood sugar in the hepatic vein (H-3a). In the perfused liver, glucagon increases also the degradation of proteins and the formation of urea (H-3b). In liver slices glucagon seems to inhibit the incorporation of amino acids into proteins (H-15).

The *hormone of the parathyroid gland* (parathormone) has been purified by counter-current distribution (H-4). The amino acid composition of bovine parathyroid hormone is shown in Table XIII-1. The hormone is free of cystine and cysteine. Oxidation of the methionine side chains inactivates the hormone (A-4). It mobilizes calcium and phosphate, particularly from the bones, and causes their excretion in the urine. The

TABLE XIII-1

AMINO ACID COMPOSITION OF SOME PROTEIN HORMONES

(Amino acid residues per protein molecule)

| Hormone | Bovine thyrotropin[a] | Prolactin[b] | Bovine somatotropin[c] | Parathyroid hormone[d] |
|---|---|---|---|---|
| Mol. wt. | 28,000 | 24,200 | 45,652 | 6943 + 300 |
| Glycine | 22.6 | 12 | 22.7 | 4 |
| Alanine | 13.7 | 11 | 29.6 | 7 |
| Valine | 8.9 | 12 | 13.9 | 8 |
| Leucine | 7.9 | 37 | 49.4 | 8 |
| Isoleucine | 6.5 | | 13.2 | 3 |
| Proline | 16.5 | 14 | 15.1 | 2 |
| Phenylalanine | 8.6 | 8 | 25.3 | 2 |
| Tyrosine | 9.4 | 7 | 11.9 | 1 |
| Tryptophan | — | 2 | 3.1 | 1 |
| Serine | 10.9 | 18 | 27.6 | 6 |
| Threonine | 14.7 | 11 | 25.7 | 1 |
| Cystine/2 | 12.4 | 6 | 8.1 | 0 |
| Methionine | 4.0 | 7 | 8.4 | 2 |
| Arginine | 6.3 | 11 | 24.0 | 5 |
| Histidine | 3.7 | 7 | 6.8 | 4 |
| Lysine | 16.2 | 10 | 22.1 | 9 |
| Aspartic acid | 14.2 | 19 | 35.9 | 9 |
| Glutamic acid | 16.5 | 22 | 40.6 | 11 |
| Ammonia | 2.12[e] | (17) | (30) | 9 |

[a] 3.6 Hexosamine residues (G-3).  [d] From (H-4).
[b] From (E-1).  [e] 2.12 gram equivalent per 100 g.
[c] From (G-8).

hormone seems to inhibit the action of the osteoblasts so that less phosphate is incorporated into the bone (H-5). It had been believed earlier that the hormone has a renal action; however, it acts also in nephrectomized animals (H-5).

We do not yet know in which organ *angiotensin* (H-16) and its precursor *angiotensinogen* are formed, and whether they are of endocrine origin. Angiotensin, previously called angiotonin or hypertensin, is formed from its precursor angiotensinogen by the action of renin, an enzyme of the kidney (H-6). Angiotensinogen, which has no pressor activity, is found in the blood plasma. Bovine angiotensinogen is a polypeptide of the following composition: Asp. Arg. Val.Tyr. Val.His.Pro.Phe.His.Leu; it has been prepared synthetically (H-7). On exposure to renin, limited proteolysis takes place and the C-terminal histidine and leucine residues are split off. The remaining octapeptide is identical with synthetic angiotensin (H-8). In horse blood the valyl residue in the Tyr.Val.His sequence is replaced by an isoleucine residue (H-8,9). Although 24 different analogues of angiotensin with shorter or longer peptide chain have been prepared, none of them has a higher pressor activity than angiotensin (H-10). The specific activity of the octapeptide is reminiscent of the analogous high pressor activity of the octapeptide vasopressin (see Chapter XIII, D). Another peptide with high biological activity is *bradykinin* which stimulates the activity of smooth muscles (H-11). Bradykinin is formed when ox blood is exposed to the action of trypsin. Its composition is: Arg.Pro.Pro.Gly.Phe.Ser.Pro.Phe.Arg (H-12). Bradykinin has been prepared synthetically before the structure of the natural product was known (H-13). Very little is known on *parotin*, a product of the parotid gland; it lowers the calcium and phosphate level in the blood of rabbits. The molecular weight of parotin is 132,000 (H-14).

REFERENCES

### Section A

**A-1.** G. Pincus and K. V. Thimann, eds., "The Hormones," 3 vol. Academic Press, New York, 1948–1955.    **A-2.** R. A. Peters, *Nature* 177: 426(1956).    **A-3.** A. Bussard, *Ann. Inst. Pasteur,* 75: 14(1948).    **A-4.** K. Hofmann, *Ann. Rev. Biochem.* 31: 213 (1962).

### Section B

**B-1.** J. Roche and R. Michel, *Physiol. Revs.* 35: 583(1955).    **B-2.** M. Heidelberger and W. Palmer, *JBC* 101: 433(1933).    **B-3.** I. Abelin and F. Bigler, *Z. physiol. Chem.* 303: 196(1956).    **B-4.** R. Pitt-Rivers, *BJ* 43: 223(1948).    **B-5.** J. Gross and R. Pitt-Rivers, *BJ* 53: 645, 652(1953).    **B-6.** C. R. Harington *et al., Nature* 179: 218(1957).    **B-7.** C. R. Harington, "The Thyroid Gland," Oxford Univ. Press, London and New York, 1938.    **B-8.** I. Snapper and A. Grünbaum, *Brit. J. Exptl. Pathol.* 17: 361(1936).    **B-9.** A. Taurogg *et al., JBC* 213: 119(1955).    **B-10.** D. M. Fawcett and S. Kirkwood, *JBC* 205: 795(1953).    **B-11.** W. Tong *et al., JBC* 227: 772(1957).    **B-12.** P. von Mutzenbecher, *Z. physiol. Chem.* 261: 253(1939).    **B-13.** E. P. Reineke and C. W. Turner, *JBC* 161: 613(1945).    **B-14.** C. H. Harington, *JCS* 1944: 193.    **B-15.** C. Rich and A. G. Bearn, *Endocrinology* 62: 687(1958).    **B-16.** K. Sterling and M. Tabachnick, *JBC* 236: 2241(1961).    **B-17.** A. L. Lehninger, *BBA* 37: 387(1960).    **B-18.** C. Martius and C. Hess, *Biochem. Z.* 326: 1(1955).    **B-19.** L. Sokoloff and S. Kaufman, *JBC* 236: 1083(1961).    **B-20.** L. J. De Groot and A. M. Davis, *BBA* 59: 581(1962).

### Section C

**C-1.** H. Jensen, "The Chemistry and Physiology of Insulin." Am. Assoc. Advance Sci., Washington, D.C., 1944.    **C-2.** F. G. Banting and C. H. Best, *J. Lab. Clin. Med.* 7: 251, 464(1922).    **C-3.** H. Dudley, *BJ* 18: 147(1924).    **C-4.** J. J. Abel *et al., J. Pharmacol. Exptl. Therap.* 31: 65(1927).    **C-5.** V. du Vigneaud *et al., J. Pharmacol. Exptl. Therap.* 32: 267(1928).    **C-6.** C. W. Pettinga, *Biochem. Preparations* 6: 28 (1958).    **C-7.** L. C. Craig *et al., JACS* 75: 5528(1953); *Biochem. Preparations,* 8: 70 (1961).    **C-8.** E. Fredericq, *BBA* 9: 601(1952).    **C-9.** I. W. Kupke and K. L. Lang,

*BBA* 13: 153(1954).   **C-10.** B. Low, *Nature* 169: 955(1952).   **C-11.** F. Sanger *et al.*, *BJ* 44: 126(1949); 49: 481(1951); 53: 266(1953); 59: 509(1955); 60: 541(1955).   **C-12.** G. H. Dixon and A. C. Wardlaw, *Nature* 188: 721(1960); D. Yu-cang *et al.*, *Scientia Sinica* 10: 84(1961).   **C-13.** R. Cecil and U. E. Loening, *BJ* 76: 146(1960).   **C-14.** F. H. Carpenter and W. E. Baum, *JBC* 237: 409(1962).   **C-15.** V. du Vigneaud *et al.*, *JBC* 94: 233 (1931).   **C-16.** M. B. Glendenning *et al.*, *JBC* 167: 125 (1947).   **C-17.** D. F. Waugh, *JACS* 70: 1850(1948); *JBC* 185: 85(1950).   **C-18.** H. C. Hagedorn *et al.*, *J. Am. Med. Assoc.* 106: 177(1936).   **C-19.** I. A. Mirsky and G. Perisutti, *JBC* 228: 77(1957).   **C-20.** H. Brown *et al.*, *BJ* 60: 556(1955); J. I. Harris *et al.*, *ABB* 65: 427 (1956); D. S. H. W. Nicol and L. F. Smith, *Nature* 187: 483(1960).   **C-21.** A. Kotaki, *J. Biochem. (Tokyo)* 50: 256(1961).   **C-22.** C. A. Villee and A. B. Hastings, *JBC* 181: 131(1949).   **C-23.** W. C. Stadie *et al.*, *Science* 110: 550(1949); *JBC* 200: 745(1953).   **C-24.** C. Park *et al.*, *Am. J. Physiol.* 182: 12(1955); *JBC* 181: 247(1949).   **C-25.** C. Villar-Palasi and J. Larner, *BBA* 39: 171(1960); *ABB* 94: 436(1961).   **C-26.** K. L. Manchester and F. G. Young, *BJ* 75: 487(1960); 81: 135(1961).   **C-27.** A. L. Lehninger and D. Neubert, *Proc. Natl. Acad. Sci. U.S.* 47: 1929(1961).   **C-28.** H. Carlin and O. Hechter, *JBC* 237: PC 1371(1962).   **C-29.** F. Sanger, *Brit. Med. Bull.* 16: 183 (1960).   **C-30.** S. Wilson *et al.*, *BBA* 62: 483(1962).

### Section D

**D-1.** R. Acher, *Ann. Rev. Biochem.* 29: 547(1960).   **D-2a.** O. Kamm *et al.*, *JACS* 50: 513(1928).   **D-2b.** R. Acher *et al.*, *JBC* 233: 116(1958); *BBA* 38: 266(1960).   **D-3.** V. du Vigneaud *et al.*, *JACS* 74: 3713(1952); *JBC* 206: 353(1954).   **D-4.** R. Acher and J. Chauvet, *BBA* 12: 487(1953).   **D-5.** H. Tuppy, *BBA* 11: 449(1953).   **D-6.** V. du Vigneaud *et al.*, *JACS* 76: 3115, 4751(1954); 79: 5572(1957); 80: 3355(1958).   **D-7.** A. Light and V. du Vigneaud, *Proc. Soc. Exptl. Biol. Med.* 98: 692(1958).   **D-8.** M. F. Bartlett *et al.*, *JACS* 78: 2905(1956).   **D-9.** M. Bodansky and V. du Vigneaud, *JACS* 81: 1258(1959).   **D-10.** H. Konzett and B. Berde, *Brit. J. Pharmacol.* 14: 133(1959).   **D-11.** W. B. Lutz *et al.*, *JACS* 81: 167(1959).   **D-12.** V. du Vigneaud *et al.*, *JBC* 235, PC 64(1960); 237: 1563(1962).   **D-13.** C. Ressler, *Proc. Soc. Exptl. Biol. Med.* 92: 725(1956).   **D-14.** S. Guttmann *et al.*, *Naturwiss.* 44: 632(1957).   **D-15.** B. Berde *et al.*, *Brit. J. Pharmacol.* 12: 209(1957).   **D-16.** I. L. Schwartz *et al.*, *Proc. Natl. Acad. Sci. U.S.* 46: 1273, 1278(1960).   **D-17.** R. D. Kimbrough and V. du Vigneaud, *JBC* 236: 778(1961).   **D-18.** R. Acher, *BBA* 42: 397(1960).   **D-19.** P. G. Katsoyanis and V. du Vigneaud, *JBC* 233: 1352(1958).   **D-20.** C. Ressler, *Science* 128: 1281(1958).   **D-21.** C. T. O. Fong *et al.*, *JACS* 81: 2592(1959).   **D-22.** H. Tuppy and H. Nesvadba, *Monatsh. Chem.* 88: 977(1957).

### Section E

**E-1.** C. H. Li, *Adv. in Protein Chem.* 12: 269(1957).   **E-2.** S. L. Steelman *et al.*, *BBA* 33: 256 (1959).   **E-3.** T. H. Lee, *JBC* 233: 917(1958).   **E-4.** J. I. Harris and P. Roos, *Nature* 178: 90(1956).   **E-5.** I. I. Geschwind *et al.*, *JACS* 78: 4494(1956).   **E-6.** I. I. Geschwind *et al.*, *JACS* 79: 620(1957).   **E-7.** J. I. Harris and P. Roos, *BJ* 71: 434ff (1959).   **E-8.** T. H. Lee *et al.*, *JBC* 236: 1390(1961).   **E-9.** S. Guttmann and R. A. Boissonas, *Helv. Chim. Acta* 42: 1257(1959).   **E-10.** R. Schwyzer *et al.*, *Helv. Chim. Acta* 47: 1702(1959).   **E-11.** K. Hofmann *et al.*, *JACS* 79: 1641, 6087(1957); 80: 1486, 6458(1958).   **E-12.** K. Hofmann *et al.*, *JACS* 82: 3732(1961).

### Section F

**F-1.** C. H. Li, *Adv. in Protein Chem.* 11: 102(1956).   **F-2.** P. S. Hench *et al.*, *Proc. Staff Meetings Mayo Clin.* 24: 281(1949); *Arch. Internal Med.* 85: 545(1950).   **F-3.** W.

F. White, *JACS* 76: 4194(1954). **F-4.** C. H. Li, *Nature* 173: 251(1954). **F-5.** P. H. Bell *et al.*, *JACS* 76: 5565(1954); 77: 2419(1955). **F-6.** H. Kappeler and R. Schwyzer, *Helv. Chim. Acta* 44: 1136(1961). **F-7.** C. H. Li *et al.*, *BBA* 46: 324(1961). **F-8.** D. Schwartz, *Nature* 183: 464(1959). **F-9.** R. G. Shepherd *et al.*, *JACS* 78: 5051(1956). **F-10.** C. H. Li *et al.*, *JBC* 190: 317(1951). **F-11.** A. V. Schally and R. Guillemin, *Proc. Soc. Exptl. Biol. Med.* 100: 138(1959). **F-12.** P. R. Dasgupta and F. G. Young, *Nature* 182: 32(1958). **F-13.** K. Hofmann *et al.*, *Proc. Soc. Exptl. Biol. Med.* 108: 559(1962).

### Section G

**G-1.** W. P. Deiss *et al.*, *J. Clin. Invest.* 38: 334(1959). **G-2.** P. G. Condliffe *et al.*, *BBA* 34: 430(1959). **G-3.** J. G. Pierce *et al.*, *JBC* 235: 78, 85(1960). **G-4.** R. D. Cole and C. H. Li, *JBC* 213: 197(1955). **G-5.** C. H. Li *et al.*, *JBC* 229: 153(1957); 233: 73(1958). **G-6.** C. H. Li *et al.*, *Science* 108: 624 (1948); *JBC* 218: 33(1956). **G-7.** U. J. Lewis and N. G. Brink, *JACS* 8: 4429(1958). **G-8.** A. J. Parcells, *Nature* 192: 971(1961). **G-9.** S. Ellis *et al.*, *Proc. Soc. Exptl. Biol. Med.* 84: 603(1953). **G-10.** P. G. Squire and C. H. Li, *JBC* 234: 520(1959). **G-11.** H. Otsuka and Y. Noda, *J. Biochem. (Tokyo)*, 41: 547(1954). **G-12.** S. Ellis, *JBC* 233: 63(1958). **G-13.** S. L. Steelman *et al.*, *Endocrinology* 56: 216(1955); 59: 256(1956). **G-14.** S. Katzman and E. A. Doisy, *JBC* 98: 739(1932); 106: 1125(1934). **G-15.** M. Reiss and F. Haurowitz, *Z. ges. exptl. Med.* 68: 371(1929). **G-16.** H. P. Lundgren *et al.*, *JBC* 142: 367 (1942). **G-17.** R. Bourrillon *et al.*, *Bull. soc. chim. biol.* 41: 493(1959). **G-18.** J. S. Evans and J. Hauschildt, *JBC* 145: 335(1942).

### Section H

**H-1.** A. Staub *et al.*, *JBC* 214: 619(1955). **H-2.** W. W. Bromer *et al.*, *JACS* 79: 2807 (1956). **H-3a.** W. C. Shoemaker *et al.*, *Am. J. Physiol.* 196: 315(1959). **H-3b.** L. L. Miller, *Nature* 185: 248(1960). **H-4.** H. Rasmussen and L. C. Craig, *JBC* 236: 759, 1083(1961). **H-5.** J. Dawson *et al.*, *BJ* 66: 116(1957). **H-6.** E. Braun-Menendez and I. H. Page, *Nature* 181: 1061(1958). **H-7.** L. Skeggs *et al.*, *J. Exptl. Med.* 106: 439(1958). **H-8.** D. F. Elliott and W. S. Peart, *BJ* 65: 246(1957); R. Schwyzer *et al.*, *Helv. Chim. Acta* 41: 1287(1958); R. M. Bumpus *et al.*, *Science* 125: 886(1957). **H-9.** W. Rittel *et al.*, *Helv. Chim. Acta* 40: 614(1957). **H-10.** R. Schwyzer, *Helv. Chim. Acta*, 44: 667(1961). **H-11.** D. F. Elliott *et al.*, *BJ* 78: 60(1961). **H-12.** H. Konzett *et al.*; G. P. Lewis; P. G. Shorley *et al.*, *Nature* 189: 998, 999(1961). **H-13.** R. A. Boissonas *et al.*, *Helv. Chim. Acta*, 43: 1349(1960). **H-14.** Y. Ito and Y. Kubota, *J. Biochem. (Tokyo)*, 47: 422(1960). **H-15.** J. Pryor and T. Berthet, *Arch. intern. physiol. et biochem.* 63: 227(1960). **H-16.** I. H. Page and F. M. Bumpus, *Physiol. Revs.* 41: 331(1961).

# Chapter XIV

## Toxins (Toxic Proteins)

Many of the toxic substances found in the animal and the plant kingdom are proteins or peptides. Some of them have been isolated in crystalline form during the last few years. The properties of these toxic substances vary considerably. They are usually classified according to their
composition and to the material from which they have been prepared.

### A. Bacterial Exotoxins (A-1,2)

The *exotoxins* pass from the bacterial cell into the culture medium or
into the organism of the host. The exotoxins of *Corynebacterium diphtheriae*, *Clostridium botulinum*, and *Clostridium tetani* have been purified
and isolated as crystals. All of these, which are extremely toxic, are typical
proteins of the globulin type.

*Diphtheria toxin* (A-3) has been purified by fractional precipitation with
ammonium sulfate. The molecular weight of the purified toxin is 74,000;
the axial ratio of the molecules is 4.7 (A-4). The toxin is rendered innocu-
ous by treatment with formaldehyde; this indicates that amino groups are
essential for the toxic action. The toxin is destroyed by heat and by
proteolytic enzymes. It combines readily with iron porphyrins; possibly,
the toxin forms the protein component of a respiratory enzyme (A-1). Its
toxic action has been attributed to its ability to split neuraminic acid
from certain serum proteins (A-15).

Crystalline *tetanal toxin* has been obtained by precipitation with
methanol at slightly acidic reaction, low ionic strength, and temperatures
of −8 to −5°C (A-5). Tetanus toxin is a very labile globulin with an isoelectric point at pH 5.1 and a sedimentation constant of 4.5S. Even at
0°C, when kept at that temperature for a long time, it is inactivated,
undergoing conversion into a nontoxic, flocculating dimer, which is still
precipitated by the homologous antibody (A-5). On reduction with sulfhydryl compounds the monomer is regenerated; it can be stabilized by
alkylation with iodoacetate (A-6). The molecular weight of the tetanal
toxin is 68,000 (A-7).

*Botulinus toxin* occurs partly as a protoxin which on incubation with trypsin is converted into the active toxin (A-8,9). The toxin has been obtained in crystals; its molecular weight is approximately 900,000 (A-10,11). The isoelectric point is 5.6, the frictional ratio, $f/f_0$, is 1.45; the axial ratio, a/b, is very high, close to 10.

The mode of action of these exotoxins is not yet entirely known. Some of them have a great affinity for nerve tissue and are able to paralyze the nervous system. The lethal dose of botulinus toxin required to kill a mouse is 0.00005 $\mu$g or approximately 20 million molecules of the toxin (A-14). This low value suggests that the toxin acts as a catalyst, producing toxic products or starting a chain reaction.

The toxin of *Clostridium welchii* (A-12) has been purified to a certain extent by fractionation with ammonium sulfate and ethanol. It seems to be an $\alpha$-lecithinase, splitting phosphocholine from lecithin or sphingomyelin (A-13).

The toxins of *Corynebacterium diphtheriae*, *Clostridium botulinum*, and *Clostridium tetani* are typical antigens. Their injection into rabbits induces the formation of powerful antitoxins, which combine with the toxins and thus abolish their toxic properties. Toxins and antitoxins neutralize each other in definite proportions and it is possible to titrate one against the other to determine their amounts (A-2).

## B. Bacterial Endotoxins

Endotoxins are found in various types of *Salmonella*, *Shigella*, and other bacteria. They are highly toxic and cause elevated temperatures in injected animals. About 0.1 $\mu$g of the endotoxin from salmonellas is pyrogenic in man (B-1). The endotoxins are extracted from the dried bacilli by dilute trichloroacetic acid or by diethylene glycol (B-2). They are resistant to trypsin and can be precipitated from the concentrated tryptic digest with ethanol. The purified endotoxins are complexes formed by a polysaccharide, a phospholipid, and a protein or peptide. They are found only in the smooth S-forms of bacteria, but not in the rough R-forms (B-1). The endotoxin complex is cleaved into its components by treatment with 90% phenol (B-3).

The serological specificity of the endotoxins is determined by their carbohydrate moiety whereas the toxicity is caused by the lipid components. During the last few years, Westphal and Lüderitz have demonstrated that the serological specificity often depends on the content of dideoxyhexoses present in the polysaccharide (B-4,8). The dideoxyhexoses found by these authors in the endotoxins of various gram-negative enterobacteriaceae are abequose, tyvelose, ascarylose, paratose, and colitose, which are the 3,6-dideoxy derivatives of D-galactose, D-mannose, L-mannose, D-glucose, and L-galactose, respectively (B-5). Each serotype corresponds to a definite chemotype; 16 different combinations of these chemotypes have been found (B-6). Very little is known about the protein or peptide component of the endotoxins; the protein may act merely as a carrier (B-1). It may be mentioned, finally, that tuberculin from tubercle bacilli is also a protein derivative; fractionation of tuberculin with ethanol and by electrophoresis furnishes 2 polysaccharides, at least 3 proteins (A, B, and C), and a nucleoprotein (D). The typical skin reaction is produced by protein A (B-7).

## C. Cyclic Peptides

Toxic cyclopeptides of a very unusual type were discovered by Hotchkiss and Dubos (C-1) in *Bacillus brevis*, a soil bacterium. They will be discussed here because they contain D-amino acids as well as other "unnatural" amino acids which have never been found in the natural proteins. The toxins of *B. brevis* are insoluble in water, but are extracted by ethanol. When ether is added to the alcoholic extract, a cyclopeptide called tyrocidin is precipitated, whereas another cyclopeptide, gramicidin, remains dissolved. As the name gramicidin indicates, the cyclopeptides are toxic for gram-positive bacteria and are therefore valuable as bactericidal agents. One of the simplest cyclopeptides of this type is gramicidin S, a decapeptide of the structure (-L-Val-L-Orn-L-Leu-D-Phe-L-Pro-)$_2$ (C-2). Like gramicidin, tyrocidin is also a cyclic decapeptide; its structure is (-L-Val-L-Orn-L-Leu-D-Phe-L-Pro-L-Phe-D-Phe-L-AspN-L-GluN-L-Tyr-) (C-3). In the similar tyrocidin B, the L-phenylalanyl residue is replaced by L-tryptophan (C-4). Neither D-phenylalanine nor ornithine have ever been found in one of the natural proteins. Other unusual components of the cyclopeptides are D-leucine, ethanolamine, $\alpha$, $\gamma$-diaminobutyric acid, D-ornithine, and D-aspartic acid. Bacitracin, a peptide from *Bacillus licheniformis*, has the following structure in which each arrow indicates a peptide bond; the direction of the arrow indicates the sequence NH·CO and not CO·NH.

$$\text{L-Ileu} \rightarrow \text{L-Cys} \rightarrow \text{L-Leu} \rightarrow \text{D-Glu} \rightarrow \text{L-Ileu} \rightarrow \text{L-Leu} \rightarrow \text{L-Lys} \leftarrow \text{L-Asp} \leftarrow \text{L-His}$$

$$\begin{array}{c} \text{D-Asp} \\ \uparrow \\ \\ \text{D-Orn} \rightarrow \text{L-Ileu} \rightarrow \text{D-Phe} \end{array}$$

Bacitracin

Unusual features of the bacitracin structure are (a) a peptide bond formed by the ε-amino group of lysine and the carboxyl group of the neighboring aspartyl residue, (b) the presence of D-ornithine, (c) the participation of the β-carboxyl group of an aspartyl residue in peptide bond formation, and (d) the formation of a thiazoline ring between the SH group of cysteine and the carbonyl C-atom of the adjacent isoleucine residue (C-5). Polymyxin from *Bacillus polymyxa* contains an octapeptide ring formed by 5 residues of L-diaminobutyric acid, 1 of L-threonine, 1 of L-leucine, and 1 of D-phenylalanine; a side chain of isopelargonyl-D-diaminobutyryl-L-threonine is attached by the carboxyl group of threonine to an amino group of diaminobutyric acid of the cyclopeptide (C-6).

Gramicidin S has been prepared synthetically (C-7). The bactericidal activity of the synthetic cyclopeptide is identical with that of the natural gramicidin S; it remains unchanged when the L-ornithine residue of gramicidin S is replaced by an L-lysine residue (C-8). The X-ray diffraction pattern indicates that the 2 pentapeptide chains which form the decapeptide ring of gramicidin S form 2 extended β-chains (see Chapter IX, B) which are linked to each other by numerous hydrogen bonds (C-8).

The toxic cyclopeptides have been discussed here because they show impressively that we must be prepared to find similar unusual components and bond types in some of the natural proteins. In spite of the high resolution of chromatographic methods, hitherto no unusual amino acids have been detected in proteins.

## D. Other Toxins

The best investigated of the animal toxins (D-1,2) are the *snake venoms*. They cause hemolysis, or have neurotoxic or cardiotoxic activity. Many of them are enzymatically active as choline esterase, phospholipase, phosphatase, proteinase, peptidase, amino acid oxidase, or as hyaluronidase (D-1). Some of the snake venoms are devoid of a measurable enzymatic activity. The snake venoms are water-soluble; they can be salted out by ammonium sulfate and have all the properties of globulins (D-5). The carefully dried venom retains its toxic activity over several years (D-3). Most of the venoms are heterogeneous mixtures; they have been fractionated by electrophoresis and chromatography. Crotoxin from *Crotalus terrificus*, the rattlesnake, has been crystallized (D-4) but is still heterogeneous. Its molecular weight is close to 30,000; it contains 4% sulfur (D-4). The toxic action is lost after exposure of crotoxin to an excess of cysteine (D-6); presumably, the toxic action is correlated with the presence of certain dithio bonds. The principal toxin of the venom from *Crotalus* is crotactin; the lethal dose of crotactin is $10^{-4}$ mg/g mouse (D-7). Russell's viper venom contains a coagulating enzyme which forms 10% of the total venom proteins. Its molecular weight is 105,000; like thrombin it has esterase activity when tested with TAME (tosyl arginine methyl ester)(D-11).

The purified venom of *Bothrops jararaca* contains no sulfur. Likewise, the venom from *Naja flava*, the african cobra, is free of sulfhydryl or dithio groups; it is inactivated by $O_2$ and by cyanides (D-8). The toxin of the venom of *Naja naja* has been purified by salting-out with sodium sulfate; the lethal dose is $6.4 \times 10^{-3}$ mg/300 g pigeon (D-9). The toxic action of the snake venoms is attributed to their enzymatic actions. Very little is known about the toxins of insects such as bees, wasps, spiders, or scorpions, which also seem to contain toxic proteins.

It may be mentioned, finally, that toxic proteins have also been found in certain plants, particularly in legumes. The best known of these toxins are abrin from *Abrus precatorius*, crotin from *Croton tiglium*, and ricin. Ricin, the toxin of castor beans, has been crystallized and found to be a protein of the molecular weight 77,000–85,000. Its isoelectric point is at pH 5.2–5.5 (D-10).

## REFERENCES

### Section A

**A-1.** A. M. Pappenheimer, Jr., *Adv. in Protein Chem.* 4: 123(1948). **A-2.** W. E. van Heyningen, "Bacterial Toxins." Blackwell, London, 1950; Hoppe-Seyler/Thierfelder, "Handbuch der physiologisch- und pathologisch-chemischen Analyse," Vol. 4, Pt. I, p. 785, Springer, Berlin, 1960. **A-3.** E. H. Relyveld, "Toxine et Antitoxine Diphthérique." Hermann, Paris, 1960. **A-4.** M. L. Petermann and A. M. Pappenheimer, Jr., *J. Phys. Chem.* 45: 1(1941). **A-5.** L. Pillemer *et al.*, *JBC* 173: 427(1948); *ABB* 22: 374(1949). **A-6.** G. Raynaud *et al.*, *Ann. Inst. Pasteur* 99: 167(1960). **A-7.** J. F. Largier and F. J. Joubert, *BBA* 20: 407(1956). **A-8.** J. T. Duff *et al.*, *J. Bacteriol.* 73: 597(1957). **A-9.** G. Sakaguchi and S. Sakaguchi, *J. Bacteriol.* 78: 1(1959). **A-10.** C. Lamanna and B. Doak, *J. Immunol.* 59: 231(1948). **A-11.** A. Abrams *et al.*, *JBC* 164:63(1946). **A-12.** M. J. Boyd *et al.*, *JBC* 167: 899(1947). **A-13.** H. G. McFarlane, *BJ* 42: 587(1948). **A-14.** F. W. Putnam *et al.*, *JBC* 176: 401(1948). **A-15.** M. D. Poulik, *Clin. Chim. Acta* 6: 493(1961).

### Section B

**B-1.** O. Westphal *et al.*, *Biochem. Z.* 330: 21, 34; *Z. Naturforsch.* 13b: 561(1958). **B-2.** A. Boivin, *Bull. soc. chim. biol.* 23: 12(1941). **B-3.** W. T. J. Morgan and S. N. Partridge, *BJ* 35: 1140(1941). **B-4.** O. Westphal and O. Lüderitz, *Angew. Chemie* 66: 407(1954). **B-5.** O. Westphal, *Ann. Inst. Pasteur* 98: 789(1960). **B-6.** F. Kauffmann *et al.*, *Zentr. Bakteriol. Parasitenk. Orig.* 178: 442; 179: 180(1960). **B-7.** F. B. Seibert, *Chem. Revs.* 34: 107(1944); *Am. Rev. Tuberc.* 59: 86(1949). **B-8.** O. Westphal *et al.*, *J. Med. and Pharm. Chem.* 4: 497(1961); *Pathologia et Microbiologia* 24: 870(1961).

### Section C

**C-1.** R. D. Hotchkiss and R. J. Dubos, *JBC* 132: 791(1940); R. D. Hotchkiss, *Adv. in Enzymol.* 4: 153(1946). **C-2.** A. N. Belozserskii and T. S. Pashkina, *Biokhimiya* 10:

352(1945); R. L. M. Synge, *BJ* 39: 351(1945).  **C-3.** A. Paladin and L. C. Craig, *JACS* 76: 688(1954).  **C-4.** T. P. King and L. C. Craig, *JACS* 77: 6624(1955).  **C-5.** L. C. Craig *et al.*, *JACS* 77: 731, 3132(1955); E. M. Lockhart and E. P. Abraham, *Biochem. J.* 62: 645(1956); E. P. Abraham, "Biochemistry of Some Peptide and Steroid Antibiotics," Wiley, New York, 1957.  **C-6.** G. Bizerte and M. Dautrevaux, *Bull. soc. chim. biol.* 39: 795(1957).  **C-7.** R. Schwyzer and P. Sieber, *Helv. Chim. Acta* 40: 624(1957).  **C-8.** R. Schwyzer and P. Sieber, *Helv. Chim. Acta* 41: 1582(1958).

*Section D*

**D-1.** W. Neumann and E. Habermann, Tierische Gifte, *in* Hoppe-Seyler/Thierfelder, "Handbuch der physiologisch- und pathologisch-chemischen Analyse," Vol. 4, p. 801. Springer, Berlin, 1960.  **D-2.** E. Kaiser and H. Michl, "Biochemie der tierischen Gifte." Deuticke, Wien, 1958; E. E. Buckley and N. Porges, eds., "Venoms, A symposium." Am. Assoc. Advance, Sci., Washington, D.C., 1956; E. A. Zeller, *Adv. in Enzymol.* 8: 459(1948).  **D-3.** K. H. Slotta, *Fortschr Chem. org. Naturstoffe*, 12: 406 (1955).  **D-4.** K. H. Slotta and H. Fraenkel-Conrat, *Nature* 142: 213(1939).  **D-5.** A. Polson *et al.*, *BJ* 40: 265(1946).  **D-6.** N. Gralen and T. Svedberg, *BJ* 32: 1375 (1938).  **D-7.** W. P. Neumann and E. Habermann, *Biochem. Z.* 327: 170(1955).  **D-8.** F. Micheel and F. Jung, *Z. physiol. Chem.* 239: 217(1936).  **D-9.** B. N. Ghosh and D. K. Chaudhuri, *J. Indian Chem. Soc.* 20: 22(1943).  **D-10.** M. Kunitz and M. McDonald, *J. Gen. Physiol.* 32: 25(1948).  **D-11.** W. J. Williams and M. P. Esnouf, *BJ* 84: 52 ff.(1962).

# Chapter XV

## Role of Proteins in Immunological Reactions[*]

### A. Antigens (A-27)

Antigens are defined as substances which on parenteral administration to the animal organism give rise to the formation of antibodies which combine specifically with the injected antigen. Historically the first antigens were red blood cells, bacteria, or viruses. The presence of antibodies in the homologous immune sera was proved by specific agglutination or lysis of the injected or invading antigens. In 1897 Kraus (A-3) demonstrated that soluble bacterial extracts and proteins can also act as antigens. These antigens are precipitated specifically by the homologous antibodies. The latter are frequently designated as precipitins. The precipitin test is the simplest method applicable for the differentiation of species-specific proteins. Its great importance for protein chemistry is obvious.

Whereas pure proteins are homogeneous antigens, bacteria and red blood cells consist of a mosaic of different substances, many of them antigenic, others non-antigenic. Thus pneumococci contain type-specific carbohydrate antigens in their capsule and species-specific protein inside the bacterial cell. Similarly red blood cells contain hemoglobin as an antigen within the cells, and various blood group substances as antigens in their membrane. Antibodies against the latter substances cause agglutination of the red blood cells whereas antibodies to hemoglobin do not agglutinate the red cells but precipitate hemoglobin from its aqueous solutions.

Most of the proteins found in nature are antigens when administered parenterally to organisms of another species. Oral administration is usually inefficient since the ingested proteins are digested by the proteolytic enzymes of the gastrointestinal tract. Their breakdown products, the peptides, are in general not antigenic. It seems that a minimum molecular weight of approximately 5,000–10,000 is a prerequisite for antigenicity. Formation of antibodies is sometimes observed after the administration of small molecules of highly reactive substances such as iodine, picryl chlo-

[*] See references (A-1,2,24).

ride, or other nitro compounds. Since all these substances combine *in vitro* with proteins, it is assumed that they also combine with proteins at the site of injection and that the conjugated proteins thus formed are the true antigens. The antigen precursors of low molecular weight have been called proantigens (A-4).

Although the formation of antibodies against bacterial antigens or viruses renders the organism immune against repeated infection, the formation of precipitins after the injection of soluble protein antigens sensitizes the organism for a second injection of the same antigen. In contrast to the first injection, which is uneventful, the second injection is followed by the rapid formation of antigen-antibody complexes in the circulating blood; it can lead to very severe *anaphylactic* symptoms and even to the death of the sensitized organism. It is clear from this example that the formation of antibodies is sometimes favorable for the organism by producing immunity, but that it also has unfavorable consequences by causing allergy, i.e., sensitization for certain antigens.

Each of the natural proteins is specific as an antigen. Antibodies produced by the injection of one protein give precipitates only with the same protein, not with other proteins. Only in those cases where the test antigen is very similar to the injected antigen are cross reactions observed. Thus antibodies to horse serum proteins are also able to precipitate donkey serum protein. Similarly, ovalbumin from ducks' eggs is precipitated by antibodies to ovalbumin from hens' eggs (A-5). On the other hand, myoglobin is serologically quite different from hemoglobin although both substances contain the same hemin (A-6). Evidently the specificity is due to the protein component and not to the hemin. Hemoglobin of man and of beef are serologically different; however a slight relationship is indicated by inhibition tests (A-7). Proteins possess not only species specificity but also organ specificity. The proteins of blood serum differ serologically from the hemoglobin or from muscular proteins of the same animal.

It was found by Landsteiner (A-8) that the specificity of proteins is not attributable to the protein molecule as a whole, but is associated only with certain *chemical groups* of the molecule. This was proved by coupling proteins with different chemical groups. The preferred method employed by Landsteiner was the coupling of proteins with diazo compounds (see Chapter VII, F). The advantages of this method are: (1) diazo derivatives of practically all substances can be prepared, so that any desired groups can be linked to the protein molecule, and (2) the coupling of diazo compounds with proteins takes place at 0°C at slightly alkaline reaction of pH 9, where most proteins are not denatured.

Amino acids or peptides can be coupled to proteins by condensing their carboxyl group with the amino group of *p*-nitraniline, reducing the nitro group of the nitranilide to $NH_2$, diazotizing with nitrous acid, and mixing the cold solution of the diazo compound with a solution of the protein at pH 8.5–9.0. The formation of an azoprotein is indicated by the appearance of an orange or red color. In the azoprotein molecule the diazotized residues are bound in o-position to the OH groups of the tyrosine residues of the protein, but also to imidazole residues and, at a large excess of the diazo compound, to amino groups of lysine (see Chapter VII, F). The antigenicity of sugars can be examined by condensing the sugars with nitrobenzyl chloride, reducing the nitrobenzyl ether of the sugar, and diazotizing and coupling to protein. Similarly the influence of optical isomers or of cis and trans isomers can be investigated. Thus Landsteiner (A-8) condensed D- and L-tartaric acid, or fumaric and maleic acid with p-nitraniline and converted the nitranilides into azoproteins. It was found in such experiments that antibodies to the D-tartranil-azoprotein are different from those to the L-tartranil-azoprotein, that cis and trans isomers, or o-, m-, and p-isomers can be differentiated by the precipitin test with the homologous antibody (A-8).

It is evident from these and many other examples that the specificity of the artificially conjugated proteins depends mainly on the chemical groups introduced into the molecule. Thus, antibodies to arsanilazo horse serum protein are able to precipitate arsanilazoovalbumin and other arsanilazo proteins. By the chemical treatment the proteins lose their species specificity. If a large number of azo groups is introduced into the horse serum globulin molecule, it is not precipitated by antibodies to horse serum globulin. Similarly the species specificity of serum globulin is lost when its molecule is iodinated or acetylated after a certain minimum number of iodine atoms or acetyl groups has been introduced into the protein molecule (A-9).

Conjugated antigens can be prepared not only by coupling of proteins with diazo derivatives of smaller molecules, but also by combination with their isocyanates or isothiocyanates and other derivatives. However none of these methods is so easily adaptable to a variety of substituents as the formation of azoproteins. Not all groups attached to proteins by means of phenylazo bonds give rise to the formation of specifically adapted antibodies. In particular, nonpolar groups such as the paraffin chains are devoid of serological specificity. It seems that the prerequisites for antigenicity are rigidity and polarity of the *determinant* groups of the injected antigen. The high serological specificity of the aromatic $N$-phenylazo derivatives produced by the diazotization method is probably due to the rigidity of their planar aromatic benzene ring. It was at first believed that only acidic groups such as $-COOH$, $-SO_3H$, $-AsO_3H_2$ induce the formation of antibodies. However, highly specific precipitating antibodies are also formed after the injection of azoproteins in which the tyrosine groups are substituted by strongly basic residues, e.g., by the azophenyltrimethylammonium cation (A-10).

The great importance of the aromatic residues of tyrosine or phenylalanine for the antigenicity of proteins has been demonstrated recently by the antigenicity of synthetic polyamino acids (A-11). If such polyamino acids are condensed with gelatin, which is a very poor antigen, very efficient antigens are obtained by using polytyrosine or polyphenyl-alanine as substituents, but not with polyalanine, polyserine, or any of the other aliphatic polyamino acids. An increase in antigenicity is also observed when mixed polymers containing the acidic glutamyl and the basic lysyl residues are attached to gelatin. Finally, a fully synthetic antigenic polyamino acid, free of gelatin, has been prepared; it consists of a chain of poly-DL-alanine-poly-L-lysine to which polypeptide chains containing L-tyrosine and L-glutamic acid are attached (A-11,12).

Very little is known on the determinant groups of the natural proteins. Since these are precipitated by the homologous antibodies without denaturation, we must conclude that the specificity of the antibodies is directed against certain groups on the surface of the globular protein antigens. The results obtained with polyamino acids suggest that here also the rigid and polar tyrosine residues play an important role. The essentiality of these groups for antigenicity has been suspected for many years because of the very weak antigenicity of gelatin which is practically free of tyrosine and cystine. The latter amino acid increases considerably the antigenicity of polyamino acids, probably owing to the formation of dithio bridges which render the structure of the polyamino acid chain more rigid (A-11). The increase in antigenicity by cystine residues and also by mixed lysyl-glutamyl polymers demonstrates that aromaticity is not a prerequisite for antigenicity. Indeed, gelatin-bound polyphenyl-alanine can be replaced by the hydroaromatic polycyclohexylalanine without loss of antigenicity (A-11).

Not all the determinant groups of the native globular protein antigens are on the surface of the globular molecules. If human serum albumin is injected into rabbits and the immune serum of the rabbits is then treated with increasing amounts of native human serum albumin until no precipitate is formed, some of the antibodies of the exhausted immune serum still combine with albumin fragments of a tryptic digest of human serum albumin (A-23,25). Evidently, some of the antibodies react with determinant portions of the human serum albumin which, in the native protein, are buried in the interior, but are made accessible to the antibodies by tryptic digestion. We have to conclude, therefore, that the injected antigen in the organism of the host is likewise degraded to fragments and that these give rise to the formation of antibodies (A-23).

Denaturation results in a change or a partial loss of the original antigenicity of the native protein. Antibodies against native ovalbumin react only weakly with denatured ovalbumin (A-13). Injection of denatured ovalbumin leads to the formation of antibodies which react specifically with denatured ovalbumin (A-14). The lower antigenicity of denatured proteins is probably due to the fact that they are hydrolyzed by proteolytic enzymes more readily than native proteins so that they are split in the organism before they arrive at the site of antibody formation.

The protein component of the azoproteins is necessary for the formation of antibodies *in vivo*, but not for the combination of the antibody with the determinant groups of the antigen. Antibodies against arsanilazoproteins also combine with arsanilazotyrosine or with arsanilic acid (A-15,16). Small molecules containing the determinant group of the antigen are designated as *haptens* (Greek: *haptein*, to bind, to hold) because they are bound by the antibody. Haptens with more than one determinant group (e.g., *tris*-arsanilazoresorcinol) form precipitates with the antibody (A-17) whereas those with only one determinant group form soluble hapten-antibody complexes. The indispensability of the protein component for the production of antibody is attributed to the carrier role of the protein macromolecule which prevents rapid excretion of the hapten and carries it to the site of antibody formation.

Insight into the fate of the injected antigen in the animal organism can be gained by labeling the antigen with radioactive isotopes and determining the isotope distribution in the organs and tissues of the host (A-1,2,18). Experiments of this type reveal that the antigens, after intravenous injection, are bound to the proteins of spleen, lymph nodes, liver, bone marrow, and other organs of the reticulo-endothelial system; immediately after injection they are found in the small cytoplasmic granules (microsomal fraction), 30–60 minutes later, in the mitochondrial fraction where they persist for several months. Autoradiography reveals that most of the antigen is present in the cytoplasm, very little if any of the antigen being found in the nuclei (A-19). It is not yet clear in which manner the injected, soluble antigens are converted into insoluble cell-bound material. They are probably incorporated into the cells by phagocytosis and degraded to fragments which form insoluble complexes with ribonucleic acid (A-20) or with other cellular components.

Antibody formation is considerably enhanced when a mixture of the antigen is injected with an *adjuvant* such as alumina, paraffin oil, or killed mycobacteria suspensions (A-21). The adjuvants protect the antigen against degradation and keep it over a long period of time in the tissue; they cause inflammation and stimulate the formation of those cell types which are able to produce antibodies (A-26). Insoluble contaminants may also be essential for the high antigenicity of bovine γ-globulin and other soluble protein antigens. If these insoluble contaminants are removed by high-speed centrifugation, bovine $\gamma$-globulin has only a very weak antigenic action (A-22).

## B. Antibodies, Their Detection and Isolation

The presence of antibodies is revealed by their specific reaction with the homologous antigen. The best investigated antibodies are in this respect the precipitins since the precipitin test can be evaluated in terms of weight units of the precipitated antibody. The methods for the detection of the nonprecipitating antibodies are much less satisfactory; they cannot give more than a rough estimate of the amount of antibody present.

The precipitin test has been refined in the last decade in such a manner that the presence of multiple antigen-antibody systems can be detected in a single test. This was first accomplished by Oudin (B-1) who mixed the immune serum with a warm agar solution in small test tubes, let the mixture cool down to form a gel, and allowed the antigen solution to diffuse into the antibody-containing gel. Two or more precipitin bands are formed when more than one antigen-antibody system is present. In the "double diffusion" method the agar gel is placed between the solutions of antigen and antibody (B-2,3). In the Ouchterlony technique (B-4) the vertical tubes are replaced by horizontal agar gels in flat dishes; the solutions of antigen and antibody are placed into holes in the gel or into bottomless glass cups. The method allows differentiation of homologous antigen-antibody pairs, each of which forms a single precipitin zone, from cross-reacting systems which form fused or branched precipitin bands. Further differentiation can be accomplished by combining the Ouchterlony agar gel diffusion with electrophoresis (see Chapter VI, C, on immunoelectrophoresis).

Early workers in the field of immunology found that the antibodies reside in the globulin fraction of the immune sera (B-5). It was assumed that antibodies are attached to the serum globulins. Analyses of the immune precipitates revealed, however, that the amount of globulin bound in the antigen-antibody complex was proportional to the antibody activity carried down with the precipitate. This led to the view that the antibodies are globulins which are complementarily adapted to the determinant groups of the antigen (B-6a,6b). This view is supported by the results of attempts to purify and isolate antibodies. Purification of antibodies can be accomplished either by nonspecific methods in which certain globulin fractions rich in antibody are separated from other fractions, or by specific methods in which the affinity of the antibody for antigen is utilized for its purification and isolation. *Nonspecific* purification can be attained by electrophoresis (B-7), by salting-out with ammonium sulfate, or by ethanol fractionation (see Chapter VIII, B). The antibodies are always found in the $\gamma$-globulin fraction, particularly in the subfractions II-1, 2, and 3 and in fraction III-1 of the ethanol fractionation procedure (B-8). Horse immune sera contain another antibody fraction, T, which migrates on electrophoresis between the $\beta$- and the $\gamma$-globulins (B-9). The blood of newborn animals is nearly or completely free of $\gamma$-globulins; this agrees with the absence of antibodies in the sera of newborn animals. Nonspecific purification of antibodies has also been accomplished by partition chromatography with potassium phosphate and cellosolve at $-3°C$ (B-10).

If a single type of antibody is to be separated from other antibodies, *specific* methods of separation must be used. These methods consist essentially of two steps: (1) formation of an insoluble antigen-antibody complex and (2) dissociation of this complex into antigen and antibody. In the first step all other proteins and also other substances are removed from the antigen-antibody complex. The first successful separation of pure antibodies was reported by Felton (B-11), who precipitated pneumococcal polysaccharide antigens by the homologous immune sera and treated the precipitate with baryta which solubilized the antibodies; the barium salts of the polysaccharide remained insoluble. Precipitates of the polysaccharide antigens have also been dissociated by concentrated saline solutions (B-12). In the author's laboratory, azoprotein precipitates were dissociated by acidification in the presence of 5% NaCl; a large portion of the antibody was solubilized whereas most of the antigen formed an insoluble complex with some of the antibody (B-13). Although it is not yet well understood why acidification should cause dissociation of the antigen-antibody complex, the method has found wide application in the dissociation of various types of antigen-antibody complexes. In many of these techniques the antibody is selectively adsorbed from the immune serum by an insoluble form of the antigen and is then split off by acidification.

Serum albumin and other protein antigens can be converted into an insoluble form by combination with carboxyl resins whose COOH groups, by treatment with thionyl chloride, have been chlorinated to form acylchloride groups (B-14, 41). In another procedure serum albumin is bound to diazotized aminobenzyl-cellulose (B-15), or to nitrated, reduced, and diazotized polystyrene resin (B-16). Particularly high yields of antibody are obtained after adsorption of antigen to cellulose powder or filter paper which has been alkylated by $m$-nitrobenzyloxy-methyl-pyridinium chloride, then reduced and diazotized (B-17). When an immune serum is passed through these columns of antigen-coated cellulose or resin, only the homologous antibody is bound; all other proteins pass through the column. Subsequently, antibody is released by acidification to pH 2–3. Azoproteins can also be eluted by a large excess of the homologous hapten which can be removed from the antibody solution by dialysis. Another technique of antibody purification is based on the coupling of antigens with $N$-acetyl-homocysteine lactone; if such sulfur-containing antigens are added to an immune serum, the homologous antibody combines with them to form a precipitate which is solubilized by acidification with glycine-$H_2SO_4$ to pH 2.4; the sulfur-containing antigen is then precipitated by a mercury salt whereas the antibody remains in solution (B-18).

The purified antibodies are almost quantitatively precipitated by the homologous antigen. The properties of purified antibodies are similar to those of the normal $\gamma$-globulins. Their molecular weight is close to 160,000; in the blood of horses, sheep, and cattle, antibodies with a higher molecular weight of about 900,000 have also been detected (B-19). The length of antibody molecules, according to electron microscopy, is approximately 250 Å (B-38). The localization of antibody molecules in electron micrographs is facilitated by coupling them with ferritin which, because of its high iron content, is easily detectable; aromatic diisocyanates (B-39) or bis-diazotized benzidine (B-40) have been used as links between antibody and ferritin molecules.

Hydrolysis of purified antibodies with concentrated HCl indicates that their amino acid composition is very similar to, if not identical with, that of normal serum $\gamma$-globulins (B-20,21). Likewise, the same $N$-terminal amino acid sequence of alanyl-leucyl-valyl-aspartyl-glutamyl- has been found in normal rabbit $\gamma$-globulins as well as in rabbit antibodies to ovalbumin (B-22) or to pneumococcal polysaccharides (B-20). Recently Porter made the important discovery that antibody, on short exposure to papain and sulfhydryl compounds, is split into three large fragments designated as I, II, and III. He also showed that fragment III is a crystalline globulin without specific affinity for the antigen, but that the fragments I and II, which are amorphous, combine specifically with the antigen (B-23). Fragment III is also found in analogous digests of normal $\gamma$-globulin. The fragments I and II, which have molecular weights close to 50,000, form soluble complexes with antigen or hapten which contain only 1 molecule of antigen or hapten per molecule of antibody fragment (B-24). If the antibody solution is exposed to pepsin instead of papain, fragment III is liberated, but the fragments I and II are obtained as a complex which is dissociated by treatment with mercaptoethylamine (B-25). A single SS bond between fragment I and II is cleaved in this reaction. On reoxidation the I-II complex is reconstituted; in contrast to the separated fragments, the I-II complex forms insoluble precipitates with the homologous antigen. If a mixture of the reduced fragments of two different antibodies such as anti-ovalbumin and anti-bovine $\gamma$-globulin is reoxidized, hybrid antibodies are formed which bind one molecule of ovalbumin and one of bovine $\gamma$-globulin (B-26). These experiments prove convincingly that the antibody fragments are univalent (i.e., have only one specifically combining group) and that the original antibody molecule is bivalent.

The author and other immunochemists had been hesitant to accept the bivalence of antibodies because antibody molecules are always adjusted to only *one* type of determinant group even if an antigen with two or more different determinant groups is injected (B-27). It was difficult, therefore, to believe that the peptide chain of an antibody molecule during the short period of its formation, which may be only a few seconds, should always react with two identical determinant groups as templates, although other determinant groups were available. The discovery of the univalence of the fragments I and II and of their combination by a dithio bond suggests that the formation of I and II precedes their combination by an SS bridge and makes it unnecessary to postulate immunological bivalence of a single peptide chain.

The mutual attraction of antigen and antibody is doubtlessly due to a close fit between the determinant group of the former and the combining group of the latter. It is not yet clear, however, whether the complementariness of the antibody molecule is brought about only by differences in conformation or whether it also involves changes in the amino acid sequence of the fragments I and II. Attempts have been made to answer this question by partially hydrolyzing purified antibodies enzymatically and subsequent "fingerprinting" of the peptide mixture, i.e., chromatography along one dimension and electrophoresis along a second. Experiments of this type reveal that the peptide mixture obtained from various types of rabbit antibodies is very similar to that of peptides obtained from normal rabbit $\gamma$-globulins, but that there are slight differences in 3 or 4 peptides which are present in only some but not in all of the antibodies examined (B-28,29). This has also been found when two antibodies of a single immune serum were compared with each other (B-30), and seems to indicate differences in the amino acid sequence. Interpretation of these experiments is made difficult by the discovery that the fragments I and II come from different antibody types in the heterogeneous antibody mixture. Some antibodies contain 2 residues of fragment I, in addition to fragment III; other antibody molecules contain two fragments of the type II (B-42). If the fractionation methods are refined, the number of separable fragments increases from three to five. The heterogeneity of peptides, obtained after tryptic digestion of antibodies, may partly be a result of this heterogeneity of antibodies.

If the disulfide bonds of antibodies are reduced by sulfhydryl compounds and the molecules then exposed to denaturing agents (B-45) or to *N*-acetic or *N*-propionic acid (B-46), the antibody molecules are broken up into 4 peptide chains. Two of these have been designated as heavy (H) or A chains, the two others as light (L) or B chains. The latter seem to contain the combining groups of the antibody molecules. Still smaller fragments may be obtained by the action of papain in urea (B-47). Papain digestion of the native antibody removes part of the A chains, but not the B chains, as fragment III (B-46). There are no indications for the presence of $\alpha$-helices in rabbit antibodies (B-48).

Very little is also known on the combining site of the antibody molecules. The high specificity of antibodies had suggested to some authors that the determinant group of the antigen is incorporated into the antibody molecule. This was disproved by the finding that antibodies to iodoproteins, arsanilazoproteins, and other conjugated proteins are free of iodine, arsenic, or other typical components of the determinant group (B-31). The assumption of a passage of the determinant group of the antigen into the antibody molecule is also irreconcilable with the fact that the number of antibody molecules formed per molecule of antigen is much higher than the number of determinant groups in the antigen molecule. Since hydrolysis of the antibodies yields only amino acids, it must be concluded that the combining groups of the antibody molecules consist of amino acid residues.

If antibodies are labeled by nonradioactive iodine in the presence of a hapten, their combining site is protected by the hapten against iodination. Subsequent removal of the hapten and iodination with $I^{131}$ leads to iodination of the combining site and demonstrates that this site contains at least 1 tyrosine residue (B-32, 43). If one portion of antibody is iodinated with $I^{131}$ in the presence of hapten and another equal portion with the radioactive isotope, $I^{125}$, in the absence of hapten, enzymatic digestion of a mixture of the two iodinated antibodies makes it possible to detect the tyrosine-containing peptide which corresponds to the combining site of the antibody; the ratio $I^{131}:I^{125}$ which is equal to 1.0 in all other peptides deviates from unity in the peptide from which the combining site is derived (B-33). Antibodies to the acidic azobenzoate group seem to contain in their combining site a positively charged group, whereas antibodies to the basic trimethylphenylammonium group contain a carboxyl group in their combining site (B-34,49). All these observations indicate that the combining sites of different antibodies differ from each other not only in their conformation but also in their amino acid composition. The size of the combining site of the antibody molecule has been estimated to have a minimum diameter of not more than 10–20 Å. In antibodies to polyglucoses the upper limit of the size of the combining site corresponds to the dimensions of a hexa- or heptasaccharide (B-35).

If antigen containing two or more different types of determinant groups is injected, different types of antibody are formed, each of them adjusted to only one type of determinant group. Thus injection of a serum globulin containing diiodotyrosine residues and also azophenylarsonate groups leads to the formation of two separable types of antibody; those directed against the diiodotyrosine groups are precipitated by iodinated ovalbumin; the supernatant solution gives additional precipitate on addition of ovalbumin-azophenyl-arsonate (B-36). No further precipitation occurs when ovalbumin containing both types of hapten groups is added. Such separation of two types of antibodies is only possible if the two determinant groups of the antigen are remote from each other in different portions of the protein molecule. If the two determinant groups are close to each other, they act as a single determinant and give rise to only one type of antibody (B-37).

Work on the structure of antibodies is made difficult by their heterogeneity. Since each injected antigen has several types of determinant groups, a multiplicity of different antibodies will always be formed. Further heterogeneity will arise by differences in the extent of complementariness. Finally, different types of antibodies may be formed in spleen, lymph nodes, bone marrow, and other lymphoid tissues.

The degradation of antibodies to univalent subunits stimulated analogous work on the normal $\gamma$-globulins of the blood serum. These too can be degraded by proteolytic enzymes to similar fragments. Dissociation of normal $\gamma$-globulins to small fragments can also be accomplished by the action of urea and reducing agents, in the absence of enzymes (B-44). Evidently, $\gamma$-globulins are not uniform molecules in the classical sense, but aggregates or micelles formed from smaller units by the closure of disulfide bonds and other types of linkage.

## C. Antigen-Antibody Interaction (C-33)

All immunological reactions are based on the same primary reaction, the combination of antigen with antibody. Precipitation, agglutination, cytolysis, or other more complicated reactions may occur secondarily. One of the simplest antigen-antibody reactions is the precipitation of a soluble antigen by the homologous antibody. If the antigen used is labeled by a colored group, by isotopes, or by an element which can be determined by chemical analysis, it is possible to carry out quantitative analyses and to examine the composition of the precipitates under different conditions. The first analyses of this kind were carried out by Wu (C-1). Investigations with more refined methods demonstrate that the antibody:antigen ratio in the precipitates increases with an increase of this ratio in the antibody:antigen mixture (B-6a; C-2,3,4). The antibody:antigen ratio in the precipitate depends on the size of the antigen employed (C-5). If cells are used as the precipitating antigen, the layer of antibody molecules bound to their surface is negligibly small compared with the volume of the cell. The reverse is true if small molecules are employed as antigens; in some precipitates only 2–3% of the precipitate is formed by the antigen (Ag), while the bulk of the precipitate consists of antibody (Ab).

The precipitate is usually considered as a lattice of alternate antigen and antibody molecules (C-6). As has been demonstrated in the preceding section, each antibody molecule seems to consist of two univalent fragments held together by an SS bond. Hence the antibody molecules are serologically bivalent. Treatment of the $AbAg_2$ precipitate with papain results in the formation of complexes of the composition FAg where F is the univalent antibody fragment mentioned in Section B of this chapter (C-34). The antigen molecules usually contain a large number of determinant groups on their surface, particularly when the molecular weight of the antigen is high. Therefore the antigen molecules are usually multivalent and bind a large number of antibody molecules if an excess of antibody molecules is present (C-7). The molar ratio antibody:antigen in thyroglobulin-antithyroglobulin precipitates is approximately 50–60 (C-7). Evidently not all of the antibody molecules in such a complex can be bound by both of their combining groups. Many of them may be bound to thyroglobulin by only one of the two combining sites. In an *excess of antigen*, all immune precipitates are soluble. This is attributed to the small size of the antigen-antibody complexes which, in the presence of an excess of antigen, consist only of a single antibody molecule and two antigen molecules. Rabbit antibody precipitates are insoluble in an *excess of antibody* whereas horse antibody precipitates dissolve in an excess of antibody. The reasons for this different behavior of rabbit and horse antibodies are not yet clear. It may be the result of differences in the physical-chemical properties of the two types of antibodies, for instance, differences in polarity or of hydration.

Since the nature of the determinant groups of the natural proteins is not known, we also do not know much about their combination with antibody. In bovine serum albumin, amino groups seem to be involved in the combination with the homologous antibody (C-37). A particularly simple immunochemical reaction is the *combination of an antibody with the free hapten* which corresponds to the determinant group of the antigen. Thus antibodies to protein-azobenzoates (prepared by coupling protein with diazotized *p*-aminobenzoic acid) combine with tyrosine azobenzoate or other substituted azobenzoates (C-8) to form soluble antibody-hapten complexes which contain 2 molecules of hapten per antibody molecule.

The affinity constant of the antibody-hapten complex can be determined by equilibrium dialysis and can be measured photometrically since the azo compounds are colored. Measurements of this type in different antibody-hapten systems indicate that $\Delta F$, the free energy of combination, is approximately $-5$ kcal/mole of hapten; $\Delta H$ is not very different from $\Delta F$, and the entropy change very low (C9 to 11). If bivalent or tervalent haptens are used instead of univalent haptens, the antibody-hapten complexes are insoluble since in such complexes alternate hapten-antibody lattices can be formed (C-12).

Similar determinations of thermodynamic data can be carried out when the antigen-antibody complex is dissolved in an excess of antigen. Ultracentrifugation has been used to determine the size of the soluble antigen-antibody complexes. Their composition, depending on the affinity constant, varies between AbAg and AbAg$_2$ where Ag and Ab are antigen and antibody molecules, respectively. In this case the $\Delta F$ values are similar to those found with haptens, close to $-5$ kcal/mole of antigen (C-13). This indicates that the combining area of the antigen is not larger than that of the hapten (C-11). Calorimetric determination of $\Delta H$ gave the value $-3.5$ kcal/mole of antigen (C-14). If insoluble antigen-antibody complexes are washed with $1\%$ saline solution, they lose small amounts of antibody but not of antigen. The insoluble complex AgAb$_n$ dissociates into insoluble AgAb$_{n-1}$ and soluble Ab. The free energy change for the reaction AgAb$_{n-1}$ + Ab = AgAb$_n$ can be determined when isotopically labeled antigen or antibody is used; it is approximately $-9$ kcal/mole; $\Delta H$ is close to 0 and $\Delta S$ positive (C-15). The increase in entropy is attributed to the release of water of hydration from the combining sites of antigen and antibody molecules.

Numerous equations have been proposed for correlating the composition of the antigen-antibody precipitates with the antigen and antibody content of the supernatant solutions. Most of these equations are based on empirical constants and hold for a limited range of Ag/Ab ratio, but fail in the regions of antigen or antibody excess. Some deficiencies have been overcome by a treatment which is based on the chain formation of branched polymers (C-16). However, even this improved theory cannot be reconciled with the different behavior of precipitates formed by rabbit and horse antibodies. The failure of all attempts to describe antigen-antibody interaction satisfactorily is principally due to the heterogeneity of the antibody population. Even if a single, well-defined antigen is used for immunization, a variety of different antibodies is formed. This multiplicity of antibody types is due to complementary adaptation of the antibodies to different determinant groups of the antigen, but also to different degrees of adaptation. Since antibodies consist of peptide chains and since these cannot be distorted beyond a certain limit, the complementary adaptation of the antibodies to determinant groups of the antigen will approach being, but will never achieve perfection. In other words, the immune serum will contain well and poorly adapted antibodies. Indeed, the immune serum may contain all intermediates between normal γ-globulins and well-adapted antibodies (C-17).

The forces involved in the mutual linkage of antigens and antibodies are the short-range van der Waals forces, possibly also electrostatic forces between groups of opposite charge and hydrogen bonds (see Fig. XV-1). As explained in Chapter X, Section A, the mutual attraction by the short-range forces increases with the sixth power of $1/r$ where $r$ is the distance between the two components. Indeed, it has been found that small

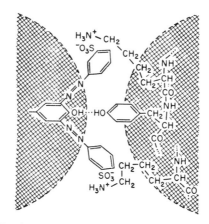

Fig. XV-1. Combination of antigen (protein-*p*-azophenylsulfonate) with the homologous antibody (C-38).

changes in the structure of a hapten cause considerable changes in the mutual affinity of hapten and antibody (C-18). It is understandable, therefore, that the antibodies against *m*-substituted haptens have much less affinity to the analogous *o*- or *p*-derivatives and vice versa. Since all antibodies and also many of the antigens are proteins, it is not surprising that their interaction depends on the pH, the salt content of the medium, its dielectric constant, and other properties. The optimum pH for the combination of antigen with antibody is close to neutrality. In acid or alkaline solutions dissociation takes place. The optimum salt content for precipitation is 1.0–1.5% NaCl for mammalian antibodies. With chicken and other avian antibodies, maximum precipitation is observed at higher salt contents (C-19). The combination of antigen with antibody is a very rapid reaction. The velocity constant is approximately $10^7$–$10^8$ per mole per second (C-39,40).

Soluble antigen-antibody complexes can be detected by paper electrophoresis since their migration velocity is different from that of free antigen or antibody (C-20). Antigen-antibody complexes of bovine serum albumin (BSA) and anti-BSA solubilized in an excess of BSA can be precipitated by half-saturation with ammonium sulfate (C-21), insulin-antiinsulin complexes by the addition of 25% ethanol and lowering the temperature to −5°C (C-22).

The combination of antigen with antibody molecules can be made visible in the electron microscope if large antigen molecules such as hemocyanin or viruses are used as antigens. The diameter of tobacco mosaic virus increases twice when the virus combines with antibody; this indicates that the orientation of the antibody molecules is perpendicular to the surface of the virus (C-23). The structure of precipitates containing smaller antigens are not resolved in electron microscopy. However, when ferritin, which contains about 20% iron, is used as an antigen, the antigen molecules can be localized in electron micrographs. The distance between adjacent ferritin molecules increases from 100 Å in the absence of antibody to 200–400 Å in its presence (C-24). Fluorescent antibody has been used to detect the appearance of actin and other muscle proteins in the growing chicken embryo and in other organisms (C-25).

If the antigens are not molecules of a definite type but bacteria or cells, the term "agglutination" is used for the formation of an insoluble antigen-antibody complex. Precipitation, agglutination, cytolysis, and other immunological phenomena are different manifestations of the same antibody; antibodies against pneumococcal polysaccharides are able to precipitate the specific polysaccharides, to agglutinate the cocci, to dissolve them in the presence of complement, or to render them susceptible to phagocytosis (C-26). The same antibodies are able to protect men and animals against infections with virulent pneumococci (C-27). Agglutination of red blood cells has a great importance in blood transfusion since the blood plasma of many individuals contains isoagglutinins, i.e., agglutinins for the red cells of persons of another blood group. The antigens responsible for this antigen-antibody reaction are called blood-group substances. They are glycoproteins with a high carbohydrate content and occur in the membrane of the red blood cells. They are discussed in Chapter XI, Section C. Agglutinins for red blood cells are also found in the extracts of lima beans and other legumes (C-28). Some of these cross-react with antibodies to the blood group substances or to antigenic bacterial polysaccharides (C-36). The nature of these agglutinins is not yet known.

The agglutination test is much more sensitive than the precipitin test, since very small amounts of antibody are sufficient to combine with the antigens in the surface of the red cells. Agglutination therefore is frequently used to detect nonprecipitating antibodies or small amounts of precipitating antibodies. In such experiments normal red blood cells are coated with the respective antigen. This is accomplished by treating the red blood cells with tannic acid (C-29), formaldehyde (C-30), or bis-diazotized benzidine (C-31), and mixing them subsequently with the respective antigen. They are then agglutinated when a serum containing antibody to the cell-bound antigen is added. In this manner less than 0.04 μg of antibody can be detected (C-32). Agglutination, like precipitation, requires the presence of bivalent antibodies (C-41).

Very little is known regarding the chemistry and the mechanism of other immunological reactions such as phagocytosis, neutralization of toxins, or anaphylactic reactions. Experimental methods used in these research areas are described in Kabat's classical book (A-24).

## D. The Mechanism of Antibody Formation

In order to gain more insight into the mechanism of antibody formation, we need to know more about the *metabolic fate of the injected antigen*. As mentioned in Section A of this chapter, isotopically labeled antigens can be used for this purpose. In view of the short half-life of $I^{131}$, the use of this isotope as a label for protein antigens is limited to short-term experiments. In long-term experiments, diazotized $S^{35}$-sulfanilic acid, 1-$C^{14}$-anthranilic acid, or tritiated aromatic amines have been coupled to various proteins (D-1 to 4). Another type of labeled protein has been obtained by the action of $S^{35}$-dichloroethylsulfone on proteins (D-5). If only traces of the isotopically labeled haptens are coupled to a protein antigen, the immunological specificity of the protein is not changed (D-6). Thus bovine serum albumin (BSA) or bovine γ-globulin (BGG) labeled by traces of $I^{131}$ or $S^{35}$-sulfanilic acid circulates in the blood plasma like BSA and BGG for several days, and is eliminated from the blood 5–7 days after the first injection at a time when antibody formation begins (D-3). If labeling is accomplished by a large amount of hapten containing traces of the isotopically labeled hapten, the serological properties of the proteins are completely changed; they are converted into heavily substituted iodo-proteins or azoproteins and are eliminated from the circulation within a few hours after injection (D-7). In both instances most of the antigen, after elimination from the blood, is found in the microsomal and later in the mitochondrial fraction of spleen, lymph nodes, liver, bone marrow, and other organs of the reticuloendothelial system.

Attempts have been made to use biosynthetically labeled proteins as antigens. Such proteins are obtained by injecting animals with radioactive amino acids containing $C^{14}$, $S^{35}$, or $H^3$. However, these internal labels are not reliable because the radioactive amino acids which are formed by the degradation of the protein *in vivo* are reutilized for the synthesis of new proteins. The different fate of internal and external labels is evident from experiments with proteins which are internally labeled with $S^{35}$-amino acids and externally with $I^{131}$; 9 days after injection of the doubly labeled proteins into rats, the ratio $S^{35}:I^{131}$ in the liver proteins is 50–100 times higher than in the injected protein (D-8).

Since immunity in some instances lasts many years, it would be important to know whether the injected antigen persists in the organism. Experiments with externally labeled antigens indicate *persistence of the injected azoproteins* in the organism for periods of at least several months (D-1 to 3). It is not yet clear in which form the injected antigens are present in the cells. Most probably, they are degraded to fragments which combine with ribonucleic acid and persist in this form in the tissues (D-9). However, we are not yet able to differentiate between nonspecifically bound fragments of the antigen molecules and those fragments which, because of their specific combination with certain intracellular components, give rise to the production of antibodies.

Very small amounts of *antibody in the serum* can be detected by the hemagglutination methods mentioned in the preceding section of this chapter or by labeling with radioactive isotopes. In investigations of the latter type internal labeling by the administration of radioactive amino acids containing $S^{35}$, $C^{14}$, or $H^3$ is very useful. The radioactive amino acids are rapidly incorporated into the newly-formed antibody molecules. These can be determined quantitatively by precipitation of the immune serum with the homologous antigen. If the concentration of the antibody is so low that no precipitation takes place, nonradioactive antibody of the same type can be added as a carrier. From the specific activity of the precipitate and that of the antibody the amount of precipitated labeled antibody can be calculated.

For the detection of *antibody in tissue sections* fluorescent antibody is at present widely used. Fluorescein isocyanate, originally used by Coons (D-10) for labeling purposes is frequently replaced by the more stable fluorescein isothiocyanate (D-11). The tissue section is first exposed to a solution of the antigen; the excess of antigen is removed by washing and the section then covered with a solution of fluorescent antibody which combines with the bound antigen and thus reveals the localization of antibody.

Internally labeled antibody containing $C^{14}$-phenylalanine has been used to determine whether internally labeled proteins exchange their amino acids with the amino acid pool of the body. If $C^{14}$-antibodies to pneumococcal polysaccharides, produced in a rabbit injected with $C^{14}$-phenylalanine, are intravenously injected into a normal rabbit, and samples of the serum of this animal are precipitated after various periods of time with pneumococcal polysaccharide, the specific $C^{14}$-activity of the antibody is found to remain unchanged (D-12). This proves that the antibody molecules do not exchange their $C^{14}$-phenylalanine residues with amino acids of the other body proteins. The great value of antibodies in experiments of this type is that antibodies, owing to specific reaction with the antigen, can be determined quantitatively in a large excess of other plasma proteins and can be isolated in a single step by the addition of the homologous antigen. Experiments of this type also make it possible to determine the biological half life of antibodies in different species (D-13, 14). The half life varies from 2 days in mice and 6 days in rabbits to 10–20 days in man and the large mammals.

Antibody formation, like the formation of other proteins is accompanied by an increase of the ribonucleic acid content of the cells. The cytoplasm of these cells has a high affinity for basic dyes such as pyronine; antibody-forming cells can be detected by means of this stain (D-15). The same cells, called plasma cells, also give a strong cytoplasmic fluorescence when they are exposed to antigen and fluorescent antibody as described above (D-10). Their antibody content is also confirmed by autoradiography after treatment with radioactive antigen (D-16). The newly formed antibody remains 10–30 minutes inside the cells and is then released into the serum (D-17). In the rabbit spleen the amount of antibody formed is approximately 32 $\mu g/g$ of spleen in 3 hours (D-17).

Antibody formation after a single injection of antigen reaches its maximum in the rabbit on the 10th to 12th day and then decreases rapidly. The precipitin test becomes negative after 3–4 weeks. If the animal is then reinjected with the same antigen, antibody production begins much earlier, increases very rapidly to a much higher maximum which is reached after 4–6 days, and persists for a long period of time. This phenomenon, called *secondary response* or *anamnestic reaction*, can be elicited only by the antigen which had been used in the primary response. Different explanations have been suggested for the intensity and rapidity of the secondary response. The author attributes it to small amounts of antibody which continue to be formed and circulate after the primary injection; these amounts of antibody are too small to be detected by the precipitin test but are easily detected by the more sensitive hemagglutination test (D-18) and remain detectable for more than a year (D-19). On reinjection of the antigen the circulating antibody disappears in a very short time from the circulation, owing to the rapid formation of antigen-antibody complexes. Such antigen-antibody complexes may also be formed in or on the surface of the antibody forming cells. These complexes are evidently a much stronger stimulus for further antibody formation than the soluble antigens of the primary reaction which circulate a long time in the blood plasma and are deposited only slowly in the reticulo-endothelial system.

Since the antibody molecules are complementarily adapted to the determinant portions of the antigen molecules, attempts have been made to manufacture antibodies *in vitro*. Normal γ-globulins were denatured under mild conditions in the presence of azoproteins and then renatured. Although the renatured globulins give precipitates with the azoproteins at pH 5, these precipitates are quite non-specific and are also given by γ-globulins denatured and renatured in the absence of the azoproteins (D-20,21). Antibody formation takes place in surviving cells or tissue cultures from sensitized animals, but not in cell homogenates or extracts. Although such cell-free preparations can still incorporate amino acids they cannot form antibodies (D-17).

Antibody formation is frequently considered as an abnormal phenomenon having nothing to do with the processes continually taking place in the normal organism. However, antibody formation is protein formation and we have no reason to assume that antibodies are formed by a mechanism differing greatly from that by which normal proteins are synthesized. We still do not know whether the amino acid sequence of antibodies differs from that of one or the other fraction of the normal γ-globulins and whether there are differences among the amino acid sequences of different antibodies. We know, however, that each antibody has a specific combining site, complementarily adjusted to the determinant group of the antigen. In other words, the antibodies differ from each other and from the normal γ-globulins by the conformation of that part of their peptide chain which forms the combining site.

Various explanations have been given for this adaptation of the antibodies to the injected antigen. The first widely accepted theory was that proposed by the German immunologist Ehrlich who in 1906 suggested that each of the antibody forming cells contained small amounts of "receptors" for the various antigens. At the time of this proposal only a few bacterial and cellular antigens were known and it was imaginable that the animal organism had a preformed receptor for each of these antigens. Ehrlich assumed that the injected antigen combined with the corresponding receptor, that this receptor was then regenerated at an accelerated rate and shed into the circulation where it appeared as antibody. In 1914, when Landsteiner discovered the large number of antigenic azoproteins, it became difficult to assume that the animal body should have preformed receptors for all these artifacts of the chemical laboratories. It was therefore suggested that the injected antigen interferes with the process of globulin formation and deranges it in such a manner that globulins are formed which are complementarily adapted to the determinant portion of the antigen (B-6a,6b; D-22,23). This theory has been designated as the *template theory of antibody formation* or as instructive theory. Its principal postulate is that antibodies can be formed only in the presence of the antigen and that antibody formation stops when all of the antigen is eliminated or destroyed (see Chapter XVI,K).

An objection to this view arises from the fact that immunity frequently lasts many years or even during the whole life of an individuum. The latter is the case when the antigen is a virus which may multiply in the invaded organism for indefinite periods of time. However, it has also been demonstrated convincingly (D-24) that the lifeless pneumococcal polysaccharides persist in the organism for a long period of time; the same has been shown for isotopically labeled proteins (D-1 to 3). This is not surprising since it is known that many of the native proteins such as bovine serum albumin or hemoglobin are highly resistant to proteolytic enzymes and undergo proteolysis only after denaturation.

According to another theory of antibody formation, each of the antibody forming cells has the capacity to form only one type or a few types of antibody and already has this capacity before injection of the antigen. If an antigen is injected, only the cells which form the homologous antibody are stimulated, and undergo multiplication so that clones of these cells are formed. A second injection of the same antigen stimulates then a large number of cells forming the corresponding antibody and thus causes the intensive secondary response (D-25). This theory, like that of Ehrlich, postulates preformed antibodies for all types of antigens and is, therefore, confronted with the same difficulty to explain the existence of preformed antibodies to the artificial conjugated proteins. In contrast to the instructive or template theory this theory has been designated as the *selective theory of antibody formation*. The antigen merely selects the antibody forming cell, but has no further function. Antibody formation, according to this theory, can continue in the absence of antigen. It would be important to know whether a cell can form only one type of antibody or several different types. Experiments with single cells of animals injected with two different antigens have shown that most of the cells form only one type of antibody; very few cells form two different antibodies (D-27).

Both theories have been modified by several authors. Thus it has been claimed that the antigen may not act as a direct matrix for the formation of antibody molecules but may rather affect earlier phases in the biosynthesis of the antibody proteins. Alternatively, the selective role of the antigen may not involve selection between cells and their clones but selection between intracellular particles such as the ribosomes. Indeed, the difficulties of describing the facts in terms of a simple clonal selection theory are considered by its authors as insuperable (D-28).

Since newborn animals have no or very little γ-globulin in their circulation, it is not surprising that they cannot form antibodies. However, if a newborn animal has been injected with a foreign serum protein, the animal remains tolerant to a reinjection of the same protein for a long period of time, usually several months. This tolerance (D-26a) is manifested by the inability of the animal to form antibodies against the previously injected antigen when it is reinjected at an age at which it has acquired the ability to form all other types of antibodies. The selective theory of antibody formation attributes tolerance to the elimination of all those cells which are receptors for the injected antigen. The instructive theory ascribes tolerance to the deposition of antigen in the immature cells where it blocks formation of the homologous antibody. This view is supported by the fact that the duration of tolerance increases with the amount of antigen injected into the immature animal (D-26b).

## E. Complement (E-1)

If red blood cells are injected into an animal of another species, antibodies are formed which cause agglutination of the injected cells and also their lysis. If the antibody-containing immune serum is "inactivated," i.e., kept for 30–50 minutes at 55°C, it loses its lytic but retains its agglutinating activity. The labile factor which is inactivated at 55°C has been called alexin or *complement*. Guinea pig serum is frequently used as complement since it is rich in the lytic factor. This factor, denoted by the symbol C' (complement), can be fractionated by dialysis against solutions of low ionic strength into a labile euglobulin fraction, which precipitates at the low ionic strength, and a labile soluble pseudoglobulin fraction in the supernatant solution. The former has been denoted as mid-piece or C'1, the latter as end-piece or C'2. The guinea-pig serum contains also a component C'3 which is inactivated by zymosan, a polysaccharide from yeast, and another component C'4 which is inactivated by small amounts of ammonia or primary amines. Kinetic analyses (E-2) reveal that the erythrocytes (E) after combination with antibody (A) combine with the four components of complement in the sequence: $EAC'1 \rightarrow EAC'1,4 \rightarrow EAC'1,4,2 \rightarrow E^*$, where $E^*$, produced by the action of C'3 on EAC'1,4,2, is the damaged red blood cell which loses hemoglobin and is converted into a "ghost." All four components of complement are protein-like substances or proteins which can be separated from each other by fractionation with ammonium sulfate (E-3). Some separation has also been accomplished by chromatography on DEAE cellulose (E-4). The latter method indicates the presence of more than one C'3 fraction called C'a and C'b (E-4). One of these seems to be thermo-labile, the other thermo-stable (E-5).

Complement is bound not only to sensitized red blood cells but also to antigen-antibody precipitates. This binding causes a small but significant increase in the weight of the precipitates (E-6,7). Experiments with iso-topically labeled C' (E-8) indicate the presence of 0.10–0.40 mg of complement per milliliter of fresh guinea-pig serum. Complement, in contrast to antibody, is not specific but is bound by all types of antigen-antibody complexes, and even by other types of insoluble protein aggregates, for instance, by heat-denatured human $\gamma$-globulin (E-9). In spite of these nonspecific phenomena, complement is of great importance for clinical diagnostics and also for our understanding of numerous pathological symptoms. In clinical laboratories complement is used for the decision whether an antigen-antibody reaction has taken place. If such a reaction is allowed to proceed in the presence of small amounts of C', this is bound and becomes unavailable for the hemolysis of sensitized red blood cells. Free C' is indicated by hemolysis. *In vivo*, complement can cause not only lysis of sensitized cells, but also lesions in the kidney and other organs in which antigen-antibody reactions have taken place. Many of the pathological symptoms are attributed to the action of complement.

Very little is known on the mechanism of C' action. The component C'1 forms the bulk of the material which is bound to antigen-antibody complexes. It is probably a proenzyme which, after combination with antigen-antibody complexes, is converted into an esterase (E-10). Hemolysis is prevented by diisopropyl-fluorophosphate (DFP) which inhibits the esterase (E-11). Experiments with isotopically labeled complement indicate that the amount of complement bound in the conversion of EAC'1 into EAC'1,4,2 is much smaller than that taken up with C'1 (E-12). Binding of C'4 has been demonstrated by means of fluorescent antibody to complement (E-13). On the basis of kinetic analyses it has been claimed that reaction of the sensitized red blood cell with a single molecule of C'4 and another of C'2 causes conversion of EAC'1 into EAC'1,4,2 and subsequent lysis by C'3 (E-14).

*Properdin* is the name given to a component of the blood serum which like C'3 is inactivated by zymosan at 37°C, but in contrast to C'3 also at 15°C (E-15). The nature of properdin is not clear. It may be an antibody rather than a factor of another nature (E-16).

## REFERENCES

**Section A**

**A-1.** F. Haurowitz, *in* "Functions of the Blood" (R. G. Macfarlane, ed.), p. 527. Academic Press, New York, 1961. **A-2.** F. Haurowitz, *Ann. Rev. Biochem.* 29: 609 (1960). **A-3.** R. Kraus, *Wien Klin. Wochschr.* 10: 431(1897). **A-4.** P. Gell *et al.*, *Brit. J. Exptl. Pathol.* 27: 267(1946). **A-5.** S. B. Hooker and W. C. Boyd, *J. Immunol.* 30: 41(1936). **A-6.** L. Kesztyüs and V. Varteresz, *Z. Immunitaetsforsch.* 105: 372 (1945). **A-7.** F. Haurowitz and P. Schwerin, *Rev. fac. sci. univ. Istanbul*, A9: 120 (1944). **A-8.** K. Landsteiner, "The Specificity of Serological Reactions," 2nd ed.

Harvard Univ. Press, Cambridge, Massachusetts, 1946. **A-9.** F. Haurowitz *et al.*, *J. Immunol.* 40: 391(1941). **A-10.** F. Haurowitz, *J. Immunol.* 43: 331(1942). **A-11.** M. Sela and R. Arnon, *BJ* 75: 91, 103; *BBA* 40: 382; *BJ* 85: 223(1962). **A-12.** T. G. Gill and P. Doty, *BBA* 60: 450(1962); *JBC* 236: 2677(1961). **A-13.** F. Haurowitz and F. Bursa, *Rev. fac. sci. univ. Istanbul* B10: 283(1945). **A-14.** C. F. C. MacPherson and M. Heidelberger, *JACS* 67: 585(1945). **A-15.** J. R. Marrack and F. C. Smith, *Brit. J. Exptl. Pathol.* 13: 394(1933). **A-16.** F. Haurowitz and F. Breinl, *Z. physiol. Chem.* 205: 259(1933). **A-17.** L. Pauling *et al.*, *JACS* 64: 2994(1942). **A-18.** F. Haurowitz, *Ergeb. Mikrobiol.* 34: 1(1961; in English). **A-19.** H. F. Cheng *et al.*, *Proc. Soc. Exptl. Biol. Med.* 106: 93(1961). **A-20.** J. S. Garvey and D. H. Campbell, *J. Exptl. Med.* 105: 361(1957); J. D. Hawkins and F. Haurowitz, *BJ* 80: 200(1961). **A-21.** J. Freund, *Am. J. Clin. Pathol.* 21: 645(1951). **A-22.** D. W. Dresser, *Nature* 191: 1169(1961). **A-23.** C. Lapresle *et al.*, *Bull. soc. chim. biol.* 41: 695(1959); *J. Immunol.* 82: 94(1959); T. Ishizaka *et al.*, *Proc. Soc. Exptl. Biol. Med.* 103: 5(1960). **A-24.** E. A. Kabat, "Experimental Immunochemistry." Thomas, Springfield, Illinois, 1961. **A-25.** E. M. Press and R. R. Porter, *BJ* 83: 172(1962). **A-26.** H. Finger, *Z. Immunitaetsforsch.* 122: 15(1961). **A-27.** L. Levine, *FP* 21: 711(1962).

*Section B*

**B-1.** J. Oudin, *Ann. Inst. Pasteur.* 89: 531(1955). **B-2.** C. L. Oakley and A. J. Fulthorpe, *J. Pathol. Bacteriol.* 65: 49(1953). **B-3.** J. R. Preer, Jr., *J. Immunol.* 77: 52(1957). **B-4.** Ö. Ouchterlony, *Arkiv Kemi* 1: 43(1949/50). **B-5.** S. Belfanti and T. Carbone, *Arch. sci. med.* 22: 9(1898). **B-6a.** F. Breinl and F. Haurowitz, *Z. physiol. Chem.* 192: 45(1930); F. Haurowitz, *Ann. Rev. Biochem.* 29: 609(1960). **B-6b.** S. Mudd, *J. Immunol.* 23: 423(1932). **B-7.** A. Tiselius, *BJ* 31: 1464(1937). **B-8.** J. L. Oncley *et al.*, *JACS* 71: 541(1949). **B-9.** J. van der Scheer and R. Wyckoff. *Science* 91: 485(1940). **B-10.** J. H. Humphrey and R. R. Porter *BJ* 62: 93(1956), **B-11.** L. D. Felton, *J. Immunol.* 22: 453(1932). **B-12.** M. Heidelberger and E. A. Kabat, *J. Exptl. Med.* 67: 181(1938). **B-13.** F. Haurowitz *et al.*, *BJ* 41: 304(1947). **B-14.** H. C. Isliker, *Ann. N.Y. Acad. Sci.*, 57: 225(1953). **B-15.** D. H. Campbell *et al.*, *Proc. Natl. Acad. Sci. U.S.* 37: 575(1951). **B-16.** L. Gyenes and A. H. Sehon, *Can. J. Biochem. and Physiol.* 38: 1235(1961). **B-17.** A. E. Gurvich *et al.*, *Biokhimiya* 24: 144(1959); 26: 934(1961). **B-18.** S. J. Singer *et al.*, *JACS* 82: 565(1960). **B-19.** M. Heidelberger and F. E. Kendall, *J. Exptl. Med.* 65: 455(1937); E. L. Smith and R. D. Greene, *JBC* 171: 355(1947). **B-20.** E. L. Smith *et al.*, *JBC* 214: 197(1955). **B-21.** S. Fleischer *et al.*, *ABB* 92: 329(1961). **B-22.** R. R. Porter, *BJ* 46: 473(1950). **B-23.** R. R. Porter, *BJ* 73: 119(1959). **B-24.** A. Nisonoff and D. L. Woernley, *Nature* 183: 1325(1959). **B-25.** A. Nisonoff *et al.*, *Nature* 189: 293(1961); *JBC* 236: 3221(1961). **B-26.** A. Nisonoff *et al.*, *ABB* 93: 460(1961); *Nature* 194: 355(1962). **B-27.** F. Haurowitz *et al.*, *J. Immunol.* 43: 331(1941). **B-28.** A. E. Gurvich and A. M. Olovnikov, *Biokhimiya* 25: 646(1960). **B-29.** D. Gitlin and E. Merler, *J. Exptl. Med.* 114: 217(1961). **B-30.** A. E. Gurvich *et al.*, *Biokhimiya* 26: 468(1961). **B-31.** F. Haurowitz *et al.*, *J. Immunol.* 43: 327(1942). **B-32.** M. E. Koshland *et al.*, *Proc. Natl. Acad. Sci. U.S.* 45: 1470(1959). **B-33.** D. Pressman and O. Roholt, *Proc. Natl. Acad. Sci. U.S.* 47: 1606(1961). **B-34.** A. L. Grossberg *et al.*, *JACS* 82: 5478 (1960). **B-35.** E. A. Kabat, *J. Immunol.* 84: 82(1960). **B-36.** F. Haurowitz and P. Schwerin, *J. Immunol.* 47: 111(1943). **B-37.** F. Haurowitz and P. Schwerin, *Brit. J. Exptl. Pathol.* 23: 146(1942). **B-38.** C. E. Hall *et al.*, *J. Biophys. Biochem. Cytol.* 6: 407(1959). **B-39.** A. F. Schick and S. J. Singer, *JBC* 236: 2477(1961). **B-40.** F. Borek, *Nature* 191: 1293(1961). **B-41.** H. C. Isliker, *Adv. in Protein Chem.* 12: 388(1957). **B-42.** J. L. Palmer *et al.*, *Proc. Natl. Acad. Sci. U.S.* 48: 49(1962).

**B-43.** A. L. Grossberg *et al.*, *Biochemistry* 1: 391(1962); L. Wofsy *et al.*, *ibid.* p. 1031.
**B-44.** G. M. Edelman *et al.*, *Proc. Natl. Acad. Sci. U.S.* 47: 1751(1961). **B-45.** G. M.
Edelman and B. Benacerraf, *Proc. Natl. Acad. Sci. U.S.* 48: 1035(1962). **B-46.** R. R.
Porter *et al.*, *Ann. Rev. Biochem.* 31: 625(1962); *ABB, Suppl.* 1: 174(1962). **B-47.** A.
Ya. Kulberg and J. A. Tarkhanova, *Fol. Biol. (Czechoslov.)* 8: 147(1962). **B-48.** H. N.
Eisen and J. H. Pearce, *Ann. Rev. Microbiol.* 16: 101(1962). **B-49.** C. C. Chen *et al.*,
*Biochemistry* 1: 1025(1962).

### Section C

**C-1.** H. Wu *et al.*, *Proc. Soc. Exptl. Biol. Med.* 25: 853(1928); 26: 737(1929). **C-2.** M.
Heidelberger and F. E. Kendall, *J. Exptl. Med.* 50: 809(1929); 62: 467, 697(1935); 65:
647(1935). **C-3.** F. Haurowitz, *Z. physiol. Chem.* 245: 23(1936). **C-4.** J. R. Marrack
and F. Smith, *Brit. J. Exptl. Pathol.* 12: 30, 182(1931). **C-5.** S. B. Hooker and W. C.
Boyd, *J. Immunol.* 30: 38(1936); *J. Gen. Physiol.* 17: 341(1934). **C-6.** J. R. Mar-
rack, "The Chemistry of Antigens and Antibodies." London (1938). **C-7.** H. E.
Stokinger and M. Heidelberger, *J. Exptl. Med.* 66: 251(1937); *JACS* 60: 247 (1938).
**C-8.** A. Nisonoff *et al.*, *J. Immunol.* 81: 126(1958); *JACS* 81: 1418(1959). **C-9.** F.
Karush, *JACS* 78: 5519(1956); 79: 3380, 5323(1957). **C-10.** S. F. Velick *et al.*, *Proc.
Natl. Acad. Sci. U.S.* 46: 1470(1960). **C-11.** P. Stelos *et al.*, *JACS* 82: 6034(1960).
**C-12.** L. Pauling *et al.*, *JACS* 64: 2994(1942). **C-13.** S. J. Singer *et al.*, *JACS* 77:
3499, 3504, 4851(1955); 78: 312(1956); 81: 3887(1959). **C-14.** R. F. Steiner and C.
Kitzinger, *JBC* 222: 271(1956). **C-15.** F. Haurowitz *et al.*, *JACS* 79: 1882(1957).
**C-16.** R. Goldberg, *JACS* 75: 312(1953). **C-17.** P. Grabar, *Bull. soc. chim. biol.* 26:
298(1944). **C-18.** L. Pauling and D. Pressman, *JACS* 67: 1003(1945); 71: 2893
(1949). **C-19.** M. Goodman *et al.*, *J. Immunol.* 72: 440(1954). **C-20.** S. Berson and
R. S. Yalow, *J. Clin. Invest.* 36: 642(1957); 38: 1996, 2017(1959); S. Nakamura *et al.*,
*Z. physiol. Chem.* 318: 115(1960). **C-21.** R. S. Farr, *J. Infectious Diseases* 103: 239
(1958). **C-22.** P. J. Moloney and M. A. Aprile, *Can. J. Biochem.* 38: 1216(1960).
**C-23.** A. Kleczkowski, *Immunology* 4: 130(1961). **C-24.** G. C. Easty and E. H.
Mercer, *Immunology* 1: 353(1958). **C-25.** A. G. Szent-Györgyi and H. Holtzer, *BBA*
41: 14(1960). **C-26.** L. D. Felton and G. Bailey, *J. Immunol.* 11: 197(1926). **C-27.**
O. T. Avery and W. F. Goebel, *J. Exptl. Med.* 54: 431(1931); C. M. MacLeod *et al.*,
*J. Exptl. Med.* 82: 445(1945). **C-28.** W. C. Boyd *et al.*, *ABB* 55: 226(1955). **C-29.**
S. V. Boyden, *J. Exptl. Med.* 93: 107 (1951). **C-30.** J. S. Ingraham, *Proc. Soc. Exptl.
Biol. Med.* 99: 452(1958). **C-31.** D. Pressman *et al.*, *J. Immunol.* 44: 101(1942).
**C-32.** A. N. Roberts and F. Haurowitz, *J. Immunol.* 89: 348(1962). **C-33.** P. Grabar,
*Adv. in Protein Chem.* 13: 1(1958). **C-34.** J. J. Cebra *et al.*, *JBC* 237: 751(1962).
**C-35.** W. C. Boyd *et al.*, *ABB* 55: 226(1955). **C-36.** G. W. G. Bird, *Experientia* 17:
408(1961). **C-37.** J. R. Marrack and E. S. Orlans, *Brit. J. Exptl. Pathol.* 35: 28(1954).
**C-38.** F. Haurowitz, in Kallos, *Fortschr. Allergielehre*, p. 19, Basle and New York,
1939. **C-39.** A. Froese *et al.*, *Canad. J. Biochem.* 40: 1786(1962). **C-40.** L. A. Day
and J. M. Sturtevant, *JACS* 84: 3768(1962). **C-41.** L. Gyenes and A. H. Sehon,
*J. Immunol.* 89: 483(1962).

### Section D

**D-1.** J. S. Ingraham, *J. Infectious Diseases* 89: 109, 114(1951). **D-2.** C. F. Crampton
*et al.*, *Proc. Soc. Exptl. Biol. Med.* 80: 448(1952). **D-3.** C. F. Crampton *et al.*, *J.
Immunol.* 71: 319(1953). **D-4.** A. N. Roberts and F. Haurowitz, *FP* 20: 21(1961)
*J. Exptl. Med.* 116: 407(1962). **D-5.** G. E. Francis *et al.*, *BJ* 60: 118, 363(1955).
**D-6.** A. S. McFarlane, *BJ* 62: 135(1956). **D-7.** F. Haurowitz and C. F. Crampton,

*J. Immunol.* 68: 73(1952).   **D-8.** S. Fleischer *et al., Proc. Soc. Exptl. Biol. Med.* 101: 860(1959).   **D-9.** J. S. Garvey and D. H. Campbell, *J. Exptl. Med.* 105: 361(1957). **D-10.** A. H. Coons *et al., J. Exptl. Med.* 93: 173(1951); 102: 49, 61(1955).   **D-11.** J. D. Marshal *et al., Proc. Soc. Exptl. Biol. Med.* 98: 898(1958).   **D-12.** J. H. Humphrey and A. S. McFarlane, *BJ* 57: 195(1954).   **D-13.** P. Gros *et al., Bull. soc. chim. biol.* 34: 1070(1953).   **D-14.** F. J. Dixon *et al., J. Exptl. Med.* 96: 313(1952).   **D-15.** A. Fagraeus, *J. Immunol.* 58: 1(1948).   **D-16.** M. C. Berenbaum, *Immunology* 2: 71 (1959).   **D-17.** B. A. Askonas and J. H. Humphrey, *BJ* 68: 252; 70: 212(1958). **D-18.** M. Richter and F. Haurowitz, *J. Immunol.* 84: 420(1960).   **D-19.** F. Haurowitz, M. Richter and S. Zimmerman, unpublished experiments.   **D-20.** F. Haurowitz *et al., ABB* 11: 515(1946).   **D-21.** J. Morrison, *Can. J. Chem.* 31: 216(1953). **D-22.** J. Alexander, *Protoplasma* 14: 296(1931).   **D-23.** L. Pauling, *JACS* 62: 2643 (1940).   **D-24.** O. K. Stark, *J. Immunol.* 74: 130(1955).   **D-25.** F. M. Burnet, *Science* 133: 307(1961).   **D-26a.** M. Hašek *et al., Adv. Immunol.* 1: 1(1962).   **D-26b.** R. T. Smith, *Ibid.*, p. 67.   **D-27.** G. J. V. Nossal and P. Mäkelä, *J. Immunol.* 87: 447, 457(1961); 88: 604(1962).   **D-28.** A. Szenberg, N. L. Warner, F. M. Burnet, and P. E. Lind, *Brit. J. Exptl. Pathol.* 43: 129(1962).

### Section E

**E-1.** M. M. Mayer, *in* E. A. Kabat, "Experimental Immunochemistry," revised enlarged edition, p. 133. Thomas, Springfield, Illinois, 1961; A. G. Osler, *Adv. in Immunol.* 1: 131(1962).   **E-2.** M. M. Mayer *et al., J. Immunol.* 73: 426(1954).   **E-3.** L. Pillemer *et al., J. Exptl. Med.* 74: 297 (1941); 76: 93(1942).   **E-4.** H. J. Rapp *et al., Proc. Soc. Exptl. Biol. Med.* 100: 730(1959).   **E-5.** J. D. Hawkins, *Nature* 186: 483 (1960); 193: 1084(1962).   **E-6.** M. Heidelberger *et al., J. Exptl. Med.* 73: 681, 695 (1941).   **E-7.** F. Haurowitz and M. M. Yenson, *J. Immunol.* 47: 309(1943).   **E-8.** N. Penn *et al., J. Immunol.* 79: 409(1957).   **E-9.** T. Ishizaka *et al., J. Immunol.* 87: 433 (1961).   **E-10.** E. L. Becker, *J. Immunol.*, 77: 462, 469(1956).   **E-11.** L. Levine, *BBA* 18: 283(1955).   **E-12.** F. Haurowitz *et al.*, unpublished experiments.   **E-13.** P. G. Klein and P. M. Burkholder, *J. Exptl. Med.* 111: 93, 107(1960).   **E-14.** T. Borsos *et al., J. Immunol.* 87: 310(1961).   **E-15.** L. Pillemer, *Ann. N.Y. Acad. Sci.* 66: 233 (1956).   **E-16.** R. A. Nelson, *J. Exptl. Med.* 108: 515(1958); H. E. Schultze *et al., Z. Immunitätsforsch.* 123: 307(1962).

# Chapter XVI

## Protein Biosynthesis

### A. Introduction

"Protein synthesis is the cardinal manifestation of life, and yet the manner in which it is achieved is still the great problem of biochemistry and of biology." These words, written 12 years ago in the first edition of this book as an introduction to this chapter, are still valid. Indeed, protein biosynthesis is at present in the center of interest of biochemists, biophysicists, geneticists, and of all scientists who are interested in the basic aspects of biology as witnessed by numerous symposia and conferences (A-1 to 7).

In the preceding chapters of this book it has been demonstrated that proteins are highly specific and that this specificity is based on a definite number and sequence of amino acids in their molecules. How is this specificity of composition, this definite order of the amino acid residues in the peptide chains accomplished? Obviously copying mechanisms or the "memory" of a computing system come to one's mind. Biochemists and biophysicists, in cooperation with geneticists, are trying feverishly to solve this problem and to clarify the mechanism by which genetic information concerning the specific amino acid sequence of proteins is transmitted from mother cell to daughter cell. Although this book deals primarily with the structure and function of proteins, it seems necessary to the author to discuss, at least in this last chapter, the central problem of present protein research, the problem of protein biosynthesis. The inclusion of this problem in a book on the structure and function of proteins is also justified by the present belief that the specificity of the amino acid sequence is dictated by specific nucleotide sequences in nucleic acids and that the mechanism by which nucleotide sequences are translated into amino acid sequences is based on intermolecular forces which operate between polynucleotides of definite conformation, enzymes, and amino acid residues which are lined up for the formation of the protein macromolecule.

Although the problem of transcribing nucleotide sequences into amino acid sequences may seem formidable, it is simplified to a certain extent (*a*) by the fact that most nucleic acids consist of only four main types of nucleotides, (b) by the absence of *free* peptide intermediates, and (c) by the fact that the influence of nucleic acids on protein biosynthesis can be investigated in simple systems such as bacteriophages during their multiplication in bacteria or in cell-free suspension of ribosomes. The raw material for protein biosynthesis are the amino acids. They are supplied to animals, plants, and bacteria by different sources. Details on the biosynthesis of amino acids, their metabolic fate, and their breakdown will be found in textbooks of biochemistry and in treatises which are devoted primarily to metabolic problems (A-8,9). The following Section B merely gives a short survey on amino acid supply and some problems which are of importance in the conversion of amino acids into proteins.

## B. Amino Acid Supply

### Amino Acid Production in Autotrophic Organism

Plants and some of the bacteria can produce the amino acids required for protein synthesis from inorganic carbon and nitrogen sources. The ultimate sources of these elements are the atmospheric carbon dioxide and molecular nitrogen. Most of the pathways leading from $CO_2$ to the carbon chains of the amino acids have been elucidated during the last decade in experiments with isotopically labeled carbon dioxide. The description of this work is beyond the purpose of this book. The results obtained and the metabolic pathways which lead from $CO_2$ to the different amino acids are described in pertinent chapters of the "Annual Reviews of Biochemistry" and in authoritative surveys on metabolic pathways (A-8,9). The first product of the assimilation of atmospheric nitrogen is ammonia which is rapidly incorporated into glutamic acid (B-1). The energy required for the assimilation of nitrogen is supplied by coupled oxidations; the assimilation of $N_2$ is retarded by carbon monoxide or by sulfides, both of which typically inhibit the respiratory enzymes (B-1). In contrast to the green plants and the photosynthetic bacteria which are able to assimilate nitrogen and produce all of their amino acids, the higher animals must rely upon a continuous supply of certain amino acids which are furnished by the food in the form of proteins. The consumed proteins are hydrolyzed in the digestive tract by proteolytic enzymes and the liberated amino acids are absorbed and incorporated into the specific body proteins.

### Proteolytic Breakdown

The most important of the **proteolytic enzymes** and their modes of action have been discussed in Chapter XII, F. Proteolytic enzymes occur not only in the secretory cells of the gastrointestinal tract but also in most of the body cells. Their presence there is revealed by the phenomenon of autolysis, i.e., the lysis of the cell following cessation of life activities. This lysis proceeds even under strictly sterile conditions and is brought about by cellular enzymes identical with or closely related to cathepsin. Autolysis is, therefore, different from putrefaction which is brought about by bacteria or molds. As mentioned in Chapter III, B, the extent of proteolysis can be measured by determining the increase in carboxyl and amino groups which are formed by the hydrolytic cleavage of peptide bonds:

$$R \cdot CO \cdot NH \cdot R' + H_2O = R \cdot COOH + R'HN_2$$

The physical properties of the protein solution are drastically changed during the first phases of proteolytic cleavage. The viscosity is lowered and the protein loses its property of being coagulated by heat. This initial phase of proteolysis is frequently accompanied by an abnormal decrease in the total volume of the solution (B-2). Possibly, the action of the proteolytic enzymes results not only in the hydrolytic cleavage of peptide bonds but also in the disaggregation of a protein polymer into its subunits.

In Chapter VII, H, it has been pointed out that many of the globular proteins in their native state are resistant to the action of proteolytic enzymes, but that they become susceptible to the enzymes after denaturation. Denatured hemoglobin is hydrolyzed by papain about 100 times more rapidly than native hemoglobin (B-3). Likewise, native collagen is hardly attacked by trypsin, while heated collagen is readily hydrolyzed (B-4). The same is true for ovalbumin and for the serum globulins (B-5). The resistance of the native proteins to the action of proteolytic enzymes is probably due to the absence of points of attack for the enzyme molecules. When the peptide chains are unfolded by denaturation, the structures required for enzyme action become accessible and proteolysis takes place. The promotion of enzymatic action by the unfolded state of the peptide chains is also shown by the behavior of monomolecular protein films in the water-air interface. Expanded protein films are readily digested whereas solutions of the same proteins are more resistant to the enzymes (B-6).

The resistance of native proteins to proteolytic enzymes may be surprising for the physiologist because animals, with the exception of man, ingest native proteins with their food. However, these proteins are rapidly denatured in the highly acidic gastric juice at a pH of 1–2. The protein exposed to the action of pepsin is, therefore, denatured protein which is easily hydrolyzed. At the end of the nineteenth century, the work of Kühne (1885) and of Hofmeister and his school had demonstrated that the treatment of proteins with the enzyme mixtures of the gastric or pancreatic juice yields intermediary breakdown products. The name *albumose* was used for products which were salted out by saturation with ammonium sulfate but, in contrast to the proteins, were not coagulated by heat. Products which were neither coagulated nor salted out were designated as *peptones*. Albumoses and peptones are mixtures of peptides of higher and lower molecular weight. They were the forerunners of the peptides whose separation by chromatography and electrophoresis has rendered possible the elucidation of amino acid sequences in numerous proteins.

As mentioned in Chapter XII, F, some of the proteolytic enzymes require the presence of manganese or other bivalent metal ions for their action. Papain and cathepsin are activated by the addition of sulfides or cyanides.

Although the proteolytic breakdown is usually formulated as the hydrolytic cleavage of peptide bonds, it is in some instances more complicated and may even involve the temporary formation of new peptide bonds. Thus, glycine anilide which is not hydrolyzed by papain undergoes hydrolysis if acetylphenylalanylglycine is added as a *cosubstrate* (B-7). Evidently an intermediary product is formed by condensation or transpeptidation; at the same time, or secondarily, one equivalent of glycine and one of aniline is released. The over-all reaction is glycine anilide → glycine + aniline. It is clear from this example that proteolysis, in some instances, involves intermediary condensation or transpeptidation.

## Essential (Indispensable) Amino Acids

Some of the amino acids can be formed in the human and animal body from other organic substances. They need not be introduced into the body with the food and have been designated as endogenous, dispensable, or nonessential amino acids. The amino acids which must be introduced into the organism from outside, usually in the form of proteins, are designated as essential or indispensable amino acids. It has been proved by the work of Rose (B-8) that valine, leucine, isoleucine, threonine, methionine, phenylalanine, lysine, and tryptophan cannot be produced in higher animals. They are essential amino acids. Histidine, tyrosine, and arginine are sometimes called semiessential amino acids since their formation in the animal body from other essential or dispensable amino acids is so slow that it does not cover the requirements of the organism (B-8).

Amino acid requirements vary from species to species. While arginine is dispensable for the dog, it is essential for the rat. Surprisingly, some of the amino acids which are dispensable for the higher animals are indispensable for certain bacteria and molds. Thus, glycine, proline, and glutamic acid can be determined microbiologically (see Chapter III, E) by means of bacterial mutants which require these amino acids for growth and multiplication.

## The Dispensable Amino Acids

The principal source of the endogenous amino acids are pyruvic acid, oxaloacetic acid, and α-ketoglutaric acid. The two last-named acids are members of the citric acid cycle whereas pyruvic acid is formed *in vivo* from carbohydrates. Transamination or amination of these three acids yield, directly, the amino acids alanine, aspartic acid, and glutamic acid. Glutamic acid is the mother substances of three other $C_5$ acids: proline, hydroxyproline, and ornithine. The pyrrolidine ring of proline is formed by ring closure between the C-atom 5 and the $\alpha$-amino group of glutamic semialdehyde. Further oxidation yields hydroxyproline. Amination of glutamic semialdehyde gives ornithine, a 2,5-diaminovaleric acid; carbamylation of ornithine and subsequent amination results in the formation of another amino acid, arginine. Pyruvic acid is also the mother substance of serine; an intermediate in this reaction is hydroxypyruvic acid. Serine, in the presence of tetrahydrofolic acid, loses readily its hydroxymethyl group and is thus converted into glycine. Reaction of serine with methionine through cystathionine leads to the formation of cystine. The interrelations between these dispensable amino acids are shown in the following diagram.

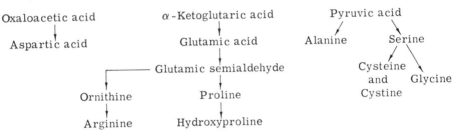

The details of these reactions have been revealed by a large number of investigations with isotopically labeled amino acids. They have been described in textbooks of biochemistry and, in more detail, in special treatises on the intermediary metabolism (B-9). The formation of the essential amino acids from inorganic material in plants and other autotrophic organisms is described in the same reference works (B-9).

## The Biological Value of Proteins

Since various proteins contain different amounts of the essential amino acids, the biological values of each of these proteins are different. Some of the vegetable proteins are deficient in lysine (see Table VIII-7). Collagen is deficient in tryptophan and cystine (see Table IX-1). Obviously it is impossible to supply all the essential amino acids by feeding one of these proteins exclusively, although a suitable mixture of two or more of these proteins may be quite satisfactory.

Although it is impossible to determine an absolute protein minimum, minima have been determined for each of the essential amino acids. They vary in the adult human being from about 0.25 g to 1.5 g of each of the amino acids per day. Since the amino acid composition of the pure proteins varies considerably, one might expect great variations in the amino acid content of different diets. Analyses of the most important foodstuffs reveal, however, that the total amino acid content of their protein mixtures is well balanced and does not vary very widely. The proteins of milk, serum, eggs, meat, fish meat, brain, corn germ, and soybean of fibrin contain 1–2% cystine and cysteine, 5–7% arginine, 2–3% histidine, 5–8% lysine, 3–5% tyrosine, 1–2% tryptophan, 4–6% threonine, 4–6% valine, 10–20% leucine, and 3–5% isoleucine (B-15). In view of this similarity in their amino acid composition, the proteins of all these foodstuffs can replace each other without any serious consequences for the organism. Since the proteins of the ingested food are hydrolyzed in the gastrointestinal tract and converted into amino acids, one might expect that proteins can be replaced by their hydrolyzates or by amino acids of the same composition. This has been confirmed by feeding experiments. Although the natural amino acids are very adequate nutrients when ingested orally, some of them are toxic when administered intravenously in large amounts. Thus, injection of glutamic acid into dogs causes vomiting (B-10). Injection of large amounts of glycine into rats produces toxic symptoms and cessation of growth (B-11). The toxic effects observed after intravenous injection of disproportionate amounts of these amino acids are, evidently, due to the abnormal high concentration of these amino acids in the blood. In the normal organism the concentration of free amino acids in the blood is extremely low. The amount of amino acid nitrogen in 100 ml of human blood serum is usually not more than 5–8 mg. The amino acid levels produced by the intravenous injection of single amino acids are much higher than the low physiological values.

The body proteins maintain their specific structure and their specific amino acid composition even under abnormal conditions. It has repeatedly been attempted to alter the composition of these proteins by an abnormal mode of nutrition, for instance, by starvation of experimental animals. Certain changes in the ratio of the amino acids have been found in such experiments and also in the serum proteins of persons suffering from hunger edema (B-12). However, it is not yet clear whether these changes are caused by the production of proteins with an abnormal amino acid content or by changes in the ratio of the protein components which form the normal protein mixture in the blood plasma, the muscles, and in other organs.

Although proteins are formed exclusively of L-amino acids, some of the unnatural D-amino acids can be utilized for the bioxynthesis of proteins. This is not surprising since deamination of the D-amino acids yields the optically inactive α-keto acids which are identical with the α-keto acids formed by the deamination of L-amino acids. Although the human organism can utilize D-lysine and D-arginine, it cannot utilize D-phenylalanine, D-tryptophan, and D-histidine (B-13). The conversion of D-amino acids into L-amino acids through deamination and reamination has been called stereonaturalization (B-14).

## C. Nonenzymatic Formation of Peptide Bonds

The hydrolytic cleavage of peptide bonds is usually a reaction in which the equilibrium is shifted toward the products of cleavage. Hence, the reverse reaction (XVI-1), the formation of a peptide from amino acids or smaller peptides is an endergonic, energy-requiring, reaction. The amount of energy required depends to a great extent on the nature of the two molecules which contribute the carboxyl and the amino group for the formation of the peptide bond.

$$R \cdot COOH + NH_2 R' = R \cdot CO \cdot NH \cdot R' + H_2O \qquad (XVI-1)$$

$$N^+H_3 \cdot CHR \cdot COO^- + N^+H_3 \cdot CHR' \cdot COO^- = NH_2 \cdot CHR \cdot CO \cdot NH \cdot CHR' \cdot COOH + H_2O \qquad (XVI-2)$$

If the two compounds are amino acids, the reaction is represented by Eq. (XVI-2). This reaction differs from reaction (XVI-1) by the charges of the amino acid molecules which in the aqueous solution occur almost exclusively in their dipolar form, i.e., as zwitterions. The pK values (see Chapter VI, A) of the amino groups in amino acids are close to 9.5 whereas those of dipeptides are approximately 8.0. Likewise, the carboxyl group of amino acids has a lower pK value than that of weak organic acids or peptides. It had been pointed out first by Linderström-Lang (C-1) that the equilibrium in reaction (XVI-2) is shifted far more to the left than in reaction (XVI-1). The difference between energies required for the condensation of $R \cdot COOH$ and $NH_2 \cdot R'$ on the one hand, and for that of $R \cdot COO^-$ and $N^+H_3 \cdot R'$ on the other, can be considered as the energy required for the conversion of the ionized into the uncharged form of the reactants. If it is intended to form a peptide bond between the carboxyl group of one amino acid or peptide and the amino group of another molecule, it is necessary to prevent participation of other carboxyl or amino groups in this reaction. This can be accomplished by the introduction of suitable substituents.

One of the most widely used methods (Bergmann and Zervas, 1932) is based on the conversion of the amino acid or peptide which carries the reactive carboxyl group into its benzyloxycarbonyl (carbobenzoxy-, cbzo-) derivative $C_6H_5 \cdot CH_2 \cdot O \cdot NH \cdot CHR \cdot COOH$. This is obtained by treatment of the amino acid or peptide with carbobenzoxy chloride $C_6H_5CH_2O \cdot COCl$. The carbobenzoxy group protects the amino group against combination with the carboxyl group of the other reactant. The carboxyl group of this second reactant (amino acid or peptide) is protected by esterification or another suitable method. In this manner, the two amino acids or peptides can combine only in one of the two possible ways.

Since this reaction is endergonic, energy must be supplied to at least one of the two reacting molecules. This can be done in different ways. One of the oldest methods is the conversion of the reacting carboxyl group into its azide which combines with the amino acid ester to yield a cbzo-peptide ester [Eq. (XVI-3)].

$$C_6H_5CH_2OCO \cdot NH \cdot CHR \cdot CON_3 + NH_2 \cdot CHR' \cdot COOC_2H_5 = N_3H + C_6H_5CH_2OCO \cdot NH \cdot CHR \cdot CO \cdot NH \cdot CHR' \cdot COOC_2H_5 \qquad (XVI-3)$$

The ester group is easily hydrolyzed at room temperature by dilute alkali, whereas the peptide bond is stable under these conditions. The carbobenzoxy group is finally split off by catalytic hydrogenation. In this manner two amino acids or peptides combine to form a new peptide bond.

The essential step is the conversion of the carboxyl group into the much more reactive azide group $-CON_3$, an acyl chloride group $-COCl$, or any other reactive group $-COX$. It will be shown later that an analogous reaction takes place when amino acids are incorporated into proteins *in vivo*. The substituent X in the biological amino acid derivative is a special form of ribonucleic acid. The conversion of the amino acid into such a reactive derivative is usually denoted as *activation*. This term is ambiguous since the term activation is frequently used for the conversion of a stable compound into a quite unstable transition state whose lifetime is only a small fraction of a second. The azides, acyl chlorides, and other activated amino acid derivatives are quite stable in the absence of water or other nucleophilic reagents. They are converted into peptides by condensation with the amino group of an amino acid ester or a peptide ester as shown by Eq. (XVI-3). With water they are hydrolyzed to the free acid and $N_3H$:

$$C_6H_5CH_2OCO \cdot NH \cdot CHR \cdot CON_3 + H_2O = C_6H_5CH_2OCO \cdot NH \cdot CHR \cdot COOH + N_3H$$
$$(XVI-4)$$

During the last few years, the methods of peptide-bond formation have been refined by means of more complicated reagents which make it possible to work under milder conditions and to obtain higher yields. The ingenious application of these methods has resulted in the synthesis of the hormones oxytocin and vasopressin (Chapter XIII, D), the adrenocorticotropic hormone (Chapter XIII, F), and of some of the bactericidal peptides (Chapter XIV, C). Doubtlessly, many more natural peptides will be synthesized in this manner in the future.

A particularly interesting reaction is the *polymerization* of amino acids and simple peptides. The most widely used method is based on the conversion of the amino acids into their *N*-carboxy anhydrides. The latter polymerize according to Eq. (XVI-5) when their solution or suspension in organic solvents is exposed to traces of water which acts as a catalyst (C-2). In this manner numerous polymers of amino acids or of small peptides have been produced (C-3).

$$n\overset{\overline{\qquad O \qquad}}{CO \cdot CHR \cdot NH \cdot CO} \rightarrow nCO_2 + (-CO \cdot CHR \cdot NH-)_n \qquad (XVI-5)$$

The *N*-carboxy anhydrides are obtained by treating amino acids with methoxycarbonyl chloride, conversion of the methoxycarbonylamino acid into its acyl chloride by means of thionyl chloride, and heating in a vacuum to 50°–60°C. Methylchloride is given off in this last step and the *N*-carboxyamino acid is obtained in good yield. The close resemblance of the polymerization product to the natural proteins is demonstrated by the hydrolytic cleavage of polylysine by trypsin (C-4), and by the identity of poly-γ-D-glutamic acid with the natural polyglutamic acid from the capsular substance of *Bacillus anthracis* and other types of bacteria (C-5). Polycystine has been obtained from polycysteine by treatment with molecular oxygen in the presence of ammonia and traces of cupric ions as catalysts (C-6). As mentioned in Chapter XV, A, some of the polyamino acids, particularly those containing tyrosine, cystine, glutamic acid, and lysine, are antigens which on injection into the animal body give rise to the formation of specific antibodies.

Polymerization of amino acids is also accomplished when the *N*-carboxy anhydrides are replaced by the analogous *N*-phosphoryl anhydrides of the amino acids (C-7). This method is of particular interest in view of the possibility that primordial proteins, in prehistoric times, may have been formed by similar mechanisms in the absence of enzymes (C-8). In the laboratory proteinlike polymers can be produced by the pyrocondensation of amino acids at elevated temperatures or by heating mixtures of organic acids and ammonia (C-8).

Peptidelike polymers are also obtained when formaldehyde, ammonia, and hydrogen cyanide are condensed to yield acetonitrile and then polymerized in the presence of kaolin at 130°C; the product on mild hydrolysis yields ammonia and polyglycine. Treatment of the polymer with formaldehyde or acetaldehyde and subsequent hydrolysis leads to the formation of serine or threonine, respectively (C-9). Formation of proteinlike macromolecules was observed when tryptic hydrolyzates were exposed to pressures of several thousand atmospheres (C-10). This is surprising since the formation of peptide bonds, owing to the release of water, is accompanied by an increase in volume. Accordingly, high pressure should facilitate the hydrolytic cleavage of peptide bonds and not their formation. An increase in temperature at the high pressures used may be responsible for the observed effects.

## D. Formation of Peptide Bonds by Proteolytic Enzymes

It has been shown in the preceding section that the equilibrium in reaction (XVI-2) is shifted far to the left side, i.e., towards hydrolysis of the peptide bond. This reaction is catalyzed by proteolytic enzymes which usually activate the carboxyl group, less frequently also the amino group (see Chapter XII, F). In the simplest of the condensation reactions, the condensation of two glycine molecules to form glycylglycine, the equilibrium constant $K$ is approximately 0.001 (C-1). It will be remembered that

$$K = \frac{[\text{glycylglycine}]}{[\text{glycine}]^2}$$

where the square brackets indicate molar concentrations. If the molar concentration of glycine is 0.1, that of glycylglycine at equilibrium is $10^{-5}$. At higher glycine concentrations more glycylglycine may be formed, but even at 1.0 $M$ glycine concentration the glycylglycine concentration is only $10^{-3}$. Evidently, very little glycylglycine is formed at equilibrium. Since the free energy change, $\Delta F$, is equal to $-RT \ln K$, where

$R = 1.98$ cal, $T$ is approximately $300°K$, and $\ln K = 2.3 \log K$, $\Delta F =$ approximately $+4.1$ kcal per mole of glycylglycine.

In the preceding Section C of this chapter, it has been pointed out that this low yield in the formation of peptides is caused by the ionization of the amino acids, and that the pK values of peptides are more favorable for condensation. Actually it has been shown that $K$ for the condensation of two molecules of glycylglycine to yield tetraglycine is 0.05 (C-1). The analogous value of $\Delta F$ is $+1.84$ kcal/mole. The $\Delta F$ value for the formation of peptide bonds inside the peptide chain is still lower when glycine is replaced by certain other amino acids. It decreases also when the terminal amino group of the peptide chain is substituted by benzoylation, carbobenzoxylation, or another similar reaction. $\Delta F$ values as low as $+0.4$ kcal/mole have been reported (D-1). Consequently, significant yields of peptide are formed under suitable conditions.

Thus, chymotrypsin catalyzes the condensation of benzoyltyrosine $C_6H_5 \cdot CO \cdot NH \cdot CH(CH_2 \cdot C_6H_4OH) \cdot COOH$ with glycylanilide $NH_2 \cdot CH_2 \cdot CO \cdot NH \cdot C_6H_5$ (D-2). The ease of peptide synthesis in this reaction is due to the lack of a free $^+NH_3$ group in benzoyltyrosine and the lack of a free $COO^-$ group in glycylanilide. The reaction is also enhanced by the low solubility of the reaction product which is continuously removed from the solution as a precipitate and thus shifts the equiiibrium toward formation of the product, benzoyltyrosyl-glycylanilide.

It is not yet quite clear whether reactions of this type proceed as direct condensations of the activated carboxyl group with the amino group of the anilide since glycylglycylanilide has been found as an intermediate in the enzymatic synthesis of benzylphenylalanyl-glycylanilide from phenylalanine glycylanilide (D-3). The finding of glycylglycylanilide indicates the intermediary transfer of a glycyl residue from glycylanilide to another molecule of glycylanilide. Transfer reactions of this type are called *transpeptidations* or *transamidations*. They require no significant energy supply since the energy required for the formation of the peptide bond is supplied by the cleavage of another peptide or amide bond. Thus, the transamidation of cbzo-glycinamide and leucyltyrosine, in the presence of papain, yields 70% of cbzo-glycylleucyltyrosine and ammonia; only 30% of the cbzo-glycinamide are hydrolyzed to cbzo-glycine (D-4). The transamidation reaction can be represented quite generally by Eq. (XVI-6):

$$R \cdot CONH_2 + NH_2 \cdot R' \rightleftharpoons R \cdot CONHR' + NH_3 \qquad \text{(XVI-6)}$$

and we can consider the amide, shown at the left side, as the activated form of the corresponding free acid $R \cdot COOH$. The mechanism of transamidation is made clear by the virtual reaction (XII-8) shown in Chapter XII. Transamidations and transpeptidations are catalyzed by papain and by the cathepsins B and C (D-5). Cathepsin B activates the carbonyl groups of *N*-acylated lysine and arginine residues whereas cathepsin C preferentially activates the carbonyl groups of *N*-substituted tyrosine, phenylalanine, and tryptophan. The amides of glycyltryptophan or glycyltyrosine are polymerized in the presence of cathepsin C, and form long peptide chains with a terminal amide group (D-6).

The requirement for *N*-substitution by benzoyl, acetyl, or other acyl groups in these reactions is reminiscent of the occurrence of terminal *N*-acetyl residues in tobacco mosaic virus protein, ovalbumin, cytochrome c, and other natural proteins and peptides, and suggests that an acetyl-amino acid might be the starting unit in the biosynthesis of some peptide chains which then grow by the stepwise addition of aminoacyl residues to the acetylated *N*-terminal amino acid. In this connection it may be re-called that some of the peptide hormones (see Chapter XIII, D and E) contain *C*-terminal amide groups and that they might be susceptible *in vivo* to transamidation reactions.

Since transamidations of the type shown in reaction (XVI-6) are catalyzed by enzymes and readily take place under physiological condi-tions, it is not surprising that the same proteolytic enzymes also catalyze *transpeptidations*, i.e., the transfer of peptide or aminoacyl residues from one peptide to another. This reaction is represented by the general equation (XVI-7).

$$R^1 \cdot CO \cdot NH \cdot R^2 + R^3 \cdot CO \cdot NH \cdot R^4 \rightleftharpoons R^1 \cdot CO \cdot NH \cdot R^4 + R^3 \cdot CO \cdot NH \cdot R^2 \quad (XVI\text{-}7)$$

Reactions of this type seem to be involved in the formation of protein-like insoluble polypeptides called *plasteins*. Their formation was first ob-served by Danilewsky (1886) who added pepsin or pancreatin to a neu-tralized concentrated enzymatic hydrolyzate of ovalbumin or other proteins. Crystalline chymotrypsin catalyzes not only plastein formation (D-7) but also the formation of the esters of di- and tripeptides of phenyl-alanine (D-7) and methionine (D-8) from the ethyl or isopropyl esters of these amino acids. The formation of plastein and of these peptide esters takes place in concentrated solutions in which the equilibrium in the reaction (XVI-8):

$$2 NH_2 \cdot CHR \cdot COOC_2H_5 \rightleftharpoons NH_2 \cdot CHR \cdot CO \cdot NH \cdot CHR \cdot COOC_2H_5 + C_2H_5OH \quad (XVI\text{-}8)$$

is shifted toward the formation of the dipeptide ester or the ester of higher oligopeptides. When aminoacyl esters are used as substrates of the enzymes, the esterified carboxyl group can be considered as "activated." The $\Delta F$ value for the hydrolysis of amino acid esters at pH 7 is very high, close to $-8.4$ kcal/mole (D-9). This $\Delta F$ value is considerably more nega-tive than the free energy change in the hydrolysis of the esters of fatty acids. Consequently, we can explain the transfer reaction by saying that the energy required for the formation of the new peptide bond is supplied by the cleavage of the ester bond.

The formation of plastein in concentrated enzymatic hydrolyzates of proteins seems also to be due to transpeptidation since addition of isotopically labeled amino acid esters causes the incorporation of phenylalanine, tyrosine, leucine, isoleucine, and of aspartic and glutamic acid residues into the insoluble plastein (D-10). None of the amino acids is incorporated if they are added to the digest as free amino acids. Although these experiments prove that amino acid residues can be incorporated into plastein by transpeptidation, it is not clear whether plastein itself is formed by transpeptidation. Insoluble "plasteins" are also formed from an octapeptide and an undecapeptide isolated from Witte peptone when these peptides are exposed to the action of pepsin at pH 4 (D-11). Moreover, a number of synthetic pentapeptides with $N$- and $C$-terminal tyrosine or phenylalanine residues, after exposure to pepsin at pH 4 or chymotrypsin at pH 8.5, form plastein-like polymers with molecular weights varying from 1500 to 4000 (D-11). It is not yet known whether their formation is due to direct condensation of the pentapeptides or whether intermediate transpeptidation takes place as described above in the condensation of benzoylphenylalanine with glycylanilide (D-3).

Although proteolysis in some instances, as shown in the preceding paragraphs, is reversible, the products formed lack the specificity and many of the other properties of the natural proteins. They owe their insolubility, probably, to their high content of amino acids with nonpolar side chains. Up to the present time none of the biologically active proteins (enzymes, hormones) has been obtained by the action of proteolytic enzymes on a digest of such a protein. Nevertheless, transpeptidations may play a role *in vivo* and may be involved in modifications of proteins, for instance, in the conversion of proenzymes (zymogens) into active enzymes. Transpeptidations may also be responsible for the occurrence of peptide fragments with identical amino acid sequences in two or more hormones formed in the same endocrine gland or in different enzymes present in the same organ (see Chapters XII,F and XIII,F).

## E. Activation of Amino Acids for Protein Biosynthesis

In the preceding section it has been shown that the first step in the peptide synthesis *in vitro* is the activation of the amino acid whose carboxyl group is involved in the formation of the peptide bond. This activation was brought about by conversion of the COOH group into COX where X was a halogen, azide, amide, or alkoxyl residue. Since many of the carbohydrates and other metabolites are activated by phosphorylation, and since dibenzoyl phosphate forms an amide bond with glycine under physiological conditions (E-1), activation of the amino acids by phosphate transfer was considered as possible for some time. Although ATP (adenosine triphosphate) is indeed indispensable for activation, phosphorylated amino acids have not been found. An important step toward the elucidation of the activation process was the discovery that homogenates of animal and bacterial cells contain enzymes which catalyze the activation of amino acids, and that this activation is accompanied by the release of stoichiometric amounts of pyrophosphate (E-2 to 4). The activation process is represented by reaction (XVI-9) where E is the amino acid-activating enzyme; PP, pyrophosphate; and AMP, adenosine monophosphate.

$$E + \text{amino acid} + ATP \rightleftharpoons E(\text{aminoacyl-AMP}) + PP \qquad \text{(XVI-9)}$$

The right side of Eq. (XVI-9) shows a complex formed by the combination of the activating enzyme with aminoacyl-AMP. The presence of this complex has been postulated because free aminoacyl-AMP cannot be detected when only small (catalytic) amounts of the activating enzyme are used. The reversibility of the reaction is proved by the formation of $ATP^{32}$ from AMP and $PP^{32}$ in the presence of enzyme and amino acids, but not in the absence of amino acids. The enzyme which catalyzes reaction (XVI-9) also catalyzes reaction (XVI-10) (E-6):

$$E(\text{aminoacyl-AMP}) + sRNA \rightleftharpoons \text{aminoacyl-sRNA} + E + AMP \quad \text{(XVI-10)}$$

Originally, it had been assumed that the amino acid is linked to RNA by an anhydride bond between the carboxyl group of the amino acid and the phosphoric acid residue of RNA. However, anhydrides of this type combine nonspecifically, even in the absence of enzymes, with native and denatured proteins (E-5). It is assumed, therefore, that the carboxyl group of the amino acid forms an ester bond with one of the OH groups at carbon atoms 2' or 3' of the ribose residues of RNA. Most probably the amino acid is first bound to the 2'-OH group, then transferred to the 3'-OH group (E-23).

sRNA is a fraction of RNA which, in contrast to the bulk of the cellular RNA, is soluble in 1.0 $M$ sodium chloride solution. The function of sRNA is to accept aminoacyl residues from the E(aminoacyl-AMP) complex and to transfer them to the ribosomes where the final steps of protein biosynthesis take place. For this reason the terms *acceptor* RNA or *transfer* RNA have been used for sRNA.

If the two reactions (XVI-9) and (XVI-10) are added up, the over-all reaction (XVI-11) is obtained:

$$\text{Amino acid} + ATP + sRNA \rightleftharpoons \text{aminoacyl-sRNA} + AMP + PP \quad \text{(XVI-11)}$$

The enzymes which catalyze the reactions (XVI-9) and (XVI-10) have been designated as *aminoacyl-RNA synthetases* (E-6). Both the enzymes and sRNA are found in the supernatant fraction of the liver homogenate after removal of the particulate fractions by centrifugation at 105,000 $g$. The aminoacyl-sRNA synthetases are precipitated from this supernatant fraction by acidification to pH 5. The precipitate, sometimes designated as the pH 5 fraction, contains also part of the sRNA.

Separation of sRNA from the enzymes can be accomplished by fractionation with phenol. sRNA is precipitated from the aqueous phase by ethanol. Its molecular weight is approximately 25,000–30,000. This corresponds to a chain of 75–90 nucleotides. sRNA forms only a small percentage of the total RNA of the cells. The bulk of the RNA is found in the microsomal fraction which is almost insoluble in 1.0 $M$ NaCl solutions. sRNA differs from the ribosomal RNA also by its high content of unusual bases, particularly pseudouridine, 5-methylcytidine, 6-$N$-methyladenosine, and smaller amounts of other bases. The nucleotide sequence of sRNA is pNpNpN...pCpCpA, where $N$ indicates any nucleotide, $C$ cytidine, and $A$ adenosine. The phosphate residues, denoted by $p$, are bound to the 5′-C-atom of the nucleotide at their right and to the 3′-C-atom at their left side. The terminal adenosine group is the acceptor group of the sRNA molecule. The aminoacyl residue is bound to the 2′- or 3′-hydroxyl group of the adenosine residue. This has been shown by the resistance of the aminoacyl-substituted terminal residue against oxidation with periodate (E-7). The reactions (XVI-9) and (XVI-10) depend on the presence of Mg$^{++}$ and K$^+$ ions. ATP can be replaced by an ATP-generating system. GTP (guanosine triphosphate) is necessary for the transfer of the aminoacyl residues to the microsome (E-8 to 10).

sRNA is not a homogeneous substance, but is a mixture of various sRNA molecules each of them *specific for one type of* L-*amino acid*. This view is based on the separation of different types of sRNA and on the independence of the activation of each of the amino acids. In other words, the different amino acids do not compete with each other for sRNA. Each of them combines only with the appropriate RNA. Since the acceptor end of all of the sRNA molecules are formed by the trinucleotide pCpCpA, and since most of them have at their other end the residue pG, the differences between the sRNA molecules must be attributed to differences in the composition and/or sequence inside the sRNA molecules. Some of these sRNA types have been purified by chromatography (E-11) or countercurrent distribution (E-12). Considerable differences in the nucleotide composition of sRNA specific for alanine, valine, and tyrosine have been found (E-12).

Attempts have been made to fractionate the aminoacyl-sRNA synthetases. It has been possible to purify some of them, for instance, an alanine-activating synthetase from pig liver (E-13) and four synthetases from *E. coli* which catalyze the activation of valine, leucine, isoleucine, and methionine, respectively (E-6). The specificity of the different purified aminoacyl-sRNA synthetases was demonstrated by their ability to catalyze (*a*) the ATP-PP exchange [reaction (XVI-9)] in the presence of the homologous amino acid, and (*b*) the formation of aminoacyl hydroxamates from amino acid and hydroxylamine. The latter test is based on the reaction: $NH_2 \cdot CHR \cdot COX + HONH_2 \rightarrow NH_2 \cdot CHR \cdot CO \cdot NHOH + HX$. The aminoacyl hydroxamates give a red color with ferric ions. The possibility of separating different aminoacyl-sRNA synthetases from each other has led to the view that there is a specific synthetase for each of the 20 amino acids and that each of these enzymes is also able to discriminate between the 20 sRNA types (E-14). Since the synthetases are inhibited by *p*-chloromercuribenzoate, they seem to contain SH groups in their active site (E-15).

It is not yet clear whether enzymes of one organism can also react with the sRNA molecules of another animal, vegetal, or bacterial species. Contradictory results have been obtained in investigations of this type. Whereas the leucine-RNA synthetases from yeast or liver catalyzed leucine transfer to yeast and liver RNA (E-16,24), the enzymes from *E. coli* had only a very low ability to form aminoacyl-sRNA in yeast or rat liver extracts (E-17).

It is not yet clear whether the activation of amino acids by the aminoacyl-sRNA synthetases and release of PP from ATP is the only mechanism of amino acid activation for protein biosynthesis. Other amino acid-incorporation enzymes have been isolated from different bacteria. The most thoroughly investigated of these is one extracted from *Alcaligenes fecalis;* it catalyzes activation not only by ATP, but also by CTP, UTP, and GTP; in contrast to the aminoacyl-sRNA synthetases, the enzymes from *A. fecalis* do not release pyrophosphate from the nucleoside triphosphates, but convert them into nucleoside diphosphates and inorganic phosphate (E-18). The same enzyme system might be active in the silk glands of the silkworms which produce the silk proteins (E-19).

When oviduct of the hen is incubated with $C^{14}$-amino acids, these are incorporated more rapidly into lipids than into the proteins (E-20). Amino acid-lipid complexes are also formed when protoplasts of *Bacillus megatherium* are incubated with $C^{14}$-amino acids (E-21). It has been suggested that amino acid-lipid complexes are physiological intermediates in the normal biosynthesis of proteins in the cells, and that the activation by means of sRNA is of importance only in cell-free systems, i.e., in homogenates or certain subcellular fractions (E-20). If this were so, the formation of aminoacyl-sRNA complexes would be an artifact. A clear answer to this problem would be very important for our evaluation of experiments on amino acid incorporation.

## F. Incorporation of Amino Acids into Ribonucleoprotein Particles (F-1)

When isotopically labeled amino acids are injected *in vivo* or when living cells or their homogenates are incubated *in vitro* with labeled amino acids, the amino acids are found first in the *ribosomes* of the microsomal fraction, i.e., in the submicroscopic particles which are sedimented by centrifuging for 1 to 2 hours at 105,000 $g$ (F-2). Ribosomes are ribonucleoprotein particles which are found in bacteria, and are also obtained from animal microsomes by treatment with deoxycholate (DOC) or certain detergents. The ribosomes contain approximately 40 to 50% RNA and 50 to 60% protein. Extraction of the microsomes with DOC removes, principally, lipoproteins rich in phospholipids and also small amounts of protein and RNA. The ribosomes *are the principal site of protein biosynthesis.* The size of ribosomes depends to a great extent on the magnesium content of the solution in which they are suspended. At high magnesium concentrations particles of the sedimentation constant 70–100 S (mol. wt. approximately 3 to 6 millions) are found; they may be polymers of 26 S units (F-3).

The prerequisites for incorporation of amino acids into microsomes are the presence of $Mg^{++}$, ATP, an ATP-generating system such as creatine phosphate and creatine phosphokinase, the pH 5 fraction (see Section E), guanosine triphosphate (GTP), and glutathione (F-4). The role of GTP is not yet understood. It does not seem to be involved in the activation of the amino acids. Although amino acids are incorporated into ribosomes by the pH 5 fraction which contains sRNA and synthetases, the incorporation is enhanced and increases about twofold when more sRNA is added to the system (F-5). This supports the assumption that the amino acid residues are transferred from their sRNA complexes to the ribosomes (F-6). The transfer seems to take place very rapidly since free complexes of aminoacyl-sRNA have never been found in the ribosomes; such complexes, if they exist, must be very unstable and short-lived (E-6). The making of nascent protein in the ribosomes is a very rapid reaction and may take only a few seconds; it has been estimated that about 200,000 amino acid residues are incorporated into protein per cell per second (F-7). Electron microscopy reveals that the ribosomes of rabbit reticulocytes (see below) in which hemoglobin is formed occur in aggregates of 5 ribosomes. These aggregates, called *polysomes*, have a sedimentation constant of approximately 170 S (F-20). They may be the true site of protein biosynthesis.

The *ribosomal* RNA forms the bulk of the cellular RNA. Its solubility is different from that of the soluble sRNA as shown in the preceding section of this chapter; its content of guanine and cytosine is lower. It was originally assumed that the ribosomal RNA might be the carrier of the code which determines the sequence of amino acids in the newly formed protein. This idea has been abandoned because there is no correlation between the base composition of the ribosomal RNA and the DNA of the same bacteria (F-8). At present, another type of RNA, called messenger RNA (mRNA) is considered as the carrier of the code. It will be discussed in Section I of this chapter. In comparison to sRNA and mRNA, the ribosomal RNA is metabolically more stable.

The newly formed protein is released from the microsomes into the soluble portion of the  homogenate (F-9, 21). This release resembles active secretion since it requires ATP, $Mg^{++}$ ions, and an enzyme which catalyzes the release (F-10). Antibodies like other proteins are also formed in ribosomes; they can be released from these by exposure to ultrasonic waves and subsequent extraction with bicarbonate at pH 9.5 (F-11).

Since the amount of protein formed in cells or in a cell-free system is much smaller than the protein present, it is not possible to decide by quantitative analyses whether the observed incorporation of isotopically labeled amino acids is due to net protein synthesis or whether it merely indicates an exchange of amino acids. Net protein synthesis can be demonstrated in the hemoglobin-forming system of the red blood cell since the newly formed hemoglobin can be purified by adding carrier hemoglobin of the same type. Hemoglobin production is particularly intense in the nucleated avian erythrocytes and in reticulocytes. The latter are immature red blood cells which are formed in animals made anemic by injections of phenylhydrazine. Rabbit reticulocytes have been widely used (F-12). According to these experiments, aminoacyl-sRNA is the only intermediate in the conversion of amino acids into hemoglobin. The amino acid chain of hemoglobin seems to grow from its amino end towards the carboxyl end (F-13). If the incubation is interrupted shortly after addition of a radioactive amino acid to a *cell-free system*, the labeled amino acid is found only in the $C$-terminal part of the peptide chains; apparently this cell-free system is not able to start new peptide chains from the $N$-terminus; it is only able to complete incomplete peptide chains by elongating them toward the carboxyl end (F-12,13).

The ribosomes of a cell seem to be a *heterogeneous* population. They can be fractionated by the addition of antibodies to the newly formed protein. Only those ribosomes which form the homologous protein are precipitated (F-14). The *specificity* of the newly formed protein depends on the ribosome-bound mRNA as well as the pH 5 fraction with the activating enzyme and sRNA. If ribosomes of rabbit red blood cells are incubated with the pH 5 fraction of human erythrocytes, both rabbit and human hemoglobin are formed (F-15). Formation of rabbit hemoglobin was also

observed when the pH 5 fraction of *Escherichia coli* was used (F-16).
These experiments would indicate the importance of the ribosomes for the
specificity of the protein formed. However, on addition of polyuridylic
acid instead of sRNA, the rabbit reticulocyte ribosomes form polyphenyl-
alanine instead of hemoglobin (F-17). Since the polynucleotide, as will be
shown later in this chapter, takes over the role of the messenger-RNA,
results obtained with poly U seem to indicate that the specificity of the
newly formed protein is determined by mRNA and that mRNA exerts
its action by being bound directly to the ribosomes. Most probably only
5 to 10% of the ribosome population is active, i.e., charged with mRNA
(F-18).

The polynucleotides organize protein synthesis through their action
not only on ribosomes but also on sRNA. This has been demonstrated by
means of poly UG, a mixed polymer of uridylic and guanylic acid. Poly
UG stimulates the incorporation of isotopically labeled cysteine into protein as shown later in Section K (Table XVI-1). When synthetic $C^{14}$-
cysteinyl-sRNA was reduced by Raney nickel and converted to $C^{14}$-
alanyl-sRNA, poly UG catalyzed the incorporation of alanine into the
ribosomal protein (F-19). Evidently, the incorporation of an amino acid
does not depend on the nature of the sRNA-bound amino acid but on the
specific structure of the sRNA. In the experiment described above, the
sRNA was evidently specific for cysteine, remained so even after conversion of the sRNA-bound cysteine to alanine, carried the alanyl residue to
the ribosome, and thus caused incorporation of alanine instead of cysteine.
These and other experiments have led to the view that sRNA has two
functions, namely, to act as acceptor and as adaptor. The acceptor role of
sRNA has been discussed in the preceding section of this chapter, where
it was mentioned that for each amino acid there is a specific sRNA which
will combine with the homologous amino acid. Each sRNA molecule
seems to possess another specific nucleotide sequence which is adapted to
definite sites of the ribosome-mRNA complex. One can imagine that these
two nucleotide sequences are complementarily adapted to each other. In
this manner each of the 20 sRNA types would be bound to different sites
on the ribosomes.

## G. The Sites and the Rate of Protein Biosynthesis

In the preceding section of this chapter the incorporation of amino
acids into ribonucleoprotein particles (ribosomes, microsomes) was
described. Ribosomes occur not only in the cytoplasm of the cell, but also
in its *nucleus* (G-1). Indeed, it has been found that $C^{14}$-labeled amino

acids are incorporated into nuclear ribosomes (G-2). In contrast to the incorporation of amino acids into cytoplasmic ribosomes, nuclear incorporation depends on the presence of sodium ions which seem to be necessary for the transport of amino acids into the nuclei. It is not yet quite clear whether the activation of amino acids in the nuclei takes place in the same manner as in the cell-free systems, namely, by means of sRNA (G-3,4). Amino acids are incorporated into both the true "acid" proteins of the nuclei which contain tryptophan, and also into the "residual" protein which is free of tryptophan (G-5).

Amino acids are also incorporated into the *mitochondrial* fraction of the cytoplasm. This incorporation requires neither the pH 5 fraction, nor the cell sap, and is unaffected by ribonuclease (G-6). The energy for this process is provided by coupling with oxidative phosphorylation (G-6). Newly formed serum albumin in the liver cells is distributed equally between mitochondria and microsomes before it is released into the blood serum (G-7). The mitochondria are small cytoplasmic granules. Their diameter is approximately 1 micron. They are much larger than the microsomes, which are invisible even at the highest optical magnification but are visible in electron micrographs. Separation of these two types of cytoplasmic granules is effected by differential centrifugation in sucrose solutions of a suitable molarity which prevents swelling and disintegration of the mitochondria.

In determinations of the *rate of protein biosynthesis* in a biological system, we have to be sure that all amino acids required are supplied in sufficient amounts and are present in adequate concentrations. In the intracellular pool the *level of free amino acids* is usually much higher than in the body fluids (G-8), and is different from organ to organ. Thus, the specific activity of $C^{14}$-glycine in the muscle, 40 to 80 days after injection, is much higher than in the liver (G-9). If nucleated red blood cells or ascites carcinoma cells are suspended in dilute solutions of $C^{14}$-glycine, the amino acid migrates into the cells against a concentration gradient until its intracellular concentration is several times higher than the extracellular concentration (G-10 to 12). We still do not know the mechanism by means of which energy is supplied for this transport against the concentration gradient. For each of the amino acids, a minimum threshold value of approximately 1.5 to 6.0 micromoles per liter, of the extracellular fluid, and of 10 to 50 micromoles per liter of the intracellular fluid is a prerequisite for protein formation in tissue cultures of human cells (G-13).

Some of the dispensable amino acids (see Section B of this chapter) can be formed in cultures of Hela cells or fibroblasts; among the amino acids which cannot be formed and which must be supplied are arginine and cystine although these are not essential for the human and animal organism (G-14). In the organism, each of the organs maintains its characteristic intracellular amino acid level. No equilibration takes place between the organs. Thus, hemoglobin and myosin do not exchange their labeled amino acids (G-9) unless their molecules are broken down completely.

Measurements of the rate of protein biosynthesis are complicated by the simultaneous breakdown of protein, and by the *reutilization of breakdown products* for the synthesis of new protein. In cultures of mammalian cells, about 1% of the protein is degraded per hour and the same amount formed in resting cells; when the cells grow, the rate of synthesis increases to fourfold the rate in resting cells (G-15). This reutilization of the administered labeled amino acids leads to fallacious high values for the half-life of the protein molecules (G-16). The half-lives determined in such experiments may be closer to the half-lives of the cells rather than to those of the cellular proteins (G-9). The difficulties arising from interference of protein catabolism with protein anabolism can be avoided by using systems in which the newly formed protein is rapidly eliminated from the cells either by secretion or by other means. In some instances it is possible to determine the extent of catabolism and to use this value for the computation of the true rate of protein biosynthesis.

Difficulties may also arise in experiments with $S^{35}$-labeled amino acids. Cystine and cysteine combine by SS-SH exchange (see Chapter VII, E and F) with SH or SS groups of the tissue proteins and thus simulate formation of peptide bonds. SS-bound amino acids can be differentiated from peptide-bound amino acids by treatment with an excess of nonradioactive cysteine, thioglycol, or other sulfhydryl compounds. Whereas peptide-bound amino acids remain protein-bound, the SS-bound cystine residues are released from the protein by exchange with the nonradioactive sulfhydryl compounds. Ambiguous results may also be obtained with $C^{14}$- or $H^{3}$-labeled amino acids, by nonspecific binding of the amino acids to proteins and other cellular constituents. Many of the amino acids have reactive side chains which can combine with other substances present in the cells. In the presence of amino acid oxidases, highly reactive keto acids are formed which may form condensation products with the amino groups of proteins and other compounds. The high sensitivity of the work with radioactive isotopes allows the detection of very small amounts of amino acids. The results are usually expressed in counts per minute. Conversion of these into weight units of protein or amino acid reveals that the extent of amino acid incorporation is sometimes extremely small. It is doubtful in such instances whether the binding of amino acids to the protein fraction indicates true protein biosynthesis. Indeed, it has been found that amino acids can be bound to proteins in the absence of enzymes (G-17,18). In view of these complications, it is very important to demonstrate unequivocally true protein biosynthesis and to differentiate it from other processes which may lead to the "incorporation" of amino acids into proteins.

The best method to prove true biosynthesis of proteins is the isolation of the newly formed protein and its identification by chemical, physical, and serological methods. This is frequently very difficult if not impossible because the amount of new protein formed is only a small fraction of the total protein present in an organ or any other biological system. Only in very few instances, unambiguous isolation and identification have been possible. Frequently, the biosynthesis of an enzyme is investigated because its presence and concentration, even in the presence of numerous other proteins, can be determined by measuring the enzymatic activity. It must be remembered, however, that the activity of enzymes depends on activating and inhibiting substances, and that an increase in activity is not a sufficient proof for an increase in the amount of enzyme.

One of the most thoroughly investigated biosyntheses is that of *serum albumin* in the liver. Investigations of the rate of formation of serum albumin in liver slices incubated with $C^{14}$-amino acids revealed that labeled serum albumin appeared in the extracellular fluid about 20 minutes after the addition of the labeled amino acids (G-19). It was at first suspected that the newly formed peptide chain might be held for some time at its template and be released only slowly. Later, it became clear that the process of serum albumin formation requires less than 1.5 minutes, but that the newly formed serum albumin remains for a long time inside the microsomes. It can be released from these much earlier by butanol (G-20), deoxycholate, or by ultrasonic treatment (G-21). If this is not done, the serum albumin remains in the microsomes for approximately 15 minutes and is then secreted into the extracellular fluid.

The rate of the biosynthesis of hemoglobin can be determined by measuring the appearance of administered radioactive iron in hemoglobin. Hemoglobin is particularly suitable for investigations of biosynthesis since it is formed in a homogenous population of red blood cells (see Section F). Another suitable object for determinations of the rate of biosynthesis is *ferritin*. Its formation can be stimulated by the oral administration of ferric saccharate to the experimental animals. Kinetic investigations on the incorporation of $C^{14}$-amino acids into ferritin show that the process of formation of a ferritin molecule cannot take more than a few minutes (G-22). Still shorter times were found for the incorporation of amino acids into the protein of *E. coli;* its formation is completed within about 5 seconds (G-23). Only a few seconds are also necessary for the formation of the protein of the mold *Neurospora crassa* (G-24). Similarly, the formation of an antibody molecule in the presence of its antigen template does not seem to require more than a few seconds (G-25).

Computations of this type indicate, also, that each pancreas cell produces about 24,000 chymotrypsinogen molecules per second (G-26) and that a plasma cell tumor forms 2 mg globulin per gram of the wet tumor in 24 hours (G-27). It may be asked whether such a high rate and extent of protein formation is reconcilable with the basic assumptions of chemical kinetics. Since the molar concentration of sRNA in rabbit reticulocytes is $6 \times 10^{-5}$ and that of the microsomes $3 \times 10^{-6}$, one can calculate the number of collisions from the kinetic theory of chemical reactions. If it is assumed that only 10% of the sRNA molecules are active, one obtains $2.2 \times 10^{19}$ collisions per second per milliliter. The rate of hemoglobin formation is sufficiently explained if only 0.01% of these collisions lead to the incorporation of an amino acid (G-28).

The difficulties arising from simultaneous anabolism and catabolism can be avoided to a certain extent by the *use of doubly labeled proteins* as a source of amino acids. If animals are injected with their homologous serum albumin, which contains $C^{14}$- and $S^{35}$-amino acids, the ratio $S^{35}/C^{14}$ in their serum proteins remains constant for a period of several days, whereas the same ratio in the tissue proteins increases immediately in proportion to the sulfur content of the tissue proteins (G-29). Hair keratin, which is particularly rich in sulfur, incorporates much more $S^{35}$ than the other proteins. If animals are injected with $S^{35}$- or $C^{14}$-serum albumin which has been iodinated *in vitro* with $I^{131}$, the ratio $S^{35}/I^{131}$ in the tissue proteins increases, and after 8 days is about 50 to 100 times higher than the $S^{35}/I^{131}$ ratio in the circulating serum albumin. This shows that the injected serum albumin, as soon as it penetrates into the cells, is degraded, that its iodine is eliminated from the body, whereas the liberated $S^{35}$-amino acids are reutilized for protein biosynthesis.

It is clear from all these examples that we are usually dealing with simultaneous biosynthesis and breakdown of proteins. If we determine merely the rate of disappearance of a radioactive label from a labeled protein, we find an *apparent half-life* and not the true biological half-life of the protein molecules. One of the few instances in which the true half-life can be determined is the metabolic breakdown of antibodies. If antibody produced in a rabbit injected with isotopically labeled amino acids, is intravenously injected into another normal rabbit, the specific activity of the circulating antibody and its ability to precipitate the antigen decrease at the same rate (see Chapter XV, D). Since no antibody protein is formed in the injected animal, the elimination rate of the labeled antibody is a direct measure of its catabolism. Consequently, the half-life measured is the true half-life of the antibody molecules. Another unusual reaction is the formation of hair keratin. Keratin is metabolically inert and does not undergo any change after its formation. If keratin is formed in an animal into which radioactive amino acids have been injected, the labeled amino acids after incorporation are trapped in the newly formed keratin. If the hair is allowed to grow for a few days and then investigated by autoradiography, the zones of incorporation of the labeled amino acid can be detected, and the rate and extent of incorporation can be estimated (see Fig. XVI-1) (G-30).

The apparent half-lives of the serum proteins are 2–5 days in the small laboratory animals, 10–25 days in the large animals and in man. In dogs, about 10% of the plasma proteins are replaced in 24 hours (G-31). No difference has been found between the apparent half-lives of newly formed

and older serum albumin molecules in rats (G-32). Evidently, the breakdown of these molecules takes place randomly without preferential degradation of the older molecules. The turnover of the muscle proteins is
much slower than that of the plasma proteins (G-33). A still lower rate of
regeneration is found in hemoglobin; only 2.5% of this protein is formed

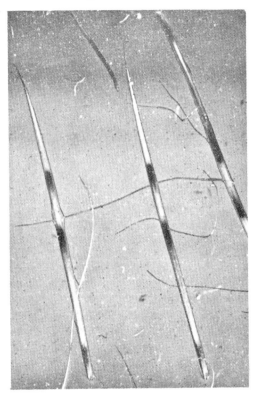

FIG. XVI-1. Autoradiogram of the hair of a rat injected intravenously with $S^{35}$-yeast
hydrolyzate. The injection was repeated 2.6 and 7.1 days later. The hair was plucked
2.5 days after the last injection. The black zones are the sites of incorporation of
$S^{35}$-amino acids (G-30).

each day (G-33). The half-life of collagen is so high that it cannot be
determined with any precision (G-34). It is quite possible that collagen
once formed is not replaced at all during the lifetime of the organism.

## H. The Problem of Peptide Intermediates

Are the amino acids of a peptide chain assembled by stepwise growth of
the chain from one end to the other, or do they combine to form free

oligopeptides which secondarily combine to form the long peptide chain? A third possibility, the simultaneous combination of hundreds of amino acids to form the peptide chain, has never been taken seriously into consideration since such a reaction would be too improbable from the point of view of kinetics. The idea of a stepwise-growing peptide chain is supported by the old experience that the body fluids do not contain any significant amounts of *free* peptide intermediates. Glutathione, carnitine, and a few other peptides which have been found, are not intermediates in the biosynthesis of proteins. Attempts have been made to use isotopically labeled amino acids for a decision between the two mechanisms of protein biosynthesis.

When living animals are injected with a mixture of several isotopically labeled amino acids, pure proteins can be isolated from their organs, and the rate of incorporation of different amino acids can be computed from their specific activities, i.e., from the counts per minute per milligram of protein-bound amino acid. The rate of incorporation depends on the dilution of the injected amino acid in the amino acid pool of the body, on the rate of formation of the protein, and on other factors. When crystalline aldolase and glyceraldehyde phosphate dehydrogenase were isolated from the muscles of rabbits injected with a mixture of 11 labeled amino acids, it was found that the ratio of incorporation into these two proteins was *the same* for all of the injected amino acids (H-1). The specific activity of each of the amino acids was 1.5 to 1.8 times higher in aldolase than in the dehydrogenase. Similar results were obtained when the specific activity of the amino acids of newly formed milk proteins was investigated (H-2). Identical specific activities were also found in peptides produced by the partial hydrolysis of milk proteins (H-2). All of these experiments seemed to indicate that the proteins are formed *directly from amino acids*, that the formation was a rapid reaction, and that free *peptides were not intermediates* in this process.

Although free peptides are present in rye grass and other plants, these peptides do not seem to be precursors of proteins; they may be produced by a side path of peptide-bond formation (H-3). In the last few years, *nucleopeptides* have been discovered in numerous bacteria (H-4,5), and also in animal tissues such as the guinea pig liver (H-6). Some of these nucleopeptides are formed in penicillin-treated staphylococci; they consist of uridine-diphospho-acetylglucosaminyl-lactyl-L-alanyl-D-glutamyl-L-lysyl-D-alanyl-D-alanine, and are formed without the participation of RNA (H-5). The composition of these peptides and particularly the occurrence of D-amino acids indicates that they are not precursors of proteins. It has been claimed that some of the other nucleopeptides are precursors of proteins (H-4,6).

When insulin or ribonuclease are formed by the incubation of pancreas slices with radioactive amino acids and the radioactive proteins are then partially hydrolyzed, the specific activity of the incorporated amino acids in different peptides is found to be *different* (H-7). This result appeared to be contradictory to the finding of equal specific activities in the peptides of milk proteins (H-2). It seemed at first to indicate the formation of peptides as intermediates and their secondary combination to form the peptide chain of the protein. However, the results of experiments of this type are also reconcilable with those reported earlier in this section if the time required for the formation of the peptide chain and the duration of the experiment is taken into consideration. Obviously, differences in specific activity can be detected only if the duration of the experiment is equal to or shorter than the time required for the synthesis of a complete peptide chain. In this case, some of the incomplete peptide chains may be completed by the incorporation of radioactive amino acids and may therefore have different specific activities in different portions of their chain. If the experiment lasts much longer than the time required for the formation of the peptide chain, most of the chains will be formed from a radioactive amino acid pool and will then have equal specific activities along their amino acid sequence.

Both types of experiments, the short-term incubations *in vitro* and the longer injection experiments *in vivo*, are therefore reconcilable with the view that the peptide chain grows by the stepwise addition of amino acids from one end to the other. This view has been supported very strongly by the analysis of *hemoglobin* formed in rabbit reticulocytes in the presence of isotopically labeled amino acids (H-8). Analysis of a partial hydrolyzate reveals that the highest specific activity is found initially in the amino acids which are close to the carboxyl end of the peptide chains of hemoglobin, that the high specific activity then extends toward the amino end, and that the activity is, finally, the same in the entire chain. This indicates growth of the peptide chain from the amino end toward the carboxyl end. The initial high activity near the carboxyl end is attributed to the completion of incomplete nonradioactive chains by the incorporation of labeled amino acids at the carboxyl end (H-8).

As mentioned in the preceding section, hemoglobin is particularly suitable for experiments of this kind since it forms the bulk of the proteins in red blood cells and is easily separated from contaminants. Although the incorporation of radioactive amino acids into hemoglobin is easily measurable, the amount of hemoglobin is still too small for the experimental proof that a net increase in hemoglobin takes place during the period covered by the experiment. However, net synthesis of hemoglobin is made very probable by the reported analysis of the peptide fragments (H-8) and may also occur in the cell-free system in which hemoglobin is formed by the ribosomes of rabbit reticulocytes (H-9). The labeled hemoglobin precursors are firmly held to the ribosomes and are not removed by washing; evidently, they are not *free* peptides but parts of the growing peptide chain which can be considered as bound peptide. Most probably, the link between ribosome and the growing peptide chain is mediated by the terminal amino group of the peptide chain (H-9).

The presence of a growing peptide chain must not be considered as a confirmation of the claim that peptides are obligatory intermediates in the formation of proteins, and that the latter are formed in a second or third step by the condensation of several peptide chains. This view has been abandoned. The incorporation of each of the amino acids seems to take place in a single step, the elongation of the peptide chain by the linkage of a single amino acid to the carboxyl end of the chain. Attempts to demonstrate the incorporation of the *N*-terminal peptide into the growing peptide chain of hemoglobin failed (H-10).

In the preceding chapters of this book, it has been pointed out repeatedly that many of the proteins consist of subunits which are held together by loose bonds. These subunits are identical with the complete peptide chains discussed in the preceding paragraphs. It seems that all of the large proteins are formed as subunits and that the molecular weight of the latter does not exceed values of 20,000–30,000. It is not yet clear whether each ribosome can form only one of these subunits (H-11) or more than one (H-12).

## I. The Role of DNA and Messenger RNA in Protein Biosynthesis

It is usually assumed that DNA (deoxyribonucleic acid) is the ultimate source of genetic information. This view is supported by the occurrence of DNA in the nucleus of the cell, particularly in the chromosomes; it is also strongly supported by the phenomenon of transformation in which DNA acts as the transforming principle. Inoculation of a strain of bacteria with the DNA fraction of another related strain may cause transformation of the former into one with some of the genetic characteristics of the latter strain (see I-1). Further support for the genetic importance of DNA is provided by the infection of bacterial cells with bacteriophages. Some bacteriophages inject their DNA into the host cells and after inactivation and destruction of *the bacterial* DNA take over the genetic apparatus of the host. The phages multiply and destroy the host cells. In both transformation and phage infection, the cellular DNA is displaced and replaced by the foreign DNA which is then replicated in subsequent generations. All of the plant viruses and many of the animal viruses are free of DNA. In these viruses RNA is the infective agent which interferes with the genetic apparatus of the infected host cell. In this text we are chiefly interested in the role of DNA and of virus RNA in the biosynthesis of proteins.

The chemical and physical properties of ribo- and deoxyribonucleoproteins have been discussed in Chapter XI, I of this book. It may be recalled that the molecular weight of DNA is several millions. Each chromosome must therefore contain 1000 or more DNA molecules. In most cells, with the exception of some viruses and phages, DNA occurs as a nucleoprotein (I-2). The presence of protein seems to be necessary for the maintenance of the conformation of the DNA molecules (I-3). Recently, considerable amounts of spermine, spermidine, and of the simpler diamines have been found in DNA preparations (I-4). The amines neutralize approximately one-third of the phosphoric acid residues of DNA, thereby stabilize DNA, and prevent the mutual repulsion of the anionic groups (I-5).

DNA is not necessary for the production of protein. When the alga *Acetabularia* is divided into a nucleated and an anucleated half, protein formation continues in both halves in spite of the absence of DNA in one of them (I-6). Mammalian nuclei continue to incorporate amino acids into their protein even after degradation of more than 75% of their DNA by DNase provided DNA is replaced by other polyanions, for instance, by polyethylenesulfonate (I-7). What is then the function of DNA in protein biosynthesis? Although it does not act directly, it may still direct protein formation by means of a mediator.

DNA is formed *in vitro* by a polymerase from the 4 deoxynucleotide triphosphates provided a small amount of DNA is present as a primer. The priming action does not depend on the native state of DNA. Heat-denatured DNA is at least as active as native DNA in this respect (I-8). The importance of primers and templates in protein biosynthesis will be discussed in Section K of this chapter. Here we are particularly interested in the role of DNA in the formation of a metabolically unstable ribonucleic acid which was first discovered in *E. coli* after infection with $T_2$ phage (I-9). The role of a messenger, a carrier of information from DNA to the ribosomes, was attributed to this unstable compound which was called messenger RNA (mRNA) or informational RNA.

mRNA forms about 5 to 15% of the total RNA and has a turnover time of about 4 to 6 seconds (I-10). The base sequence of mRNA seems to be complementary to that of the homologous DNA. This means that for each deoxyadenosine (dA) of the DNA molecule, the mRNA molecule contains one uridine (U) residue, for each T one A, for each dG one C, and for each dC one G. For example, if the DNA contains 30% A, 30% T, 20% G, and 20% C, the mRNA will contain 30% U, 30% A, 20% C, and

20% G. Although this still does not prove that the sequence of the nucleotides is complementary to that in DNA, such complementariness is strongly indicated by the observation that mRNA and the DNA which acted as a template in its formation form stable hybrids when their solution is carefully heated *in vitro* (I-11). The formation of the complementary RNA is catalyzed in the presence of DNA by an RNA polymerase which has been isolated from *E. coli* (I-12). In the DNA-mRNA hybrid, a single strand of mRNA combines either with a single strand or with a double strand of DNA, in the latter case forming a triple strand (I-13,14). As mentioned in a preceding section of this chapter, mRNA combines with the ribosomes and may thereby transmit the genetic information from DNA to the ribosomes (I-14,22). mRNA may also interact with the aminoacyl-sRNA complex or with the activating enzyme and thereby direct each of the amino acids to the appropriate site on the ribosome template (I-15,16). Each mRNA molecule seems to act only once or a few times before it is metabolized (I-17,21).

Frequently, it has been found that protein biosynthesis is accompanied by an increase in the total RNA (I-18,19), particularly the ribosomal RNA. If RNA is destroyed by the action of RNase, the formation of proteins is severely affected (I-18). However, formation of protein and total RNA do not always run parallel. Thus, no RNA is formed in *Tetrahymena pyriformis* when it is raised in a culture free of uracil, although division and multiplication of cells and protein formation continue under these conditions (I-20).

The observations recorded in this and the preceding sections lead to the conclusion that DNA is not directly involved in protein biosynthesis, that sRNA has a role in the activation of amino acids, ribosomal RNA in the assembly of the amino acids, and that mRNA may act as a carrier of genetic information from DNA to the activated or ribosome-bound amino acids.

## J. The Action of Analogs, Mutagens, and Inducers (J-24).

If one of the essential amino acids in the culture medium of bacteria or molds, or in the food of animals is replaced by an *amino acid analog*, the latter may be incorporated into newly formed proteins. Analogs of this type include *p*-fluorophenylalanine (J-1) or thienylalanine for phenylalanine, ethionine or selenomethionine for methionine (J-2), and tryptazan derivatives for tryptophan. Although incorporation of some of these amino acid analogs has been observed, only a small number of the new proteins formed were in all respects identical with the normal proteins and few of these had the same biological activities. None of the analogs can replace the natural amino acid for a long period of time. Their incorporation indicates merely that the specificity of some of the enzymes involved in protein biosynthesis is not absolute.

The action of *purine or pyrimidine analogs* (J-3) is more interesting than that of amino acid analogs because it gives us some insight into the mechanism of protein biosynthesis. The best-investigated of these analogs are 5-fluorouracil, 6-mercaptopurine, 8-azaguanine, 6-azauracil, 5-bromouracil, and 2-thiouracil. Some of these are incorporated into DNA or RNA and may in this manner interfere with protein biosynthesis. Most of these analogs inhibit the growth of bacteria, molds, or mammalian tissues. Interest in the purine and pyrimidine analogs was particularly stimulated by the demonstration that some of them inhibit the growth of tumors in experimental animals.

5-Fluorouracil seems to be incorporated, instead of uracil, into the messenger RNA (J-4) and causes the formation of an anomalous alkaline phosphatase (J-5) and galactosidase (J-6) in *E. coli*. 5-Fluorouracil also inhibits the synthesis of DNA in *E. coli* (J-7), probably by preventing the conversion of uracil into thymine (5-methyl uracil) (J-8). 5-Bromouracil is an analog of thymine; indeed it replaces thymine in the DNA molecule. If nucleic acids are exposed to the action of nitrous acid, mutations are observed (J-9). They are attributed to the conversion by nitrous acid of adenine into hypoxanthine, guanine into xanthine, and cytosine into uracil (J-9,10). Mutations have also been produced by the administration of 5-bromouracil and 2-aminopurine (J-10). Protein biosynthesis is blocked by chloramphenicol although this drug does not affect the formation and accumulation of RNA (J-11).

Information on the mechanism of protein biosynthesis has also been obtained by the study of the regulation of enzyme formation (J-23,24). The rate of formation of certain microbial enzymes depends on the presence in the culture medium of the substrates of these enzymes. Thus the rate of β-galactosidase formation in *E. coli* increases several hundred-fold if lactose replaces glucose as source of carbon. Originally this phenomenon was designated as *enzymatic adaptation*. The finding, however, that non-metabolizable analogs of the substrate such as thiomethylgalactoside in the case of β-galactosidase also induced the formation of the enzyme, and that formation of the enzyme was not necessarily of adaptive value to the organism, led to the use of the term *enzyme induction*. Substances which cause increased enzyme production are denoted as *inducers*.

More recently it has been shown that the rate of formation of other microbial enzymes is slowed down or *repressed* by the presence in the medium of the product of enzyme action or by analogs of the product. Both may serve as *repressors*. For example, the formation of the enzyme tryptophan synthetase which catalyzes the formation of tryptophan is repressed in *E. coli* by tryptophan (J-12) and also by the non-metabolizable analog 5-methyltryptophan. A common mechanism may underlie enzyme induction and enzyme repression (J-15).

The site and mode of action of inducer and repressor are not yet quite clear. Both may interact with an "aporepressor" produced by the cell as a result of the activities of so-called regulator genes. Alternatively, the inducer may combine with the traces of non-induced constitutive enzyme which is formed at about $\frac{1}{100}$ to $\frac{1}{1000}$ the rate of the inducible enzymes. The inducer may in some manner secure release of the enzyme from its template under conditions where the bound enzyme would repress further enzyme formation in the absence of inducer (J-13,14,17). In other words, the inducer would act as a "derepressor" (J-16). A comparison of inducible and constitutive yeast β-galactosidase yielded only small differences with respect to the affinity of the two enzymes for a series of natural substrates (J-19). In similar experiments, induced penicillinase in *Bacillus cereus* appeared to be identical with the constitutive penicillinase (J-20). The induced enzymes were formed from amino acids directly and not by a modification of precursor proteins (J-21). The time required for the assembly of the amino acids is approximately 10 seconds. The lag frequently observed in induction has been attributed to the time during which the repressor is degraded or inactivated (J-22).

## K. Templates and the Coding Problem

The principal problem, with which we are confronted in protein biosynthesis, is that of the *highly specific sequence of amino acids* in the protein molecules. In the preceding sections of this chapter it has been shown that the tissues contain free amino acids as precursors of the proteins, that these amino acids are activated by ATP in the presence of an activating enzyme, that the amino acids combine with the terminal adenosine group of sRNA, are transferred to the ribosomes, and that there they assemble in the growing peptide chain. The order of the amino acids in the peptide chain seems to be determined by genetic elements through the mediation of a particular ribonucleic acid, called template RNA or messenger RNA (mRNA). This last step poses serious difficulties to our understanding. We still do not know the mechanism by which a certain sequence of nucleotides should determine a sequence of amino acids. We speak, at present, of templates, primers, and of the coding problem and it might be useful to describe first how and why these ideas were developed.

*The Mechanism of Replication.* The problem of protein specificity and its transmission from generation to generation was intriguing even before the experimental proof that each protein had its definite amino acid sequence. It was generally accepted that the species specificity and identity of the protein molecules was brought about by a copying process. Two different mechanisms were taken into consideration. According to one of these, the sequence of the amino acids in the peptide chain was determined by two factors: (a) the protein forming enzymes and (b) the ability of each of the incorporated amino acids to influence the enzymatic incorporation of the next amino acid into the growing peptide chain. According to the other view, the sequence of amino acids in the nascent peptide chain was determined by another macromolecule, sometimes called a matrix or a template, in which an analogous sequence of small building elements was present.

*The Zymosequential Hypothesis.* As for the factor (a) in the first of these two hypotheses, it is difficult to imagine that the information for the entire amino acid sequence could be stored in a single enzyme whose own amino acid sequence usually is quite different from that of the protein produced. It was therefore assumed that each protein is synthesized by a series of proteinases of different specificities (K-1). Concerning factor (b) listed above, we know that the stability of a peptide bond depends on the nature of the two amino acid residues involved in its formation. Since there are approximately 20 natural amino acids, each of them can combine with one of 20 other amino acids. Some of these 400 possible peptide bonds are more stable than others. It can be reasonably assumed that the tendency to form the stable bonds is greater than the tendency to form labile bonds and that, therefore, certain sequences may be favored (K-2). This might have had particular importance for the formation of peptides in the primordial world where condensation of the amino acids may have taken place at elevated temperatures, in the absence of enzymes (K-2).

*Sequential Patterns.* In the preceding chapters of this book, it has been pointed out repeatedly that the amino acids in the peptide chains of the proteins are not randomly aligned but that certain sequences are repeated more frequently than others. Numerous other regularities have been observed and discussed, particularly by Šorm (K-3), but also by other investigators (K-4 to 6).

Thus many of the 124 dipeptide units of ribonuclease (see Fig. XII-4) appear twice or more frequently in the molecule (K-3). In the cytochrome c molecule, 10 out of 17 lysyl residues occur as basic dipeptides linked to other lysyl, arginyl, or histidyl residues (see Fig. XII-6). Sometimes an amino acid sequence is repeated in the same molecule in the reverse order. An example can be found in the amino acid sequences 69–79 and 29–19 of the ribonuclease molecule (K-3). Another striking pattern has been shown in corticotropin (see Chapter XIII, F) where a chain of 12 aminoacyl residues is repeated in such a manner that the five acidic aspartyl and glutamyl residues of one chain are replaced by five basic lysyl or arginyl residues in the other (K-7).

Frequently an amino acid is replaced preferentially by a related amino acid. Thus arginine in vasopressin may be replaced by lysine (K-8). Keil and Šorm (K-9) have denoted this "lysine-arginine group" by the letter I; other similar groups are A = aspartic acid, glutamic acid, asparagine, glutamine; S = serine and threonine; L = glycine, alanine; U = leucine, isoleucine, and valine; F = phenylalanine and proline; T = tyrosine and tryptophan. Regular arrangements of amino acids are frequently found immediately adjacent to cystine bridges; these seem to play a particular role in the arrangement of the amino acid sequences (K-10).

Although some doubts have been raised about the universality and the extent of these regular patterns (K-11), there can be hardly any doubt that some of the patterns occur more frequently than expected in a random sequence of the aminoacyl residues. It has been suggested that similarity of amino acid sequences in different proteins may indicate biological evolution from a common precursor (K-3). Whatever the ultimate cause of the occurrence of regular patterns in the amino acid sequence may be, it points to the action of some regulating or organizing mechanism.

*Organizers.* The postulate for an organizer was made long before the discovery of the role of nucleic acids in protein biosynthesis. Since specific long-range forces which would act over distances of several hundred angstrom units are not known, it was assumed that the peptide chain of the nascent protein molecule was expanded in the same manner as in monomolecular protein films (Chapter VII, J) and that replication took place in close contact with the organizer which was probably *an expanded macromolecule containing all the information required for the assembly of the amino acids* (K-12 to 14).

It may be recalled that the diameter of a globular protein molecule is in the range of 30 to 120 Å and that we do not know of any specific forces, which in an aqueous solution, would operate over distances of this magnitude. The forces which operate between ionic, polar, and nonpolar groups (see Chapter VII, E) decrease very rapidly with increasing distance between the groups involved (see Chapter X, A) and become negligibly small when this distance is more than 5–6 Å. For this reason it was assumed (a) that replication would take place in direct contact with an organizing macromolecule, (b) that a globular protein cannot be copied directly in its three-dimensional conformation, and (c) that this conformation is produced in a second phase of biosynthesis by folding of the peptide chain.

*The Action of Primers.* Before going further, it may be useful to discuss the general action of organizers. The terms "template" and "primer" are used at present, sometimes interchangeably, for this important material. The term *primer* had been applied by Cori and Cori (K-15) when they discovered that small amounts of glycogen or amylose were necessary for the enzymatic synthesis of glycogen or starch from glucose-1-phosphate in the presence of phosphorylase. The primer participates in the polymerization by forming the origin of the growing polysaccharide. If the primer is a straight-chain polyglucoside, a straight chain is formed; if it is branched, a branched polysaccharide is produced by phosphorylase. In all these reactions, the primer is incorporated into the newly formed polymer. Its role is limited to that of a starter of the reaction (see Chapter XII, G).

In other instances, the primer acts not only as a starter, but also determines the sequence of the assembling monomers (K-16). The most important example of this type is the priming role of DNA or of other polynucleotides in the polymerization of deoxyribonucleotides which is catalyzed by Kornberg's enzyme, DNA polymerase (K-17). The newly formed DNA seems to have the same nucleotide composition and nucleotide sequence as the DNA which is used as a primer. If the DNA primer is replaced by a mixed polymer of deoxyadenylic and thymidylic acid, this primer will prime its own replication. Evidently, this type of primer has two functions: (a) to start the polymerization, and (b) to act as a template for its own replication.

*Templates.* The term *template* is used in a twofold sense, namely (a) for a molecule which serves as a matrix for the formation of its *replica*, and (b) for molecules which serve as matrices for the formation of a *complementary* product. (K-18,19). For instance, the determinant group of an antigen acts as a template for the formation of antibodies (K-20) (see Chapter XV, B and D). Or DNA, according to Watson and Crick, (K-34) acts as a template for the formation of other DNA molecules in which each adenine, thymine, guanine, and cytosine of the template causes formation of thymine, adenine, cytosine, and guanine, respectively, in the newly formed DNA. As a result of this process, the newly formed strand is not a replica of the template DNA, but is complementarily adjusted in such a manner that each of its bases can combine by means of hydrogen bonds with the opposite base in the template DNA. Similarly, DNA may act as a template for the formation of messenger-RNA. The base pairs in the DNA-RNA pair would then be adenine → uracil, thymine → adenine, cytosine → guanine, and guanine → cytosine (see Section I of this chapter).

Neither templates nor primers are catalysts in the common sense of the word since they do not act in the absence of enzymes. The primer, in contrast to typical enzymes, is consumed in the reaction and built into the chain of the polymer. The template differs from an enzyme or a primer by the wealth of detailed information which it transmits to the biosynthetic system. It may be useful in this context to recall the old finding of Bredig (K-21) that the synthesis of mandelonitrile $C_6H_5 \cdot CHOH \cdot CN$ from benzaldehyde $C_6H_5 \cdot CHO$ and HCN proceeds asymmetrically and yields optically active mandelonitrile when the catalyst, diethylamine, is bound to the asymmetric molecule of cellulose. In the absence of cellulose, equal amounts of D- and L-mandelonitrile are formed. Diethylamine acts in this reaction as a catalyst. Cellulose acts merely as an organizer.

*The Idea of Protein Templates.* What is the chemical nature of the organizer which supplies the information for the sequence of the growing peptide chain? Could we not assume a process analogous to Kornberg's polymerase where the DNA primer supplies all necessary information for the sequence of the newly formed DNA? Could not a protein in the expanded state of monomolecular film act as a template for its own replication? Such a view was advocated by the author in the first edition of this book when next to nothing was known on the role of nucleic acids in the biosynthesis of proteins (K-22). In contrast to many of his colleagues, the writer is not yet convinced that this idea is obsolete. Protein templates may still have a role in addition to that of nucleic acid templates. Therefore they are discussed later in this section.

Originally it was suggested that free amino acids in the vicinity of the protein template would be loosely bound to those amino acid residues of the template which have the same side chains and that a nonspecific peptidase or protease might then link up the amino acids to yield a peptide chain alongside the identical peptide chain of the expanded protein film of the template. We have learned, in the meantime, that the immediate precursors of the newly formed peptide chain are not free amino acids but activated amino acids in the form of aminoacyl-sRNA complexes. This would still be compatible with the simple assumption that the sequence of amino acids in the nascent peptide chain is determined by a protein template.

The linear peptide chain formed under the directing action of either RNA or any other type of template folds up when the end product is a globular protein. Some investigators assume that this folding is determined only by the amino acid sequence and that the final conformation is a state of the protein in which the energy content of the molecule is particularly low. This view is based on the phenomenon of spontaneous renaturation of some of the denatured proteins (see Chapter XII, G). According to another view, the folding of the newly formed peptide chain takes place in a second phase of protein biosynthesis, and depends on various external factors.

*Interference of Antigens with γ-Globulin Biosynthesis.* The postulate of *two phases of protein biosynthesis* is principally based on the role of the antigen in antibody formation. The antibodies, as shown in Chapter XV, B, are typical globular proteins of the γ-globulin type. Their amino acid composition is identical with that of the normal γ-globulins or very similar to it. Their amino acid sequence must also be very similar since only very small differences have been discovered when partial hydrolyzates of different antibodies and normal γ-globulins were "fingerprinted." Yet the conformation of the antibodies must be quite different from that of normal γ-globulins and also different from antibody to antibody since each of them is complementarily adjusted to a determinant group of its homologous antigen. Breinl and the author (see Chapter XV) originally suggested that mutual adaptation of antigen and antibody is accomplished by changes in the amino acid composition and/or sequence in the peptide chains of the antibody molecules. Pauling (K-23) later proposed that the complementariness of the antibodies is brought about merely by the appropriate folding of their peptide chains and not by changes in the primary structure.

The author (K-20), accepting this view, has therefore suggested that the antigen molecule interferes only with the second phase of protein biosynthesis, with the folding of the peptide chain and not with the first phase in which the amino acids are assembled. This view is not shared by those who believe that the conformation of proteins is definitely determined by their amino acid sequence, and that different conformations indicate the presence of different amino acid sequences in each of the different antibodies. The experiments published at the time of this writing do not yet allow us to decide whether the small differences discovered by Gurvich (K-25) and Gitlin (K-26) in the partial hydrolyzates of different rabbit antibodies are caused by interference of the antigen molecules with the sequential assembly of amino acids, or whether they are an expression of the heterogeneity of the antibodies which always consist of a mixed population of molecules formed in lymph nodes, spleen, and in other organs.

*Nucleic Acid Templates.* During the last decade, important arguments have been advanced in favor of the view that the amino acid sequence of proteins is determined by a *nucleic acid template* rather than by a protein template. The first incentive to this idea appears to have been the finding of Brachet (K-27) that intensive protein biosynthesis is usually accompanied by simultaneous formation of ribonucleic acid, and that protein synthesis is inhibited when the protein-forming system is exposed to the action of ribonuclease. At the same time, Caspersson (K-28) by means of ultraviolet microscopy succeeded in localizing nucleic acids intracellularly, in both nuclei and cytoplasm. It had been known for a long time that nucleic acids occur in the chromosomes, the site of genetic information. It was therefore suggested as early as 1940 by Jansen (K-29) and by Friedrich-Freksa (K-30) that nucleic acids act as templates for the biosynthesis of proteins. It was impossible at that time to prove or disprove this view since nothing was known on the sequence of nucleotides in nucleic acids, nor on the sequence of amino acids in any protein. Ten years later, Chargaff (K-31) made the important discovery that the ratios adenine/thymine and guanine/cytosine in DNA were always 1.0 and that also in RNA the ratio of 6-amino to 6-keto groups in the bases was close to 1.0. The first X-ray diffraction analyses (K-32,33) of DNA revealed that its structure was that of long chains, possibly helical like those discovered earlier in proteins (Chapter VII, B). These discoveries were interpreted in an ingenious manner by Watson and Crick (K-34) who attributed the base ratio of unity to the formation of hydrogen bonds (a) between adenine and thymine residues and (b) between guanine and cytosine residues in two antiparallel DNA strands of a twin DNA helix, and suggested that this helix was the carrier of genetic information in the chromosomes.

The dominant role of the nucleic acids was indicated by their role in the transformation phenomenon and in the infection of bacterial cells by bacteriophages. Both phenomena have been discussed earlier in Section I of this chapter. Both strongly support the view that the transfer of DNA from one organism to the other involves penetration of the DNA molecules of the donor into the acceptor organism where the foreign DNA takes over the role of the genetic elements (K-35 to 38). However, no *direct* influence of DNA on protein formation was detected. Most observations pointed to a role of RNA in protein biosynthesis. As a consequence of these observations, it was suggested that DNA governed the formation of RNA, and RNA that of the proteins. The mechanism of these controls has been discussed in the Sections E, F, and I, of this chapter and need not be repeated here.

*The Nucleotide Code.* Shortly after the enunciation of the view that the amino acid sequence of the proteins was determined by nucleic acids, attempts were made to reconcile the 4-letter code of the nucleic acids with the fact that at least 20 code letters were required for the differentiation of 20 amino acids. On purely mathematical basis, Gamow (K-39) suggested that combinations of two or three adjacent nucleotides might be the true coding elements. The four purine and pyrimidine bases present in DNA or RNA can form 16 combinations of two and 64 combinations of three bases. This would give a sufficient number of triplets as code words for each of the amino acids. It would be easy to test this hypothesis if we were able to compare the amino acid sequence in a protein with the nucleotide sequence of the nucleic acid of the same cell or tissue. Unfortunately this is impossible at present because we have no satisfactory methods to determine the nucleotide sequence in the nucleic acids. Another difficulty arises from the heterogeneity of the nucleic acids. Most cells produce more than one type of protein and contain mixtures of nucleic acids. New methods will be necessary for the complete separation of nucleic acids.

Let us, however, assume that there indeed exists *colinearity* between the sequence of trinucleotides in a nucleic acid and the sequence of amino acids in a protein chain. Would a sequence of $n$ amino acids require a chain of $n$ trinucleotides, i.e., $3n$ mononucleotides, or would the trinucleotides overlap? In the latter case, $n$ amino acids would require only $n + 2$ mononucleotides. Some investigators have proposed an overlapping code (K-40). If the code were overlapping, each nucleotide would be part of three overlapping trinucleotides. Changes in a single nucleotide would therefore affect three amino acids. The investigation of different types of hemoglobins has revealed that in most of them only a *single* amino acid is replaced by another amino acid (see Chapter XI, G). Deletion of one or of two adjacent nucleotides causes mutation whereas deletion of a third neighboring nucleotide restores the original state by bringing it back into register. On the basis of this and other evidence, Crick *et al.* (K-41) conclude that the code consists of triplets or, less probably, groups of six nucleotides, which *do not overlap*. It is most probably a *degenerate code*. This means that some of the amino acids are coded by more than one triplet. Some of the triplets may not act as code words. They are called nonsense triplets. Those which have sense have been called *codons* (K-60).

It can easily be seen that two adjacent triplets may contain more information than that needed for two adjacent amino acids. For example, if UUG and UCA were the codes for valine and threonine, then the nucleotide sequence U*UGUCA* would also include the triplet G*UC* which might be the code for arginine. One would then be dealing with an overlapping code. Overlapping of this type can be avoided by excluding some of the triplets. A code, in which no overlapping is possible, has been called a comma-less code (K-42). It cannot contain more than 20 out of the possible 64 triplets, and is sometimes called a "dictionary." Only five of the possible comma-free codes contain the "magic number" of 20 triplets (K-42). The others have less than 20 triplets.

*The Role of DNA, sRNA, and mRNA.* Although we cannot yet answer the question why and how the trinucleotide residues of polynucleotides stimulate the incorporation of the appropriate amino acids, the discovery of the specific stimulating action of synthetic polynucleotides, which was made while this book was written had an enormous impact on our views on protein biosynthesis. It is too early to evaluate fully the importance of these experiments. However, there is no doubt that they strongly support the claim that nucleic acids supply information for the alignment of the amino acids in definite sequences. The picture which emerges from this work and from the results described in the preceding sections of this chapter is the following.   (1) DNA can act as a template for the formation of complementary DNA or RNA strands; RNA formed in this manner has been called messenger RNA or informational RNA (Section I); RNA of this type occurs also in some viruses where it seems to replace DNA. (2) The amino acids are activated by ATP in the presence of specific-activating enzymes and sRNA, are coupled to sRNA, and are then transferred to the ribosomes (see Section F). (3) The assembly of the amino acids in the nascent peptide chain takes place in the ribosomes and is directed there by the action of messenger RNA.

There are still many unsolved problems in this fascinating picture. Some of them have been discussed earlier in this section, for instance, the possibility that amino acid incorporation into proteins in some systems may take place without the participation of sRNA and an aminoacyl-sRNA synthetase (E-18), or that it may be preceded by the formation of aminoacyl-lipid complexes (E-20,21), or of nucleopeptides (H-4 to 6). Some of the other problems will be discussed in the following paragraphs.

*Differences in Base Ratios.* One of the principal difficulties was the quite different base ratio in RNA as compared to DNA of the same cell. However, if we keep in mind the high molecular weight of DNA, which may amount to several millions, and the much smaller molecular weight of sRNA or the informational RNA, it is clear that the portion of the DNA molecule which is complementary to sRNA or mRNA would be only a very small part of the large DNA molecule. The complementary base ratio of these two small pieces would be overshadowed by different base ratios in other parts of the DNA molecule. It has been shown by Spiegelman *et al.* (K-61) that hybridization of sRNA or mRNA with small portions of the coding DNA takes place. Hence we have every reason to assume that the base ratio in the small RNA molecules is exactly complementary to that of the combining piece of the giant DNA molecule.

*The Dilemma of Nonsense Chains in DNA.* If duplication, as postulated, takes place in such a manner that the base pairs in the two strands are A-T, T-A, C-G, and G-C, *the two antiparallel strands will usually be different.* If we show the phosphate residues by arrows which point from the 3'-hydroxyl group of one nucleotide to the 5'-hydroxyl residue of the adjacent nucleotide, duplication of a chain of the composition $A \rightarrow G \rightarrow G \rightarrow T \rightarrow C$ will yield the complementary chain $T \leftarrow C \leftarrow C \leftarrow A \leftarrow G$ which, since its orientation is antiparallel to the mother chain, can also be written as $G \rightarrow A \rightarrow C \rightarrow C \rightarrow T$. This is quite different from the mother chain. We still do not know which of the two chains acts as a template for the production of mRNA and, subsequently, for the assembly of the amino acids in the nascent peptide chain. It is usually claimed that one of the chains is a nonsense chain. However, the triplets UUU and AAA in the synthetic polynucleotides poly U and poly A cause the incorporation of the amino acids phenylalanine and lysine, respectively. Since UUU is complementary to the triplet dAdAdA in the DNA molecule (dA = deoxyadenylic acid), and AAA complementary to TTT, it is clear that *both* chains of the DNA double strand contain triplets which have sense and can act as code words.

It is easy to construct a DNA molecule which on duplication will reproduce an *identical antiparallel chain*. The prerequisit for this is that one-half of the DNA chain must be a complementary replica of the other half. Taking, as an example, the chain $A \rightarrow G \rightarrow G \rightarrow T \rightarrow C$ shown above, and assuming that it folds back on itself and forms the chain

$$A \rightarrow G \rightarrow G \rightarrow T \rightarrow C$$
$$\downarrow$$
$$T \leftarrow C \leftarrow C \leftarrow A \leftarrow G$$

which we can write as $A \rightarrow G \rightarrow G \rightarrow T \rightarrow C \rightarrow G \rightarrow A \rightarrow C \rightarrow C \rightarrow T$, we find that replication of the latter chain yields an *identical* chain. In this single chain the ratio A/T and G/A is equal to 1.0. Since these ratios in single-strand DNA are different from 1.0, the composition of the single-strand DNA cannot be of this type. In the double-strand complementary DNA preparations, the ratios A/T and G/C are obviously 1.0. The assumption of a chain which replicates itself in such a manner that the antiparallel replica is identical with the mother chain, would eliminate the difficulty of differentiating between a sense and a nonsense chain. Although genetic arguments have been advanced against the idea of such self-replicating chains (K-51), recent X-ray work indicates that sRNA indeed coils back on itself as a double helix with complementary base sequences in its two halves (K-62). We may designate such a chain which is complementary to itself as *homocomplementary* in contrast to all other *heterocomplementary* chains which are complementary to but not identical with their parent chain.

*Coding by Synthetic Polynucleotides.* The strongest support for the coding role of the nucleic acids came from recent startling experiments of Nirenberg *et al.* (K-43) which were confirmed and extended by Ochoa *et al.* (K-44). In these experiments, the incorporation of C[14]-amino acids into protein in a cell-free system was investigated. The system (K-45) consisted of the ribosomes of *E. coli* and the supernatant fluid, but was free of cell debris which sediment at 30,000 *g*. The ability to incorporate amino acids into protein was not lost when this system was frozen. The incorporation of amino acids into proteins depended, in this system, on the presence of ribosomal RNA which was called "template RNA" but may be identical with the messenger RNA from *E. coli* (Section I of this chapter). The template RNA could be replaced by RNA from yeast or tobacco mosaic virus, *but also by polynucleotides* prepared *in vitro* by the enzymatic polymerization of nucleotides.

When polyuridylic acid was used instead of the template RNA, only the incorporation of phenylalanine was stimulated; the product of this reaction was identified as polyphenylalanine. Analogous experiments with various mixed uridine polynucleotides (see Table XVI-1) revealed that each of them preferentially stimulated the incorporation of only one or a few amino acids in addition to phenylalanine. Since at the present time, no methods are available to prepare polynucleotides of a definite nucleotide sequence, the composition of the polynucleotides is expressed by indicating only their over-all content of A (adenylic acid), U (uridylic acid), G (guanylic acid), and C (cytidylic acid). Thus the symbol poly UG$_2$ indicates that the polynucleotide consists of approximately one-third of uridylic and two-thirds of guanylic acid residues. For statistical reasons, it can be assumed that it contains predominantly the triplets of the formula UG$_2$, namely UGG, GUG, and GGU, and to a lower percentage triplets of the over-all formulas U$_3$, U$_2$G, and G$_3$. Nothing is known about the precise frequency and the sequence of these in the poly UG$_2$ molecule.

TABLE XVI-1
AMINO ACIDS AND THEIR CODING RNA TRIPLETS[a]

| Amino acids | Triplets | Amino acids | Triplets |
|---|---|---|---|
| Alanine | CUG,CAG,CCG,(CG) | Leucine | UAU,UUC,UGU,(UU) |
| Arginine | GUC,GAA,GCC,(GC) | Lysine | AUA,AAA,(AA) |
| Asparagine | UAA,CUA,CAA,(CA,AA) | Methionine | UGA |
| Aspartic acid | GUA,GCA,(GA) | Phenylalanine | UUU,(UU) |
| Cysteine | GUU | Proline | CUC,CCC,CAC,(CC) |
| Glutamic acid | AUG,AAG,(AG) | Serine | CUU,ACG |
| Glutamine | ? ,AGG,AAC | Threonine | UCA,ACA,CGC,(CA) |
| Glycine | GUG,GAG,GCG,(GG) | Tryptophan | UGG |
| Histidine | AUC,ACC,(AC) | Tyrosine | AUU |
| Isoleucine | UUA,AAU,(UU) | Valine | UUG |

[a] From (K-64).

We know, however, from the experiments of Nirenberg (K-43) and Ochoa (K-44) and their co-workers that poly UG$_2$ stimulates the incorporation of glycine and tryptophan. It can be suspected, therefore, that one of the three triplets UGG, GUG, and GGU is the code for glycine and another the code for tryptophan. The third may be a nonsense triplet.

The first set of coding triplets or codons was based on the observation that incorporation of each of the 20 amino acids required the presence of the nucleotide U (uridylic acid) in the synthetic polynucleotides (K-43,44). Shortly thereafter it was found, however, that some of the U-free polynucleotides are also able to act as codons (K-63). Thus, poly CA causes incorporation of proline and, to a lesser extent, of threonine and histidine. The present picture emerging from these and similar investigations (K-64,65) is shown in Table XVI-1.

The *first triplet* recorded in Table XVI-1 for each of the amino acids is the triplet which contains uridylic acid. The only amino acid for which such a triplet has not yet been found is glutamine. The original finding of U in all triplets led to the view that one of the U nucleotides in each triplet was necessary as such and not involved in the coding process, and that the code was indeed a doublet code (K-66). The doublets are shown in parentheses in Table XVI-1.

It can be seen from the table that the code is highly degenerate. Thus both phenylalanine and leucine are coded by poly U. If both amino acids are present, phenylalanine is incorporated predominantly. In the absence of phenylalanine leucine is incorporated. The degeneracy of the code has been further demonstrated by the isolation of two different sRNA preparations, both of them activating leucine; one of these preparations activated for the $U_2G$ code, the other for $U_2C$ (K-67). It is evident from all these data that the relation between amino acids and their coding triplets is more complicated than originally assumed. It may depend not only on the sequence of nucleotides in the triplet or doublet but also on the secondary structure of the polynucleotide chain and its conformation.

One of the strongest arguments in favor of the code is based on the fact that most replacements of amino acids in mutations of hemoglobin and other proteins require the change of only *one* of the three nucleotides present in the coding triplet (K-47). Thus, the change from valine to isoleucine in hemoglobin would require a change from $U_2G$ to $U_2A$. Transitions of this type have been designated as *primary transitions* (K-48). However, the exchange of arginine (code GUC) for lysine ($UA_2$) in vasopressin (Chapter XIII, D) does not follow this rule. On the other hand, poly U causes the formation of polyphenylalanine not only in the cell-free system of *E. coli* but also in rabbit reticulocytes (F-17). Moreover, a protein similar to the TMV protein is formed when the RNA of tobacco mosaic virus is added to the cell-free system of *E. coli* (K-46). Experiments of this type indicate that the code may be of universal validity.

The first insight into the *sequence* of nucleotides in a triplet was gained by the observation that the trinucleotide AUU, if it was attached to poly U, caused the incorporation of tyrosine; similarly GUU in an analogous experiment led to the incorporation of cysteine (K-64). If it is further assumed that the observed mutational changes of amino acids in hemoglobin or TMV protein involve the exchange of only *one* of the three nucleotides of a triplet, a complete set of sequences for each of the coding triplets can be derived (K-68). In Table XVI-1 these sequences are shown by the underlined values. The other triplets shown in the table do not indicate any particular sequence but indicate merely the frequency of the recorded mononucleotides in the coding polynucleotide. Thus CCG means a polynucleotide in which approximately one third of the nucleotides consists of guanylic acid and two thirds of cytidylic acid residues.

Although the results obtained with polynucleotides are of extreme importance, and although they open up new avenues to the elucidation of protein biosynthesis, we still do not understand the underlying mechanism. We cannot yet answer the question why $U_3$ should stimulate the incorporation of phenylalanine and not that of say glycine or arginine.

*The Central Problem.* The principal difficulty, as mentioned repeatedly in the preceding paragraphs, arises when we try to understand the coordination of each of the amino acids with a specific trinucleotide. It is not difficult to visualize highly specific affinity between DNA and mRNA, or between mRNA and sRNA. In both instances we may be dealing with complementary nucleotide sequences in parts of these molecules. Even if only a few trinucleotides were complementarily constructed, the mutual forces of attraction, caused by hydrogen bonds between A and T (or U) and between G and C should be sufficient to explain the specificity of mutual attraction and the *transcription* of the base sequence of DNA into the base sequence of RNA. It is much more difficult to visualize the *translation* of the RNA transcript into the amino acid sequence of the growing peptide chain, particularly in view of the fact that all sRNA molecules combine with the aminoacyl residues by means of the same bond, namely an ester bond formed by the amino acid and the terminal cytidylcytidyladenyl residue of sRNA.

Since there is a specific sRNA for each of the amino acids, there must be differences in the nonterminal portion of the sRNA molecules. These differences in the nucleotide sequence should then determine which of the amino acids is bound to the terminal CCA residue. In this process, like in all other steps of protein biosynthesis, proteins are involved, namely the amino acid-activating enzymes which seem to be specific for each of the amino acids. *It is this step, the combination of the aminoacyl residue with a specific sRNA molecule, in which,* according to the view prevailing at present, *the nucleotide code is translated into the amino acid code* since the further fate of the aminoacyl-sRNA complex depends on its sRNA moiety and not on the aminoacyl residue. It will be recalled that the alanyl-sRNA complex obtained from cysteyl-sRNA by reduction with Raney nickel is incorporated into the cysteine sites and not into alanine sites (F-19). The transfer RNA (sRNA) which is specific for cysteine delivers its amino acid load as directed by the template RNA(mRNA). The mutual "recognition" of the sRNA and mRNA is evidently based on the mutual combination of complementary trinucleotide sequences (K-52). Hence, it is not too difficult to visualize a mechanism in which mRNA would line up the aminoacyl-sRNA complexes in the sequence in which they should deliver their aminoacyl residues to the growing peptide chain.

It is easy to postulate that each amino acid combines with a specific

sRNA molecule. However, formidable difficulties arise when we attempt
to understand this reaction in terms of molecular specificity. How should
we visualize a nucleotide sequence in the sRNA molecule which would
direct only 1 of the 20 amino acids into the terminal cytidylcytidyladenyl
trinucleotide? How can we imagine such a sequence which would exclude
all but 1 amino acid from the formation of an ester bond with the CCA
trinucleotide? Since specific synthetases are involved in the combination
of the amino acid with sRNA, it is reasonable to assume that both the
enzyme and sRNA are involved in the specific formation of the ternary
E(aminoacyl-sRNA) complex, and that a special conformation of the
trinucleotides rather than the chemical properties of the constituent
mononucleotides is responsible for their specific coding action. It would be
easier to understand the different coding action of the different triplets if
there were any chemical or physical relations between their properties
and those of the activated and incorporated amino acids. We do not know
of any relation of this type. Moreover, the degeneracy of the code (see
Table XVI-1) makes such a relation extremely improbable. We may ask
whether there is any other chemical or biochemical relation between
nucleotides and amino acids. Before answering this question it may be
useful to discuss briefly chemical and biochemical differences between
the four mononucleotides.

*The Chemical Specificity of Nucleotides.* The code listed above does not reveal
any connection between the acidic and basic properties of nucleotides and amino
acids. The basic properties of adenine, guanine, cytosine, uracil, and thymine are
weak. In the nucleotides the basic character is more than compensated by the
acidic phosphoric acid residues which confer on the nucleic acids their acidic
properties. Therefore, at pH 7 all nucleotides and nucleic acids are present as
anions of alkali salts. In contrast to the mono- and polynucleotides, the amino
acids display a wide array of acidic and basic groups. Their isoelectric points
range from pH 3.0 to 10.8 (see Table VI-1). The monotony of the nucleotides
does not make it probable that they should differ from each other in their affinity
for the different amino acids. In view of the chemical properties of the nucleotides, one would rather expect that all of them would have the highest affinity
for arginine, which is the most basic amino acid, and the lowest affinity for
glutamic and aspartic acid. Since we do not know of such a correlation, the conformation rather than the ionization and polarity of the trinucleotides may govern
their relations to the amino acids. Unfortunately, nothing is known on their
conformation nor on any close fit between specific pairs of trinucleotides and
amino acids.

*Biochemical Specificity of Nucleotides.* In biochemical reactions the nucleotides,
in spite of their similarity, have revealed a surprising specificity as coenzymes in
metabolic reactions. The important role of ATP in oxidative phosphorylation
has been known for many years. More recently, Leloir (K-49) discovered the
paramount role of uridylic acid residues in the transfer of carbohydrate derivatives. Kennedy (K-50) found a similar role of cytidylic acid in the transfer of

lipid residues. Finally, guanosine triphosphate, as mentioned earlier, is involved in the activation of amino acids. At present it is still an enigma why each of the four nucleotides has a specific role as a coenzyme and why they cannot replace each other in all instances. Most probably, uridylic acid fits somehow to the surface of carbohydrate residues and/or to that or their enzymes, and cytidylic acid to the analogous substrates and/or enzymes involved in lipid transfer.

Since the relatively simple mononucleotides act as specific coenzymes in definite metabolic reactions (see preceding paragraph) one might reasonably assume that the trinucleotides have a similar more specific role, even if they are present as trinucleotide residues in a polynucleotide macromolecule. According to this view we would have to consider the coding RNA chain as a chain of coenzymes for those enzymes which catalyze the activation of amino acids by sRNA and their transfer to the ribosomal sites of protein synthesis (see Section F of this chapter). Although this view does not clarify the mechanism by means of which information is transmitted from the nucleotide chain to the peptide chain, it points to analogous coenzyme functions of nucleotides in carbohydrate and lipid metabolism, and may thus facilitate our understanding of the coding process.

The assumption that the conformation of the trinucleotides may be more important than their chemical composition is supported by the fact that the code shown in Table XVI-1 is degenerate. Thus the incorporation of leucine is stimulated not only by a polynucleotide of the average composition $U_2A$, but also by $U_2C$ and $U_2G$. Similarly, there are multiple code words for some of the other amino acids. The degeneracy of the code shows clearly that there cannot be a one-to-one relation between amino acids and trinucleotides. This is also demonstrated by the fact that poly U, as mentioned earlier, acts as a code not only for the incorporation of phenylalanine but also for that of leucine. Since the polynucleotide chain is flexible, the conformation of the nucleotide triplets may vary to a certain extent and may depend on its environment.

It may be affected particularly by basic groups which would neutralize the phosphoryl residues and cause changes in the "microscopic ionization constants" (see Chapter VI, B). The interpretation of the coding process is complicated by the observation that a large part of the coding polynucleotides is degraded to small oligonucleotides before the incorporation of the amino acids begins (K-70). The life time of the mRNA molecules seems to be only 2 or 3 minutes (K-71).

*The Central Dogma.* As mentioned earlier, it has been claimed that information goes from RNA to protein but not in the reverse direction. According to this "Central Dogma" (K-59), the proteins act merely as catalysts and not as templates. In an earlier part of this discussion, the role of templates, primers, and catalysts has been discussed, and it has been shown that we cannot always clearly differentiate these factors. We know that proteins are indispensable for replication, duplication, and polymerization. Protein formation precedes the multiplication of the phage DNA in the invaded bacteria (K-53 to 56). All this points to the fact that an "early protein" is required before the phage DNA can multiply. Could this protein not have the role of a template? Is information, as claimed by the Central Dogma, transmitted only from RNA to protein and not also in the reverse direction? It seems to the author strange indeed that nature should not have made use of the transfer of information from protein to protein and also from protein to nucleic acids.

The view that proteins may be active as templates may seem a heresy in the present era of the Central Dogma. It may appear less heretic when we remember that most of the nucleic acids *in vivo* are bound to an excess of protein. In Chapter XI, I, it was pointed out that we still do not know whether the "nucleoproteins" isolated from tissues are artifacts formed during the preparation from protein and nucleic acid, or whether they exist in the cells as such. If they are present in the cells, we would like to know whether there is a definite correlation between the peptide chain and the adjacent nucleotide chain. Nothing is known in this respect. However, we know very well that proteins and nucleic acids combine with each other *in vitro*. It is reasonable to assume that combination will also occur *in vivo* and that this combination will affect the conformation of the nucleotide triplets. In this manner proteins might affect the coding process.

The assumption of a strict colinearity of nucleotide and amino acid sequences has been criticized as an oversimplification according to which life would be merely an expression of the chemistry of nucleic acids (K-57). In a lighter vein, an audience recently was warned that nucleic acids alone cannot determine whether an organism becomes a *coli* bacillus, a mouse, an elephant, or a scientist. It seems to the author quite possible that information may be transmitted not only from nucleic acids to proteins but also in the opposite direction, from proteins to nucleic acids and to proteins. This idea is supported by the action of antigens in antibody formation. The high affinity of antibodies to the homologous antigens is certainly based on mutual complementariness which is proved beyond any doubt by the binding to antibody of the determinant groups of the antigen and of small hapten molecules of the same structure. It is easy to understand the mutual complementariness of antigen and antibody if we assume that the antigen acts as a template and that the peptide chains of the antibody molecule are folded in such a manner that complementariness results. It would be much more difficult to imagine a mechanism in which information from the antigen would pass first to the nucleic acids and from these to the nascent antibody molecule (see Chapter XV).

Whatever the mechanism of antibody formation may be, it is clear that the information for their specific structure is supplied by the antigen molecule which may be a protein, a polysaccharide, or some other molecule with suitable determinant groups and with the ability to interfere with $\gamma$-globulin production. If the folding of $\gamma$-globulins is affected by the administration of antigens, how will this process proceed in the absence of antigens? We do not know the answer to this question, but it is reasonable to assume that the folding and the final shape of the $\gamma$-globulin molecule will then depend on other components in the cell, proteins, polysaccharides, or nucleic acids which may act as "normal templates" for the formation of $\gamma$-globulin molecules. Although we may, for the sake of simplification, speak of two phases of protein biosynthesis, the formation of the peptide chain and its folding, the two processes may occur almost simultaneously. It is imaginable that interference of a template with the folding pattern may affect the sequential pattern and thereby may prevent or favor the incorporation of certain amino acids into particular geometrical patterns of the three-dimensional conformation of the globular molecule.

The immunochemical observations show clearly that not all information required for the biosynthesis of proteins is supplied by nucleic acids, and that proteins and other substances may act as templates. Life may then be more than merely the "expression of the chemistry of nucleic acids" (K-57).

Admittedly, we ended up in speculations. However these may be permissible or even necessary in the field of protein biosynthesis in which, according to one of the investigators "workers are on a crest of a wave of discovery, speculation, reviewing, and frantic exploitation unparalleled in the history of biological research" (K-58) and about which, a few years ago, the following statement was made: "It is remarkable that one can formulate principles such as the Sequence Hypothesis and the Central Dogma, which explain many striking facts and yet for which proof is completely lacking" (K-59).

In view of these statements it may be useful to keep in mind that the principal result of the important discoveries described in the preceding paragraphs is the recognition that the nucleotide sequence in polynucleotides acts as a code which is translated into a definite sequence of amino acids in the growing peptide chain, but that we are still completely ignorant concerning the mechanism of this translation. This was clearly expressed by one of the principal contributors to the coding problem who in a survey lecture illustrated the coding polynucleotide by a straight chain of mononucleotides, the resulting peptide chain by a parallel chain of amino acid residues, and then inserted on the blackboard a mysterious rectangle between the two chains. The rectangle, as he explained, was a "black box" which contained the unknown mechanism for the translation of the nucleotide code into the amino acid sequence (K-69). In other words, we see the phenomena but do not yet understand the mechanism by means of which they are produced. We do not yet know the content of the "black box."

## REFERENCES

### Section A

**A-1.** *Brookhaven Symposia in Biol.* 12.  **A-2.** R. C. J. Harris, ed., "Protein Biosynthesis." Academic Press, New York, 1961.  **A-3.** *Intern. Congr. Biochem. 5th Congr., Moscow* (1961) Symposium No. 2.  **A-4.** "Solvay Conference on Chemistry. Nucleoproteins" (11th Conference). Wiley (Interscience) New York, 1959.  **A-5.** *Cold Spring Harbor Symposia Quant. Biol.* 26 (1961).  **A-6.** H. Chantrenne, "Biosynthesis of Proteins," Pergamon Press, New York, 1961.  **A-7.** J. S. Fruton, *in* "The Proteins" (H. Neurath, ed.), 2nd edition. Academic Press, New York, in press.  **A-8.** D. M. Greenberg, ed., *in* "Metabolic Pathways," Vol. 2, p. 173. Academic Press, New York, 1960.  **A-9.** A. Meister, "Biochemistry of the Amino Acids." Academic Press, New York, 1957.

### Section B

**B-1.** R. H. Burris and P. W. Wilson, *Ann. Rev. Biochem.* 14: 685(1945).  **B-2.** K. Linderstrøm-Lang and C. F. Jacobsen, *Compt. rend. trav. lab. Carlsberg, Sér. chim.* 24: 1(1943).  **B-3.** H. Lineweaver and S. R. Hoover, *JBC* 137: 325(1941).  **B-4.** E. Cherbuliez and K. H. Meyer, *Compt. rend. trav. lab. Carlsberg, Sér. chim.* 22: 118 (1938).  **B-5.** F. Haurowitz *et al.*, *JBC* 157: 621(1945).  **B-6.** E. K. Rideal, *Proc. Roy. Soc.* B116, 200(1934); J. H. Schulman and E. K. Rideal, *BJ* 27: 1581(1933).  **B-7.** O. K. Behrens and M. Bergmann, *JBC* 129: 587(1939).  **B-8.** W. C. Rose *et al.*, *JBC* 143: 115; 146: 683(1942); 148: 457(1943); *Proc. Am. Phil. Soc.* 91: 1(1947).  **B-9.** D. M. Greenberg, ed., "Metabolic Pathways," 2 vols. Academic Press, New York, 1960; A. Meister, "Biochemistry of the Amino Acids." Academic Press, New York, (1957).  **B-10.** S. C. Madden *et al.*, *J. Exptl. Med.* 81: 439(1944).  **B-11.** R. Gingras *et al.*, *Rev. can. biol.* 6: 802(1947).  **B-12.** A. Roche, *Bull. soc. chim. biol.* 16: 270 (1934); M. Florkin and G. Duchateau, *Acta Biol. Belg.* 2: 219(1942).  **B-13.** A. A. Albanese, *Abstr. 114th Meeting Am. Chem. Soc.* C37(1948).  **B-14.** Y. Kotake and S. Goto, *Z. physiol. Chem.* 270: 48(1941).  **B-15.** R. J. Block and D. Bolling, *Arch. Biochem.* 3: 217(1944).

### Section C

**C-1.** K. Linderström-Lang, "Proteins and Enzymes." Stanford Univ. Press, Stanford, California, 1952.  **C-2.** H. Leuchs, *Chem. Ber.* 39: 857(1906).  **C-3.** E. Katchalsky and M. Sela, *Adv. in Protein Chem.* 13: 244(1958).  **C-4.** E. Katchalsky *et al.*, *JACS* 69: 1551(1947).  **C-5.** W. Hanby *et al.*, *Nature* 161: 132(1948); 163: 483(1949).  **C-6.** A. Berger *et al.*, *JACS* 78: 4483(1956).  **C-7.** A. Katchalsky and M. Paecht, *JACS* 76: 6042(1954).  **C-8.** S. W. Fox *et al.*, *JACS* 77: 1048(1955); 82: 3745 (1960); *Science* 132: 200(1960).  **C-9.** S. Akabori, *Kagaku* 25: 54(1958); cited in *Angew. Chem.* 70: 52(1958).  **C-10.** S. E. Bressler *et al.*, *Biokhimiya* 20: 463(1955).

### Section D

**D-1.** A. Dobry *et al.*, *JBC* 195: 149(1952).  **D-2.** M. Bergmann and J. S. Fruton, *JBC* 124: 321(1938).  **D-3.** F. Janssen *et al.*, *JACS* 75: 704(1953).  **D-4.** M. J.

Mycek and J. S. Fruton, *JBC* 226: 165(1957). **D-5.** J. S. Fruton *et al.*, *JBC* 226: 173 (1957); 218: 59(1956). **D-6.** J. S. Fruton *et al.*, *JBC* 195: 645(1952); 204: 891(1953); *Biochemistry* 1: 19(1962). **D-7.** H. Tauber, *JACS* 71: 2952(1949); 73: 1288(1951); 74: 847(1952). **D-8.** M. Brenner *et al.*, *Helv. Chim. Acta* 33: 568(1950); 40: 937(1957). **D-9.** W. P. Jencks *et al.*, *JBC* 235: 3608(1960). **D-10.** J. Horowitz and F. Haurowitz, *JACS* 77: 3138(1955); *BBA* 33: 231(1959). **D-11.** Th. Wieland *et al.*, *Ann. Chem.* 633: 185(1960); 651: 172(1962).

*Section E*

**E-1.** H. Chantrenne, *BBA* 4: 484(1950). **E-2.** F. Lipmann *et al.*, *JACS*, 74: 2384 (1952). **E-3.** M. Hoagland *et al.*, *BBA* 24: 215; 26: 215(1957); *JBC* 231: 241(1958); E. B. Keller and P. C. Zamecnik, *JBC* 221: 45(1956). **E-4.** K. Ogata *et al.*, *BBA* 25: 659; 26: 656(1957). **E-5.** P. Castelfranco *et al.*, *JACS* 80: 2335(1958); K. Moldave *et al.*, *JBC* 234: 841(1959). **E-6.** P. Berg *et al.*, *JBC* 236: 1726ff.(1961). **E-7.** H. G. Zachau *et al.*, *Proc. Natl. Acad. Sci. U.S.* 44: 885(1958). **E-8.** P. C. Zamecnik *et al.*, *J. Cellular Comp. Physiol.* 47: Suppl. 1, p. 81(1956). **E-9.** R. J. Mans and G. D. Novelli, *BBA* 50: 287(1961). **E-10.** S. Lacks and F. Gros, *J. Mol. Biol.* 1: 301(1959). **E-11.** M. L. Stephenson and P. C. Zamecnik, *Proc. Natl. Acad. Sci. U.S.* 47: 1627 (1961). **E-12.** R. W. Holley *et al.*, *JACS* 83: 4861(1961). **E-13.** G. C. Webster, *BBA* 49: 151(1961). **E-14.** P. Berg, *Ann. Rev. Biochem.* 30: 293(1961). **E-15.** J. W. Davie *et al.*, *ABB* 65: 21(1956). **E-16.** R. Rendi and S. Ochoa, *Science* 133: 1367(1961). **E-17.** P. Berg and E. J. Ofengand, *Proc. Natl. Acad. Sci. U.S.* 44: 78(1958). **E-18.** M. Beljanski *et al.*, *Compt. rend.* 250: 624(1960); *Proc. Natl. Acad. Sci. U.S.* 44: 494, 1157(1958); *BBA* 56: 559(1962). **E-19.** I. Suzuke *et al.*, *J. Biochem. (Tokyo)* 49: 81(1961). **E-20.** R. W. Hendler, *JBC* 234: 1466(1959); *Nature* 193: 821(1961). **E-21.** G. D. Hunter and R. A. Goodsal, *BJ* 78: 564(1961). **E-22.** J. Apgar *et al.*, *JBC* 237: 796(1962). **E-23.** P. C. Zamecnik, *BJ* 85: 237(1962). **E-24.** R. Rendi and S. Ochoa, *JBC* 237: 3707(1962).

*Section F*

**F-1.** R. C. J. Harris, ed., "Protein Biosynthesis." Academic Press, New York, 1961. **F-2.** J. W. Littlefield *et al.*, *JBC* 217: 111(1955); E. B. Keller *et al.*, *J. Histochem. and Cytochem.* 2: 378(1954). **F-3.** M. L. Petermann and M. G. Hamilton, *in* "Protein Biosynthesis" (R. C. J. Harris, ed.), p. 235. Academic Press, New York, 1961. **F-4.** M. G. Hoagland *et al.*, *JBC* 231: 241(1958); A. Korner, *BJ* 81: 168(1961). **F-5.** E. H. Allen and R. S. Schweet, *BBA* 44: 55(1960). **F-6.** F. H. C. Crick, *Brookhaven Symposia in Biol.* 12: 35(1959); W. Zillig *et al.*, *Z. physiol. Chem.* 318: 100(1960). **F-7.** K. McQuillen, *in* "Protein Biosynthesis" (R. C. J. Harris, ed.), p. 263. Academic Press, New York, 1961. **F-8.** A. N. Belozersky and A. S. Spirin, *Nature* 182: 111 (1958). **F-9.** J. L. Simkin and T. S. Work, *BJ* 65: 307; 67: 617(1957). **F-10.** G. Webster and J. B. Lingrel, *in* "Protein Biosynthesis" (R. C. J. Harris, ed.), p. 301. Academic Press, New York, 1961; J. T. Lett and W. N. Takahashi, *ABB* 96: 569 (1962). **F-11.** B. A. Askonas, *in* "Protein Biosynthesis" (R. C. J. Harris, ed.), p. 363. Academic Press, New York, 1961. **F-12.** R. S. Schweet *et al.*, *Proc. Natl. Acad. Sci. U.S.* 44: 1029(1958); 46: 1030(1960); *JBC* 237: 760(1962). **F-13.** H. M. Dintzis, *Proc. Natl. Acad. Sci. U.S.* 47: 247, 899(1961). **F-14.** S. Spiegelman *et al.*, *Proc: Natl. Acad. Sci. U.S.* 47: 114, 137, 1135, 1564(1961). **F-15.** J. Kruh *et al.*, *BBA* 55: 690(1962). **F-16.** G. v. Ehrenstein and F. Lipmann, *Proc. Natl. Acad. Sci. U.S.* 47. 941(1962). **F-17.** H. R. V. Arnstein *et al.*, *Nature* 194: 1042(1962). **F-18.** R. W. Risebrough *et al.*, *Proc. Natl. Acad. Sci. U.S.* 48: 430(1962). **F-19.** F. Chappeville *et al., Proc. Natl. Acad. Sci. U.S.* 48: 1086(1962). **F-20.** J. R. Warner *et al., Science* 138: 1399(1962). **F-21.** J. L. Simkin, *BJ* 70: 305(1958).

## Section G

**G-1.** J. Bonner, *in* **F-1,** p. 323. **G-2.** V. G. Allfrey and A. E. Mirsky, *in* "Protein Biosynthesis (R. J. C. Harris, ed.), p. 49. Academic Press, New York, 1961. **G-3.** V. A. Gvostev, *Biokhimiya* 25: 920(1960). **G-4.** R. B. Khesin, *Proc. Intern. Congr. Biochem. 5th Congr., Moscow* (1961) Symposium 2, Abstr. No. 7. **G-5.** I. B. Zbarsky, *Biokhimiya* 16: 112, 390(1951). **G-6.** D. B. Roodyn *et al., BJ* 80: 9(1961). **G-7.** A. H. Gordon and J. H. Humphrey, *BJ* 78: 551(1961). **G-8.** J. H. Humphrey and B. D. Sulitzeanu, *BJ* 68: 146(1958). **G-9.** J. Kruh *et al., Compt. rend. soc. biol.* 150: 1356 (1956); *Rev. franç. études clin.* 2: 820(1957). **G-10.** H. N. Christensen *et al., JBC* 194: 41, 57(1952). **G-11.** E. Heinz, *JBC* 211: 781(1954). **G-12.** A. Lietze *et al., ABB* 76: 255(1958). **G-13.** H. Eagle *et al., JBC* 236: 2039(1961). **G-14.** H. Eagle, *JBC* 214: 839(1955). **G-15.** H. Eagle *et al., JBC* 234: 592(1959). **G-16.** S. Margen and H. Tarver, *J. Clin. Invest.* 35: 1161(1956). **G-17.** A. S. Konikova *et al., Biokhimiya* 19: 440(1954). **G-18.** M. Luck *et al., JBC* 197: 869(1952); *ABB* 73: 391(1958). **G-19.** T. Peters, Jr., *JBC* 200: 461(1953); 229: 659(1957); 237: 1186(1962). **G-20.** R. B. Khesin, *Biokhimiya* 18: 462(1957). **G-21.** P. N. Campbell, *BJ* 74: 107(1960). **G-22.** R. B. Loftfield *et al., JBC* 219: 151(1956); 231: 925(1958). **G-23.** R. J. Britten *et al., Proc. Natl. Acad. Sci. U.S.* 41: 863(1955). **G-24.** M. Zalokar, *BBA* 46: 423(1961). **G-25.** F. Haurowitz *in* "Molecular Structure and Biological Specificity" (L. Pauling and H. A. Itano, eds.), p. 18. Am. Inst. Biol. Sci., Washington, D.C., 1957. **G-26.** Z. Berankova *et al., Collection Czechoslov. Chem. Communs.* 24: 3476(1959). **G-27.** B. A. Askonas, *in* "Protein Biosynthesis" (R. C. J. Harris, ed.), p. 363. Academic Press, New York, 1961. **G-28.** P. O. P. Ts'o and A. Lubell, *ABB* 86: 19(1960). **G-29.** S. Fleischer *et al., Proc. Soc. Exptl. Biol. Med.* 101: 860(1959); H. Walter *et al., JBC* 224: 107(1957). **G-30.** S. Fleischer *et al., JBC* 234: 2717(1959). **G-31.** L. L. Miller, *J. Exptl. Med.* 90: 297(1949). **G-32.** H. Walter and F. Haurowitz, *Science* 128: 140 (1958). **G-33.** R. Schoenheimer, "The Dynamic State of Body Constituents." Harvard Univ. Press, Cambridge, Massachusetts, 1932. **G-34.** R. C. Thompson and J. E. Ballou, *JBC* 223: 795(1956). **G-35.** K. McQuillen *et al., Proc. Natl. Acad. Sci. U.S.* 45: 1437(1959).

## Section H

**H-1.** M. V. Simpson and S. F. Velick, *JBC* 208: 61(1954); 216: 179(1955). **H-2.** T. S. Work *et al., BJ* 58: 326(1954); 61: 105(1955); 63: 69(1956). **H-3.** R. L. M. Synge and M. A. Youngson, *BJ* 78: 708(1961). **H-4.** J. Cerna *et al., Collection Czechoslov. Chem. Communs.* 26: 1212(1961). **H-5.** E. Ito and J. L. Strominger, *JBC* 235: PC 5(1960). **H-6.** P. Szafranski and M. Bagdasarian, *Nature* 190: 719(1961). **H-7.** M. Vaughan and C. B. Anfinsen, *JBC* 211: 367(1954). **H-8.** H. M. Dintzis, *Proc. Natl. Acad. Sci. U.S.* 47: 247(1961). **H-9.** R. Schweet *et al., Proc. Natl. Acad. Sci. U.S.* 44: 1029 (1961); 46: 1030(1960); *JBC* 237: 760(1962). **H-10.** J. P. Burnett, Jr., and F. Haurowitz, *Z. physiol. Chem.*, in press (1963). **H-11.** K. McQuillen, *in* "Protein Biosynthesis" (R. C. J. Harris, ed.), p. 260. Academic Press, New York, 1961. **H-12.** D. Schwartz, *Proc. Natl. Acad. Sci. U.S.* 48: 750(1962).

## Section I

**I-1.** S. Zamenhof, "The Chemistry of Heredity." Thomas, Springfield, Illinois, 1959. **I-2.** H. Ris, *in* "Chemie der Genetik." Springer, Berlin, 1959. p. 3. **I-3.** L. F.

Cavalieri and B. H. Rosenberg, *Biophys. J.* 1: 337(1961). **I-4.** H. Tabor, *Biochem. Biophys. Research Communs.* 3: 382(1960); H. R. Mahler *et al.*, *Biochem. Biophys. Research Communs.* 4: 79(1961). **I-5.** P. T. Mora *et al.*, *JBC* 237: 157(1962). **I-6.** J. Brachet, *Arch. biol.* (*Liège*) 53: 207(1941); H. Chantrenne, *Ann. Rev. Biochem.* 27: 35(1958). **I-7.** V. G. Allfrey and A. E. Mirsky, *Proc. Natl. Acad. Sci. U.S.* 45: 1325 (1959). **I-8.** A. Kornberg, *in* "Molecular Control of Cellular Activity" (J. A. Allen, ed.), p. 245. McGraw-Hill, New York (1962); R. L. Sinsheimer, *Ibid.* p. 174. **I-9.** E. Volkin and L. Astrachan, *Virology* 2: 149(1956). **I-10.** F. Gros *et al.*, *Cold Spring Harbor Symposia Quant. Biol.* 26: 111(1961). **I-11.** S. Spiegelman *et al.*, *Proc. Natl. Acad. Sci. U.S.* 47: 1135, 1564(1961). **I-12.** M. Chamberlin and P. Berg, *Proc. Natl. Acad. Sci. U.S.* 48: 81(1962). **I-13.** H. M. Schulman and D. M. Bonner, *Proc. Natl. Acad. Sci. U.S.* 48: 53(1962). **I-14.** G. Zubay, *Proc. Natl. Acad. Sci. U.S.* 48: 456 (1962). **I-15.** J. R. Fresco, *Am. Scientist* 50: 158(1962). **I-16.** S. Brenner *et al.*, *Nature* 190: 576(1961). **I-17.** A. Tissieres and J. D. Watson, *Proc. Natl. Acad. Sci. U.S.* 48: 1061(1962). **I-18.** J. Brachet, "Un Symposium sur les Proteines," p. 123. Desoir, Liège 1946. **I-19.** T. Caspersson, *Symposia Soc. Exptl. Biol.* 1: 137(1947). **I-20.** S. Lederberg and D. Mazia, *Exptl. Cell Research* 21: 590(1961). **I-21.** H. Matthaei *et al.*, *Proc. Natl. Acad. Sci. U.S.* 48: 666(1962). **I-22.** I. M. Eisenstadt *et al.*, *Proc. Natl. Acad. Sci. U.S.* 48: 652, 659(1962).

### Section J

**J-1.** D. Steinberg *et al.*, *Science* 124: 389(1956). **J-2.** D. Gross and H. Tarver, *JBC* 217: 169(1955). **J-3.** R. E. Handschumacher and A. D. Welch, *in* "The Nucleic Acids" ed. by (E. Chargaff and J. N. Davidson, eds.), Vol. 3, p. 477. Academic Press, New York, 1960. **J-4.** S. Champe and S. Benzer, *Proc. Natl. Acad. Sci. U.S.* 48: 532 (1962). **J-5.** S. Naono and F. Gross, *Compt. rend.* 250: 3889(1960). **J-6.** A. Bussard *et al.*, *Compt. rend.* 250: 4049(1960). **J-7.** J. Horowitz *et al.*, *Nature* 184: 1213(1958); *BBA* 29: 222(1958). **J-8.** C. Heidelberger *et al.*, *BBA* 20: 445(1957). **J-9.** A. Gierer and K. W. Mundry, *Nature* 182: 1457(1958). **J-10.** E. Freese, *Brookhaven Symposia in Biol.* 12: 63(1959). **J-11.** F. E. Hahn *et al.*, *BBA* 26: 469(1957). **J-12.** G. N. Cohen and F. Jacob, *Compt. rend.* 248: 3490(1959). **J-13.** H. V. Rickenberg, *Nature* 185: 240(1960). **J-14.** B. G. Haughton and H. K. King, *BJ* 80: 268(1961). **J-15.** J. Monod *et al.*, *Compt. rend.* 254: 4214(1962). **J-16.** G. N. Cohen and F. Gros, *Ann. Rev. Biochem.* 29: 525(1960). **J-17.** H. G. Vogel, *Proc. Natl. Acad. Sci. U.S.* 43: 491 (1957). **J-18.** J. Monod and F. Jacob, *Cold Spring Harbor Symposia Quant. Biol.* 26: 389(1961). **J-19.** A. S. L. Hu *et al.*, *ABB* 91: 210(1960). **J-20.** M. R. Pollock *et al.*, *BJ* 70: 665(1958); *BJ* 62: 391(1956). **J-21.** D. S. Hogness *et al.*, *BBA* 16: 99 (1959). **J-22.** A. B. Pardee and L. S. Prestidge, *BBA* 49: 77(1961). **J-23.** J. Monod and F. Jacob, *Cold Spring Harbor Symposia Quant. Biol.* 26: 193; J. Monod, *Biochem. Soc. Symposia* 21: 104(1961). **J-24.** M. R. Pollock *in* "The Enzymes" (P. D. Boyer *et al.*, eds.), 2nd ed. Vol. 1, p. 618, Academic Press, New York, 1959.

### Section K

**K-1.** J. S. Fruton, *Harvey Lectures* 51: 64(1957). **K-2.** S. W. Fox *et al.*, *JACS* 75: 5539(1953); *Science* 132: 200(1960). **K-3.** S. Sorm *et al.*, *Collection Czechoslov. Chem. Communs.* 22: 1310(1959). **K-4.** F. Lanni, *Proc. Natl. Acad. Sci. U.S.* 46: 1563(1960). **K-5.** H. Gibian, *Z. Naturforsch.* 16b: 18(1961). **K-6.** F. Sorm, *et al.*, *Collection Czechoslov. Chem. Communs.* 26: 531, 1174, 1180(1961). **K-7.** D. Schwartz, *Nature* 183: 464(1959). **K-8.** M. F. Bartlett *et al.*, *JACS* 78: 2905(1956). **K-9.** B. Keil and F. Sorm, *Collection Czechoslov. Chem. Communs.* 27: 1310(1962). **K-10.** F. Lanni and

F. Sorm, *Collection Czechoslov. Chem. Communs.* 27: 469(1962). **K-11.** J. Williams *et al., J. Mol. Biol.* 3: 532(1961). **K-12.** L. T. Troland, *Am. Naturalist* 51: 321(1917). **K-13.** H. J. Muller, *Sci. Monthly* 44: 210(1937). **K-14.** J. B. S. Haldane, *in* "Perspectives in Biochemistry" (J. Needham and D. E. Green, eds.), p. 1. Cambridge Univ. Press, London and New York, 1937. **K-15.** C. F. Cori and G. T. Cori, *JBC* 135: 641(1940); *FP* 4: 234(1944). **K-16.** S. Ochoa, *in* "Solvay Conference on Chemistry, Nucleoproteins" (11th Conference), p. 241. Wiley (Interscience), New York, 1959. **K-17.** A. Kornberg, *Science* 131: 1503(1960). **K-18.** A. G. Passinsky, *Biofizika* 5: 16(1960). **K-19.** G. Pontecorvo, *Symposia Soc. Exptl. Biol.* 12: 1(1960). **K-20.** F. Haurowitz, *Biol. Revs.* 27: 247(1952). **K-21.** G. Bredig and F. Gerstner, *Biochem. Z.* 250: 414(1932). **K-22.** F. Haurowitz, *Quart. Rev. Biol.* 24: 93(1949). **K-23.** L. Pauling, *JACS* 62: 2643(1940). **K-24.** A. E. Gurvitch and A. M. Olovnikov, *Biokhimiya*, 25: 646(1960). **K-25.** A. E. Gurvitch *et al., Biokhimiya* 26: 468(1961). **K-26.** D. Gitlin and E. Merler, *J. Exptl. Med.* 114: 217(1961). **K-27.** J. Brachet, *Arch. biol. (Liège)* 53: 207(1941). **K-28.** T. Caspersson, *Naturwiss.* 29: 33(1941). **K-29.** L. Jansen, *Protoplasma* 33: 410(1939). **K-30.** H. Friedrich-Freksa, *Naturwiss.* 38: 376(1940). **K-31.** E. Chargaff, *Experientia* 6: 201(1950). **K-32.** R. Franklin *et al., Trans. Faraday Soc.* 50: 298(1954). **K-33.** M. H. F. Wilkins *et al., Nature* 172: 759(1953). **K-34.** J. D. Watson and F. H. C. Crick, *Nature* 171: 737(1953). **K-35.** H. Ephrussi-Taylor, *Exptl. Cell Research* 6: 94(1954). **K-36.** R. D. Hotchkiss and J. Marmur, *Proc. Natl. Acad. Sci. U.S.* 40: 55(1954). **K-37.** J. Marmur and R. D. Hotchkiss, *JBC* 214: 383(1955). **K-38.** J. Lederberg, *Physiol. Revs.* 32: 403(1954). **K-39.** G. Gamow, *Nature* 174: 318(1954); with M. Ycas, *Proc. Natl. Acad. Sci. U.S.* 41: 1011(1955). **K-40.** C. R. Woese, *Nature* 194: 1114(1962); *Biochem. Biophys. Research Communs.* 5: 88(1961); R. Wall, *Nature* 193: 1268(1962). **K-41.** F. H. C. Crick *et al., Nature* 192: 1227(1961). **K-42.** C. Levinthal, *in* "Physical Science" (J. L. Oncley, ed.), p. 249. Wiley, New York. **K-43.** M. Nirenberg *et al., Proc. Natl. Acad. Sci. U.S.* 47: 1588(1961); 48: 104, 410, 666(1962). **K-44.** S. Ochoa *et al., Proc. Natl. Acad. Sci. U.S.* 48: 63, 1936(1962). **K-45.** H. Matthaei and M. W. Nirenberg, *Proc. Natl. Acad. Sci. U.S.* 47: 1580(1961). **K-46.** A. Tsugita *et al., Proc. Natl. Acad. Sci. U.S.* 48: 846 (1962). **K-47.** E. L. Smith, *Proc. Natl. Acad. Sci. U.S.* 48: 667 (1962). **K-48.** J. M. Diamond and G. Braunitzer, *Nature* 194: 1287(1962). **K-49.** L. F. Leloir, *in* "Currents in Biochemical Research" (D. E. Green, ed.), p. 585. Wiley (Interscience), New York, 1956. **K-50.** E. P. Kennedy, *JBC* 209: 525(1954). **K-51.** D. Schwartz, *Proc. Natl. Acad. Sci. U.S.* 48: 750(1962). **K-52.** J. Fresco and D. B. Strauss, *Am. Scientist* 50: 158(1962). **K-53.** G. Kellenberger, *Adv. in Virus Research* 8: 1(1961). **K-54.** H. V. Aposhian and A. Kornberg, *JBC* 237: 519(1962). **K-55.** G. Koch and A. D. Hershey, *J. Mol. Biol.* 1: 260(1959). **K-56.** L. Astrachan and E. Volkin, *BBA* 32: 449(1959). **K-57.** B. Commoner, *Science* 133: 1745(1961). **K-58.** M. B. Hoagland, *Chem. Eng. News*, April 30, 1962, p. 70. **K-59.** F. H. C. Crick, *Symposia Soc. Exptl. Biol.* 12: 138(1958), p. 160. **K-60.** F. H. C. Crick, *Science* 139: 461(1963). **K-61.** S. Spiegelman *et al., Proc. Natl. Acad. Sci. U.S.* 48: 1069, 1466(1962); *FP* 22: 36(1963). **K-62.** M. Spencer *et al., Nature* 194: 1014(1962). **K-63.** M. Bretscher and M. Grunberg-Manago, *Nature* 195: 283(1962). **K-64.** A. J. Wahba *et al., Proc. Natl. Acad. Sci. U.S.* 48: 1683(1962). **K-65.** O. W. Jones, Jr., and M. W. Nirenberg, *Proc. Natl. Acad. Sci. U.S.* 48: 2115(1962). **K-66.** R. B. Roberts, *Proc. Natl. Acad. Sci. U.S.* 48: 1245(1962). **K-67.** B. Weisblum *et al., Proc. Natl. Acad. Sci. U.S.* 48: 1439(1962). **K-68.** T. H. Jukes, *Proc. Natl. Acad. Sci. U.S.* 48: 1809(1962). **K-69.** S. Brenner, Lecture at the AAAS Meeting, Philadelphia, December 28, 1962. **K-70.** S. H. Barondes and M. H. Nirenberg, *Science* 138: 810, 813(1962). **K-71.** J. E. M. Midgley and B. J. McCarthy, *ABB* 99: 696(1962).

# Index

## A

Abrin, 367
Absorbancy of proteins, 140
Acceptor-RNA, 281, 406
Acetone, use in preparation of proteins, 8
$N$-Acetyl groups in proteins, 160, 404
$N$-Acetylation, 154
Ac-Globulin, 194
$\alpha_1$-Acid glycoprotein (orosomucoid), 188, 253
ACTH, 355
Actin, 225–227
Activated states, 314
Activation energy, 166, 172
Actomyosin, 225, 227
N,O-Acyl shift, 27
Adaptive enzymes, 422
Adrenocorticotropic hormone, 355
Agar gel diffusion, 374
Agglutination, 383
Alanine, 35, 96
$\beta$-Alanine, 40
Albumins, 3, 184, 185, 192
Alcohol dehydrogenase, 302, 305, 329, 330
$N$-Alkylation, 53, 154
Alkylation of SH, 57
Amandin, 67, 203
Amidination, 40, 155
Amino acids
    activation, 401, 405
    acylation, 52
    analogs, 421
    buffer action, 96
    chromatography, 29–31, 33
    color reactions, 34
    dielectric increments, 125
    dispensable, 397
    electrochemistry, 92
    fractionation, 29, 33
    ionization constants, 96
    isolation, 29, 33
    microbiological determination, 34, 43
    optical resolution, 33

paper chromatography, 29, 31, 32
polymerization, 401
quantitative determination, 42–48
sequence, 51–61
supply, 395
Amino acid-lipid complexes, 408
Amino acid oxidases, 304, 332
D-Amino acids, 399
Amino groups, determination, 27, 28
Aminoacyl hydroxamates, 408
Aminoacyl-sRNA synthetase, 406, 407
$\gamma$-Aminobutyric acid, 40
S-Aminoethylcysteine, 58
Ammonia formation, 26
Ammonium sulfate precipitation, 12
Amylases, 301, 302, 306, 324–326
Amyloid, 240, 255
Analogs, action, 421
Anamnestic reaction, 386
Anaphylactic symptoms, 370
Angiotensin, 359
Angiotonin, 359
Anhydrohemoglobin, 265
Anionic detergents, 241
Anomalous pK values, 101
Antibodies, 374–379
    complementariness, 438
    denaturation, 165
    $\Delta F$ of combination, 381
    half-life, 386, 415
    formation, 384–389, 414, 427, 439
    release, 410
Antibody-hapten complex, 380
Antigens, 369–374
    adjuvants, 373
    metabolic fate, 384
    persistence, 385
    template role, 426, 428
Antigen-antibody complexes, 379, 382
Antipathin, 222
Apoenzymes, 296–298
Apparent specific volume, 120
Archibald method, 72
Arginine, 26, 29, 39, 97, 98, 101
Ascorbic acid oxidase, 260